THE
# ELEMENTS
OF
# GENETICS

# THE ELEMENTS OF GENETICS

*Irwin H. Herskowitz*

HUNTER COLLEGE / THE CITY UNIVERSITY OF NEW YORK

MACMILLAN PUBLISHING CO., INC.
*New York*

COLLIER MACMILLAN PUBLISHERS
*London*

Copyright © 1979, Irwin H. Herskowitz

Printed in the United States of America

All rights reserved. No part of this book may be reproduced or transmitted in any form or by any means, electronic or mechanical, including photocopying, recording, or any information storage and retrieval system, without permission in writing from the Publisher.

Macmillan Publishing Co., Inc.
866 Third Avenue, New York, New York 10022

Collier Macmillan Canada, Ltd.

Library of Congress Cataloging in Publication Data

Herskowitz, Irwin Herman, (date)
  Elements of genetics.

  Bibliography: p.
  Includes index.
  1. Genetics. I. Title. [DNLM: 1. Genetics. QH430 H572e]
QH430.H45       575.1       78-7295
ISBN 0-02-353950-X

Printing: 2 3 4 5 6 7 8      Year: 9 0 1 2 3 4 5

# To the New Additions to Our Pedigree

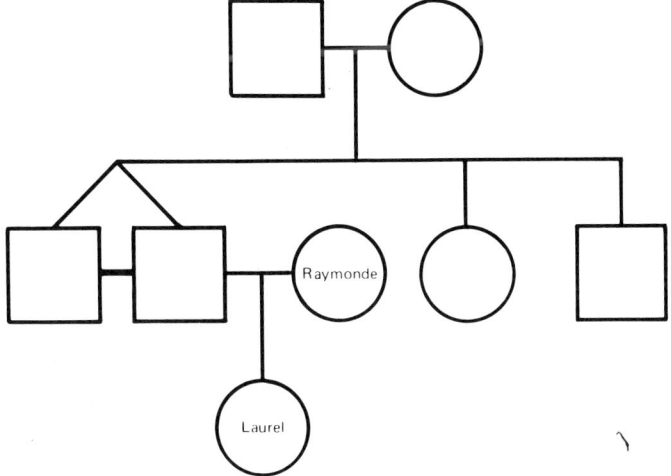

# Preface

THIS BOOK WAS WRITTEN to fulfill the author's need for a college text in genetics that can be used (a) in a one quarter or a two or three credit one semester course, (b) by students in nursing or health science programs, in addition to (c) biology majors who have had only one year of college biology or human physiology. The overall philosophy of the approach taken here is the same as was stated in the preface of my *Principles of Genetics*.

> Most first courses in college biology provide a reasonably good introduction to genetics. Accordingly, students starting their first course in college genetics not only have some background in the origins and early advances in genetics, but also have some knowledge of the recent progress made through biochemical and microbial studies. Because of this prior exposure, the students also come to the course with enthusiasm and interest. It is feasible, therefore, to approach the subject in a highly structured manner.
>
> This book aims to elucidate the principles of genetics, many of which were recently discovered in molecular and microbial studies. Since principles are dealt with rather than history, no distinction is made between "classical" and "modern" genetics, and the presentation aims to be logical rather than chronological.

As before, each chapter starts with a brief introduction followed by a series of numbered conclusions or postulates, each of which is then proved, supported, or discussed. Each chapter ends with a summary and questions and problems. Also, as before, (a) the few names in the text—Watson, Crick, Wilkins, Mendel, Barr, Hardy and Weinberg—are there simply because they are uniquely important, widely known, or commonly used; (b) a glossary; and (c) answers to selected questions and problems are included.

The present text differs from *Principles of Genetics* mainly in the following respects: (a) eukaryotic principles are illustrated with examples from human beings whenever possible; (b) the chemistry of genetics is less detailed; (c) the applications and implications of genetics are

## PREFACE

discussed in much greater detail; (d) the main body of the chapters in all but the last part of the text has been shortened by about 40 per cent; (e) the technical terminology has been reduced more than 20 per cent; (f) much of the text remaining has been rewritten for clarity; (g) many new diagrams have been added; (h) all but less than a dozen questions and problems are new; (i) the appropriate bibliography follows the glossary; and (j) the supplementary sections and biometrical appendix have been omitted.

I wish to thank my wife, Reida Postrel Herskowitz, for her help with the typescript, and for her support and encouragement.

# Abbreviated Contents

### PART I WHAT THE GENETIC MATERIAL IS

*Chapter 1* Genetic material is nucleic acid  3
*Chapter 2* Structural organization of nucleic acids and chromosomes  16

### PART II WHAT THE GENETIC MATERIAL DOES

*Chapter 3* Chromosome replication  43
*Chapter 4* Transcription  56
*Chapter 5* Translation and its code  67

### PART III HOW THE GENETIC MATERIAL IS VARIED, PACKAGED, AND DISTRIBUTED

*Chapter 6* Mutation  87
*Chapter 7* Genetic recombination between viruses  101
*Chapter 8* Genetic recombination between bacteria: I. Transformation and generalized transduction  113
*Chapter 9* Genetic recombination between bacteria: II. Restricted transduction and conjugation  123
*Chapter 10* Genetic recombination in eukaryotes: I. Mitosis, meiosis, and segregation  137

*Chapter 11* Genetic recombination in eukaryotes:
II. Sex linkage and crossing over                158
*Chapter 12* Gross changes in nuclear chromosomes    172
*Chapter 13* Nonmendelian genes in eukaryotes        190

## PART IV  HOW THE GENETIC MATERIAL CHOOSES WHICH PARTS ARE PRESENT AND FUNCTIONAL

*Chapter 14* Programmed gene synthesis, destruction, and mutation    207
*Chapter 15* Regulation of gene action               221
*Chapter 16* Heterochromatization in eukaryotes      236

## PART V  HOW GENE PRODUCTS INTERACT AND THE PHENOTYPIC CONSEQUENCES OF GENE ACTION

*Chapter 17* Phenotypic effects of environment, genotype, and single loci    249
*Chapter 18* Phenotypic interactions of two or more loci    262
*Chapter 19* Determination of sex in eukaryotes      275
*Chapter 20* Differentiation and development         289

## PART VI  HOW THE PRECEDING CAME ABOUT IN INDIVIDUALS AND POPULATIONS

*Chapter 21* The origin and evolution of genetic material    307
*Chapter 22* Population genotypes and mating systems    315
*Chapter 23* Factors that affect gene frequencies in populations    323
*Chapter 24* Genetic variability of populations and speciation    332

## PART VII  THE PRESENT AND FUTURE CONSEQUENCES OF GENETICS

*Chapter 25* Applications to agriculture and ecology    343

*Chapter 26* Applications to behavior, and social
and political issues 357
*Chapter 27* Applications to medicine and public
health 373

Glossary 395
Bibliography 409
Answers to selected questions and
problems 421
Index 427

# Detailed Contents

## PART I  WHAT THE GENETIC MATERIAL IS

### Chapter 1  Genetic Material Is Nucleic Acid     3

| | |
|---|---:|
| 1.1 Evolution of the Universe | 3 |
| 1.2 Features of Organisms | 4 |
| 1.3 Need for Genetic Material | 8 |
| 1.4 RNA and DNA Genetic Material | 9 |
| 1.5 Meaning of Genetic Material | 13 |
| 1.6 DNA as Main Genetic Material | 13 |
| 1.7 Scope of Genetics | 14 |
| Summary and Conclusions | 15 |
| Questions and Problems | 15 |

### Chapter 2  Structural Organization of Nucleic Acids and Chromosomes     16

| | |
|---|---:|
| Nucleic Acid Structure | 16 |
| 2.1 RNA Structure | 16 |
| 2.2 DNA Structure | 18 |
| 2.3 The DNA Double Helix | 19 |
| 2.4 Watson, Crick, and Wilkins | 20 |
| 2.5 Information Content of Nucleic Acids and Organisms | 22 |
| 2.6 Determination of Base Sequence | 24 |
| 2.7 Double-helical Regions | 25 |
| 2.8 Making and Breaking Double Helixes | 27 |
| 2.9 Usefulness of Manipulating Double Helixes | 28 |
| Chromosome Organization | 31 |
| 2.10 In Viruses | 31 |
| 2.11 In Bacteria | 33 |
| 2.12 In Nuclear Chromosomes | 36 |
| Summary and Conclusions | 38 |
| Questions and Problems | 40 |

# PART II  WHAT THE GENETIC MATERIAL DOES

## *Chapter 3*  Chromosome Replication — 43

- RNA Replication — 43
- 3.1 *De novo* Synthesis — 43
- 3.2 Template Synthesis — 44
- 3.3 Genes Defined — 45
- DNA Replication — 46
- 3.4 Semiconservative Replication — 46
- 3.5 In Prokaryotes and Eukaryotes — 46
- 3.6 DNA Modification — 51
- 3.7 DNA Restriction Nucleases — 52
- Summary and Conclusions — 53
- Questions and Problems — 54

## *Chapter 4*  Transcription — 56

- 4.1 Genetic Material Redefined — 56
- Transcription in Prokaryotes — 57
- 4.2 In *E. coli* — 57
- 4.3 Tailoring Transcripts — 59
- Transcription in Eukaryotes — 60
- 4.4 General and Unique Features — 60
- Transcription in Viruses — 62
- 4.5 Relation to Host — 62
- 4.6 One or Two Sense Strands — 64
- Reverse Transcription — 64
- 4.7 RNA Viruses — 64
- Summary and Conclusions — 66
- Questions and Problems — 66

## *Chapter 5*  Translation and Its Code — 67

- 5.1 General View of Translation — 68
- 5.2 The Common Amino Acids — 70
- The Genetic Code — 70
- 5.3 Degeneracy of the Code — 70
- 5.4 Universality of the Code — 73
- Translation — 73
- 5.5 Ribosomes — 74
- 5.6 Ribosomal Functions — 74
- 5.7 tRNA's and Their Functions — 76
- 5.8 Basis of Degeneracy — 78
- 5.9 Starting, Continuing, and Ending Polypeptide Synthesis — 79
- 5.10 Polyribosomes — 79
- 5.11 Polyribosomes in Prokaryotes vs. Eukaryotes — 79
- 5.12 Signal Polypeptides and Membranes — 81
- Summary and Conclusions — 83
- Questions and Problems — 83

# PART III  HOW GENETIC MATERIAL IS VARIED, PACKAGED, AND DISTRIBUTED

## Chapter 6  Mutation — 87

- 6.1 Types of Mutation — 88
- 6.2 Detection of Mutations — 89
- 6.3 Undetected and Protein-modifying Mutations — 89
- 6.4 Transitions and Transversions — 91
- 6.5 Physical Mutagens — 93
- 6.6 Repair of Mutations — 94
- 6.7 Errors of Repair — 96
- 6.8 Spontaneous Mutations — 97
- Summary and Conclusions — 99
- Questions and Problems — 99

## Chapter 7  Genetic Recombination Between Viruses — 101

- 7.1 T-even Phage Life Cycle — 102
- 7.2 Genetic Recombination Expected — 105
- 7.3 Genetic Recombination Observed — 106
- 7.4 Sequencing Loci from Recombination Frequencies — 107
- 7.5 Circular Recombination Map of $\phi$T4 — 109
- 7.6 Recombination Between RNA Viruses — 110
- Summary and Conclusions — 111
- Questions and Problems — 112

## Chapter 8  Genetic Recombination Between Bacteria: I. Transformation and Generalized Transduction — 113

- 8.1 Clones — 113
- Genetic Transformation — 114
- 8.2 DNA Transforms — 114
- 8.3 Transformation as Genetic Recombination — 115
- 8.4 Occurrence in Various Organisms — 117
- Generalized Genetic Transduction — 117
- 8.5 Temperate Phage as Mediator — 117
- 8.6 DNA in Generalized Transducing Phage — 119
- 8.7 Occurrence in Various Organisms — 120
- Summary and Conclusions — 121
- Questions and Problems — 122

## Chapter 9  Genetic Recombination Between Bacteria: II. Restricted Transduction and Conjugation — 123

- Restricted Transduction — 123
- 9.1 Life Cycle of $\phi\lambda$ — 123
- 9.2 Transduction by $\phi\lambda$ — 126

## DETAILED CONTENTS

| | |
|---|---:|
| Bacterial Conjugation | 128 |
| 9.3 Sex Particles | 128 |
| 9.4 Plasmid and Episomal Sex Particles | 130 |
| 9.5 Hfr and Generalized Transduction | 130 |
| 9.6 F Derivatives | 133 |
| 9.7 Other Plasmids and Sex Particles | 134 |
| Summary and Conclusions | 134 |
| Questions and Problems | 135 |

### *Chapter 10* Genetic Recombination in Eukaryotes: I. Mitosis, Meiosis, and Segregation     137

| | |
|---|---:|
| Mitosis | 137 |
| 10.1 DNA Replication and Nuclear Cycle | 137 |
| 10.2 Mitotic Stages | 138 |
| 10.3 Genetic Program for Mitosis | 141 |
| 10.4 Nuclear Chromosome Characteristics | 142 |
| 10.5 Complete and Partial Redundancy | 143 |
| Meiosis | 144 |
| 10.6 The Stages of Meiosis | 145 |
| 10.7 Special Features in Females | 148 |
| 10.8 Segregation | 149 |
| 10.9 Independent Segregation | 150 |
| 10.10 Genetic Program for Meiosis | 151 |
| Summary and Conclusions | 153 |
| Questions and Problems | 155 |

### *Chapter 11* Genetic Recombination in Eukaryotes: II. Sex Linkage and Crossing Over     158

| | |
|---|---:|
| Sex Linkage | 158 |
| 11.1 Sex Chromosomes and Autosomes | 158 |
| 11.2 Common and Unique Sex Chromosome Loci | 159 |
| 11.3 Albinism and Colorblindness | 159 |
| 11.4 Phenotypic Ratios and Sex Linkage | 163 |
| 11.5 Nondisjunction | 164 |
| Crossing Over | 166 |
| 11.6 Crossovers | 166 |
| 11.7 Genes Tend to Remain Linked | 167 |
| 11.8 Crossover Recombination Maps | 168 |
| Summary and Conclusions | 169 |
| Questions and Problems | 170 |

### *Chapter 12* Gross Changes in Nuclear Chromosomes     172

| | |
|---|---:|
| 12.1 Detection of Gross Changes | 172 |
| Unbroken Chromosome Changes | 173 |
| 12.2 Polyploidy | 173 |
| 12.3 Loss or Gain of Individual Chromosomes | 174 |
| 12.4 Detriment Due to Chromosome Loss or Gain | 175 |
| 12.5 Monosomic and Trisomic Human Beings | 176 |

| Broken Chromosome Changes | 178 |
| --- | --- |
| 12.6 Ligation of Broken Ends | 178 |
| 12.7 One Nonrestituted Break | 179 |
| 12.8 Chromatid Breaks | 180 |
| 12.9 Two Nonrestituted Breaks | 180 |
| 12.10 Reciprocal and Half-translocation | 183 |
| 12.11 Reciprocal and Half-translocation in Human Beings | 185 |
| Summary and Conclusions | 187 |
| Questions and Problems | 188 |

### *Chapter* 13 Nonmendelian Genes in Eukaryotes — 190

| 13.1 Foreign Nonmendelian Genes | 190 |
| --- | --- |
| 13.2 Nonmendelian Genes Changed to Mendelian Genes | 191 |
| 13.3 Mendelian Genes Changed to Nonmendelian Genes | 193 |
| 13.4 Chloroplast DNA is Genetic Material | 193 |
| 13.5 Nonmendelian Recombination of chl DNA | 195 |
| 13.6 Mitochondrial DNA is Genetic Material | 196 |
| 13.7 Mapping of rDNA and tDNA in mit DNA | 198 |
| 13.8 Nonmendelian Recombination of mit DNA | 199 |
| 13.9 Other Normal Cytoplasmic DNA's | 200 |
| Summary and Conclusions | 202 |
| Questions and Problems | 203 |

## *PART IV* HOW THE GENETIC MATERIAL CHOOSES WHICH PARTS ARE PRESENT AND FUNCTIONAL

### *Chapter* 14 Programmed Gene Synthesis, Destruction, and Mutation — 207

| Gene Synthesis | 207 |
| --- | --- |
| 14.1 Replicons in Prokaryotes | 207 |
| 14.2 Regulator Genes and Replicons | 208 |
| 14.3 Replicons in Eukaryotes | 209 |
| 14.4 Amplification | 211 |
| Gene Destruction | 216 |
| 14.5 Programmed in Eukaryotes | 216 |
| Mutation | 218 |
| 14.6 Programmed in Prokaryotes | 218 |
| 14.7 Programmed in Eukaryotes | 218 |
| Summary and Conclusions | 219 |
| Questions and Problems | 219 |

### *Chapter* 15 Regulation of Gene Action — 221

| In Prokaryotes | 221 |
| --- | --- |
| 15.1 The Promoter Region | 221 |
| 15.2 The Operator | 224 |
| 15.3 The Transcription Terminator | 226 |
| 15.4 Translation and mRNA | 226 |
| 15.5 Other Translation Regulators | 228 |

## DETAILED CONTENTS

|  |  |
|---|---|
| In Eukaryotes | 228 |
| 15.6 Heterochromatin | 228 |
| 15.7 Histones | 229 |
| 15.8 Histones as Repressors | 231 |
| 15.9 Nonhistone Proteins as Activators | 232 |
| 15.10 Regulation of Translation | 233 |
| Summary and Conclusions | 234 |
| Questions and Problems | 235 |

### Chapter 16 Heterochromatization in Eukaryotes 236

|  |  |
|---|---|
| Human Beings and Other Mammals | 236 |
| 16.1 Dosage Compensation | 236 |
| 16.2 Suppression of One of Two Alleles | 237 |
| 16.3 Heterochromatization Causes Suppression of Alleles | 237 |
| 16.4 Position Effect | 238 |
| 16.5 Permanency of Heterochromatization | 239 |
| *Drosophila* | 240 |
| 16.6 Dosage Compensation | 240 |
| 16.7 V-type Position Effects | 241 |
| 16.8 Heterochromatization Causes V-type Position Effects | 241 |
| 16.9 Factors Affecting Heterochromatization | 242 |
| Maize | 243 |
| 16.10 Controlling Genes | 243 |
| Summary and Conclusions | 245 |
| Questions and Problems | 245 |

## PART V  HOW GENE PRODUCTS INTERACT AND THE PHENOTYPIC CONSEQUENCES OF GENE ACTION

### Chapter 17 Phenotypic Effects of Environment, Genotype, and Single Loci 249

|  |  |
|---|---|
| 17.1 Twin Studies | 249 |
| 17.2 Genetic and Nongenetic Environment | 251 |
| 17.3 Penetrance and Expressivity | 252 |
| 17.4 Phenocopies | 253 |
| 17.5 Multiple Phenotypic Effects of Single Genes | 254 |
| 17.6 Dominance and Recessiveness | 255 |
| 17.7 Adaptiveness and Recessiveness of Mutants | 258 |
| 17.8 Allele Dosage and Optimum Phenotypic Effect | 258 |
| 17.9 Alleles and Viability | 259 |
| Summary and Conclusions | 259 |
| Questions and Problems | 260 |

### Chapter 18 Phenotypic Interactions of Two or More Loci 262

|  |  |
|---|---|
| 18.1 Proteins Act Alone or in Combination | 262 |
| 18.2 Dominance and Phenotypic Classes | 263 |

| | |
|---|---|
| 18.3 Epistatic–Hypostatic, Complementary, and Duplicate Genes | 266 |
| 18.4 Continuous (Quantitative) Traits | 268 |
| 18.5 Genetic Basis for Quantitative Traits | 268 |
| 18.6 Variability of Quantitative Traits | 269 |
| 18.7 Dominance and Quantitative Traits | 271 |
| 18.8 Selection and Quantitative Traits | 271 |
| Summary and Conclusions | 272 |
| Questions and Problems | 273 |

## *Chapter 19* Determination of Sex in Eukaryotes — 275

| | |
|---|---|
| Yeast | 275 |
| 19.1 Mating Type and Transposable Genes | 275 |
| *Drosophila* | 276 |
| 19.2 Sex Types and Chromosomal Balance | 276 |
| 19.3 Mosaicism and Gynandromorphs | 278 |
| Hymenoptera | 279 |
| 19.4 Sex Type and Number of Chromosome Sets | 279 |
| Human Beings | 280 |
| 19.5 Genotypic and Phenotypic Sex | 280 |
| 19.6 Chromosome Number and Abnormal Sex Types | 281 |
| 19.7 Sex Chromosome Rearrangements and Abnormal Sex Types | 283 |
| Environmentally Determined Sex | 285 |
| 19.8 Based on Single Genotypes | 285 |
| Summary and Conclusions | 286 |
| Questions and Problems | 286 |

## *Chapter 20* Differentiation and Development — 289

| | |
|---|---|
| 20.1 Differential Gene Transcription | 289 |
| Survival During Early Development | 291 |
| 20.2 Stockpiling in Oocytes | 291 |
| 20.3 Use of Stockpiled Materials | 292 |
| The Coordination of Body Parts | 292 |
| 20.4 Intracellular Gene Action | 293 |
| 20.5 Hormones | 294 |
| THE SYNTHESIS OF SPECIFIC TISSUE PROTEINS | |
| Hemoglobin Synthesis | 295 |
| 20.6 Hemoglobins and Development | 295 |
| 20.7 Regulation of Transcription and Translation | 297 |
| Antibody Synthesis | 298 |
| 20.8 Antibodies and Plasma Cells | 298 |
| 20.9 Regulation of Transcription and Translation | 300 |
| Summary and Conclusions | 301 |
| Questions and Problems | 302 |

# *PART VI* HOW THE PRECEDING CAME ABOUT IN INDIVIDUALS AND POPULATIONS

## *Chapter 21* The Origin and Evolution of Genetic Material — 307

| | |
|---|---|
| 21.1 Origin of Proteins and Nucleic Acids | 307 |

| | |
|---|---|
| 21.2 The First Organism | 308 |
| 21.3 The Genetic Code and Translation Machinery | 309 |
| 21.4 Selection and Biochemical Complexity | 309 |
| 21.5 The Quantity of Genetic Material | 310 |
| 21.6 Evolutionary Trees | 312 |
| 21.7 Evolution of Gene Functions | 312 |
| Summary and Conclusions | 313 |
| Questions and Problems | 314 |

## Chapter 22 Population Genotypes and Mating Systems 315

| | |
|---|---|
| Random Mating | 315 |
| 22.1 Allele and Genotypic Frequencies | 315 |
| 22.2 The Hardy–Weinberg Principle | 316 |
| 22.3 Dominance | 317 |
| Nonrandom Mating | 318 |
| 22.4 Rare Alleles | 318 |
| 22.5 Inbreeding | 318 |
| 22.6 Consequences of Homozygosity | 320 |
| Summary and Conclusions | 321 |
| Questions and Problems | 321 |

## Chapter 23 Factors That Affect Gene Frequencies in Populations 323

| | |
|---|---|
| Selection | 323 |
| 23.1 Differential Conservation of Genotypes | 323 |
| 23.2 Selection Coefficients of Detrimental Alleles | 324 |
| 23.3 Heterosis Due to a Single Locus | 325 |
| 23.4 Heterosis Due to Many Loci | 326 |
| Mutation | 326 |
| 23.5 Effect on Allele Frequencies | 326 |
| 23.6 Equilibrium between Mutation and Selection | 327 |
| Migration and Genetic Drift | 329 |
| 23.7 Effect on Allele Frequencies | 329 |
| Summary and Conclusions | 330 |
| Questions and Problems | 330 |

## Chapter 24 Genetic Variability of Populations and Speciation 332

| | |
|---|---|
| 24.1 Great Genetic Variability in Populations | 332 |
| 24.2 Genetic Load and Genetic Death | 333 |
| 24.3 Adaptiveness of a Mutational Load | 334 |
| 24.4 Race | 334 |
| 24.5 Adaptiveness of Races | 335 |
| 24.6 Partial Reproductive Isolation | 335 |
| 24.7 Speciation of Races | 336 |
| 24.8 Speciation from Founder Individuals | 337 |
| 24.9 Speciation by Single Species Polyploidy | 337 |
| 24.10 Speciation after Interspecific Hybridization | 337 |
| Summary and Conclusions | 338 |
| Questions and Problems | 338 |

# PART VII THE PRESENT AND FUTURE CONSEQUENCES OF GENETICS

## Chapter 25 Applications to Agriculture and Ecology — 343

| | |
|---|---|
| Genetics and Agriculture | 343 |
| 25.1 Crops as a Source of Nutrition | 343 |
| 25.2 Controlled Breeding and Artificial Selection of Plants | 345 |
| 25.3 Polyploidy, Introgression, and Mutagenesis of Plants | 348 |
| 25.4 Selective Breeding of Animals | 350 |
| 25.5 Maintaining Genetic Variability | 351 |
| Genetics and Ecology | 352 |
| 25.6 Unintended Ecological Changes | 352 |
| 25.7 Intentional Ecological Changes | 354 |
| Summary and Conclusions | 355 |
| Questions and Problems | 356 |

## Chapter 26 Applications to Behavior, and Social and Political Issues — 357

| | |
|---|---|
| Applications to Behavior | 357 |
| 26.1 Molecular Basis of Memory and Learning | 358 |
| 26.2 Behavior Affected by One, Two, or More Loci | 359 |
| 26.3 Abnormal Human Behavior and Genetic Changes | 361 |
| APPLICATIONS TO SOCIAL AND POLITICAL ISSUES | |
| Races and Intelligence | 364 |
| 26.4 Adaptiveness of Human Races | 364 |
| 26.5 Equalizing Different Human Races | 364 |
| 26.6 Black and White Races | 365 |
| Politics | 366 |
| 26.7 Stalin's Suppression of Genetics | 366 |
| 26.8 Hitler's "Aryan" Master Race | 367 |
| Law and Religion | 369 |
| 26.9 Ecclesiastical and Civil Laws on Marriage | 369 |
| 26.10 Determination of Parentage | 370 |
| Summary and Conclusions | 371 |
| Questions and Problems | 371 |

## Chapter 27 Applications to Medicine and Public Health — 373

| | |
|---|---|
| Applications of Immunology | 373 |
| 27.1 ABO Blood Types and Transfusions | 373 |
| 27.2 Mother–Fetus Blood Incompatibility | 374 |
| 27.3 Tissue Incompatibility | 375 |
| Applications to Aging, Death, and Cancer | 375 |
| 27.4 Somatic Aging and Death is Programmed | 375 |
| 27.5 Protein Damage as Cause of Aging and Death | 376 |
| 27.6 Genetic Basis of Cancer | 378 |
| Applications to Environmental Mutagenesis | 379 |
| 27.7 Physical Mutagens | 379 |
| 27.8 Chemical Mutagens | 381 |

## DETAILED CONTENTS

| | |
|---|---|
| Applications to Genetic Diseases and Genetic Engineering | 383 |
| 27.9 Detection of Genetic Diseases | 383 |
| 27.10 Combatting Genetic Diseases at Present | 384 |
| 27.11 Combatting Genetic Diseases in the Future | 385 |
| 27.12 Cell, Nuclear, and Molecular Cloning | 387 |
| Summary and Conclusions | 390 |
| Questions and Problems | 390 |
| | |
| Glossary | 395 |
| Bibliography | 409 |
| Answers to Selected Questions and Problems | 421 |
| Index | 427 |

# PART I

# WHAT THE GENETIC MATERIAL IS

# Chapter 1

## Genetic Material Is Nucleic Acid

EACH TYPE OF ORGANISM on earth today possesses both unique characteristics and the capacity to reproduce itself. This book deals with the information-bearing material within organisms that is responsible for an organism's characteristics and self-reproduction. The first chapter will identify the general nature and general properties of this informational material in several simple organisms; the remainder of the book will go into details of the properties, functions, and significance of such material.

**1.1** The universe has evolved physically, chemically, and organismally.

Our universe. has been undergoing nonrandom, directional transformations, or evolution, for billions of years. The unfolding story or history of the universe can be likened to a very long motion picture in which human beings first appear on earth in the last few frames. Despite the disadvantage that we can see only the last few frames, our study of the geology and astronomy of these frames has already told us a great deal about what has happened earlier in the motion picture. We believe, for example, that the universe is probably more than 10 billion years old, during which time galaxies, stars, and planets—our own galaxy and solar system among them—were formed. We have also been led to believe that for the first 5 billion years or so of this chemical and physical evolution of the universe, no organisms—that is to say, no life—existed anywhere. Then, starting about 5 billion years ago, organisms began appearing spontaneously on a small fraction of numerous planets where conditions were suitable. The first organisms on earth, however, did not appear until about 4 billion years ago (Figure 1-1), a delay that is generally attributed to earth's relative poverty in the chemical raw materials needed for the creation of organisms.

# 4 WHAT THE GENETIC MATERIAL IS

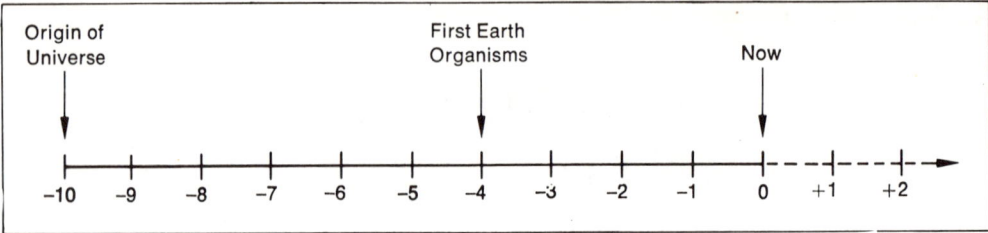

FIGURE 1-1. The march of time. Each unit on the scale represents 1 billion years. Note that 1 billion = 1,000,000,000 = 1 thousand millions = 1 million thousands.

The spontaneous formation of an organism is expected to be a rare event on any planet; and the first organism produced is expected to be very simple. Accordingly, we expect that life on our planet had a single origin and that all life on earth has descended from a single, simple organism. The 2 million kinds or species of organisms in existence on earth today—those which are visible in the last frame of our motion picture—are the product of an organismal evolution that has taken place over the past 4 billion years. Today's organisms occur in a bewildering variety that ranges from viruses and bacteria through protozoans, sponges, corals, and jellyfish; flat, round, and segmented worms; shellfish and starfish; spiders and insects; finned fishes; amphibians; reptiles; birds; mammals; algae and fungi; mosses; ferns; and seed plants.

## 1.2 Organisms have common structural, physiological, and molecular features.

Despite the great diversity of present-day organisms and the likelihood that some of them may not have had an ancestor in common for as long as 4 billion years, a survey of organisms reveals (1) structural, (2) physiological, and (3) molecular features they hold in common.

1. STRUCTURE. With the exception of the simplest organisms, the viruses, all organisms are composed of one or more cells and cell products. Cells vary in size and complexity from the tiny and relatively simple cell of a bacterium, about 100 of which can fit across the dot of an i, to the giant and relatively complex yolk, which is the single cell of a chicken or an ostrich egg.

Features of a typical cell can be suggested by imagining a plastic bag containing a very porous sponge that is saturated with a thick vegetable soup. This sponge, in turn, surrounds a smaller plastic bag containing noodle soup. The outer plastic bag represents the *cell membrane*, or outside limit of the cell (Figure 1-2). The inner plastic bag represents the *nuclear membrane*, the outside limit of the *nucleus*. The sponge represents the *endoplasmic reticulum*, a network of membranes which form channels that are often interconnected and which also connect the two other membranes. The soups and membranes make up *protoplasm*,

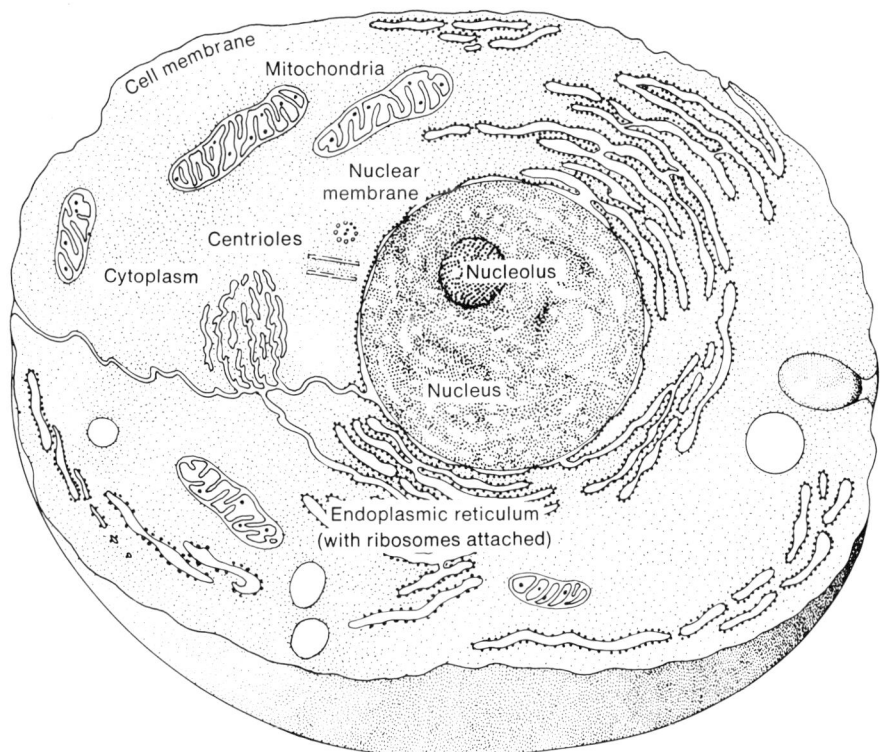

FIGURE 1-2. Diagram of a eukaryotic cell, sliced through its nucleus, showing some of the organelles present.

which is called *cytoplasm* outside the nucleus. The vegetables in the cytoplasm include various types of bodies (those surrounded by membranes are called *organelles*), such as *ribosomes, mitochondria,* and, in green cells, *chloroplasts.* These bodies will be described and discussed in detail in later chapters. The noodles of the nucleus represent the *chromosomes* ("colored bodies," called such because they stain with certain dyes). One or a few chromosomes are also present in each mitochondrion and chloroplast. Nucleus-containing cells are *eukaryotic* cells, and organisms composed of one or more of such cells are *eukaryotes.*

The cells of bacteria (and blue-green algae), on the other hand, can be likened to a single plastic bag, the size of a mitochondrion or smaller, containing soup with relatively few vegetables and only a noodle or two. Such cells consist largely of a small mass of protoplasm bounded by a cell membrane, containing ribosomes and one or a few chromosomes (Figure 1-3). Lacking a nucleus (the chromosomes are not bounded by a nuclear membrane), the cells are said to be *prokaryotic*; the cells also lack an endoplasmic reticulum and other organelles, such as mitochondria and chloroplasts.

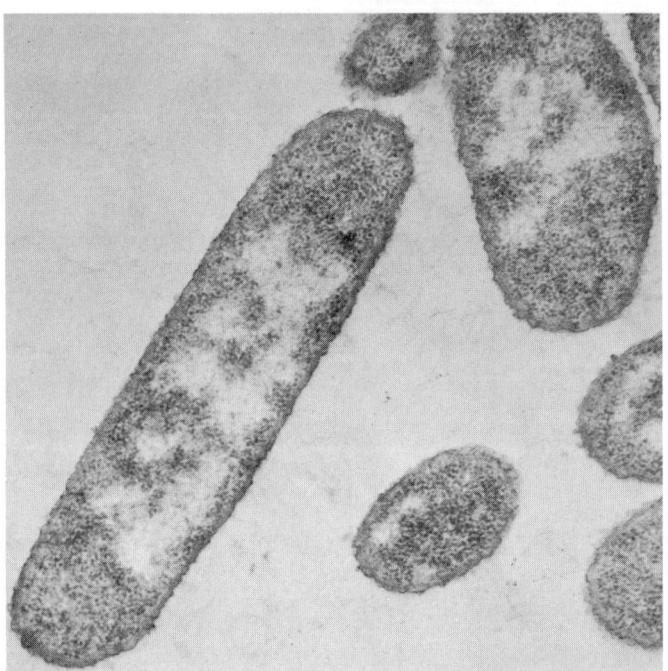

FIGURE 1-3. Prokaryotic cell. Thin slices of a rod-shaped bacterium as seen in the electron microscope. The numerous dark dots are ribosomes; the lighter areas contain chromosome fibers. Original magnification 20,000×. Present magnification about 13,000×. (Courtesy of E. Kellenberger.)

2. PHYSIOLOGY. The common features of cell structure are accompanied by common features of cell physiology or function. Cells undergo thousands of chemical reactions and physical changes that comprise the cells' *metabolism*. The function of this metabolism is to maintain the cell or organism (by repairing and/or replacing its parts), to grow, and, most important, to produce more cells or organisms of the same kind—that is, to *reproduce*. Metabolism occurs primarily at sites and surfaces provided by membranes, organelles, and large molecules (including chromosomes). The reactions that synthesize more energy-providing substances or protoplasm are *anabolic*; the reactions that degrade energy-providing substances or protoplasm are *catabolic*.

A free-living virus is not a cell and does not metabolize. When a virus enters a host cell, however, the metabolic machinery of its host is commandeered to reproduce the virus. Thus, like all cellular organisms, a virus requires cell metabolism in order to reproduce.

3. MOLECULES. All organisms are similar in that each contains unique giant molecules of two classes. Each of these molecules consists of a long chain composed of many units (of several discrete kinds) joined to each other. The units join together like children's building blocks, each with a single knob and a hole of the same size (Figure 1-4A). This type of structure permits the knob of piece 2 to fit into the hole of piece 1, the knob of piece 3 to fit into the hole of piece 2, and so forth. The individual structural units, or *monomers*, are added stepwise to produce first the

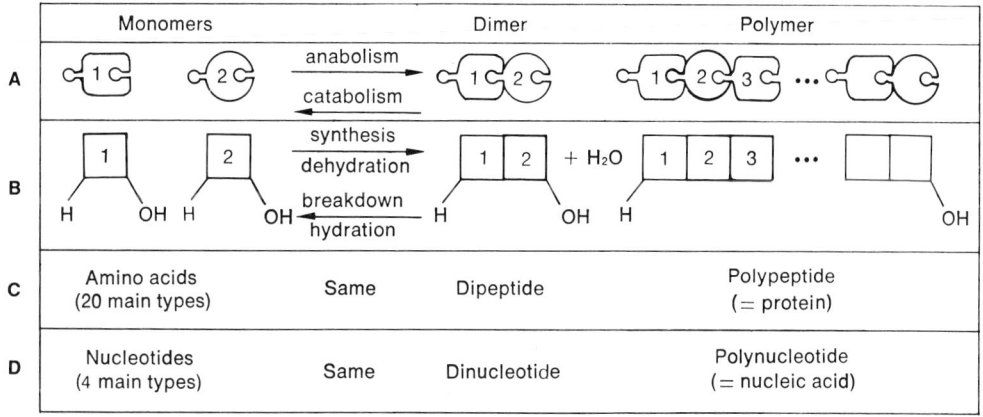

FIGURE 1-4. Similarities in the synthesis and breakdown of the two main types of biological giant molecule.

*dimer*, then the *trimer*, and so forth, and eventually the *polymer*. In chemical terms, the anabolic reaction of *polymerization* is a synthesis involving the release or removal of one molecule of water for each monomer joined to the growing chain (Figure 1-4B). Conversely, the catabolic reaction of *depolymerization* is a breakdown involving the addition of one molecule of water for each monomer removed from the polymer.

The two classes of giant polymers common to all organisms are *polypeptides* and *polynucleotides*. Each class of giant molecule has a different set of monomers.

a. *Polypeptide.* A polypeptide is a polymer of *amino acid* monomers each containing carbon (C), oxygen (O), hydrogen (H), nitrogen (N), and sometimes sulfur (S). Twenty main types of amino acid are found in organisms; the structure and characteristics of these will be described in Chapter 5. It is sufficient to note here that the union between (two) amino acids is made by a *peptide bond* to produce a (di)peptide. A *protein* is a compound of one (or more) polypeptide chain(s) (Figure 1-4C).

The cell's proteins have two general functions: one structural, the other metabolic. Proteins (together with other compounds, lipids) make up the cell's membranes; these are necessary for maintaining the cell's physical integrity, for compartmentalization, and for the attachment and support of various bodies and molecules. Individual proteins (including sometimes those that are in membranes) may function metabolically as *enzymes*. An enzyme is like a marriage counselor—it can help in the formation or dissolution of unions without becoming permanently involved. Chemically, enzymes are proteins (sometimes combined with other substances) whose surface contains one or more sites (Figure 1-5), on which other chemical substances can react more readily than they would in the absence of the enzyme. The enzyme acts as a catalyst for a reversible chemical reaction and is left unchanged when such a reaction takes place. Almost all reactions in a cell are facilitated by enzymes, so

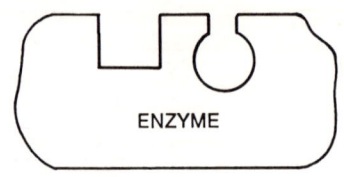
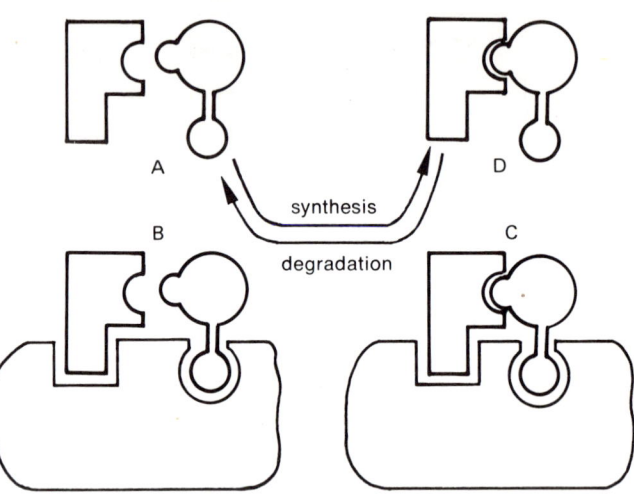

FIGURE 1-5. The catalytic action of a single enzyme. Synthesis: Two separate molecules (A) attach to the enzyme at a special site (B), where they are relatively readily joined (C) and released as a single molecule (D). Degradation: The single complex molecule (D) attaches to the special site on the enzyme (C), where it is relatively readily degraded (B) and its two component molecules are released (A). Note that in acting as a catalyst, the enzyme is not permanently changed by the chemical reaction proceeding in either direction.

that metabolism that would occur slowly in the absence of enzymes is accelerated tremendously when they are present. Note that additional proteins must be synthesized to supply the membranes and enzymes needed to replace proteins that are continuously being worn out or destroyed through molecular accidents, as well as those needed for growth and reproduction.

b. *Polynucleotide.* A polynucleotide, also called a *nucleic acid*, can be of two types: the *polyribonucleotide* or *ribonucleic acid* (*RNA*) and the *polydeoxyribonucleotide* or *deoxyribonucleic acid* (*DNA*). Each type is a polymer of nucleotide monomers (Figure 1-4D) containing C, O, H, N, and P (phosphorus); each contains only four main types of nucleotide. The chemical structure of DNA and RNA and their components will be discussed further in Chapter 2.

Like protein, the nucleic acid of each organism is unique in kind and amount. Since organisms reproduce, their unique nucleic acid must be replicated so that each of the progeny can have the characteristic nucleic acid content of the species.

**1.3  Each organism contains genetic material.**

The preceding discussion has shown that all organisms have unique requirements at the structural level (cells), at the physiological level

# GENETIC MATERIAL IS NUCLEIC ACID

(metabolism), and at the molecular level (proteins and nucleic acids). What does fulfilling these requirements entail? Each organism must possess a facility or factor that (1) persists during the entire existence of the organism, (2) is repeated in each of its progeny, (3) is different in different organisms, (4) can change and still be reproduced (so that organisms can evolve into new types), and (5) contains the information needed for synthesizing characteristic proteins and nucleic acids. For purposes of scientific investigation, we assume that we are dealing with a material factor. Since this material factor contains the instructions for the creation, or genesis, of an organism, we can call it *genetic material*. The single most important feature of all organisms is, with this view, genetic material. The study of the properties and functions of genetic material thus comprises the core of the study of all organisms, that is, of the science of biology.

## 1.4 The genetic material is RNA in some viruses and is DNA in other viruses.

Some mature viruses are composed only of nucleic acid and protein in combination. Clearly, their genetic material must be one or the other or a combination of these two types of giant molecule. A determination can be made through infection experiments.

1. RNA VIRUSES. *Tobacco mosaic virus (TMV)* is a virus composed entirely of protein and ribonucleic acid (Figure 1-6); it attacks tobacco leaves. Infection can be caused at will by exposing rubbed tobacco leaves to TMV. The rubbing produces a lesion through which the virus enters the host cell. After a single TMV particle enters a host cell and an incubation period has elapsed, hundreds of TMV progeny are formed. Suitable analysis reveals that the uninfected host cell normally contains neither the protein nor the RNA that are characteristic of TMV. Since there are many different strains of TMV and progeny TMV are always the same strain as the parent, the information for TMV protein and RNA must reside solely in TMV's genetic material.

The protein and RNA of TMV can be separated from each other by placing the viruses in a phenol (carbolic acid) and water mixture. The RNA enters the water and the protein enters the phenol. When the phenol and water fractions are separated, and the phenol and water removed, the protein and RNA are obtained in the pure state. Under appropriate conditions the separate protein and RNA can be rejoined, and the reconstituted virus will infect and produce progeny just as well as the original virus. This result indicates that the isolation process does not damage either component so that it cannot function normally.

When rubbed tobacco leaves are exposed to pure viral protein, the host cell synthesizes neither complete viral progeny nor any viral protein or viral RNA (Figure 1-7). When rubbed tobacco leaves are exposed to pure TMV RNA, however, hundreds of TMV progeny, composed of typical viral protein and viral RNA, are produced. Different strains of TMV and

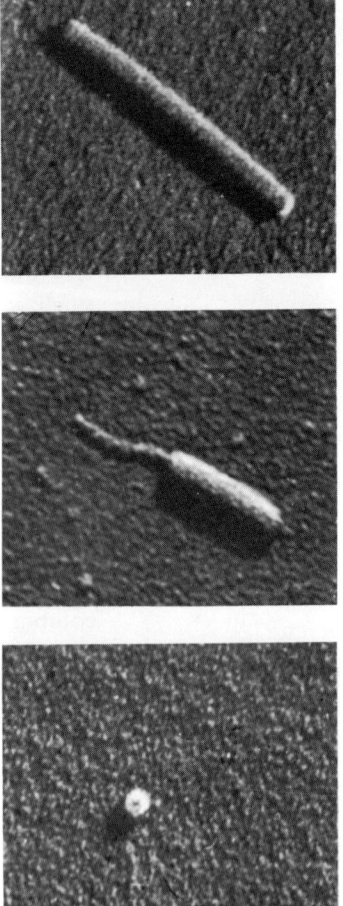

FIGURE 1-6. Tobacco mosaic virus as seen in the electron microscope. The particle is a rod (top) whose center is hollow (bottom). The walls of the cylinder are composed of about 2200 identical protein subunits (each containing 158 amino acids in a single polypeptide chain) arranged in a gentle spiral or helix through which is threaded a single molecule of RNA. The middle photograph shows a particle whose protein has been partially removed by treatment with detergent. (Courtesy of R. G. Hart.)

FIGURE 1-7. The infectivity of the separate components of TMV and of reconstituted TMV.

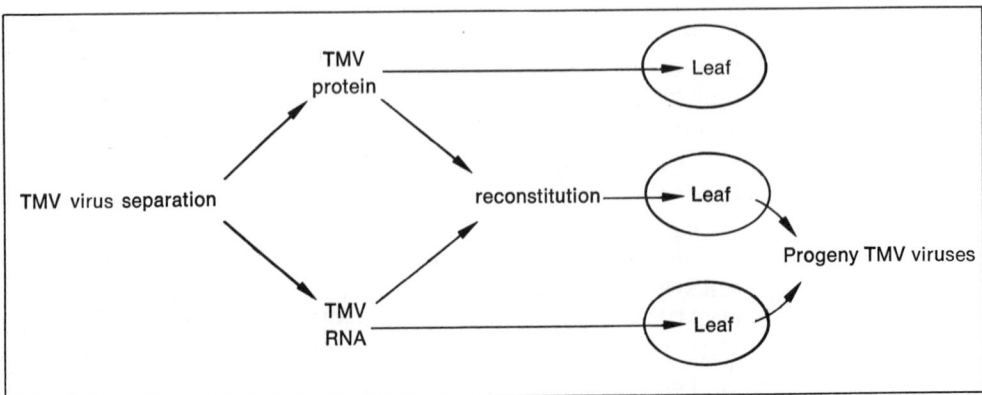

different strains of tobacco have been tested with the same result: viral progeny identical to the parent type are produced when the RNA fraction is used. When the viral RNA enters the tobacco cell, the host's metabolic machinery is somehow directed by the information in the RNA to synthesize viral RNA and protein. Since the RNA of TMV contains all the information needed to direct the synthesis of both viral protein and viral RNA in the host, RNA is the genetic material of TMV. The preceding results also prove that although TMV protein may function to protect the RNA and increase the efficiency of infection, it plays no essential role in the synthesis of either the RNA genetic material or of itself.

Other *RNA viruses* are also composed only of RNA and protein. These include viruses that attack animal cells, such as poliomyelitis, influenza, and encephalitis viruses, as well as some viruses that attack bacteria, called *bacteriophages*, or *phages*. In these cases, too, RNA isolated from the virus is infective and gives rise to complete progeny viruses. Thus, RNA is also the genetic material of these, and presumably all other, RNA viruses.

2. DNA VIRUSES. There are many *DNA viruses* composed of characteristic protein and deoxyribonucleic acid. One example is the small, rather simple, spherical bacteriophage X174 (Figure 1-8) that has the colon bacterium, *Escherichia coli*, as its host. The DNA and protein of DNA viruses can be separated and isolated by the phenol–water treatment and tested for infectivity as in the case of RNA viruses. When DNA isolated from $\phi$X174 ($\phi$ = Greek lowercase letter phi = phage) is mixed with *E. coli* cells, whose cell wall (located outside the cell membrane) is partially removed to facilitate the entry of polymers, numerous $\phi$X174 progeny viruses are produced. Since *E. coli* normally contains neither the DNA nor the protein of $\phi$X174, and since neither of these polymers is

FIGURE 1-8. The simple, spherical structure of phage X174, as seen in electron microscope photographs and, at the right, a model. The phage seems to consist of a protein shell or coat composed of 12 identical subunits arranged symmetrically around a core of DNA. (Courtesy of R. W. Horne.)

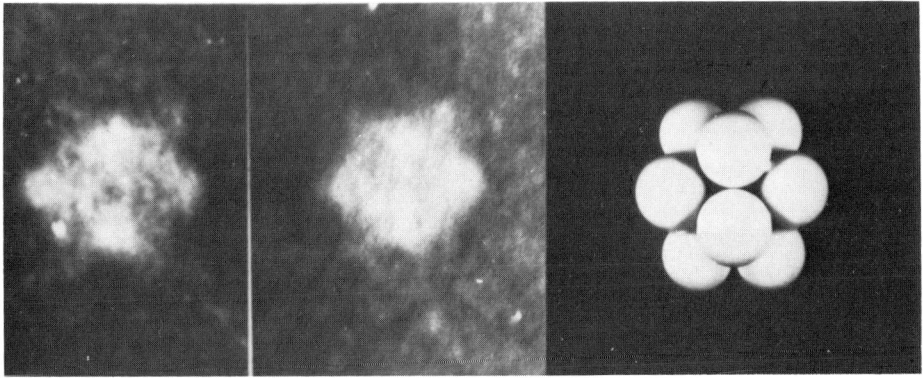

synthesized in suitably prepared *E. coli* exposed to φX174 protein, we conclude that DNA is the genetic material of φX174.

Other DNA phages of *E. coli* are larger and structurally more complex than φX174. Many of these phages look like a cellophane-covered lollipop with a tubular stick that is frayed at the free end. This is true, at least in part, of phage lambda (φλ) and of phages of the T series (Figure 1-9)—phages that have proved to be particularly useful for genetic investigation. The cellophane and the candy represent the protein coat and the DNA, respectively, of the *phage head*. The tubular stick represents the hollow *phage tail*; and the frayed ends, the *tail fibers*; both of these structures and the coat are composed of unique proteins. Infection is accomplished through attachment of the phage, tail first, to

FIGURE 1-9. The structure and form of T-even phage. **A**: Diagrammatic representation of an intact phage. **B**: Electron microscope photograph of φT4. The head (packed with DNA) has a tail structure (containing 24 striations) attached. At the base of the tail is the base plate, to which are attached six long tail fibers, kinked in the middle. These are the structures of primary attachment to the host cell of this phage, *Escherichia coli*. Magnification 300,000×. (Courtesy of T. F. Anderson.)

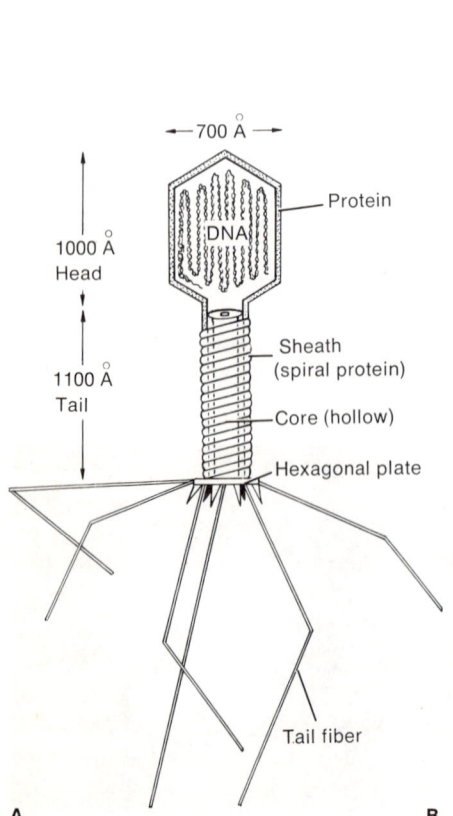
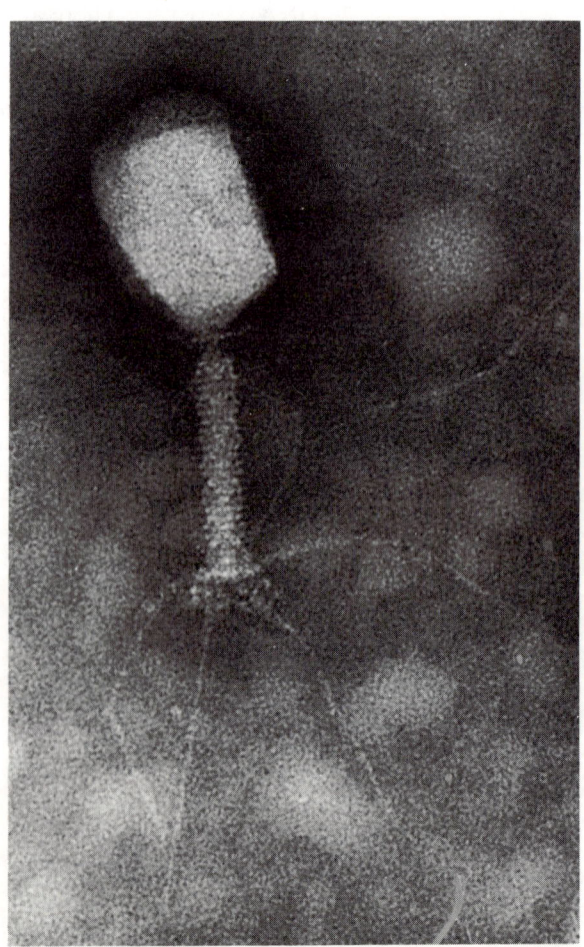

A

B

# GENETIC MATERIAL IS NUCLEIC ACID

the *E. coli* cell wall; a tail protein contracts, driving the core of the tail through the bacterial wall, after which all the DNA and a small amount of protein are injected into the host's protoplasm. After a period of incubation (whose events we consider in detail in Section 7.1), the host cell bursts, liberating numerous progeny viruses.

It has been possible to remove the relatively large molecule of DNA from λ and T phages in a pure and intact condition, and by infecting *E. coli* with it, to obtain progeny λ and T phages, respectively. This result proves conclusively that DNA is also the genetic material in the larger DNA phages. We presume that DNA is the genetic material in all DNA viruses.

**1.5** Any organismal nucleic acid that contains information used for its own reproduction is considered to be genetic material.

The term *chromosome* will henceforth refer to each separate thread or fiber (our cellular noodle) composed completely or partially of genetic material of DNA or RNA type. Some organisms contain one chromosome and others two to many chromosomes in their *chromosome sets*. A chromosome set is a group of one or more chromosomes composed of one chromosome of each kind normally present. In the human being, for example, a nuclear chromosome set contains 23 chromosomes.

In every organism we shall find some chromosomes or parts of chromosomes to contain information used for the synthesis of characteristic protein, and other chromosomes or parts of chromosomes to contain information used for other organismal purposes. When only a *portion* of the nuclei acid content of an organism is under consideration or investigation, we shall identify it as genetic material if it contains information used for its own replication—that is, if it contributes to *self-replication*.

**1.6** DNA is the main genetic material in cellular organisms.

Both DNA and RNA are present in all uninfected prokaryotic and eukaryotic cells (and in certain exceptional viruses). In organisms that contain both types of nucleic acid, the genetic material is mainly or exclusively DNA. We shall present proof in later chapters that bacteria, mitochondria, and chloroplasts contain a single DNA chromosome as their main or exclusive genetic material and that nuclei contain one or more DNA chromosomes as their main or exclusive genetic material.

DNA has been reported to occur in a variety of other structures found in eukaryotic cells—kinetoplasts, cell membrane, yolk platelets, centrioles. As we shall discuss in Section 13.9, we already know that kinetoplast DNA is self-replicating and hence, by definition, is genetic material.

RNA is detected wherever large amounts of DNA are found in the cell, whether in the protoplasm of prokaryotes or in the cytoplasm, nucleus, or organelles of eukaryotes. Most, perhaps all, of the RNA in uninfected cells is *not* genetic material.

**1.7** **The remainder of this book deals with genetics—the study of the properties, functions, and significance of self-replicating nucleic acid.**

The overall aim of this book is to summarize our basic knowledge about the genetic material of present-day organisms—those occurring in the last few frames of our motion picture—in order to understand the role of genetic material in past, present, and future biological evolution—that is, in the earlier, present, and future frames of the motion picture. Having identified genetic material as RNA in certain viruses and DNA in other viruses and all uninfected cells, we can pose a series of general questions whose answers will comprise the remainder of the subject matter of this book on *genetics*. Genetics can be defined as the study of the properties, functions, and significance of organismal nucleic acids that specify their own replication. The subject matter will emphasize the genetics of human beings and of organisms of great theoretical or practical importance to human beings. The overall goal of this treatment is to answer the following kinds of questions:

What are the chemical and physical characteristics of RNA and DNA? How is the genetic material organized in different organisms? Where and how do nucleic acids store information? (Part I—What the Genetic Material Is.)

How is the information in genetic material used in the synthesis of more nucleic acid (including itself and also, as it turns out, all the nongenetic RNA of a cell) and of protein? (Part II—What the Genetic Material Does.)

What kinds of abnormal changes occur in genetic nucleic acids? Can some of these errors in nucleic acid be repaired? What are the organismal consequences of those errors that are not repaired? Does the genetic material itself program the occurrence of variations? How is the genetic material shuffled (1) within an organism and (2) when it is transmitted to its progeny or other organisms? (Part III—How Genetic Material Is Varied, Packaged, and Distributed.)

What other kinds of information are contained in genetic material? How does the genetic material regulate its own synthesis, destruction, and variability? How does the genetic material select and regulate those portions of it that are functional in a given cell at a given time? (Part IV—How the Genetic Material Chooses Which Parts Are Present and Functional.)

How do the products of the functioning of genetic material affect each other? What are the consequences of such interactions on the structure, functioning, differentiation, and development of organisms? (Part V—How Gene Products Interact and the Phenotypic Consequences of Gene Action.)

How did genetic material originate and evolve to its present condition? How does genetic material behave in populations? What are the evolutionary consequences of this behavior? (Part VI—How the Preceding Came About in Individuals and Populations.)

What are the past, present, and future applications and implications of genetics to agriculture, ecology, behavior, social and political issues, medicine, and public health? (Part VII—The Present and Future Consequences of Genetics.)

## SUMMARY AND CONCLUSIONS

All present-day organisms on earth are probably descendants of a single ancestor. They also have similar structural requirements (one or more cells), physiological requirements (metabolism of protoplasm), and molecular requirements (proteins and nucleic acids, in particular) for maintenance, repair, growth, and reproduction. These characteristics of organisms depend ultimately upon a nucleic acid genetic material that (1) is stable, (2) contains information used in replicating itself and its modifications, and (3) carries the specifications for the organism's protein. Although the genetic material is RNA in some viruses and DNA in others, as proved in this chapter, the main genetic material of cellular organisms is DNA, as will be shown in later chapters.

This book is devoted to the study of the properties, functions, and significance of organismal nuclei acids that assist in their own replication—that is, the study of genetics. It is hoped that a knowledge of the genetics of present-day organisms will help us understand the role of genetics in past, present, and future biological evolution.

## QUESTIONS AND PROBLEMS

**1.1** *Anagrams*: Unscramble and define each of the following: **a.** sceeting. **b.** schroomcome. **c.** meroply. **d.** depipet. **e.** zenemy.

**1.2** *Riddle*: What functional organism has a head that cannot think and a tail that cannot wag?

**1.3** *Movies*: **a.** "Invisible Man." **b.** "Topper." **c.** "Heaven Can Wait." **d.** "Blithe Spirit." What genetic problems are entailed in these movies?

**1.4** The earth has just undergone a "close encounter of the third kind." How can we identify the genetic material of an extraplanetary visitor?

**1.5** Is the proof of the existence of genetic material a disproof of the existence of God? Explain.

**1.6** Explain your opinion of the statement: "Viruses are sometimes alive and other times not."

**1.7** State one evidence that the protein of TMV is not genetic material.

# Chapter 2

# Structural Organization of Nucleic Acids and Chromosomes

*[Handwritten notes:]*
*Polypeptide - polymer of amino-acid.*
*Polynucleotide - also called nucleic acid*
*Nucleotide - 1 organic base + 1 sugar + 1 phosphate*
*Purines - large organic base A + G*
*Pyrimidine - small organic base C + U (T in DNA)*

SINCE THE SCOPE OF GENETICS has been defined or delimited in terms of the properties, functions, and significance of genetic material, we will now consider the chemical and physical characteristics of nucleic acids. After this is done we shall describe the organization of nucleic acids in different kinds of chromosomes in various organisms.

# NUCLEIC ACID STRUCTURE

**2.1** Most RNA is a single-stranded polymer of nucleotides.

Many of the characteristics of RNA as it is ordinarily found in organisms can be represented by the single chain-like structure shown in Figure 2-1. The chain is made up of two kinds of alternating links, P's and R's. Attached only to the R links are rectangular plates of four kinds and two sizes. The larger plates are labeled A or G; the smaller plates are labeled C or U.

Chemically, the backbone is made up of alternating *phosphate* ($PO_4$) groups (shown as P in Figure 2-1) and *ribose* sugar groups (shown as R). To each ribose is attached one *organic base*. This base may be one of two larger types called *purines*, *adenine* (A) and *guanine* (G) (Figure 2-2), or one of two smaller types called *pyrimidines*, *cytosine* (C) and *uracil* (U) (Figure 2-3).

One organic base, in combination with one sugar and one phosphate, comprises a *nucleotide*. The RNA in Figure 2-1 can be considered a polymer obtained by joining nucleotides together into a single chain or strand.

16

# STRUCTURAL ORGANIZATION OF NUCLEIC ACIDS 17

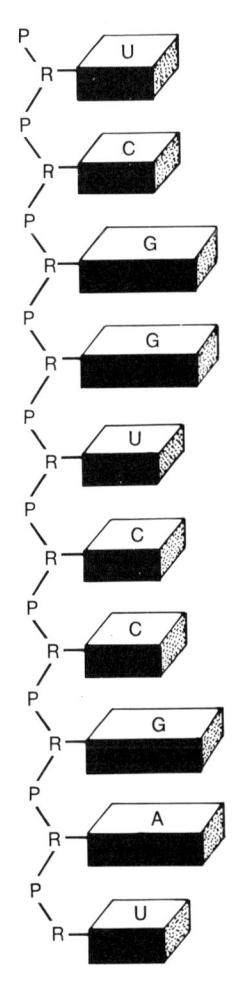

FIGURE 2-1. The RNA polymer. P = phosphate; R = ribose sugar; U = uracil; C = cytosine; G = guanine; A = adenine.

FIGURE 2-2. Purines commonly found in DNA and RNA.

Adenine

Guanine

FIGURE 2-3. Pyrimidines commonly found in nucleic acid. In DNA: cytosine and thymine; in RNA: cytosine and uracil.

Cytosine

Uracil

Thymine

## 2.2 Most DNA is a double-stranded polymer of nucleotides.

Many of the characteristics of DNA as it is ordinarily found in organisms can be represented by the ladder-like structure shown in Figure 2.4. The side chains of the ladder are made up of two kinds of alternating links, P's (*phosphates*) and D's (*deoxyribose* sugars). The rungs, attached only to the D links, are composed of rectangular plates of four kinds and two sizes, joined so that all rungs of the ladder are of uniform width. Specifically, each rung must be either C and G (or G and C) joined by three H's (*hydrogen* or *H bonds*), or A and T (*thymine*) (or T and A) joined by two H's.

The DNA ladder is actually composed of two half-ladders or two chains. Each of these chains can be considered a polymer of nucleotides. The two chains in double-stranded DNA are *complementary* to each other. Any base sequence is possible in one strand of DNA, provided that the other strand has a base sequence that correctly fills out the ladder. In other words, the bases in one chain must complement those in the

FIGURE 2-4. A chain ladder as a two-dimensional representation of double-stranded DNA. D = deoxyribose sugar; P = phosphate; A = adenine; T = thymine; G = guanine; C = cytosine; H = hydrogen.

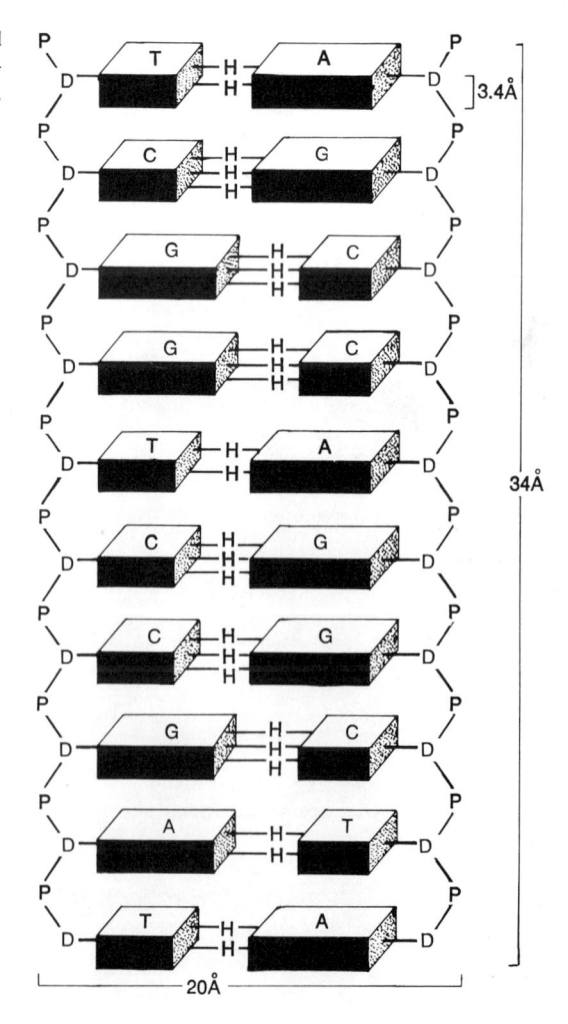

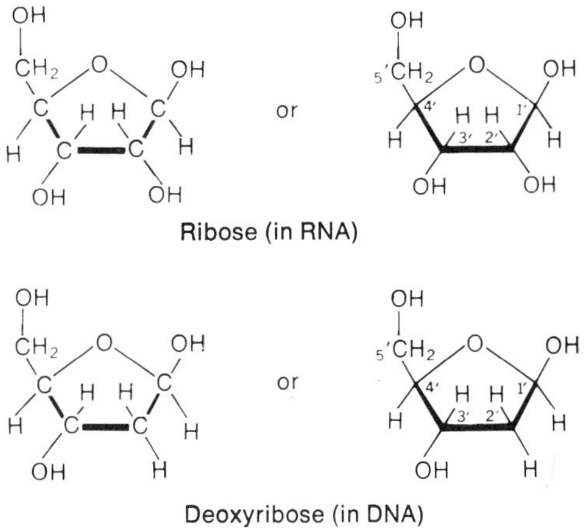

FIGURE 2-5. Simple sugars found in nucleic acids. Note that both diagrams to the right have been abbreviated by omitting the 4 C's in the ring.

Ribose (in RNA)

Deoxyribose (in DNA)

other—A must complement T, and G must complement C. Thus, if we know the base sequence in one chain of the ladder, we also know the base sequence in the complementary chain. For example, if one chain contains the sequence ATTGC, the other chain will have TAACG in the corresponding region. Therefore, in double-stranded DNA, the number of A's equals the number of T's and the number of G's equals the number of C's.

We see that RNA and DNA are similar in being polymers of nucleotides in which A, G, and C can be bases. RNA and DNA differ in three respects: (1) RNA is usually single-stranded, DNA is usually double-stranded. (2) Each nucleotide in RNA has a ribose and each one in DNA has a deoxyribose. These two sugars are very similar (Figure 2-5), since removal of one O (from the 2' position) in ribose yields deoxyribose. (3) RNA has U, whereas DNA has T as one of its possible bases. These two pyrimidines are also very similar (Figure 2-3), since the addition of a *methyl* ($CH_3$) group (to the 5 position) converts U to T. In other words, T is methyl U.

## 2.3 Double-stranded DNA is twisted to form a double helix whose strands face in opposite directions.

The final structural features of double-stranded DNA to be considered are dictated by the sugars, which face one way (with the O in the ring up) in one strand and the opposite way (with the O in the ring down) in the other (Figure 2-6). Because of this opposite arrangement of the sugars in the two strands, and because the sugar binds to an off-center position of its attached base, the whole DNA molecule is caused to twist or coil, forming a *double helix*. This necessity for coiling is not apparent in a two-dimensional representation, but is clear in a three-dimensional arrangement of the atoms. One can visualize a DNA double helix by imagining in our chain ladder that (1) the chains attach to off-center positions in a rung, so that there is more space on one side of the ladder than on the other, and (2) one end of the ladder is fixed while the other

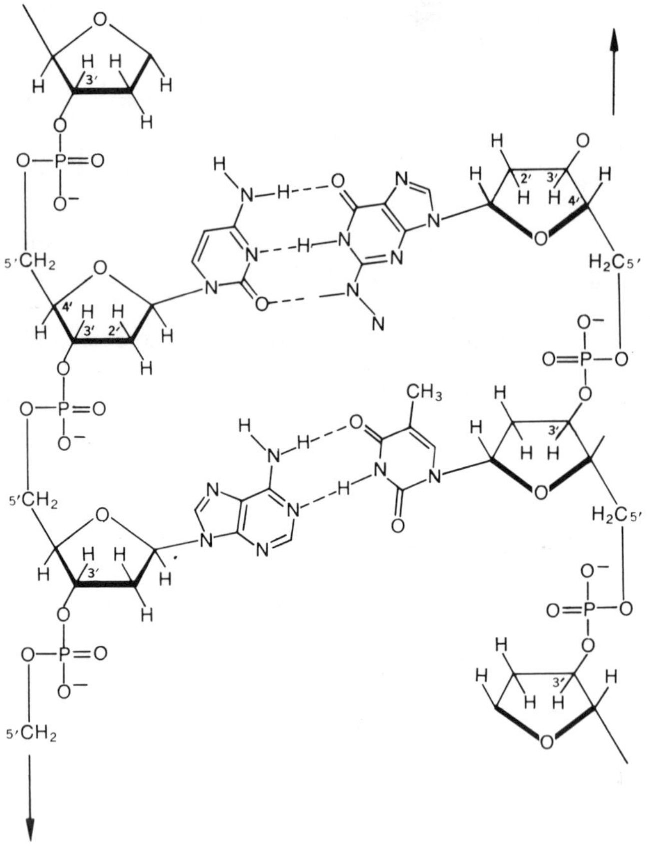

FIGURE 2-6. Two-dimensional representation of double-stranded DNA showing the opposite orientation of the sugars in the two strands and, therefore, the opposite direction in which the two strands run.

end is twisted in a clockwise direction. The result is a double helix in which the spaces on opposite sides are unequal in size. Figure 2-7 shows the larger or major and the smaller or minor spaces in the DNA double helix. Notice that since the sugar molecule has distinguishable right-side-up and upside-down ends, and since all the sugars in a strand of DNA (or RNA) face the same way, the polynucleotide is said to be *polarized* and the two strands in a double helix are said to have opposite *polarity*, that is, to run in opposite directions or *antiparallel* to each other (Figure 2-6).

**2.4** **Double-helix DNA was first described by J. D. Watson and F. H. C. Crick, who interpreted the results of M. H. F. Wilkins and coworkers.**

The solution of how DNA is chemically organized to form a double-stranded helix is probably the single most important advance in biology since the discovery of cells. As we shall see in Chapter 3, the requirement that bases bond in complementary pairs gives us immediate insight into the mechanism of DNA replication—the most important function of genetic material. This concept of DNA structure, proposed in 1953 by James D. Watson and Francis H. C. Crick, was based on a three-

STRUCTURAL ORGANIZATION OF NUCLEIC ACIDS 21

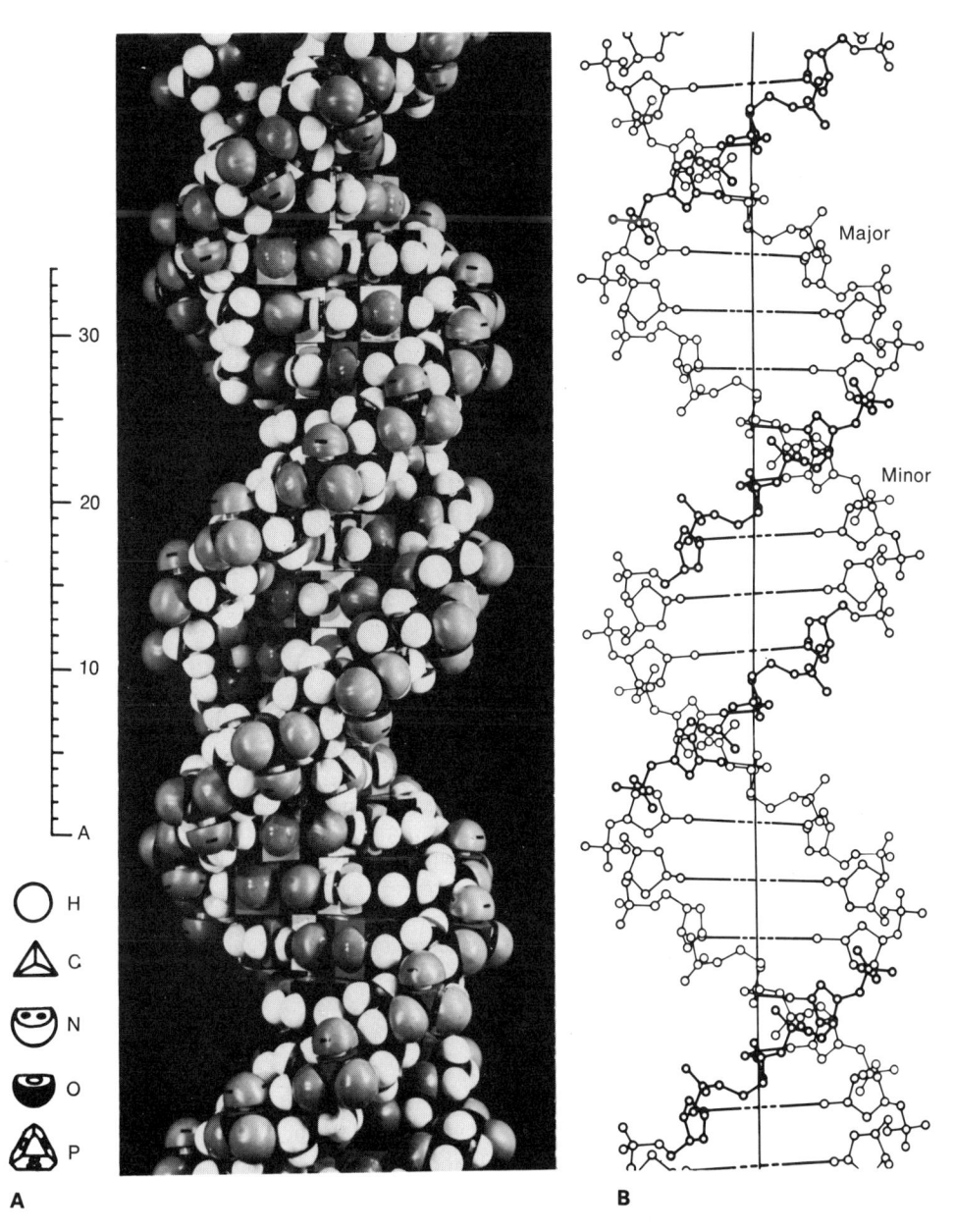

FIGURE 2-7. Molecular model of double-helical DNA. **A**: Space-filling model whose units represent different atoms. **B**: Its corresponding sugar–phosphate backbone, indicating the major and minor spaces. (Courtesy of M. H. F. Wilkins.)

dimensional molecular model they constructed. This model contained the correct number of chemical components arranged so as to fulfill certain dimensional requirements for DNA that had been discovered by Maurice H. F. Wilkins and coworkers. Wilkins had studied the patterns produced when X-ray radiation is diffracted (deflected or broken into components) by a group of DNA molecules arranged in parallel. These patterns give

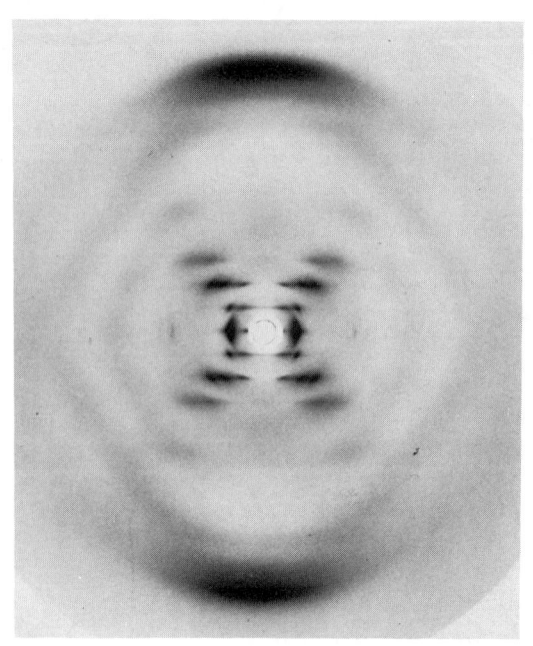

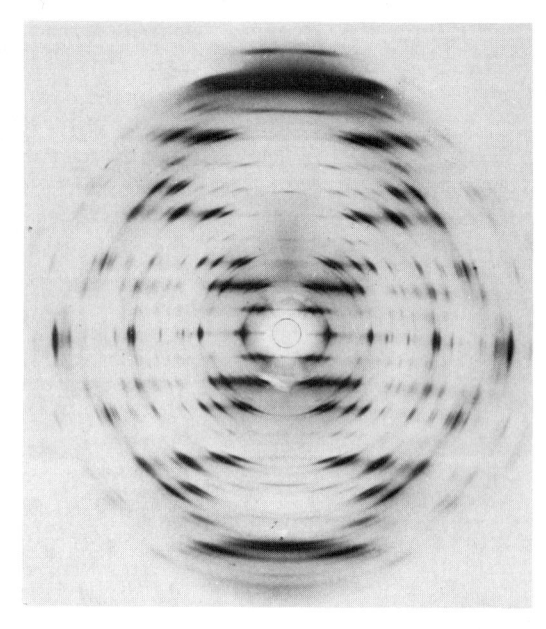

FIGURE 2-8. X-ray diffraction photographs of suitably hydrated fibers of DNA. **A**: Pattern obtained using the sodium salt of DNA. **B**: Pattern obtained using the lithium salt of DNA. The black spots located symmetrically near the upper and lower edges in both photographs correspond to nucleotides regularly spaced along a DNA chain. Other symmetrically arranged spots indicate other, larger, spacing regularities. (Courtesy of Biophysics Research Unit, Medical Research Council, King's College, London.)

information about distances between repeated features of such molecules (Figure 2-8). Three repeat distances—3.4 Å (Angstrom units), 20 Å, and 34 Å—were found (see Figure 2-4), which Watson and Crick correctly interpreted as representing (1) the distance between successive nucleotides in the polymer, (2) the diameter of each double-stranded DNA in the sample, and (3) the distance between complete turns of the helix each forms. In honor of this work Watson, Crick, and Wilkins were awarded Nobel prizes in 1962, and as a result, one strand of double-stranded DNA is sometimes called *Watson* (*W*) in genetic discussions and its complement is called *Crick* (*C*).

**2.5** The information in a genetic nucleic acid resides in the number, kind, and sequence of its nucleotides. Different organisms differ in such information.

Genetic nucleic acids contain the information needed to make organisms. In DNA or RNA genetic material, the information resides in the number, kind, and sequence of nucleotides present; or, more precisely, the number, kind, and sequence of bases the nucleotides contain (since the

# STRUCTURAL ORGANIZATION OF NUCLEIC ACIDS 23

| Organism | Nucleotide Pairs |
|---|---|
| Viruses | |
| $\phi$X174 | 4500* |
| T2 coliphage | $1.9 \times 10^5$ |
| Polyoma | $0.5 \times 10^4$ |
| Bacteriophage $\lambda$ | $0.5 \times 10^5$ |
| Bacteria | |
| Aerobacter aerogenes | $1.9 \times 10^6$ |
| Escherichia coli | $4 \times 10^6$ |
| Bacillus megaterium | $3 \times 10^7$ |
| Fungi | |
| Saccharomyces cerevisiae | $7 \times 10^7$ |
| Aspergillus | $4 \times 10^7$ |
| Porifera | |
| Tube sponge | $0.05 \times 10^9$ |
| Coelenterate | |
| Jellyfish, *Cassiopeia* sperm | $0.3 \times 10^9$ |
| Echinoderm | |
| Sea urchin, *Lytechinus* sperm | $0.8 \times 10^9$ |
| Annelid | |
| Nereid worm, sperm | $1.4 \times 10^9$ |
| Mollusks | |
| Limpet, *Fissurella bandadensis* sperm | $4.7 \times 10^9$ |
| Snail, *Tectorius muricatus* sperm | $6.3 \times 10^9$ |
| Insecta | |
| Drosophila | $0.8 \times 10^8$ |
| Crustacean | |
| Cliff crab, *Plagusia depressa* sperm | $1.4 \times 10^9$ |
| Chordate | |
| Tunicate, *Asidea atra* sperm | $0.15 \times 10^9$ |
| Vertebrates | |
| Dipnoan | |
| Lungfish, *Protopterus* | $47 \times 10^9$ |
| Amphibia | |
| Frog | $23 \times 10^9$ |
| Toad | $7 \times 10^9$ |
| Necturus | $3.4 \times 10^9$ |
| Amphiuma | $90 \times 10^9$ |
| Elasmobranch | |
| Shark, *Carcharias obscurus* | $2.6 \times 10^9$ |
| Teleost | |
| Carp | $1.7 \times 10^9$ |
| Reptiles | |
| Green turtle | $2.5 \times 10^9$ |
| Alligator | $2.4 \times 10^9$ |
| Birds | |
| Chicken | $1.1 \times 10^9$ |
| Duck | $1.2 \times 10^9$ |
| Mammals | |
| Dog | $2.5 \times 10^9$ |
| Man | $2.8 \times 10^9$ |
| Horse | $2.8 \times 10^9$ |
| Mouse | $2.4 \times 10^9$ |

FIGURE 2-9. DNA content per chromosome in some viruses and some prokaryotic or eukaryotic cells.

*For double-stranded replicative form

sugars and phosphates alternate monotonously to make up the backbone of the polymer). As expected, different species differ in the amount of genetic nucleic acid they contain (Figure 2-9). It is generally true that the more complex the organism, the larger the amount of DNA per chromosome set. For example, the number of pairs of nucleotides (equal

| Species | Base, % | | | |
|---|---|---|---|---|
| | Adenine | Thymine | Guanine | Cytosine |
| Man (sperm) | 31.0 | 31.5 | 19.1 | 18.4 |
| Chicken | 28.8 | 29.2 | 20.5 | 21.5 |
| Salmon | 29.7 | 29.1 | 20.8 | 20.4 |
| Locust | 29.3 | 29.3 | 20.5 | 20.7 |
| Sea urchin | 32.8 | 32.1 | 17.7 | 17.7 |
| Yeast | 31.7 | 32.6 | 18.8 | 17.4 |
| Tuberculosis bacillus | 15.1 | 14.6 | 34.9 | 35.4 |
| *Escherichia coli* | 26.1 | 23.9 | 24.9 | 25.1 |
| Vaccinia virus | 29.5 | 29.9 | 20.6 | 20.3 |
| E. coli bacteriophage T2 | 32.6 | 32.6 | 18.2 | 16.6* |

*Hydroxymethylcytosine, a derivative of cytosine.

FIGURE 2-10. Base composition of DNA from various organisms.

to base pairs) is roughly $10^5$ in T phages, $10^6$ in *E. coli*, $10^8$ in yeast, and $10^9$ in human beings.

Different organisms with the same quantity of DNA genetic material, such as a worm and a crab (see Figure 2-9), must differ in the quality of their DNA information—that is, in the relative amounts of the four types of bases (Figure 2-10) and/or their sequence in the polymer. Different members of a single species with a large DNA content ordinarily have no detectable differences in base composition or amount. (In human beings, however, a net shift of several hundred base pairs from AT to GC, or the reverse, would shift the percentages of bases so little as to go undetected.) Within a species many of the genetic differences are probably due to differences in base sequence.

**2.6 Techniques are now available to sequence the bases in DNA and RNA.**

Single-stranded pieces of RNA and DNA up to 100 nucleotides long are now readily base-sequenced. When the DNA or RNA is longer than this, it can be fragmented into suitably sized pieces by enzymes called *nucleases*, which preferentially break, nick, or scission a polynucleotide at specific places in its backbone. Different nucleases have different molecular requirements. For example, some nucleases degrade only DNA, others degrade only RNA; some work only on double-stranded regions, others work only on single-stranded regions; some can attack a nucleic acid only at its terminus, removing one end nucleotide at a time; others can attack at a more internal position when they encounter certain bases or combinations of bases. In short, nucleases are available to degrade a nucleic acid in many, and specific ways.

Sequencing of bases can be accomplished by treating nucleic acid with successive nucleases of specific types and then analyzing the fragments produced. Another, relatively simple, way to sequence DNA is to treat a

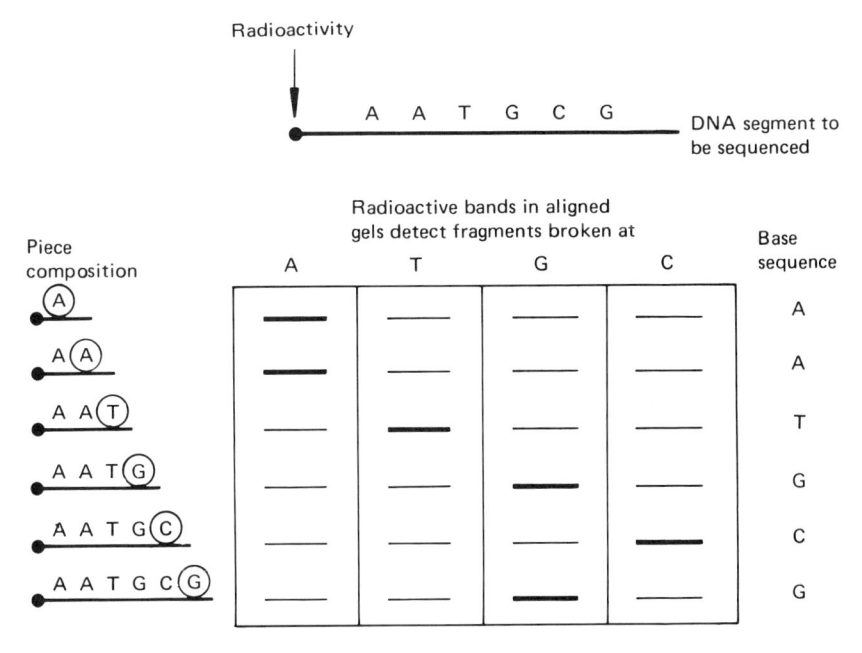

FIGURE 2-11. Sequencing bases in single-stranded DNA. See the text for details.

sample containing many copies of a DNA strand with a specific chemical reagent that preferentially breaks different strands at a single A or T, C, or G. When the same end of the DNA copies is radioactive, the result is a series of radioactive fragments of different sizes that will migrate in a gel in strict accordance with their length. When four gels produced from the four different chemical cleavages are aligned (Figure 2-11), the base sequence can be read directly from the pattern of radioactive bands.

**2.7 Single-stranded nucleic acids can have double-helical regions if folding brings complementary regions together.**

If single-stranded DNA folds on itself so that the bases in one region are brought adjacent to complementary bases in another region and the regions are antiparallel, base pairing will occur and produce a double-helical region within the single strand (Figure 2-12, part 2). Several naturally occurring, single-stranded RNA's also have double-helical regions that are composed of antiparallel sequences that have formed base pairs—G and C being complements, as in DNA, and A and U being complements (as expected, since U is so similar to T). Figure 2-13 shows a "t" RNA where three regions fold so that adjacent sequences can base-pair and where the two terminal regions base-pair; Figure 2-14 shows a "5S r" RNA, where two regions fold so adjacent sequences can base-pair and where the terminal regions base-pair; Figure 2-15 shows the many double-helical regions present in a 387-nucleotide-long segment of the RNA chromosome of $\phi$R17, which give the whole chromosome segment the appearance of a flower.

26    WHAT THE GENETIC MATERIAL IS

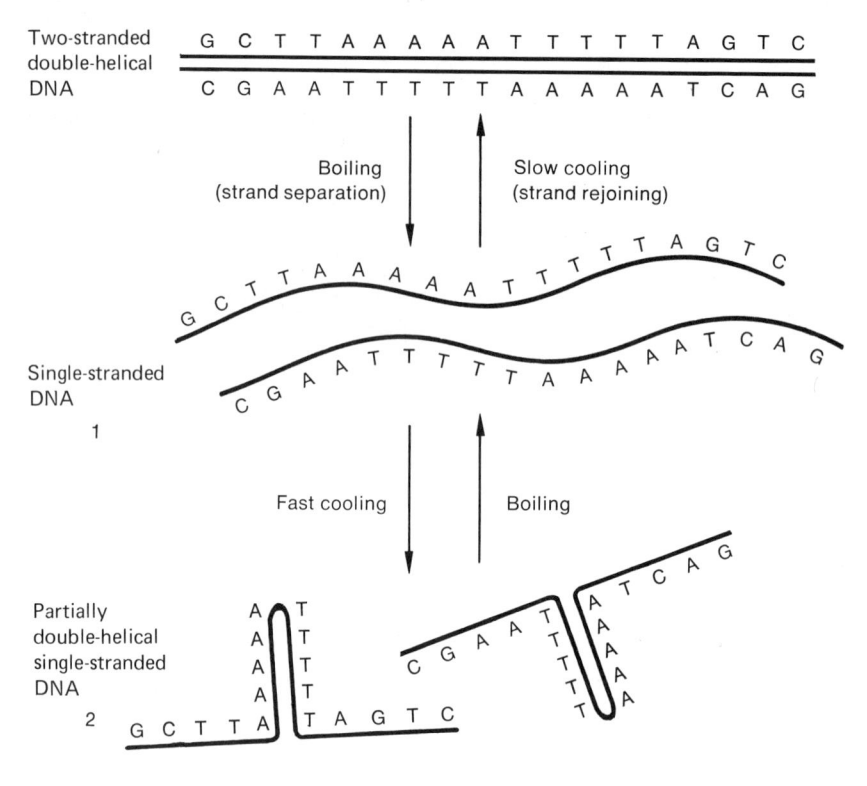

FIGURE 2-12. Strand separation and joining of two-stranded, double-helical DNA: (1) has no base pairs and, therefore, the maximum absorption of ultraviolet light; (2) has some base pairing and, therefore, less than the maximum absorption of UV light.

FIGURE 2-13. The base sequence and cloverleaf structure of a tRNA molecule from yeast. DiMe, dimethyl; Me, methyl; DiH, dihydro; $\psi$, pseudouracil (rearranged U); I, hypoxanthine.

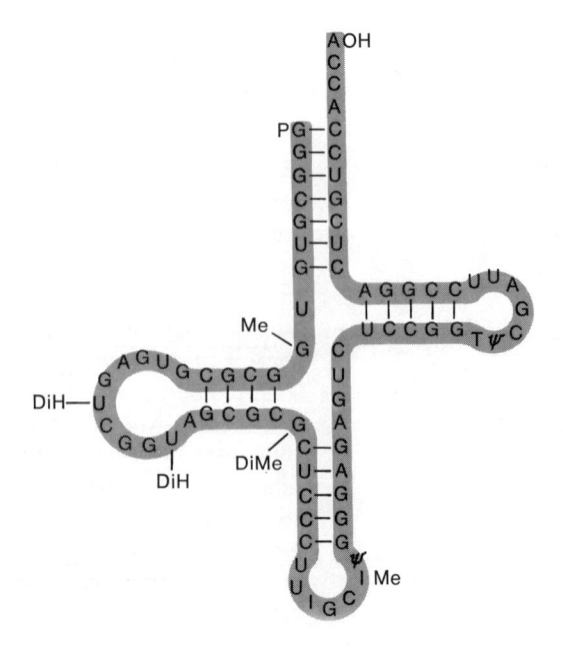

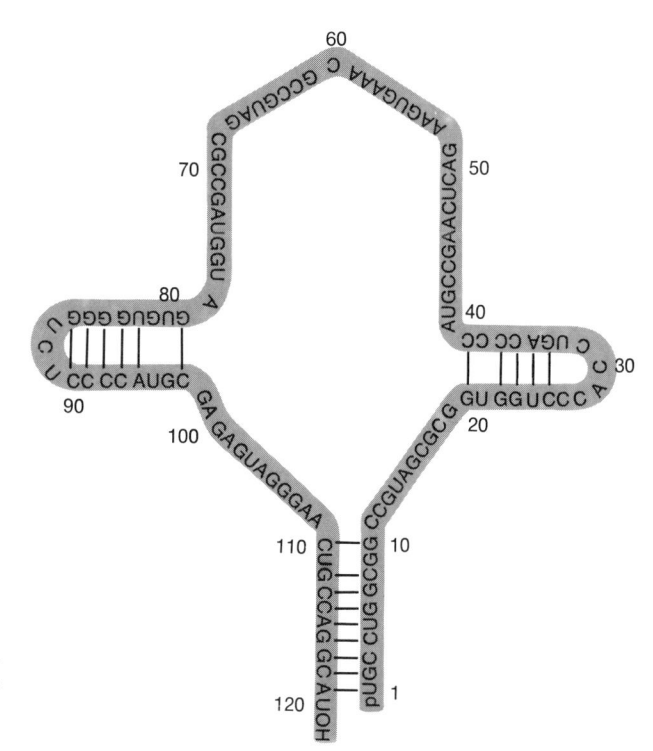

FIGURE 2-14. Base sequence and pairing in 5S rRNA of *E. coli*. Solid line, standard base pair (G—C or A—U). (After G. G. Brownley, F. Sanger, and B. G. Barrell.)

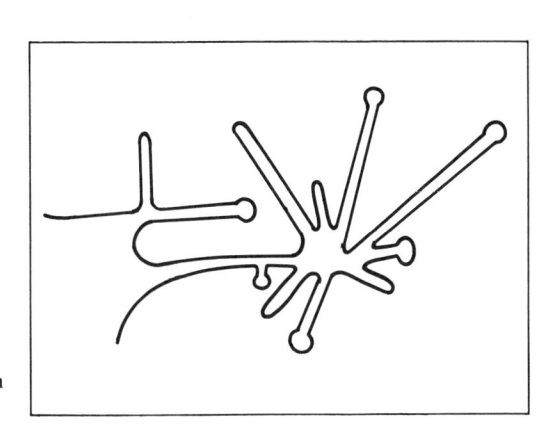

FIGURE 2-15. The flower configuration of a segment of an RNA phage chromosome.

**2.8 Double helices are destroyed by heating and may be formed upon cooling.**

When partially double-helical, single-stranded RNA is heated sufficiently, energy is provided for the breaking of H bonds. The members of each base pair separate from each other, and the RNA becomes single-stranded throughout. When the solution is cooled, however, the molecule will tend to re-form the double-helical regions.

When completely double-stranded DNA is heated sufficiently, all H bonds will break and the two complementary strands will separate, provided that one or both strands have ends that make the necessary unwinding possible (Figure 2-12). (The two strands in double-stranded DNA are twisted around each other like the strands of a two-stranded rope and can only be separated if at least one of the two strands has ends that are free to spin.) When the heated mixture is cooled slowly, there is adequate time for Watsons to find their Cricks and the double-stranded condition is often re-formed (Figure 2-12). When the heated mixture is cooled quickly, however, the strands generally remain single. Like single-stranded RNA, a single strand of DNA can base-pair with itself if folding brings together complementary parts of the strand in an antiparallel manner. This pairing may occur between adjacent sequences (Figure 2-12, Part 2) or nonadjacent ones, as occurs in single-stranded RNA.

**2.9** The ability to destroy and re-form double-helical nucleic acids experimentally has important applications in genetic studies.

1. MAPPING AT-RICH REGIONS IN DOUBLE-STRANDED DNA. It is possible to map the location of regions in double-stranded DNA that contain runs of successive AT base pairs. The DNA is heated to a temperature sufficient to break the two H bonds joining each AT pair but insufficient to break the three H bonds joining each GC pair. The result is observed in the electron microscope. The magnification of the electron micrograph is insufficient to show individual base pairs. It is sufficient, however, to show that a duplex has separated into its two component strands wherever it has runs of AT pairs and that it has not done so elsewhere (Figure 2-16).

FIGURE 2-16. Strand-separation mapping of duplex DNA. Above: Electron micrograph of partially strand-separated φλ DNA. Below: The corresponding map of its AT-rich regions. (Courtesy of R. B. Inman.)

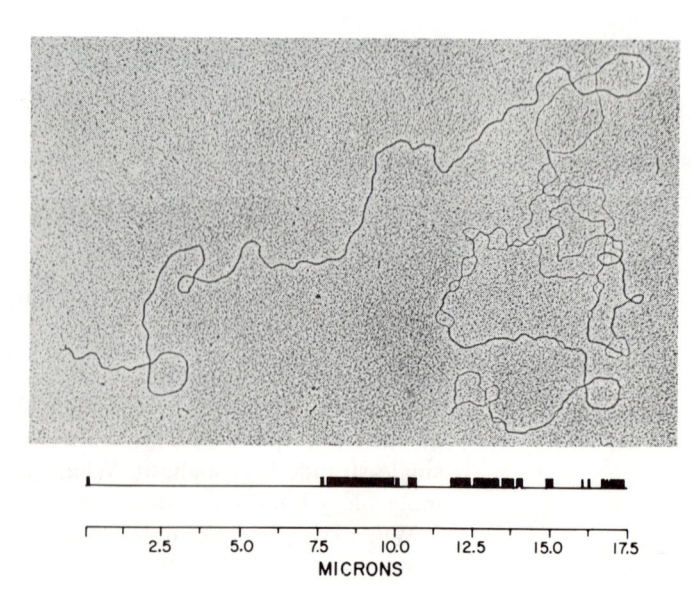

2. BASE-RATIO DETERMINATION. The organic bases in nucleic acids characteristically absorb light in the *ultraviolet* (*UV*) range at a wavelength of about 2600 Å. DNA absorbs as much as 40 per cent more at this wavelength, however, in its single-stranded condition than it does in its double-stranded condition. This difference can be used to determine the base ratio of duplex DNA, by recording the amount of UV light absorbed while the DNA is heated. DNA relatively rich in A and T will show its greatest increase in UV absorption at a comparatively low temperature (since the AT base pair has only two H bonds to break), while DNA that is rich in G and C will show this phenomenon at a higher temperature (since the GC base pair has three H bonds to break). One therefore can determine the base composition of a sample of unknown duplex DNA by comparing its absorption profile with that determined for duplex DNA's of known base compositions.

3. FORMATION OF MOLECULAR HYBRIDS. Double-stranded hybrid nucleic acids can be formed by combining single strands that are sufficiently complementary.

   a. *Hybrid DNA–RNA duplexes* (each composed of one DNA strand and one RNA strand held together by CG, AT, and AU pairs). Hybrid DNA–RNA duplexes can be made by filtering single-stranded RNA through a jelly-like semisolid made of agar and water in which single-stranded DNA from the same organism has been trapped. When the DNA is in excess, almost no RNA passes through the filter, showing that all or almost all RNA from a given organism is complementary to its own DNA. When the DNA and RNA are derived from widely different organisms, however, much of the RNA passes through the filter because its base sequences are not complementary to those in the DNA and does not form DNA–RNA hybrid molecules.

   b. *Hybrid DNA–DNA duplexes.* Hybrid DNA–DNA duplexes made up of single DNA strands from two different species can also be formed using the filtration technique, if the genetic material of these species contains similar base sequences. Molecular hybridization, therefore, can be a particularly useful technique in determining the evolutionary similarity of DNA's from different organisms.

   c. *Mapping by hybridization.* Since an individual's RNA is complementary to its DNA, it is possible to map the place in DNA that is complementary to a given segment of RNA. This is done by forming a DNA–RNA hybrid molecule that is detected in one of two ways. If the duplex DNA is relatively short, its strands are separated and permitted to hybridize with the RNA. The double-stranded regions formed can be differentiated in electron micrographs from the single-stranded regions, and, therefore, can be mapped (Figure 2-17). If the duplex DNA is long and can be attached to a slide, the DNA is heated and cooled quickly to separate the complementary strands, which will still remain attached to the slide, and is then exposed to a given segment of RNA that is radioactive. The slide is stained for DNA so that the chromosome can easily be seen in the ordinary light microscope. It is then washed to remove nonhybridized RNA and exposed to a photographic emulsion

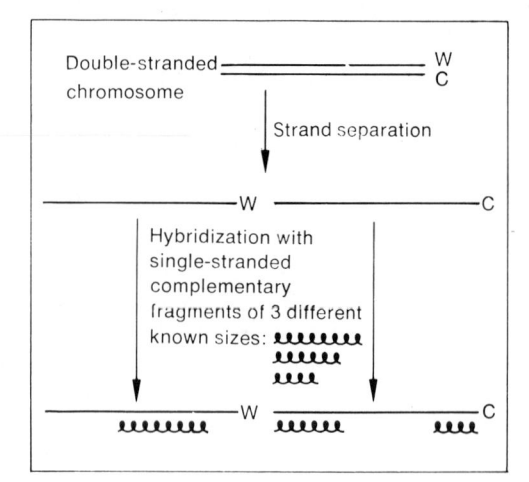

FIGURE 2-17. Mapping by hybridization between a short strand-separated chromosome and complementary nucleic acid fragments. The longest fragment is complementary to W, the other two are complementary to regions in C that are relatively far apart.

FIGURE 2-18. Mapping by hybridization between a long chromosome, which has been strand-separated but kept in place on a slide, and complementary radioactive fragments.

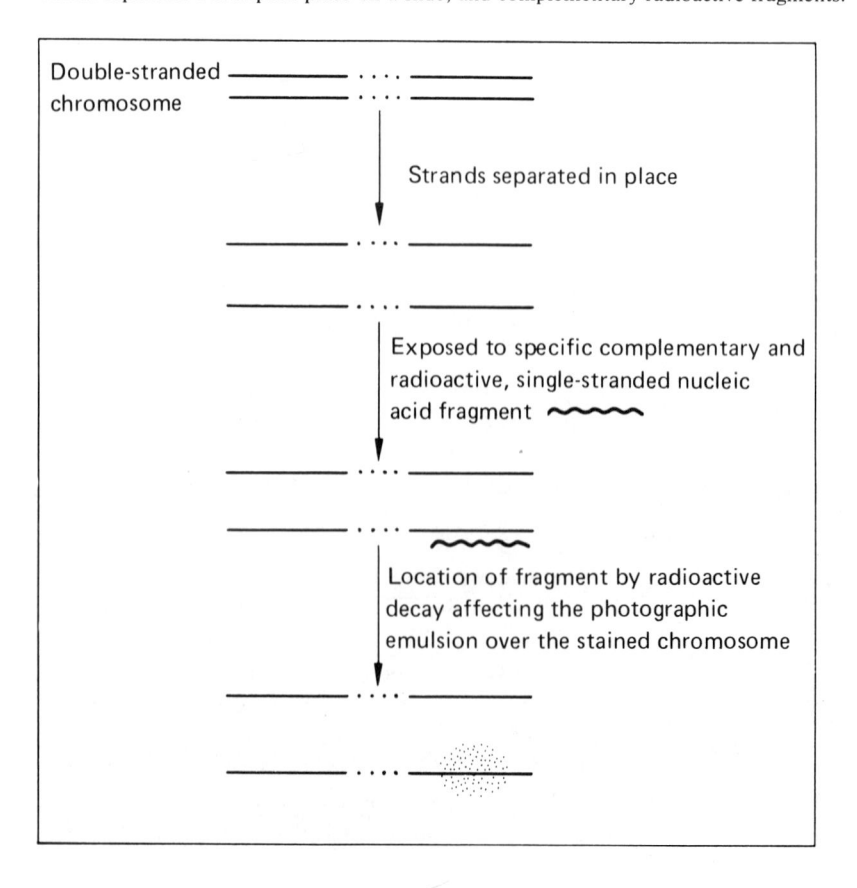

# STRUCTURAL ORGANIZATION OF NUCLEIC ACIDS 31

that detects the position of radioactivity and hence the RNA that has hybridized (Figure 2-18).

# CHROMOSOME ORGANIZATION

**2.10** RNA and DNA virus chromosomes may be single-stranded rods or rings or double-stranded rods. DNA virus chromosomes may also be double-stranded rings.

RNA VIRUSES. Some observations have already been made on the way RNA chromosomes are packaged. In the case of TMV, for example, the chromosome occurs as a single strand of RNA in the form of a *rod* (or nonring) that does not seem to be extensively folded in the mature virus (part A, Figure 2-19). In small RNA phages, however,

FIGURE 2-19. The nucleic acid conformation and content of viral chromosome sets.

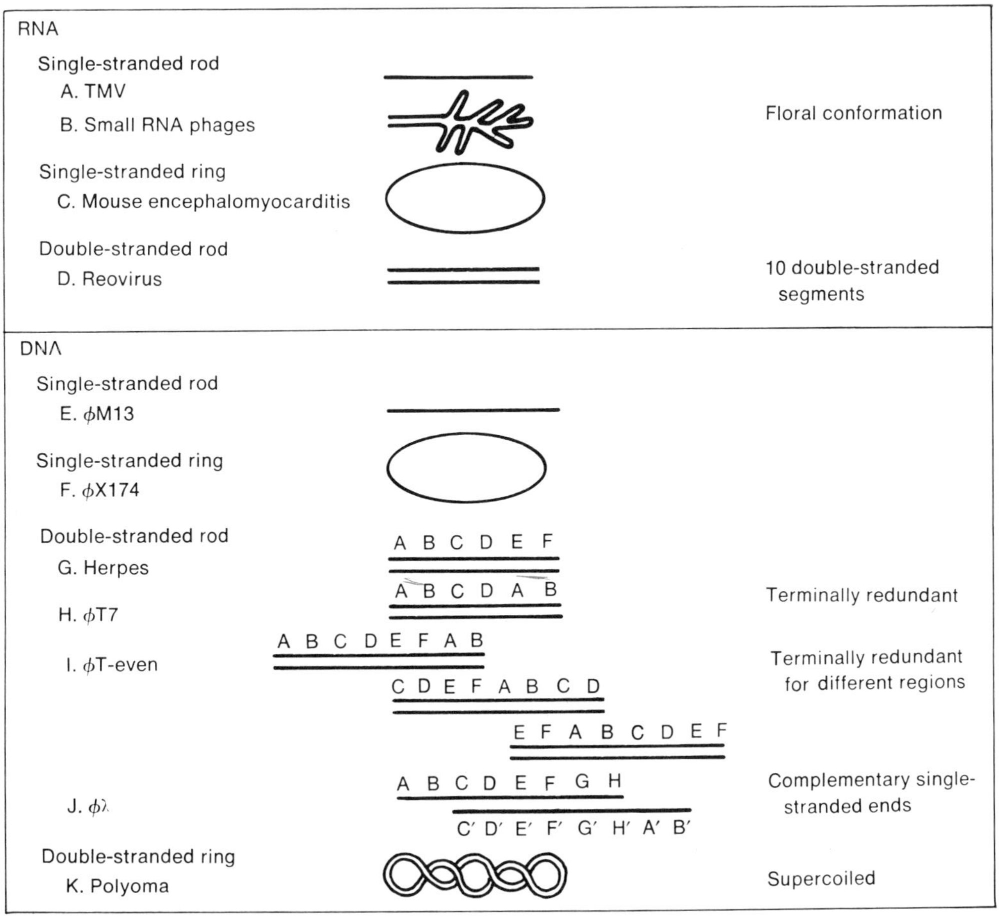

```
RNA
  Single-stranded rod
    A. TMV
    B. Small RNA phages                              Floral conformation
  Single-stranded ring
    C. Mouse encephalomyocarditis
  Double-stranded rod
    D. Reovirus                                      10 double-stranded
                                                     segments
DNA
  Single-stranded rod
    E. φM13
  Single-stranded ring
    F. φX174
  Double-stranded rod
    G. Herpes            A B C D E F
    H. φT7               A B C D A B                 Terminally redundant
    I. φT-even    A B C D E F A B
                  C D E F A B C D                    Terminally redundant
                      E F A B C D E F                for different regions
                  A B C D E F G H                    Complementary single-
    J. φλ         C' D' E' F' G' H' A' B'            stranded ends
  Double-stranded ring
    K. Polyoma                                       Supercoiled
```

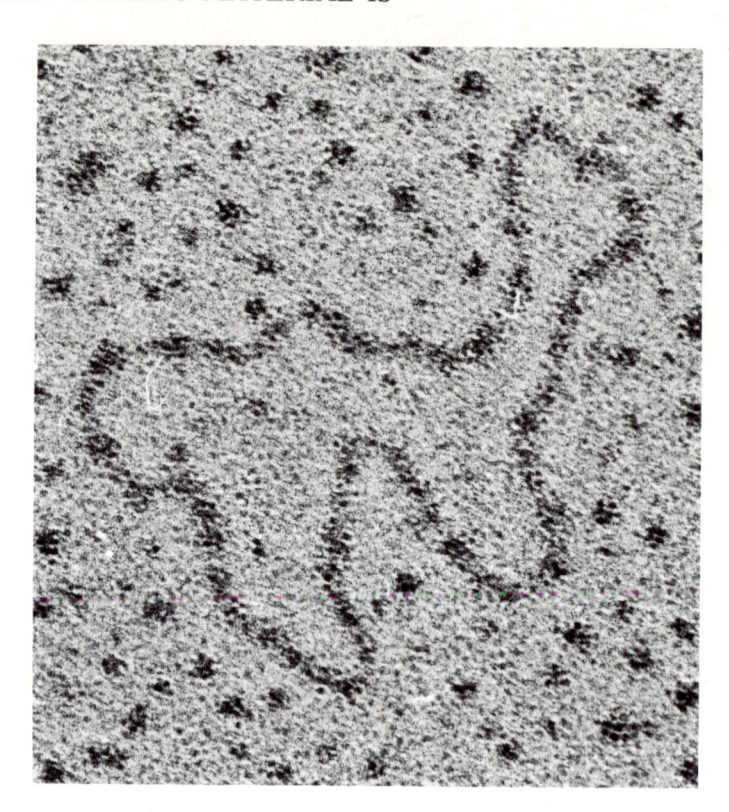

FIGURE 2-20. Single-stranded circular DNA chromosome of φX174. (Courtesy of D. Dressler and J. Wolfson.)

the single-stranded RNA chromosome seems to be extensively folded in the mature virus, to pack the RNA in a small volume, and to have a floral appearance in the host (part B, Figure 2-19). The RNA chromosome is sometimes a single-stranded *ring* (part C, Figure 2-19). Several viruses, including a phage, have double-stranded RNA as their genetic material, which in the particular case of reovirus is composed of 10 double-stranded chromosomes (part D, Figure 2-19).

DNA VIRUSES. Some DNA phages have a single-stranded rod chromosome (part E, Figure 2-19), others such as φX174 have a single-stranded ring chromosome (part F, Figure 2-19; Figure 2-20). Many viruses have double-stranded rod DNA. These chromosomes are of four types:

1. In some viruses there is nothing special about the sequences at the ends of the chromosome (part G, Figure 2-19, where each different part of the chromosome is designated by a different letter).
2. In some phages, however, the base sequence at one end is repeated in the same sequence at the other end, so that each chromosome, besides containing all sequences, has the same repeat or *redundancy* (part H, Figure 2-19, where the AB region is *terminally redundant*).
3. Other phages are also terminally redundant for a certain length of the chromosome, which, however, is from different regions in different individuals of the same type of phage (as in certain T phages; part I, Figure 2-19). Such an array of chromosomes can be obtained in the following way. One

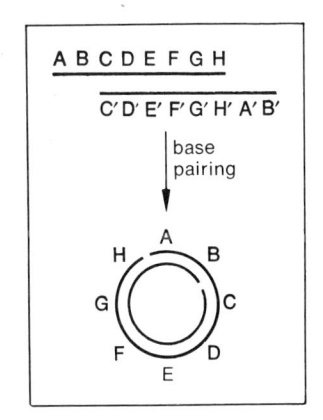

FIGURE 2-21. Formation of a doubly nicked double-stranded ring by means of base pairing between complementary single-stranded ends of an otherwise double-stranded rod nucleic acid.

long duplex DNA is formed that is composed of repeats of a six-letter chromosome (ABCDEFABCDEFABCDEF...). Mature phage chromosomes are then obtained by cutting the long DNA chromosome into successive lengths of eight letters, starting at the left. This produces, as in the figure, chromosomes ABCDEFAB, CDEFABCD, and EFABCDEF, which are terminally redundant for successive segments of a chromosome.

4. Still other mature phage chromosomes that are not redundant have complementary single-stranded ends (as in $\phi\lambda$; part J, Figure 2-19, where region A is complementary to A' and B to B'). These ends, if allowed to base-pair, would form a double-stranded ring, each of whose strands is broken or nicked once at nearby places (Figure 2-21).

Finally, there are DNA chromosomes that are double-stranded rings (as in polyoma virus; part K, Figure 2-19). Such chromosomes have additional twists locked into their structure that make the molecule supercoil and collapse on itself (part A, Figure 2-22). (A rubber band becomes supercoiled as it continues to be twisted.) When such supercoiled DNA is exposed to a nuclease, the nicks produced furnish free ends that permit the molecule to unwind these extra twists and assume an expanded circular configuration (parts B and C, Figure 2-22).

The supercoiled form of polyoma DNA must help in packing the double ring (with its protein) into the relatively small volume of the mature virus (Figure 2-23), just as folding RNA does in the case of the smaller RNA phages. In the case of T phages (Figures 2-23 and 2-24), where the DNA enters the empty phage head during the production of mature phage, the packing of the DNA seems to be accomplished either by folding the DNA back and forth, in the fashion of a firehose, or by spinning it into a ball from the outside in (which, like a ball of twine, can be unwound from the inside out without being turned). The result of this regular packing is that the DNA can pass through the hollow core of the tail and into the new host without breaking or forming knots.

## 2.11 Bacterial chromosomes are folded and supercoiled in a regular manner.

The problem of packing DNA exists on the same scale, if not a greater one, in prokaryotes. Figure 2-25 shows the double-stranded ring

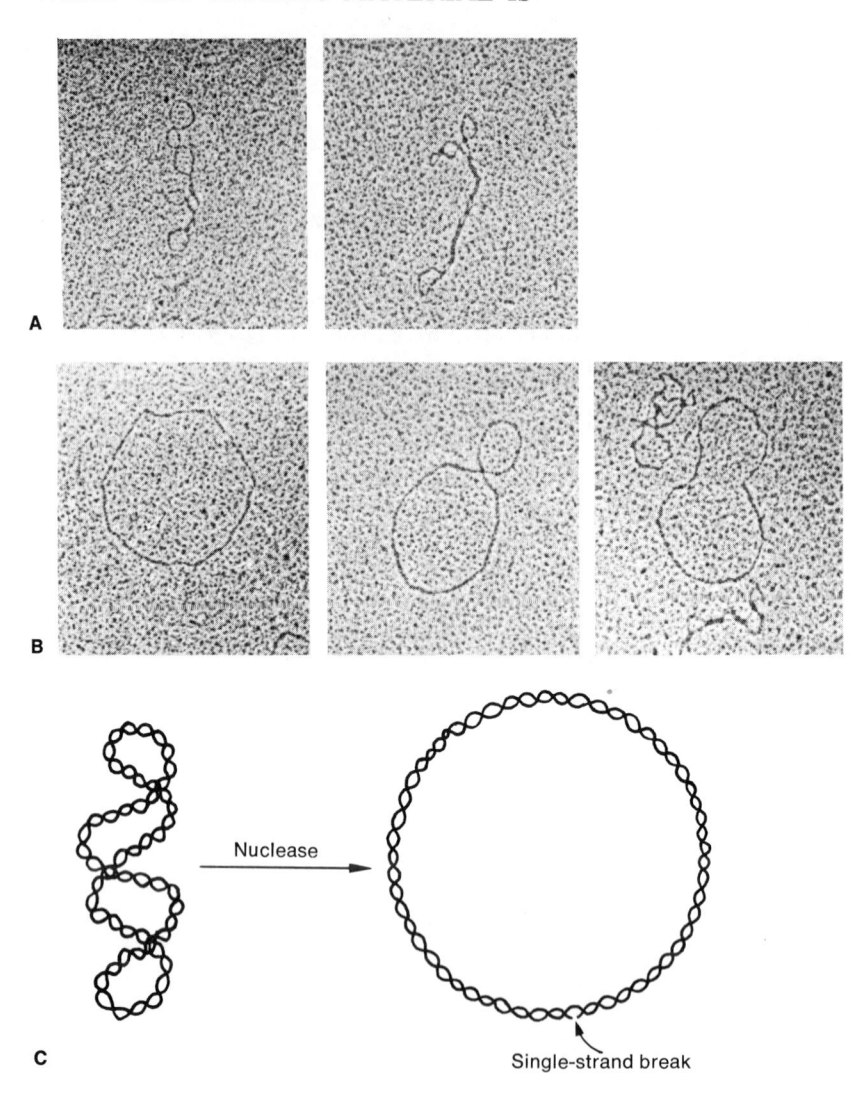

FIGURE 2-22. Electron micrographs of the supercoiled (**A**) and untwisted (**B**) forms of the ring duplex DNA of polyoma (33,000×). (Courtesy of J. Vinograd.) **C**: The conversion of the supercoiled to the untwisted form.

chromosome that is normally included in a bacterium. This chromosome is able to fit inside a bacterium because it is folded and supercoiled (as in *E. coli*, Figure 2-26). Supercoiling produces spirals with four to seven turns. How the folds and supercoils produce the condensation of the bacterial chromosome is shown in the model of Figure 2-27. Such a chromosome organization must not only serve to pack the DNA within the cell, but it must be advantageous for chromosome functioning, and permit the chromosome to replicate and be distributed to daughter cells efficiently without breaking or tangling.

# STRUCTURAL ORGANIZATION OF NUCLEIC ACIDS 35

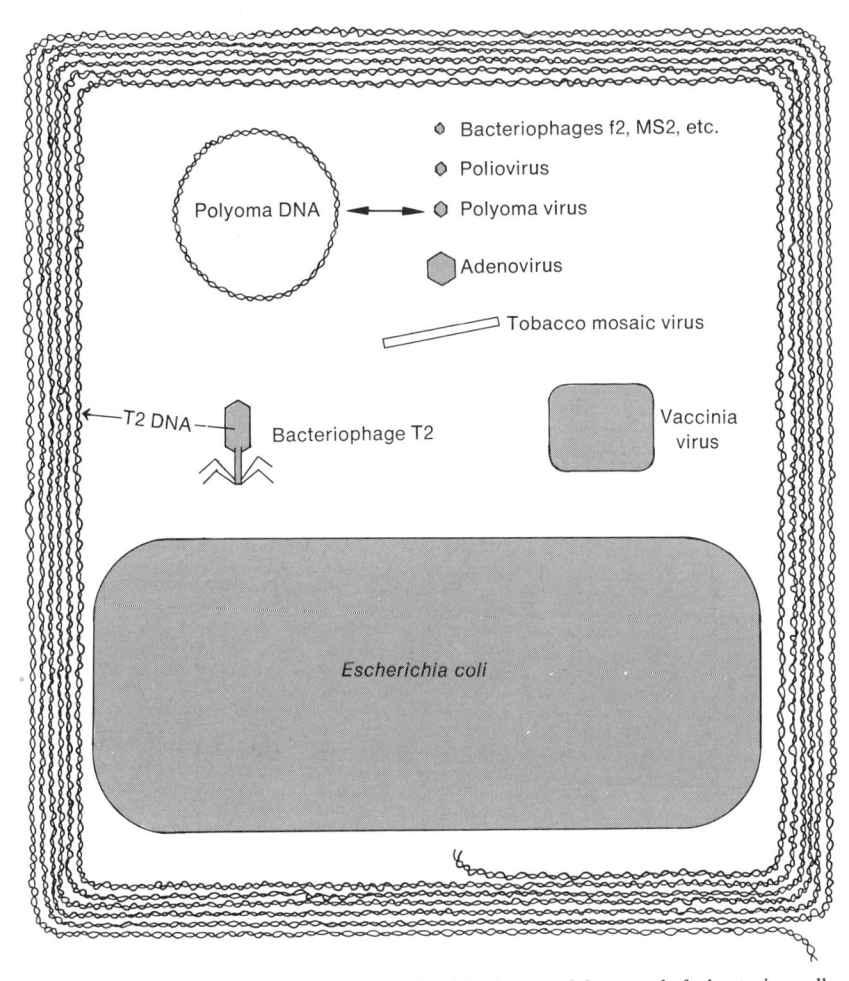

FIGURE 2-23. Sizes of viruses, of the nucleic acids of some of them, and of a bacterium, all drawn to the same scale.

FIGURE 2-24. Electron micrograph of φT2 with its DNA extruding from the head. The double-stranded chromosome is a single filament. (Courtesy of A. K. Kleinschmidt, 1962. *Biochim. Biophys. Acta*, **61**:861.)

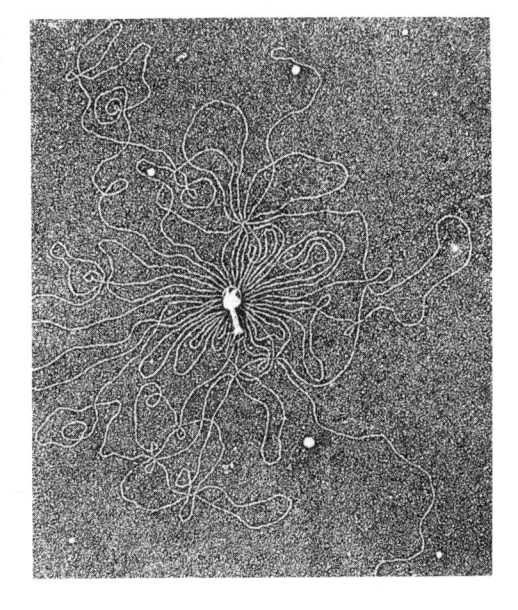

FIGURE 2-25. The double-stranded DNA chromosome extruding from a partially disrupted bacterium (*Hemophilus influenzae*). (Courtesy of L. A. MacHattie and C. A. Thomas, Jr.)

**2.12 A nuclear chromosome set is composed of two or more supercoiled and folded chromosomes whose DNA is combined with basic protein.**

The DNA in the nucleus of a eukaryote is ordinarily divided into two or more (usually rod) chromosomes. Such chromosomes usually contain a single duplex of DNA prior to replication. The amount of DNA per chromosome is so great—a human chromosome is at least 20 times as long as a bacterial chromosome—that it must be coiled and folded extensively in some pattern not yet fully understood. Figure 2-28 shows a replicated and condensed human chromosome in which one can see the tremendously large number of folds the chromosomal fiber makes. The apparent diameter of this fiber is about 250 to 300 Å, whereas one double helix has a diameter of 20 Å. This difference is explicable, at least in part, in terms of the properties of nucleic acid.

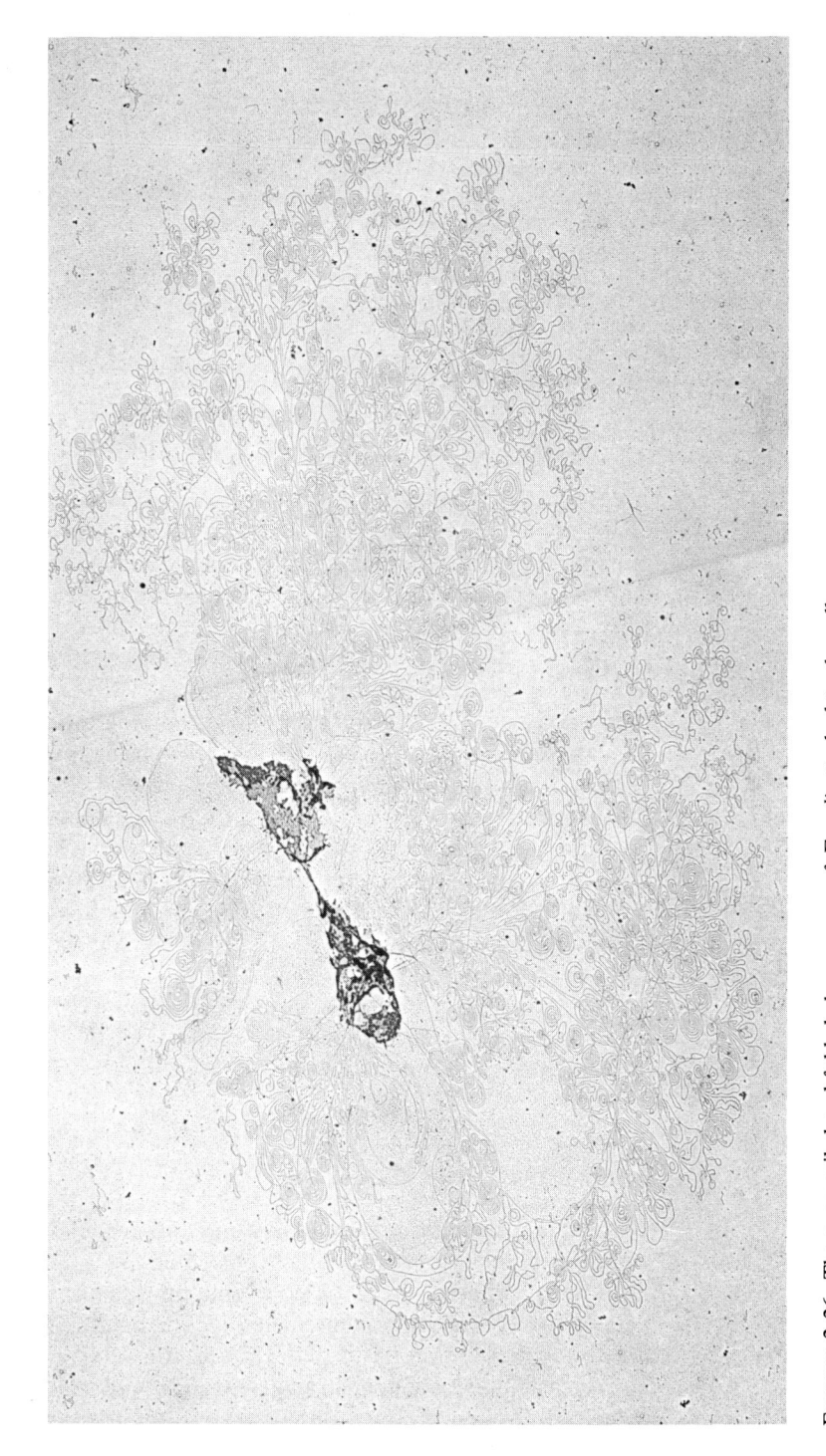

FIGURE 2-26. The supercoiled and folded chromosome of *E. coli* attached to the cell membrane (central dark area). 11,000×. (Courtesy of H. Delius and A. Worcel, 1974. *J. Mol. Biol.*, **82**:107–109. Copyright by Academic Press (London) Ltd.)

38                    WHAT THE GENETIC MATERIAL IS

FIGURE 2-27. A two-dimensional model of the condensed bacterial chromosome. Steps **A** to **C** illustrate the two configurational changes which condense the DNA. **A**: The circular, unfolded chromosome. **B**: The folded chromosome containing seven domains of supercoiling. The actual number of domains is probably greater (12 to 80) but is reduced for simplicity. **C**: The folded and supercoiled chromosome. RNA molecules, in binding to the DNA, define the positions of the folds and also segregate the DNA into domains of supercoiling. (Courtesy of D. E. Pettijohn and R. Hecht, 1973. *Cold Spring Harbor Sympos. Quant. Biol.*, **38**:39.)

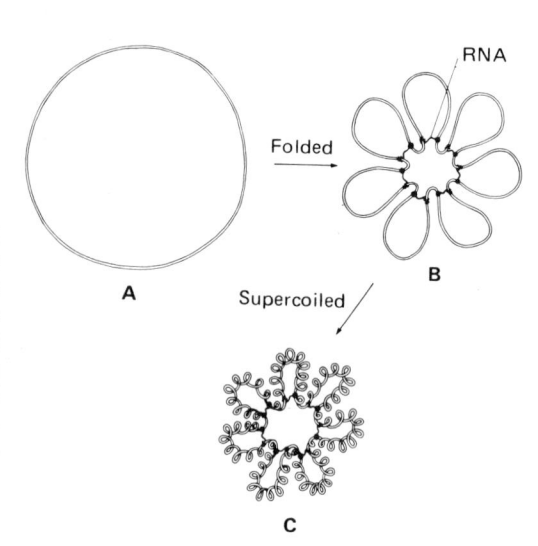

Pure nucleic acid is not easy to fold or coil because the acidic phosphates are negatively charged. These groups repel each other and tend to prevent the association of different parts of the same (or different) duplexes of nucleic acid. Such associations are made possible, however, when these acidic groups are neutralized by positively charged basic groups. This neutralization must be one of the functions of the proteins of the smaller spherical viruses.

In eukaryotes DNA is usually found in combination with basic proteins such as *histone* or *protamine*, which are rich in basic amino acids. (Nuclear chromosomes also contain varying amounts of nonhistone protein.) Further information about all these proteins, their chemistry, origin, and function is presented in Chapter 15. In histone-containing cells, one DNA duplex is part of a unit chromosomal fiber 30 to 40 Å thick. The DNA in this fiber can coil (like the filament in a light bulb) around the histone to produce a fiber about 125 Å thick, and then fold into a fiber 250 to 300 Å thick.

There are, however, some special chromosomes that are known to contain several to many (sometimes over 1000) DNA duplexes. Further discussion of these many-threaded chromosomes is postponed to later chapters.

## SUMMARY AND CONCLUSIONS

This chapter deals with the composition and properties of nucleic acids and how these are organized in chromosomes.

Nucleic acid structure is based on the phosphate, sugar, and organic base content of its component nucleotides. When the sugar is ribose, the polymer is RNA; when it is deoxyribose, it is DNA. Since these sugars have an asymmetric shape, all single-stranded nucleic acid molecules are polarized. Inasmuch as any one of the organic bases can be attached to any sugar, the single-stranded molecule can have any base sequence. The four usual bases in RNA are C, G, A, and U; in DNA they are C, G, A, and T.

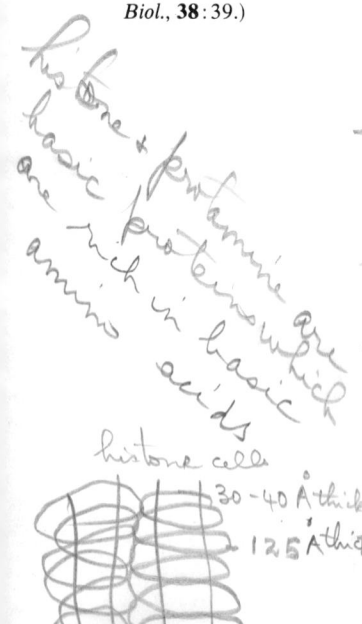

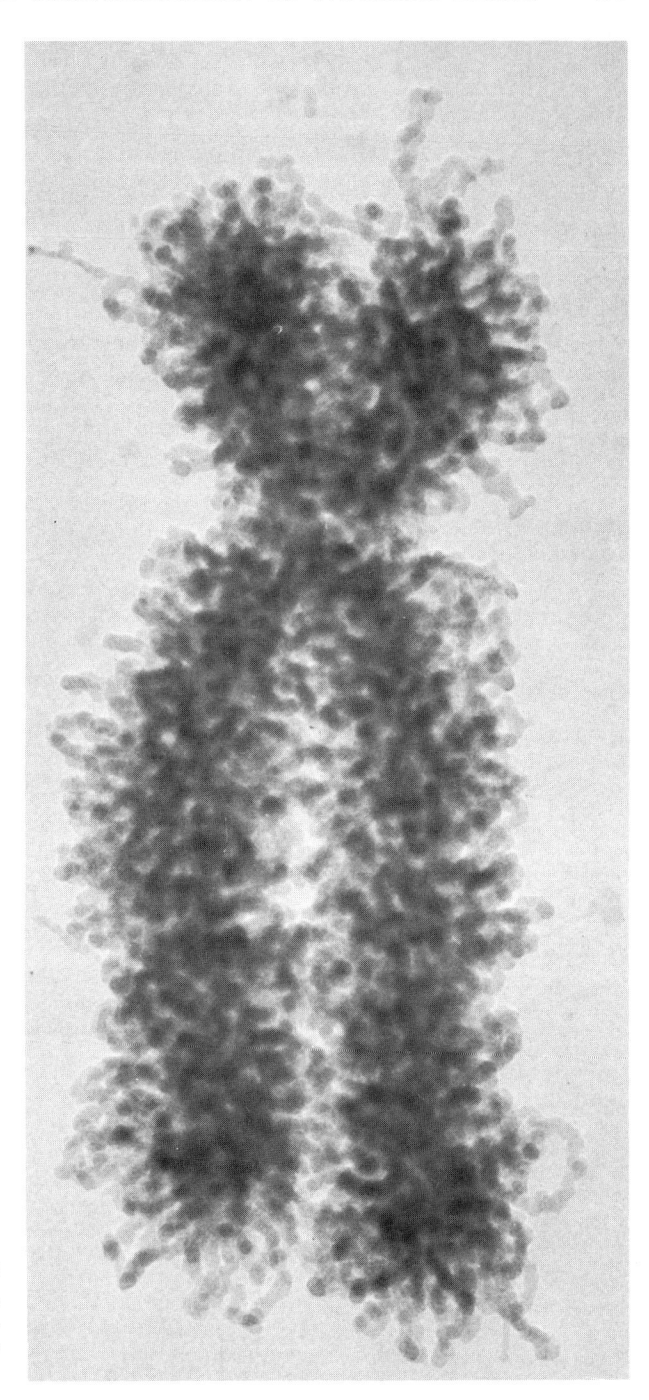

FIGURE 2-28. Divided and compacted human chromosome composed of two greatly folded, DNA-containing fibers. (Courtesy of E. J. DuPraw, 1970. *DNA and Chromosomes.* New York: Holt, Rinehart and Winston.)

The information in a nucleic acid resides in the number, kind, and sequence of its nucleotides (or, simply, its organic bases). The genetic nucleic acids of different organisms differ in such information.

The organic bases in nucleic acids can pair by means of H bonds—C can pair only with G, and A can pair only with T (or U). When two strands or two parts of a single strand are in an antiparallel alignment, basepairing produces a double helix or a double-helical region. Most DNA is double-stranded, whereas most RNA is single-stranded.

Since H bonds broken by heating can re-form upon cooling, double helixes or double-helical regions can be destroyed and re-created. This technique is useful for mapping chromosomes, determining their base ratio, and producing hybrid double-stranded nucleic acids.

Genetic RNA and DNA may be organized in chromosomes that are single-stranded rods or rings, or double-stranded rods. Genetic DNA may also occur as double-stranded rings as well. Many genetic nucleic acids are folded and/or supercoiled. Nuclear DNA chromosomes usually contain a single duplex whose folding and supercoiling is aided by joining to basic protein.

## QUESTIONS AND PROBLEMS

**2.1** *Anagrams*: Unscramble and define each of the following: **a.** toneelucid. **b.** deripinimy.  **c.** rinupe.  **d.** seencula.  **e.** unredcandy.
**2.2** Given the sequence A A T G C C G A T: **a.** Write the complementary DNA sequence.  **b.** Write the complementary RNA sequence.
**2.3** Distinguish between a double-stranded helix and a double-helical strand.
**2.4** How does the DNA of $\phi$X174 differ from the DNA of its *E. coli* host?
**2.5** How does the RNA of reovirus differ from the RNA of TMV?
**2.6** Suppose that you have three unlabeled test tubes of DNA, two from *E. coli* and one from human beings. How would you proceed to identify the two from *E. coli*?
**2.7** You determined that T comprises 30 per cent of all bases in whale DNA when the test tube drops and the sample is lost. What can you conclude about the other bases present?
**2.8** Name a DNA phage whose chromosome is: **a.** single-stranded throughout.  **b.** double-stranded throughout.  **c.** partly single- and partly double-stranded.  **d.** begins and ends with the same base sequence.
**2.9** Discuss the mechanisms used to compact the chromosomes of a: **a.** virus.  **b.** prokaryote.  **c.** eukaryote.
**2.10** State five differences between the DNA's of $\phi$X174 and human beings.
**2.11** An unlabeled test tube contains either TMV viruses or human nuclear chromosomes. How could you determine which type it contained?
**2.12** *Balloons*: Fill appropriately.

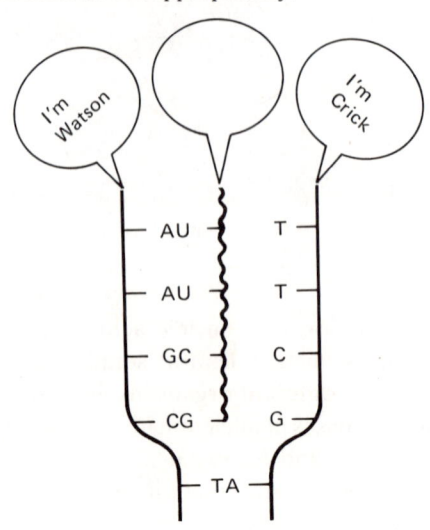

*PART II*

# WHAT THE GENETIC MATERIAL DOES

# Chapter 3

# Chromosome Replication

*[Handwritten notes: in vitro — (in the test tube) biological processes studied outside the organism; in vivo — biological processes studied within the living organism; de novo synthesis — made without using a template]*

WE HAVE SAID that the single most important feature of genetic material is its ability to direct its own replication. This chapter starts with a brief discussion of the synthesis of one type of nongenetic RNA and the replication of genetic RNA. The bulk of the chapter deals with the replication of genetic DNA, which comprises the bulk of all genetic material.

## RNA REPLICATION

**3.1** A small portion of all cellular RNA is synthesized independently of preexisting nucleic acid.

Enzymes that combine nucleotides into polymers of nucleic acid are called *nucleic acid polymerases*. Those that combine ribose-containing nucleotides are *RNA polymerases*. For example, an RNA polymerase isolated from *E. coli* can join ribose-containing nucleotides each of whose base is adenine (A) into an RNA polymer called *poly A* (Figure 3-1). This synthesis can be observed in the test tube—that is, *in vitro*. Such syntheses also occur in organisms—that is, *in vivo*. In these cases RNA synthesis does not require the presence of other nucleic acids, and the RNA made is new and is said to have been synthesized *de novo*. Only a small portion of the total organismal RNA, however, is made *de novo*. Some of it occurs free of, and some of it becomes attached to, other RNA strands.

FIGURE 3-1. The enzymatic synthesis of poly A from ribose nucleotides, each of whose base is A.

```
|A                          A|
|A      E. coli             A|
|A      RNA polymerase      A|
|A         ———————→         A|
|A                          A|
|A                          A|
|A                          A|
```

43

## WHAT THE GENETIC MATERIAL DOES

<u>Since *de novo*-synthesized RNA is made independently of all nucleic acids, including those made *de novo*, it never repeats any previously existing nucleic acid information, and is therefore nonreplicational nucleic acid.</u> In other words, <u>*de novo*-synthesized RNA does not function as genetic material.</u>

### 3.2 RNA is replicated by RNA polymerases that use existing RNA strands as molds or templates to synthesize complementary strands

Other RNA polymerases require the presence of RNA strands in order to synthesize new RNA. Such enzymes use a strand of RNA as a *mold* or *template* to attract and base-pair with free complementary nucleotides (Figure 3-2A). When the free nucleotides are base-paired with the template bases, the enzyme can then join them together into a new, antiparallel, complementary strand of RNA (Figure 3-2A, where the half-arrowheads indicate the direction of polarity). If the new strand is then used as a template to synthesize an antiparallel, complementary

FIGURE 3-2. The replication of RNA.

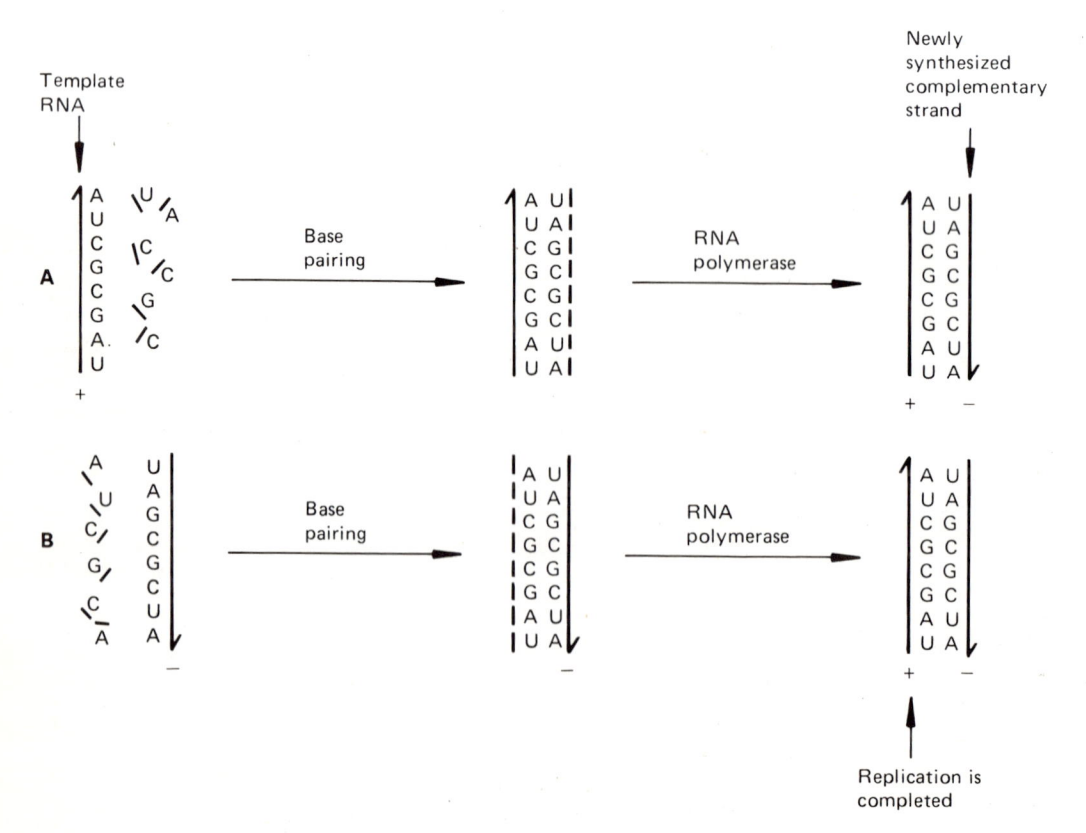

# CHROMOSOME REPLICATION

strand (Figure 3-2B), the next complement formed will be an exact copy of the original strand, and replication of the original strand will have occurred.

We see, therefore, that replication of an RNA strand requires *two* successive syntheses. For example, consider the replication of the single-stranded RNA chromosome of $\phi Q\beta$. The virus RNA strand is called the + strand (Figure 3-2A). It is used as a template to synthesize a complementary − strand; the − strand is then used to enzymatically synthesize a complementary + strand (Figure 3-2B), which results in the replication of the phage chromosome. Since the information in the + strand is used (eventually) to produce another + strand, such RNA is genetic material and the RNA polymerase employed can be called an *RNA replicase.*

RNA polymerases that require a template do not start the synthesis of a complementary strand at a random position on the template strand. Each different polymerase must recognize a particular base sequence before it can start working. This recognition sequence is different for different polymerases. In the case of $\phi Q\beta$ replicase, for example, the recognition sequence is near one end of the + strand (say AUCG in Figure 3-2A). Note (as in the figure) that the − strand also has this sequence at its corresponding end, permitting it to be recognized by the replicase.

Finally, it should be noted that some of the nucleotides in certain RNA chromosomes are not part of the genetic material. For example, the last nucleotide in the + and in the − strand of $\phi Q\beta$ is apparently not used as a template to make a complement. This nucleotide can be removed and the strand can function normally *in vivo*. More striking is the situation in the case of reovirus, which has 10 +− double-stranded RNA chromosomes for its chromosome set. Each of the 10 + strands has poly A attached, the poly A containing about 20 per cent of all the bases present. Replication is accomplished without using these + strands as templates. Specifically, the 10 − strands are used as templates to make 10 poly-A-free + strands; these new + strands are then used as templates to make 10 new − strands, completing the replication of the genetic material. Subsequently, poly A made *de novo* is attached to each of the 10 progeny + strands. Accordingly, this poly A is not genetic material, and its function is not yet known.

**3.3** A *gene* is a base sequence in genetic nucleic acid that contains information used to perform a specific function for an organism.

The base sequence in an RNA chromosome that recognizes an RNA polymerase serves a specific function for the organism. Any base sequence of genetic nucleic acid that contains information that is used independently to perform a specific function for an organism is called a *gene.* Since genetic material contains information for the synthesis of specific proteins, there must be genes for proteins. Whereas the *RNA polymerase recognition gene* of $\phi Q\beta$ is relatively short, the genes for proteins, which vary in length, are all much longer. We shall also see (in Chapter 4) that other kinds of gene are known. (All presently-known genes function either to bind proteins or to serve as templates for the synthesis of complementary nucleic acids.) As we would expect, because different tasks require different amounts of information, the number of bases varies in different kinds of gene. Because there is a considerable amount of genetic material of unknown function in many organisms, especially eukaryotes, it is possible that some types of gene are still to be discovered. It should be noted that a given sequence of nucleotides may be part of two genes of the same or different types. This economy of nucleotides has been observed in certain viruses and in prokaryotes, and will be illustrated later.

# DNA REPLICATION

**3.4** Replication *in vivo* of double-stranded DNA results in two duplexes, each of which contains a parental strand and a newly synthesized complementary strand.

Replication of double-stranded DNA is essentially identical in all organisms. We can obtain an overall picture of the results of DNA replication by observing the distribution of parental DNA that is labeled (having been synthesized from heavy or from radioactive nucleotides) relative to progeny DNA that is unlabeled (because it is synthesized from light or from nonradioactive nucleotides). The first replication produces two half-labeled duplexes (Figure 3-3); when each of these half-labeled duplexes replicates, each produces one half-labeled duplex and one unlabeled duplex. This distribution is obtained because after replication, each duplex contains a parental strand and a newly synthesized complementary strand. In the diagram, the first replication produced one duplex composed of a parental (labeled) Watson that had been used as a template to synthesize a complementary (unlabeled) Crick, and one duplex composed of a parental (labeled) Crick that had been used as a template to synthesize a complementary (unlabeled) Watson. The second round of replication produced, from each half-labeled duplex, one half-labeled duplex (the labeled strand was used to make an unlabeled complement) and one unlabeled duplex (the unlabeled strand was used to make an unlabeled complement). In other words, replication conserves in each progeny duplex one of the two strands in its parent duplex, and is therefore said to be *semiconservative*.

**3.5** The process of semiconservative replication in prokaryotes and eukaryotes is described.

Having described the molecular *product* of the semiconservative replication of DNA, we are ready to consider the molecular *process* by which

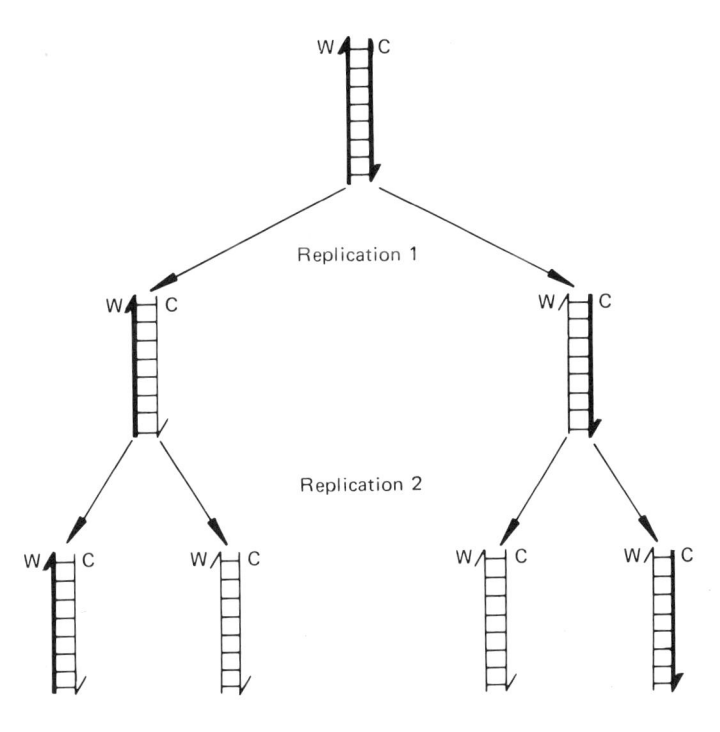

FIGURE 3-3. The semiconservative replication of double-stranded DNA. Thick line = labeled parental complements; thin line = unlabeled progeny complements. The half-arrowhead represents the direction of polarity of the DNA strand.

this replication is accomplished. We will describe the process as it occurs in prokaryotes and then tell how it differs in eukaryotes.

Replication starts at a single, particular region in the single, circular double-stranded chromosome of the prokaryote. The nucleotide sequence in this region serves, therefore, as the *replication origin gene, o* (Figure 3-4A). The H bonds holding the complementary strands of this gene together are broken, permitting the separation of the two complements in this region of the chromosome. This separation produces a *replication eye* (Figure 3-4B). At one or more times during replication a nuclease nicks one of the strands so the duplex can untwist its supercoils as well as the helical turns that accumulate when the strands in a segment of the duplex are separated. The two separated complements are prevented from reuniting by a protein that preferentially binds to single-stranded DNA just as soon as it is produced at both corners of a replication eye.

Whereas all RNA polymerases can start the synthesis of a new strand *in vivo*, for some unknown reason no DNA polymerase can. All DNA polymerases therefore require a *primer*, that is, a nucleic acid strand that can be lengthened by newly made DNA. In the replication of cellular DNA, RNA is used as primer. This RNA is made by an RNA polymerase that synthesizes short strands of RNA from ribose nucleotides of A, U, C, and G, according to the template instructions of the DNA in the replication eye (Figure 3-4C). After these segments of RNA are suitably long, a *DNA polymerase* will add to them deoxyribose nucleotides of A, T, C, and G according to the instructions of the template DNA in the replication eye (Figure 3-4D). The replication eye enlarges or opens further at

48   WHAT THE GENETIC MATERIAL DOES

A. Segment of parent duplex

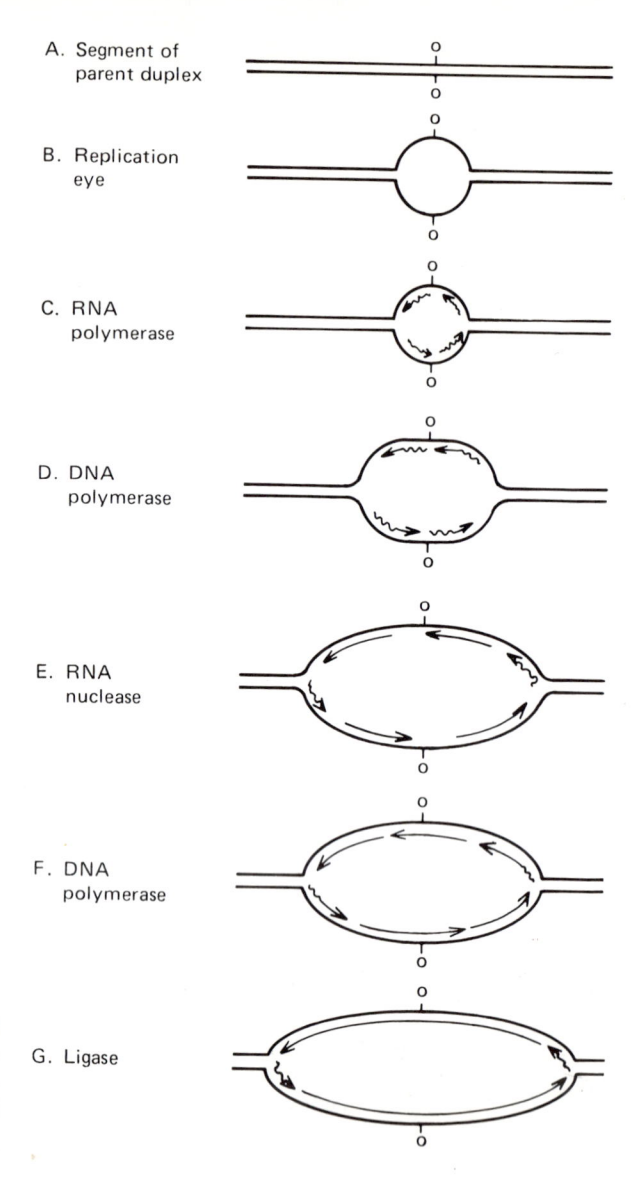

B. Replication eye

C. RNA polymerase

D. DNA polymerase

E. RNA nuclease

F. DNA polymerase

G. Ligase

FIGURE 3-4. Sequence of events in the process of semiconservative replication of DNA. Wavy line = complementary RNA; straight line = complementary DNA; *o* = replication origin gene. See the text for details.

both corners, so that the replication process can be continued in two directions.

In order to complete the replication of the DNA, the RNA primers are degraded by means of an *RNA nuclease*, starting first in the region of *o* (Figure 3-4E) and then extending in both directions. This produces gaps that can be filled in by DNA polymerase lengthening the DNA fragments that remain after RNA removal (Figures 3-4F). (Notice that some sequences of DNA are used as templates twice in the replication process, once to produce complementary RNA primers and later, after this RNA is removed, to produce complementary DNA.) Once the gaps in the DNA are repaired, the result is duplex DNA with nicks. The nicks are

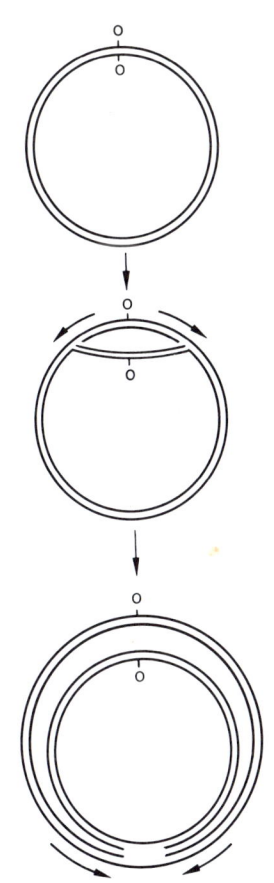

FIGURE 3-5. The bidirectional replication of the *E. coli* chromosome starting from a single origin, *o*.

removed by an enzyme, called *ligase*, that joins or ligates the adjacent ends of two DNA fragments (Figure 3-4G).

The bidirectional replication process continues (Figure 3-5) for about 30 minutes, at which time the two growing points of the eye meet, nicks are removed, and the replication of the chromosome is complete. In the present case of semiconservative replication, one parental duplex ring produces two duplex rings.

The process of semiconservative DNA replication is essentially the same in eukaryotes as described for prokaryotes, with one major exception. Because nuclear chromosomes are usually very much longer than a prokaryote chromosome, replication using a single eye would take very much longer. It is no surprise, therefore, that a nuclear chromosome has many replication start points, and hence many replication eyes that are enlarging bidirectionally (Figure 3-6). We can imagine a multieyed nuclear chromosome as being equivalent to the product of breakage and union of a series of single-eyed ring chromosomes, each the size of a bacterial chromosome (Figure 3-7). Replication is completed when the eyes meet each other and their newly synthesized DNA's are ligated to remove all nicks.

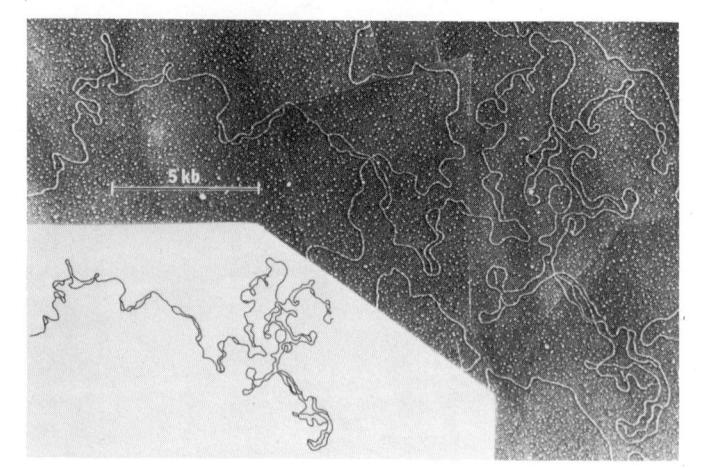

FIGURE 3-6. Bidirectionally replicating eyes in a nuclear chromosome of the fruit fly, *Drosophila*. The chromosome fragment shown contains 23 eyes in a length of 119 kilobases (kb). (1 kb = 1000 base pairs in duplex nucleic acid.) (Courtesy of H. J. Kriegstein and D. S. Hogness, 1974. *Proc. Nat. Acad. Sci., U.S.*, **71**: 136.)

FIGURE 3-7. Diagrams representing the conversion of four nonreplicating ring chromosomes (**A**) via breakages (at arrows) and unions (**B**) into one replicating (four-eyed) rod chromosome (**C**).

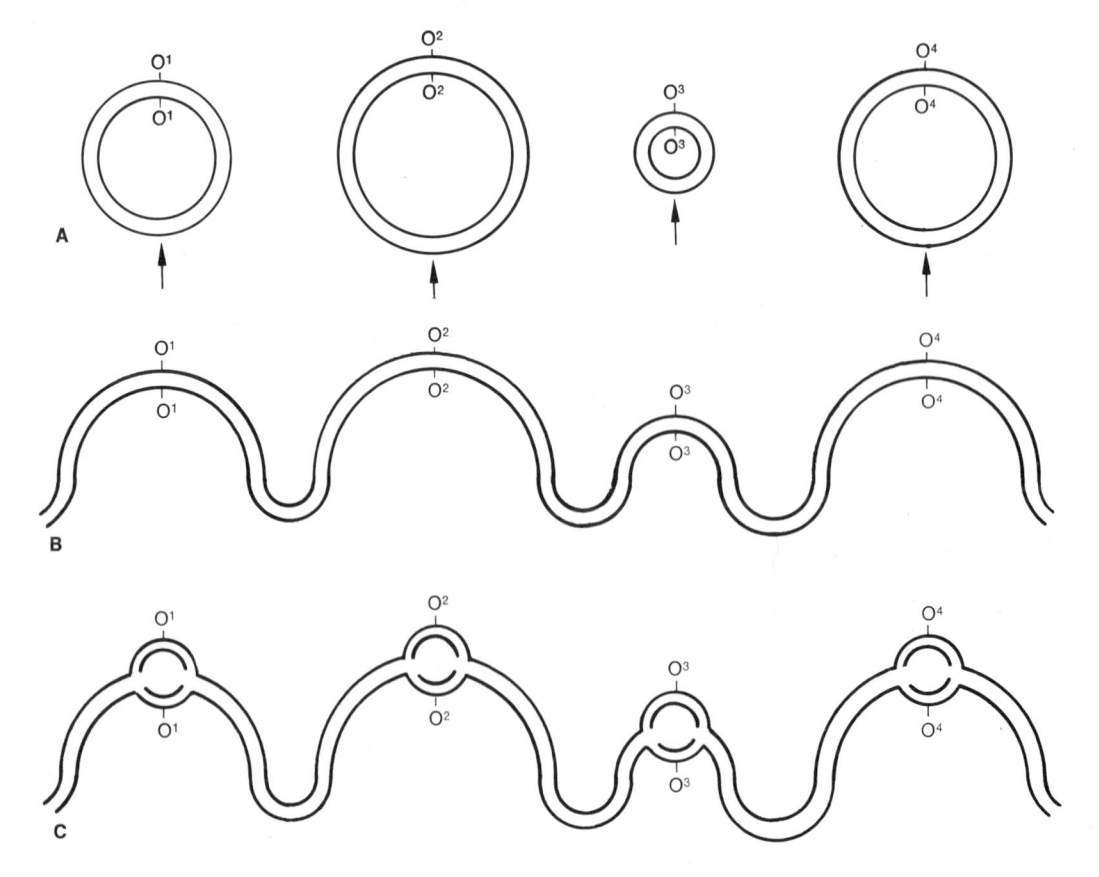

**3.6** The DNA's of many, probably most or all, organisms attain characteristic conformations by being modified enzymatically.

Of the four bases, A, T, C, and G, only T has a methyl group—T being *methylated* U. When naturally occurring DNA is analyzed, however, small amounts of methylated A, C, and G are found in almost all species. This methylation occurs in a part of the base that has no effect on its base-pairing ability (just as T and U are equally good base-pairing partners of A). This modification of DNA occurs during DNA synthesis or very soon thereafter through the action of *DNA methylases*. These enzymes work only on newly made complements close to the growing points of a replication eye. Methylation is carried out to different extents and in different patterns in different species, the methylases being species-specific. As a result, the DNA of different species has a surface configuration or conformation that is characteristic—equivalent, so to speak, to a set of fingerprints. Although *DNA modification* probably occurs in all organisms, we shall give specific examples of it only in bacteria and phages, where the phenomenon is best known.

1. BACTERIA. The bacterium *Hemophilus influenzae* has four different methylases that methylate A's. Each of these modification enzymes requires one or more different sequences of three bases, the middle one being A, as a template before it can act to methylate the middle base (Figure 3-8). Of the 16 possible sequences of three bases with A in the center, 8, or one half, are methylated by these DNA methylases. Since C's are methylated by other DNA methylases with similar, or perhaps lengthier, template requirements, we can understand how a large number of different methylation patterns are generated by species-specific DNA methylases, thereby fingerprinting DNA conformation.

2. VIRUSES. As viral DNA's replicate in their host, the DNA modification enzymes of the host give the viral DNA the host's fingerprint.

FIGURE 3-8. Modification of *Hemophilus* DNA by DNA methylases.

| DNA Methylase | DNA Modification |
|---|---|
| I | C Å C → |
| II | Pu Å C → |
| III | B Å A → |
| IV | G Å T → |

*The amino group (at position 6) is methylated; Pu, A or G; B, any usual base.

# WHAT THE GENETIC MATERIAL DOES

|          | Pyrimidine |  |
|----------|------------|--|
| Replaced | Substitute | Virus |
| T | U<br>hydroxymethyl U<br>dihydroxypentyl U*<br>bromo U | φPBS1 and φPBS2 of *Bacillus subtilis*<br>φSP8 of *B. subtilis*<br>φSP15 of *B. subtilis*<br>infectious bovine rhinotracheitis |
| C | methyl C<br>hydroxymethyl C | φ of *Xanthomonas oryzae*<br>φT2, φT4, φT6 of *E. coli* |

*Replaces one half of the T's.

FIGURE 3-9. Pyrimidine substitutions used in synthesizing DNA's of different viruses.

Viral DNA may, in addition, establish its own uniqueness in two general ways. (a) Viral DNA may substitute for some or all of one of the pyrimidines (T or C) a derivative of that pyrimidine which has the same base-pairing characteristics. Figure 3-9 gives specific examples of such substitutions. (b) The virus chromosome may contain information for the synthesis of virus-specific enzymes that modify newly made viral DNA in a virus-specific pattern. For example, some virus chromosomes specify their own types of DNA methylase. In the case of certain phages, glucose sugars are added to some of the bases in species-specific amounts and patterns by other phage-specified enzymes (Figure 3-10).

## 3.7 Restriction enzymes restrict the DNA persisting in a cell to normally modified DNA.

It is advantageous to a cellular organism for its DNA to be modified in a characteristic pattern. When a portion of the organism's own DNA is defectively modified (because it failed to be modified or became abnormally modified in the course of aging), it can be recognized by the organism's nucleases and degraded. (The degraded piece can sometimes be replaced by a normal piece as the result of a repair mechanism, to be

FIGURE 3-10. The kind and amount of glucoses added to hydroxymethyl C in T-even-phage DNA.

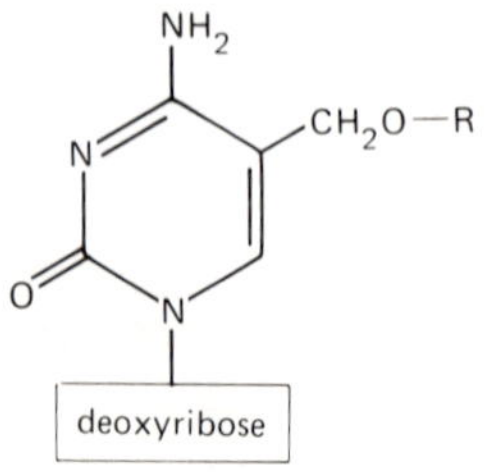

| R = | T2 | T4 | T6 |
|-----|----|----|----|
| H | 25 | 0 | 25 |
| α-glucosyl | 70 | 70 | 3 |
| β-glucosyl | 0 | 30 | 0 |
| β-(1,6)-glucosyl-<br>α-glucosyl | 5 | 0 | 72 |
|  | 100 | 100 | 100 |

# CHROMOSOME REPLICATION

```
A    ... G T Py│Pu A C ...
     ... C A Pu│Py T G ...


B    ... G T Py Pu Å C ...
     ... C A Pu Py T G ...
              *
*Methylated at the 6-amino position.
```

FIGURE 3-11. A restriction enzyme in *Hemophilus* degrades the sequence in **A** (arrows) but not the one in **B** (the result of DNA methylase II of Figure 3-8 acting upon the sequence in **A**). Py = T or C; Pu = A or G.

discussed in Chapter 6.) When DNA synthesized in one species infects another species, the host's nucleases may recognize this DNA as being improperly modified, and degrade it. Such nucleases are called *restriction enzymes*, since the DNA persisting in an organism is restricted to normally modified DNA. Restriction enzymes, therefore, must not be able to degrade correctly modified DNA.

*Hemophilus*, for example, has a restriction enzyme that recognizes a particular set of base sequences (Figure 3-11), and breaks both strands at the same level at a particular place in the base sequence. Note that the base-pair sequence recognized by this restriction enzyme, and by many other nucleases, too, is symmetrical; that is, the base sequence in one strand is reversed in the other. In other words, the region contains a *palindrome* and is nicked once in each strand. Had this particular sequence been synthesized in *Hemophilus*, it would have had two of its A's methylated by one of the organism's DNA methylases (methylase II in Figure 3-8), and therefore would have been rendered immune to degradation by the restriction enzyme.

At least some, if not all, bacterial restriction enzymes act by recognizing nonmodified base sequences, as is the case for the restriction enzyme already discussed. Accordingly, as mentioned in the last section, the DNA of normally infecting phage is appropriately modified and is, therefore, immune to its host's restriction enzymes. If the phage DNA specifies additional modifications of the phage DNA, this DNA will still be immune to the host's restriction enzymes. The resultant phage DNA will also be immune to any phage-specified restriction enzymes that recognize nonmodified base sequences, whereas the bacterial DNA, which does not have the additional phage-specific modifications, may be degraded.

## SUMMARY AND CONCLUSIONS

Although some *in vivo* RNA is synthesized *de novo*, most of it is synthesized by RNA polymerases that use nucleic acid as a template to synthesize complementary RNA. With a template of genetic RNA, synthesis starts at a particular base sequence—the RNA polymerase recognition gene.

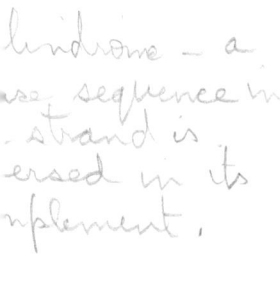

A gene is any base sequence of genetic nucleic acid that contains information that is used to perform a function for an organism. Genes with different functions have different lengths.

Replication of the single-stranded and double-stranded RNA chromosomes of $\phi Q\beta$ and reovirus require two successive template syntheses. These syntheses are followed by the nontemplate addition of one and many nongenetic ribose nucleotides to progeny chromosomes.

Double-stranded DNA chromosome replication is semiconservative. Replication starts at one replication origin gene in a prokaryote and at many such genes in a nuclear chromosome, and proceeds bidirectionally in the replication eye formed at each start point. RNA-primed pieces of DNA are synthesized through the actions of RNA polymerase and DNA polymerase; after the RNA is removed and the gaps produced are filled in with DNA, the DNA pieces are ligated.

The DNA's of probably most or all cellular organisms attain characteristic conformations by being modified by species-specific methylases. Such modifications render the organism's DNA immune to the cell's restriction enzymes. In addition to their host's modifications, viral DNA's sometimes have virus-specific modifications. These are due to the incorporation of modified pyrimidines during viral DNA synthesis or to the addition of methyl groups or glucoses to newly made viral DNA. Viral DNA fingerprinted in this manner is protected from the host's restriction enzymes that recognize and degrade unmodified sequences.

## QUESTIONS AND PROBLEMS

3.1 *Anagrams*: Unscramble and define each of the following: **a.** petemalt. **b.** indelompar. **c.** silage. **d.** empirr. **e.** heatsmely.

3.2 *Riddle*: What eyes are not for seeing?

3.3 *Song*: "Stay as Sweet as You Are." Has this application in this chapter? Explain.

3.4 Distinguish among *in vivo*, *in vitro*, and *de novo*.

3.5 In what ways are genetic RNA and genetic DNA modified or tailored?

3.6 Name four different proteins needed for replication of the *E. coli* chromosome.

3.7 Which is more correct: The two strands in double-stranded DNA are like two bedsprings pushed together, or are like the two strands of a two-stranded rope? How would semiconservative replication differ in these two cases?

3.8 DNA synthesis *in vitro* is permitted to double the initial amount of template DNA. When the initial DNA is from a cow it will have been replicated but not when it is from $\phi$X174. Explain.

3.9 State three different chemical features that give a T-even phage DNA a unique (species-specific) conformation.

3.10 What are the advantages of restriction enzymes specified by genes in *E. coli*? In an *E. coli* phage?

3.11 What are the possible advantags of enzymes that require a palindrome in nucleic acid in order to function?

# CHROMOSOME REPLICATION

**3.12** *Balloons*: Fill appropriately.

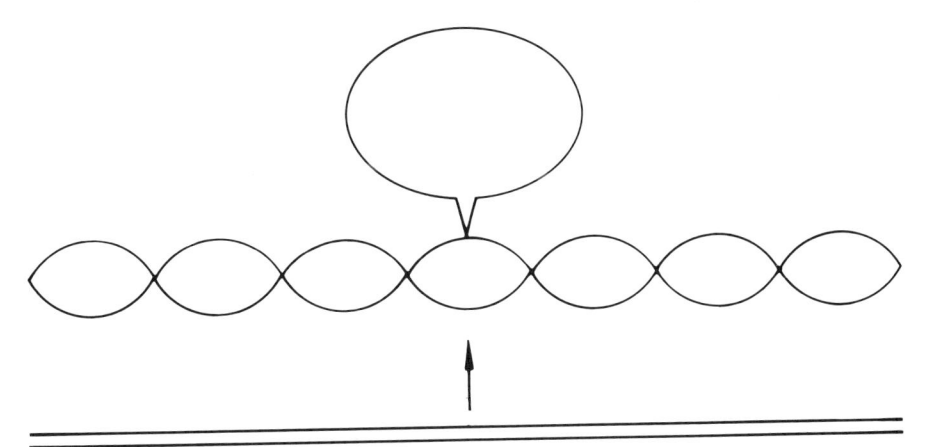

# Chapter 4

# Transcription

ALTHOUGH AN ORGANISM carries information for the synthesis of its characteristic proteins in either RNA or DNA genetic material (Chapter 1), only RNA information is used by the protein-synthesizing machinery in a cell. Thus, while proteins can be made directly using the information either in the + chromosome strand or in its − complement in the case of RNA viruses, whenever DNA is the genetic material, they must be made indirectly, by means of the information in an RNA intermediate. In other words, when DNA is the genetic material, the protein-containing information must first be copied or transcribed into RNA. We have already noted in Chapter 3 that an RNA polymerase can use single-stranded DNA as a template to synthesize complementary RNA that serves as primer for DNA replication. The RNA polymerase that catalyzes the synthesis of RNA using a DNA template is said to catalyze *transcription*, and is called *transcriptase*. In view of the equivalence of the base information in DNA and RNA, it is no surprise that RNA can be used to synthesize complementary DNA in a process called *reverse transcription*, catalyzed by a DNA polymerase called *reverse transcriptase*.

This chapter discusses transcription in prokaryotes, eukaryotes, and viruses, and ends with a brief discussion of reverse transcription. The next chapter will deal with the protein-synthesizing machinery and how the RNA *transcripts*, whose synthesis is described in this chapter, are used in protein synthesis.

**4.1** Genetic material is redefined as any nucleic acid that replicates or that was synthesized using a nucleic acid template and is known to be used as a template to synthesize nucleic acid.

Up to this point only nucleic acids that replicate have been considered to be genetic material. This definition needs to be modified and expanded

# TRANSCRIPTION

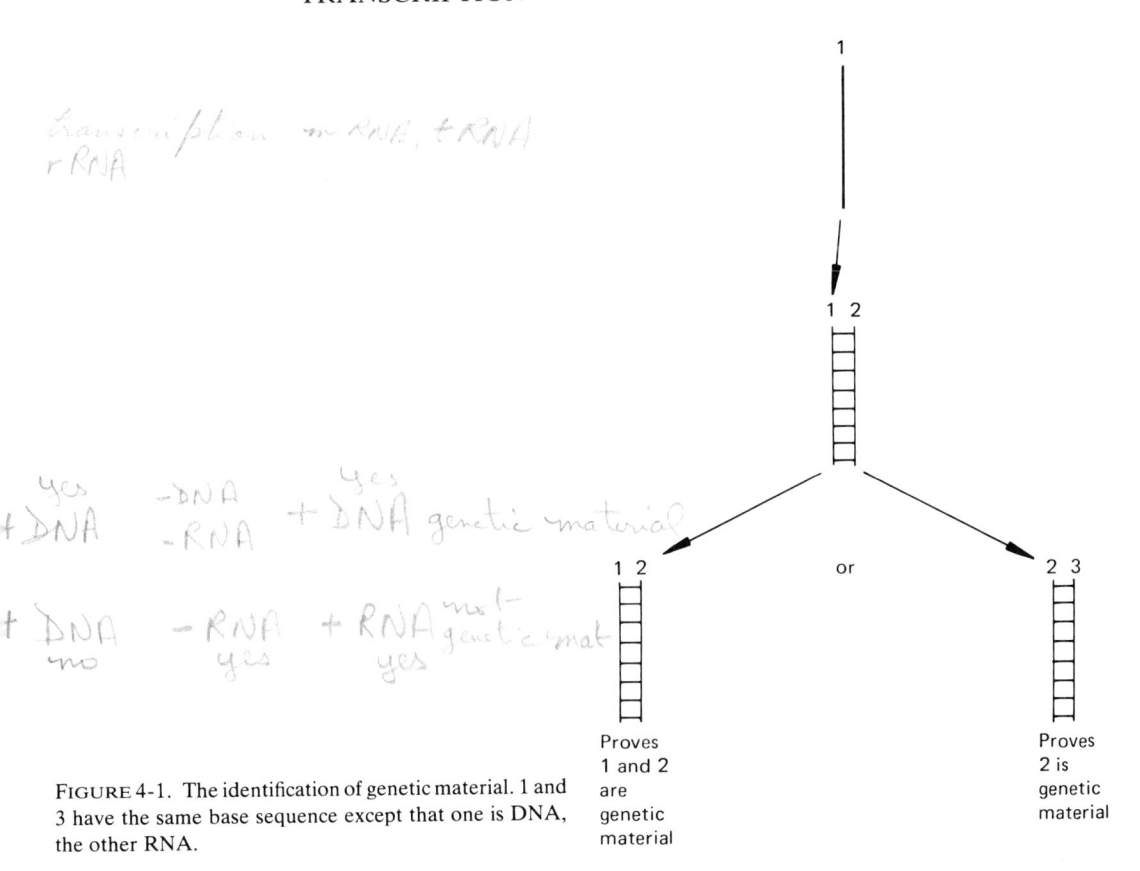

FIGURE 4-1. The identification of genetic material. 1 and 3 have the same base sequence except that one is DNA, the other RNA.

Diagram: 1 → (1 2 duplex) → splits into (1 2, Proves 1 and 2 are genetic material) or (2 3, Proves 2 is genetic material)

for two reasons. (1) It does not include DNA or RNA that is an intermediate template in replication. For example, a + strand RNA may be reverse-transcribed to − strand DNA, which is transcribed to replicate + strand RNA. According to our present definition, the − strand DNA is not genetic material because it is not replicated. (2) Cells that do not divide and organisms that do not reproduce obviously contain, although they do not replicate, genetic material. Their genetic material is identified because it is a nucleic acid that was made using a nucleic acid as a template, and that is known to serve as a template in transcription. Accordingly, our definition of *genetic material* is expanded to include any nucleic acid that replicates or that was synthesized using a nucleic acid template and is known to be used as a template to synthesize nucleic acid (Figure 4-1).

# TRANSCRIPTION IN PROKARYOTES

**4.2** Transcription is best understood, hence most easily described, in *E. coli*.

Transcription is best understood in *E. coli*. The main characteristics of the transcription process are:

transcription - synthesis of RNA using a DNA template



1. It requires ribose nucleotides of A, U, C, and G as the raw materials for making RNA.

2. It requires a transcriptase. *E. coli* transcriptase is a protein composed of six polypeptide chains ($\alpha_2\beta\beta'\omega\sigma$), five of which are different. Hence, five different genes are needed to make *E. coli* transcriptase.

3. In any region of the *E. coli* chromosome undergoing transcription, the two complementary strands separate and only one of them is used as a template for RNA synthesis (Figure 4-2). The strand that is the template for transcription is called the *sense strand.*

promoter - transcriptase recognition gene

4. Only a portion of the *E. coli* chromosome is transcribed as a unit. The *E. coli* chromosome contains many transcriptional units, many of which may be undergoing transcription at the same time. Many transcriptional units are *operons* that contain information for the synthesis of specific proteins (Figure 4-3). Such an operon contains the following sequence of genes: (a) the *promoter*, which binds transcriptase, but is not transcribed; (b) the *operator*, which when unbound permits transcription to start, but when bound by a particular protein prevents transcription from starting; (c) usually two or more *genes for protein* then follow; (d) the *terminator*, which transcriptase recognizes as a signal to end transcription and release the transcript. Some operons may contain information for a single protein or for the synthesis of RNA transcripts used as part of the machinery for protein synthesis. The unique feature of the operon is its operator gene. Nonoperon transcriptional units also occur which, of course, also start and end with a promoter and terminator. In some transcriptional units the sense strand is Watson, in others it is Crick.

FIGURE 4-2. One-complement transcription of a two-complement template. Straight line = single-stranded DNA; wavy line = single-stranded RNA.

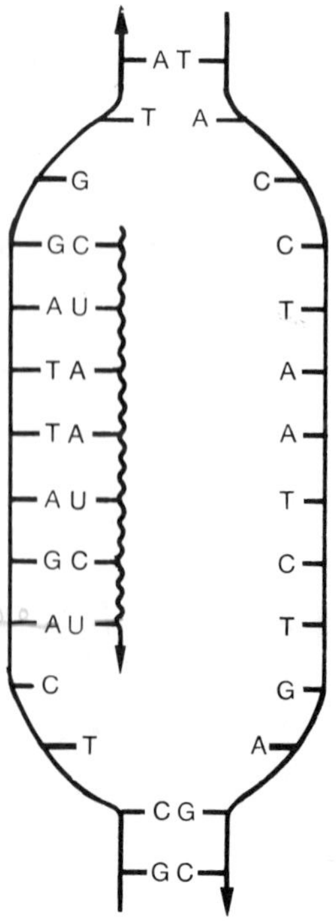


operon - duplex DNA sequence in prokaryotes which contains an operator gene that regulates the transcription of one or more polypeptide-coding genes.

sense strand - DNA strand used as a template for transcription

transcriptase - DNA dependent RNA polymerase

TRANSCRIPTION 59

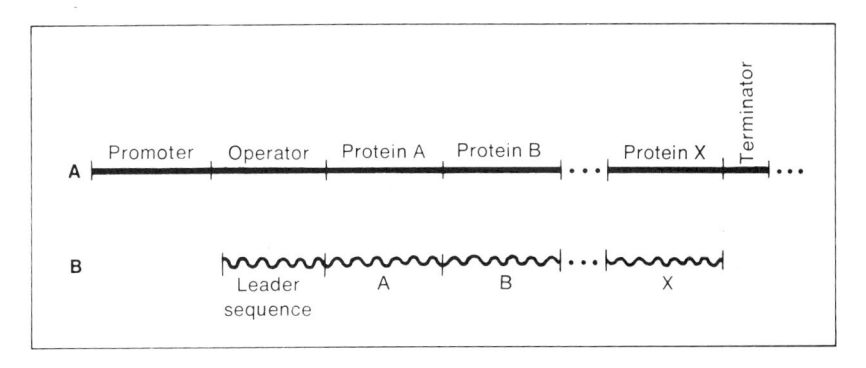

```
A  |Promoter|Operator|Protein A|Protein B| ··· |Protein X|Terminator
B       ~~~~~~~~~~~~~~~~~~~~~~~~ ··· ~~~~~~
        Leader      A        B              X
        sequence
```

FIGURE 4-3. An operon unit of transcription for protein synthesis in the *E. coli* chromosome (**A**) and its mRNA transcript (**B**). Short vertical lines indicate imaginary boundaries.

## 4.3 RNA transcripts are of three types; each is modified or tailored after synthesis.

RNA transcripts are of three types:

1. They may contain information for the synthesis of one or more specific proteins, as is true of the transcript of the operon in Figure 4-3. Since such RNA serves as a messenger of protein information, it is called *messenger RNA* or *mRNA*.

2. Other RNA transcripts are used to make three lengths of RNA which, together with proteins, make up a ribosome, a structure used as part of the machinery for protein synthesis. These *ribosomal RNA's* or *rRNA's* are of different sizes—16S, 23S, and 5S. (The larger an RNA or other polymer is, the faster it will fall when ultracentrifuged, and the larger its S value will be.)

3. The third type of transcript is also used to make an RNA that functions as part of the machinery for protein synthesis, and is called *transfer RNA* or *tRNA*. There are 30 to 40 different types of tRNA in *E. coli*, all about 4S in size.

All RNA transcripts are modified or tailored after synthesis before becoming mature m, r, or t RNA's. Such tailoring includes the modification of certain nucleotides (for example, by methylation or by internal rearrangement of their bases), the removal of nucleotides, and the addition of a nucleotide sequence produced *de novo*. The specific tailoring the three types of transcript undergo is summarized as follows:
*mRNA.* The beginning or first-synthesized end of the mRNA transcript does not carry protein information and is called the *leader sequence* (Figure 4-3). A few nucleotides in the leader sequence are methylated, and some of the sequence may be lost by way of nuclease action. Some mRNA's have *de novo* synthesized poly A, of unknown function, attached at their end.

*rRNA.* The initial transcript is selectively methylated, then trimmed of extra bases, and broken into its three main components.

*tRNA.* The initial transcript is selectively methylated, and some bases have chemical groups added or are internally rearranged to other forms. The molecule is then trimmed of extra bases, and the *de novo* synthesized nucleotide sequence CCA is added enzymatically to its end.

# TRANSCRIPTION IN EUKARYOTES

**4.4** Transcription in eukaryotes is similar to that in prokaryotes, although it differs in several respects.

As in prokaryotes, transcription of nuclear DNA in eukaryotes produces transcripts that are tailored into mRNA's, rRNA's, and tRNA's. Tailoring of eukaryotic transcripts differs, however, in that it often, but not always, includes cutting out internal sequences of bases, and splicing together pieces that specify a single polypeptide (Figure 4-4) or make up a single molecule of rRNA or tRNA. The function of the discarded portions of the transcripts is unknown. The process and products also differ in the following respects:

1. TRANSCRIPTASE. Eukaryotes have several nuclear transcriptases, each kind transcribing a different group of genes (Figure 4-5).

FIGURE 4-4. A unit of transcription for protein synthesis in a nuclear chromosome (**A**), its giant RNA transcript (**B**), and the mRNA produced by cutting and splicing (**C**).

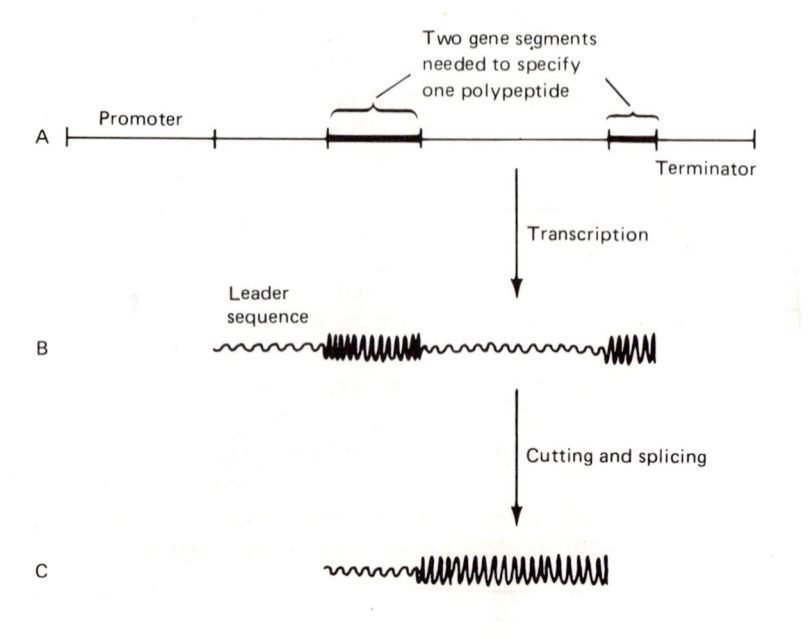

TRANSCRIPTION 61

| Transcriptase | α-Amanitin Concentration | Template Transcribed |
|---|---|---|
| I | Insensitive to high | Nucleolus organizer rDNA |
| II | Sensitive to low | Giant DNA mDNA |
| III | Sensitive to high | tDNA 5S rDNA |

FIGURE 4-5. Types of eukaryotic transcriptases, their sensitivities to α-amanitin, and their templates.

2. mRNA. Whereas the transcript that yields mRNA in prokaryotes is about 30S, it is a giant molecule of up to 200S in eukaryotes. Tailoring produces mRNA's only one fifth to one tenth this size. Moreover, unlike much prokaryotic mRNA, each nuclear mRNA carries the information for a single polypeptide only (Figure 4-4). Finally, most nuclear mRNA's have poly A, 50 to 200 bases long, added to their end.

3. rRNA. In prokaryotes, there are a few repeats of the DNA sequence whose transcript yields the three types of rRNA, so that the genes for rRNA, called *rDNA*, are redundant. In nuclei, rDNA is much more highly redundant, although it makes up a similar fraction of the total DNA. Moreover, the rDNA that forms 5S rRNA is separate from the genes for the two larger pieces of rRNA. Groups of 5S rDNA genes are scattered throughout the chromosome set. There are, moreover, two different kinds of 5S rDNA. Groups of one kind produce *somatic 5S rRNA*, the only kind found in ordinary body, or *somatic*, cells; most groups produce *oocyte 5S rRNA*, which is also produced in *oocytes*—the female cells that develop into eggs. The sequences of these two types of 5S rRNA differ by about six bases. The 5S rDNA genes in any group are separated by segments of DNA that are not transcribed. The function of such *transcription-silent* or *spacer DNA* is unknown.

The nuclear rDNA genes whose transcript will be tailored into the two larger pieces of rRNA are all ordinarily clustered on one chromosome. Such rDNA comprises a region called the *nucleolus organizer*, which together with its transcripts and proteins comprise the nuclear inclusion called the *nucleolus*. Many nuclei contain two sets of chromosomes, and thus generally contain two nucleoli. If their nucleolus organizers are close to each other, there may be a single, larger, nucleolus that mingles the substances of the two. As is the case in a group of 5S rDNA genes, the rDNA genes in the nucleolus organizer are separated by transcription-silent DNA (Figures 4-6 and 4-7). Since many transcriptases can work on the same transcription-active region, starting at a promoter, each has progressively larger pieces of rRNA attached to it. (In the case of transcription of genes for mRNA, *mDNA*, and of genes for tRNA, *tDNA*, the progressively longer pieces are growing mRNA's and tRNA's.) Finally, it should be noted that the two larger pieces of nuclear

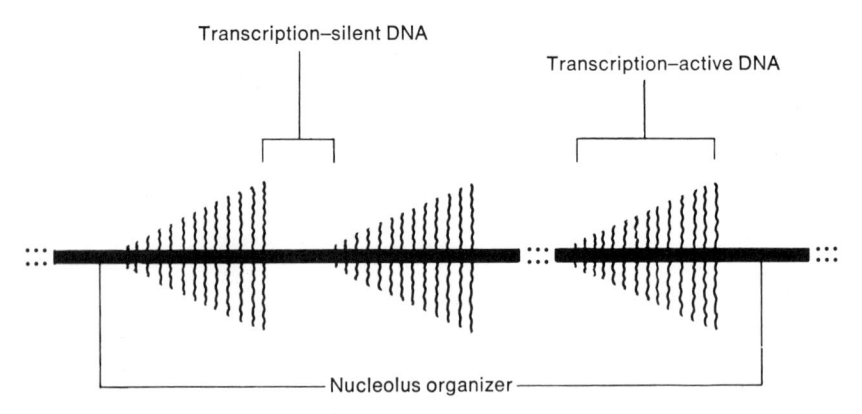

FIGURE 4-6. The nucleolus organizer portion of an amphibian chromosome, showing the alignment of 3 of about 450 repeated transcription-active and transcription-silent regions. Heavy line = double-stranded DNA; wavy line = single-stranded RNA.

rRNA are longer than their counterparts in prokaryotes and differ somewhat in length and/or base sequence in different eukaryotes.

4. tRNA. Although tDNA's are not redundant in prokaryotes, they are redundant in eukaryotes, where they comprise a similar fraction of the total DNA. Nuclear tDNA's occur in clusters, and the genes within a cluster are also separated by transcription-silent DNA. The tRNA's of different species differ, the differences being more in base modification and base sequence than in base number.

## TRANSCRIPTION IN VIRUSES

**4.5** The chromosomes of viruses are adapted to serve as mRNA's or to make mRNA's that simulate the mRNA's of their hosts.

Most mRNA's seem to contain a leader sequence that is lightly methylated. Prokaryotic mRNA's, however, often contain information for several polypeptides and lack poly A at their ends, whereas nuclear mRNA's contain information for only a single polypeptide and almost always carry poly A at their ends. The RNA and DNA viruses that attack prokaryotes or eukaryotes are adapted either to serve as mRNA's that conform with these characteristics of the mRNA's of their hosts or to make mRNA's that do so.

RNA VIRUSES. The RNA chromosomes of RNA viruses are adapted to serve as mRNA's in their hosts. In RNA phages, for example, the chromosome contains a leader sequence, no modified nucleotides in the remainder of the molecule, and information for more than one polypeptide, but does not have poly A at its end. In eukaryotic RNA viruses,

TRANSCRIPTION 63

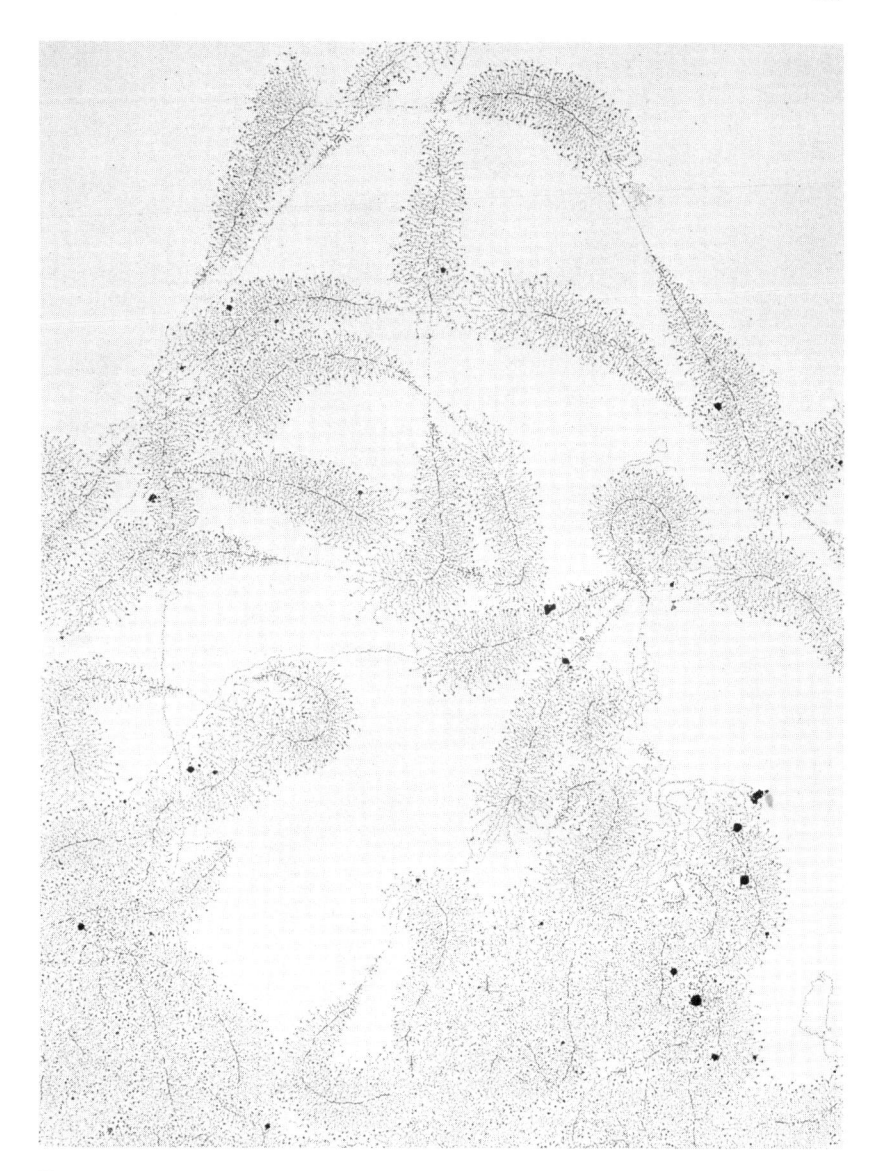

FIGURE 4-7. Electron micrograph of a portion of the nucleolus organizer rDNA of an amphibian. (Courtesy of O. L. Miller, Jr., and B. R. Beatty, Biology Division, Oak Ridge National Laboratory. From the cover photo of *Science*, 164 (May 23, 1969). Copyright 1969 by the American Association for the Advancement of Science.)

the beginning of the chromosome to be used as mRNA is lightly methylated, otherwise contains no modified bases, contains information for a single polypeptide, and carries poly A at its end (as it true, for example, in reovirus). In the case of the Newcastle disease virus, the RNA is a continuous + strand, but the − complement occurs as short pieces, and it is these that serve as mRNA's.

DNA VIRUSES. The DNA of DNA phages is largely organized into operons each of whose mRNA's specifies more than one polypeptide. On

                    D protein information
    ┌─────────────────────────────────────┐
    ↓                                     ↓
────────────────────────────────────────────── φX174 DNA
        ↑                           ↑
        └───────────────────────────┘
               E protein information

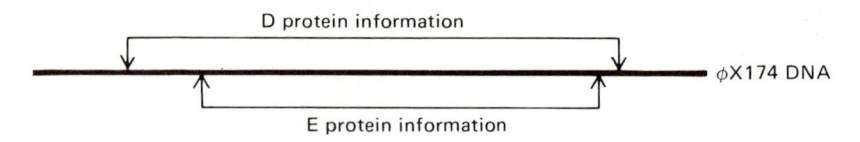

FIGURE 4-8. Nucleotides are sometime shared by two genes.

the other hand, the DNA of DNA viruses attacking eukaryotes is used to synthesize mRNA's that are lightly methylated at their start, specify but a single polypeptide, and end in poly A. DNA viruses have another genetic feature that is important for transcription of their own chromosomes. The DNA virus usually contains information for either a completely new transcriptase or a protein factor that combines with the host's transcriptase. In either event the result is a transcriptase that is virus-specific; that is, it recognizes virus promoter genes and not those of its host, so that it exclusively transcribes virus DNA.

**4.6** In certain small viruses only one strand is transcribed; in certain other viruses both strands in a given region are transcribed.

In most larger DNA viruses, as in cellular organisms, transcription switches from one strand to the other in different transcriptional units. In the case of certain small DNA viruses, however, only one strand is transcribed. For example, in the case of φX174, which contains information for only about eight proteins, the phage + strand is used as a template to make a complementary − strand within 10 seconds after infecting *E. coli*, and only the − strand is transcribed. φX174 is interesting in another respect. The nucleotide sequence that specifies protein E is included within the nucleotide sequence that specifies protein D (Figure 4-8), illustrating that the same nucleotides can be part of two genes (Section 3.3).

Although *Simian virus 40 (SV40)* switches its sense strand in different transcriptional units, late in an infection of monkey cells that will produce virus progeny, SV40 DNA is extensively transcribed in a given region from both strands. Some of this RNA is subsequently degraded. Transcription of both DNA complements has also been reported in the polyoma virus. (It also occurs in the mitochrondrial chromosome, as will be described in Chapter 13.)

# REVERSE TRANSCRIPTION

**4.7** Probably all tumor-producing RNA viruses as well as certain non-tumor-producing RNA viruses use reverse transcription as part of their reproductive cycle.

Many RNA viruses use RNA replicase to replicate their chromosomes. Certain non-tumor-producing RNA viruses and perhaps all tumor-

# TRANSCRIPTION

producing RNA viruses, however, seem to require reverse transcriptase as one of the polymerases needed for replication. Reverse transcriptase is found in the mature virus particle of many RNA tumor viruses and this enzyme is probably made using information contained in viral RNA. Evidence that such enzymes function as reverse transcriptases *in vivo* is that after infection, but not before, the nucleus contains DNA complementary to the viral RNA chromosome. This DNA is double-stranded, inserted into the normal DNA of the nucleus, and seems to be required for viral RNA replication.

Figure 4-9 offers a somewhat speculative view of the replication of *Rous sarcoma virus* (*RSV*), one of the RNA tumor viruses. The mature virus enters the host cell nucleus, where its poly A-carrying chromosome is uncoated from its protein coat, and the reverse transcriptase is liberated (A). A complementary − DNA strand is synthesized (B), using as primer a tRNA, or the poly A-containing end of RSV RNA, or an mRNA. Reverse transcriptase or DNA polymerase then synthesizes a + DNA complement, forming +− double-stranded DNA, using one of the RNA primers already mentioned. A nuclease then breaks both strands of a host chromosome, after which a ligase inserts the viral DNA duplex into the host chromosome (D). Using the − complement of the inserted DNA as template, transcriptase would then produce progeny + RNA chromosomes (E), to whose ends poly A would be attached, completing the replication of mature RSV RNA.

It is not known how infection of an RNA tumor virus causes the transformation of a host cell from the normal to the cancerous state. This may entail the action of one or more genes of the virus or host, or both. Cells that contain the viral DNA in their chromosomes can reproduce cells of the same type that retain the cancer potential, however, even when they do not contain tumor virus RNA chromosomes.

FIGURE 4-9. A speculative view of RSV replication. See the text for descriptive details.

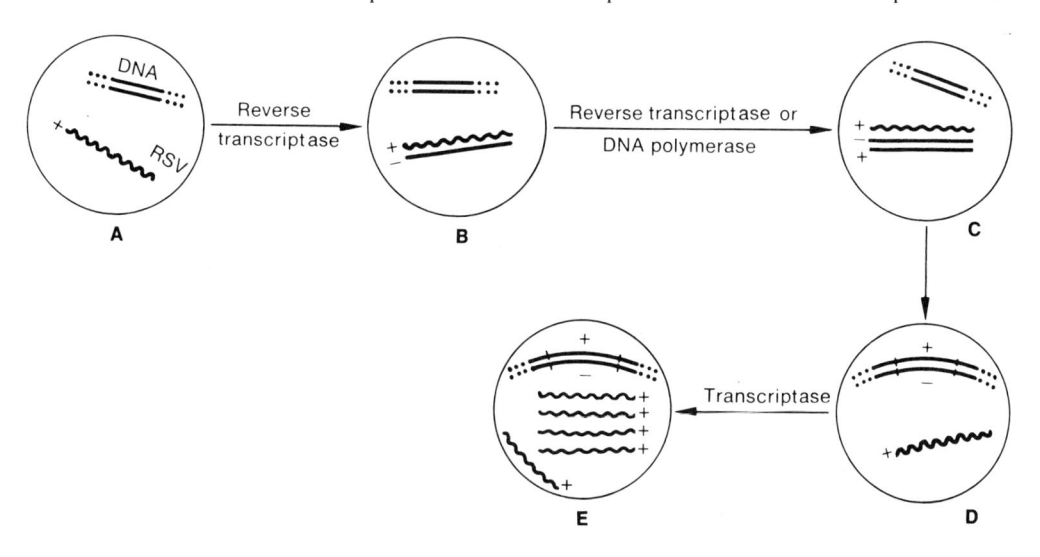

# SUMMARY AND CONCLUSIONS

Genetic material is redefined as any nucleic acid that replicates or that was synthesized using a nucleic acid template and is known to be used as a template to synthesize nucleic acid.

Transcription produces RNA complementary to a DNA template; reverse transcription produces DNA complementary to an RNA template.

In prokaryotes, transcription of mRNA is usually accomplished by operon transcriptional units that contain the following linear sequence of genes: promoter gene, operator gene, usually two or more protein-specifying genes, terminator gene. In eukaryotes, nuclear transcription produces mRNA's that specify single polypeptides and almost always have poly A attached at the end, these mRNA's having been derived from tailored giant RNA transcripts. All mRNA's seem to start with a leader sequence that does not specify protein.

Transcription in both prokaryotes and eukaryotes also produces rRNA's and tRNA's that are tailored from longer transcripts. Both types of RNA are needed in the protein-synthesizing machinery to be discussed in the next chapter. With few exceptions, only one of the two strands of DNA in a given region is transcribed.

Some DNA is not transcribed. This includes the promoter gene, at least in prokaryotes, and the transcription-silent regions that precede nuclear rDNA's and tDNA's. Viruses that are to serve as mRNA's, or are to generate mRNA's, are adapted to the mRNA requirements of their prokaryotic or eukaryotic hosts.

Reverse transcription occurs in eukaryotic cells infected with RNA tumor viruses. The DNA product is later transcribed to genetic RNA.

# QUESTIONS AND PROBLEMS

**4.1** *Anagrams*: Unscramble and define each of the following: **a.** pritton-crains. **b.** pooner. **c.** cullosune. **d.** roilingat. **e.** moropret.

**4.2** What does transcriptase have as its template? Its product?

**4.3** How is the tailoring that produces a tRNA and a dress similar?

**4.4** What three types of gene are present in all operons?

**4.5** How many other genes are necessary in order to transcribe the genes in *E. coli* operon X?

**4.6** List five different functions performed by five different kinds of genes.

**4.7** In what respects are transcription in prokaryotes and eukaryotes similar? Different?

**4.8** List four types of RNA normally or abnormally found in human beings and state one specific way each is tailored.

**4.9** What differences do you expect between a virus whose host is a prokaryote and a virus whose host is a eukaryote?

**4.10** What does reverse transcriptase have as its template? Its product?

**4.11** Describe two different mechanisms for the replication of RNA viruses.

# Chapter 5

## Translation and Its Code

DNA AND RNA INFORMATION is copied into new nucleic acid through replication or transcription by means of a single biological "language" whose "alphabet" consists of four organic bases—A, G, C, and T or U. Proteins, however, make use of a different biological language whose alphabet consists of 20 amino acids. Since proteins are synthesized according to nucleic acid information (present as, or transcribed into, RNA), a *translation* of information is required from the language of nucleic acid into the language of protein (Figure 5-1). Translation proceeds only in the direction from DNA to RNA to protein.

The translational information in RNA resides in its sequence of organic bases. Since there are 20 amino acids and only four bases, there must be some combination or group of base "letters" that uniquely specifies each amino acid (Figure 5-2). We find that the "word," coding unit, or *codon* of mRNA that translates into a single amino acid is a sequence of three successive bases—a triplet.

FIGURE 5-1. The flow of information between nucleic acids and from them to protein.

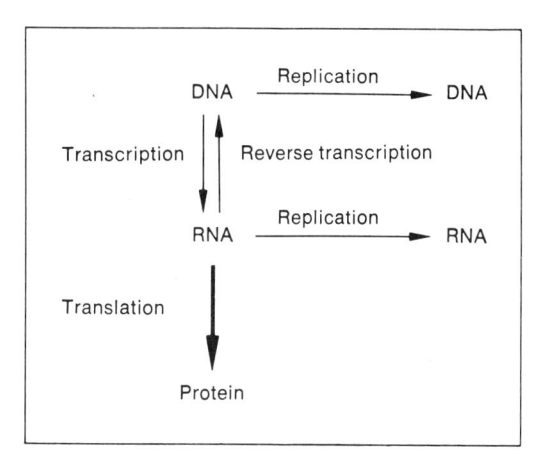

68 WHAT THE GENETIC MATERIAL DOES

| Language (Polymer) | Alphabet (Monomers) | Equivalent Words |
|---|---|---|
| Nucleic acid ↓ Protein (polypeptide) | 4 nucleotides (organic bases) ↓ 20 amino acids | 1 codon = group of 3 letters ↓ 1 amino acid |

FIGURE 5-2. Translation from the language of nucleicacid into the language of protein.

### 5.1 Translation of mRNA occurs at the ribosome by tRNA's carrying amino acids.

We will take a general view of the mechanism of translation before discussing the details of the genetic code and some details about the machinery for translation. Translation occurs only in cellular organisms whose DNA is transcribed into m-, t-, and rRNA's. The mRNA transcript carrying the triplet codons (Figure 5-3) is translated on a ribosome by tRNA's, each of which carries a single amino acid. The first translated codon in mRNA, the *initiator codon*, is almost always AUG. This codon base-pairs with the complementary triplet sequence, or *anticodon*, UAC, which is present on the particular tRNA that carries the amino acid methionine (Met) or a derivative of it (fMet). The second codon binds the anticodon of a tRNA that is carrying, or charged with, its appropriate amino acid (in the case shown in Figure 5-3, codon CCC base-pairs with the anticodon GGG of a different tRNA charged with the amino acid proline, Pro. The portion of the (f)Met attached to its tRNA is then joined enzymatically by *polypeptide polymerase* to the free end of the second amino acid (Pro), forming a chain of two amino acids. The now uncharged tRNA is released, permitting the mRNA to be moved relative to the ribosome (upward in the figure) so the third codon (GGG) can bind its appropriate charged tRNA (glycine, Gly, tRNA). The portion of the second amino acid (Pro) bound to its tRNA is then joined by polypeptide polymerase to the free end of the third amino acid (Gly), producing a chain of three amino acids, and releasing the now-uncharged tRNA.

In this way, each polypeptide is synthesized stepwise, beginning with (f)Met and continuing one amino acid at a time. Although all growing polypeptides have (f)Met at their free end, the end synthesized first, no completed polypeptides do. This change is accomplished by the enzymatic removal, sometime prior to the completion of the polypeptide, of either the f portion, leaving Met, or the entire (f)Met, depending upon whether or not the completed polypeptide is to start with Met. All internal AUG codons are translated by a different tRNA charged with Met, which has a different base sequence from the initiator tRNA charged with (f)Met.

# TRANSLATION AND ITS CODE

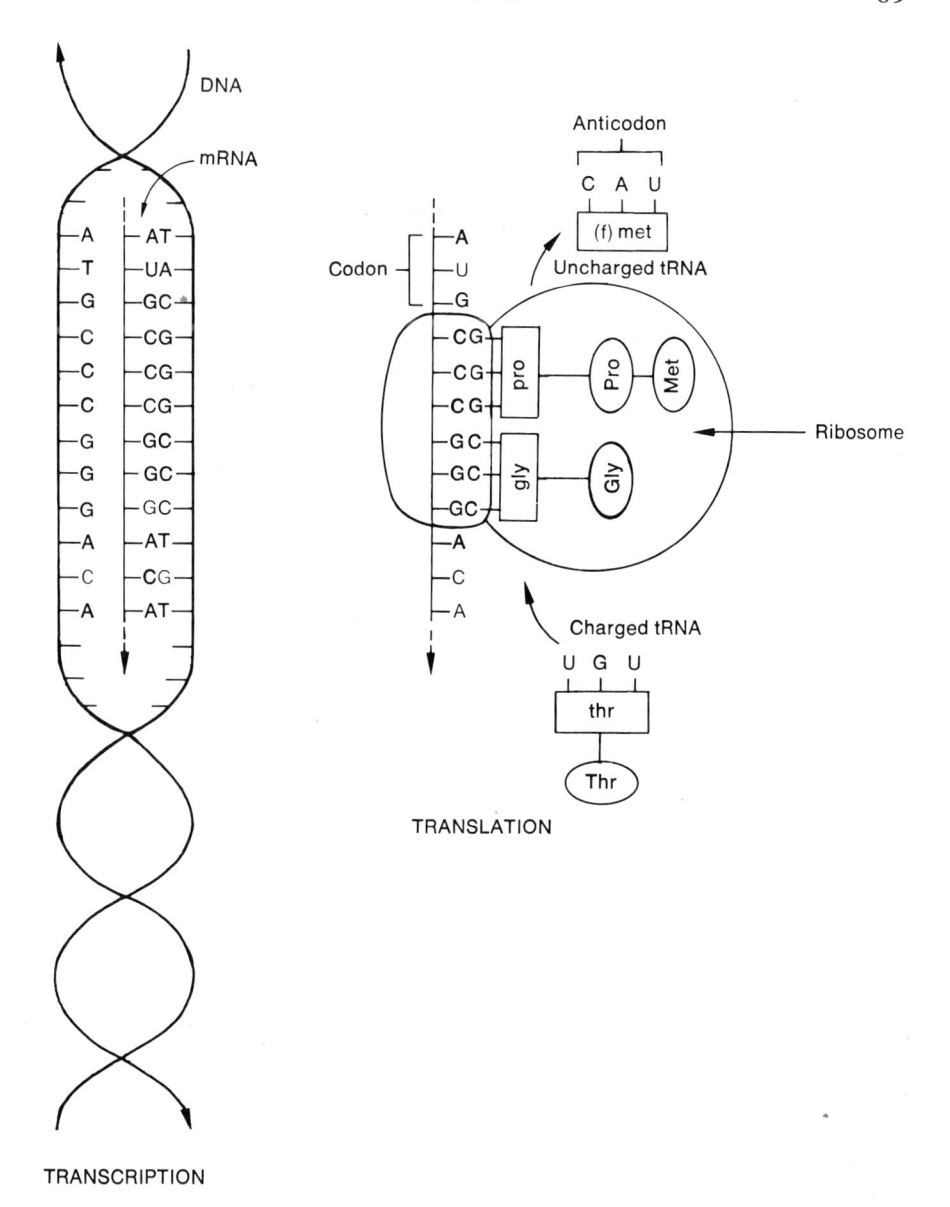

FIGURE 5-3. General view of transcription and translation. Rectangle with anticodon = uncharged tRNA; oval = amino acid; rectangle + oval = loaded or charged tRNA.

We see, therefore, that tRNA's function not only as carriers of amino acids to the ribosome, but as adapter molecules whose anticodons place amino acids in the sequence dictated by the codons in mRNA. In other words, the charged tRNA's act as the dictionary that translates the words of nucleic acid into the words of protein.

70    WHAT THE GENETIC MATERIAL DOES

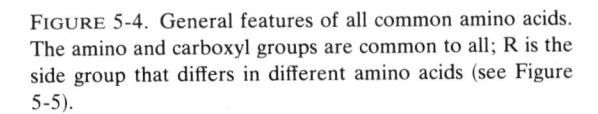

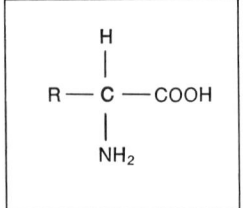

FIGURE 5-4. General features of all common amino acids. The amino and carboxyl groups are common to all; R is the side group that differs in different amino acids (see Figure 5-5).

$$\begin{array}{c} H \\ | \\ R-C-COOH \\ | \\ NH_2 \end{array}$$

**5.2  There are 20 common amino acids which, when joined by peptide bonds, produce a polypeptide.**

We are now ready to consider the chemical nature of amino acids and the chemical mechanism of their polymerization. All amino acids are similar (Figure 5-4) in having a C atom to which are attached a *carboxyl*, COOH, group; an *amino*, $NH_2$ (one exception has an NH), group; and an H atom. The amino acids differ in the fourth group attached to this atom, this side group being designated R.

The names and structures of the 20 common amino acids found in protein are given in Figure 5-5, where the unshaded portions represent their R groups. The amino acids in the figure are grouped according to their behavior under metabolic conditions, foremost among which is the fact that protoplasm is mostly water. There are two classes of amino acids: those that have a relatively strong affinity for water and so are said to be *hydrophilic*, and those that do not and are said to be *hydrophobic*. Hydrophobic amino acids are neutral compounds since they carry no charge on their R side group. Hydrophilic amino acids are neutral, acidic, or basic, depending on their R groups. Most proteins contain most types of amino acid.

Amino acids polymerize, as described briefly in Section 1.2 and Figure 1-4, at the ribosome. This polymerization starts with the union between the carboxyl group (the group by which all amino acids are attached to their tRNA's) of the first amino acid, (f)Met, with the free (unattached) amino group of the second (Figure 5-6). Formation of this peptide bond releases a molecule of water and produces a dipeptide. The tripeptide is made by a peptide bond between the carboxyl group of amino acid 2 and the amino group of amino acid 3. The growing polypeptide assumes the three-dimensional conformation dictated by the amino acids it already contains. In the water-rich cell, the hydrophobic amino acids tend to "hide" in the interior of the polypeptide, while the hydrophilic, charged (acidic or basic) amino acids tend to lie at the periphery of the molecule.

# THE GENETIC CODE

**5.3  Two or more codons have the same meaning in the translation process.**

Since codons are part of an mRNA molecule which is polarized, they are always read by charged tRNA's in the same direction. Since each of the

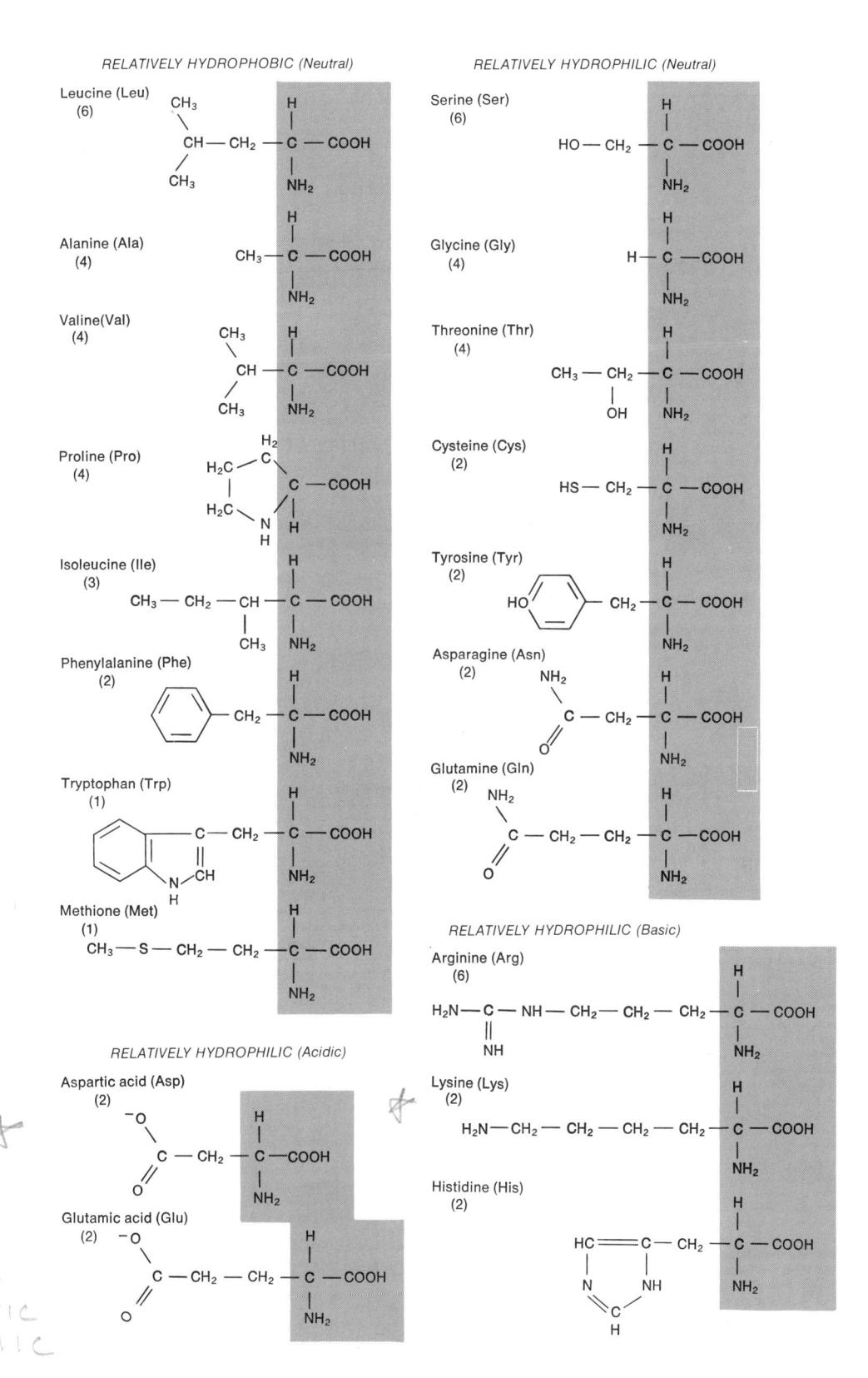

FIGURE 5-5. The 20 common amino acids, grouped according to their behavior under metabolic conditions. The number of codons that each has is shown in parentheses. The shaded portion is common to all amino acids except Pro.

71

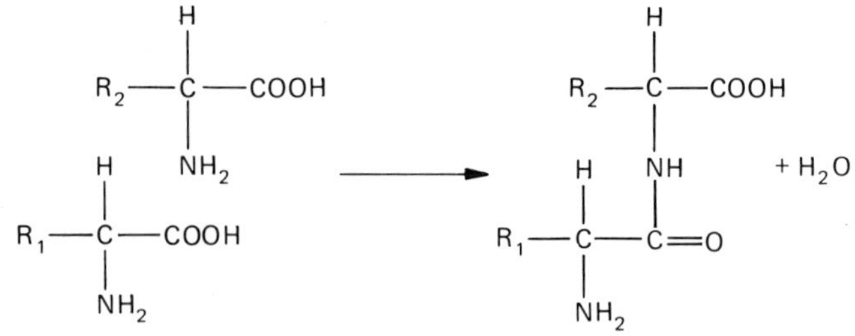

72        WHAT THE GENETIC MATERIAL DOES

$$R_2-\overset{\overset{\displaystyle H}{|}}{\underset{\underset{\displaystyle NH_2}{|}}{C}}-COOH \qquad\qquad R_2-\overset{\overset{\displaystyle H}{|}}{\underset{\underset{\displaystyle NH}{|}}{C}}-COOH$$

$$\longrightarrow \qquad + H_2O$$

FIGURE 5-6. Formation of a peptide bond. The carboxyl end of amino acid 1 joins the amino end of amino acid 2, releasing $H_2O$.

$$R_1-\overset{\overset{\displaystyle H}{|}}{\underset{\underset{\displaystyle NH_2}{|}}{C}}-COOH \qquad\qquad R_1-\overset{\overset{\displaystyle H}{|}}{\underset{\underset{\displaystyle NH_2}{|}}{C}}-\overset{}{\underset{\underset{\displaystyle O}{\|}}{C}}$$

three bases in a codon triplet has four possible alternatives, there are 4×4×4, or 64, possible unidirectional triplet combinations. All 64 possible triplet codons occur in mRNA. The codon equivalents for the amino acids are presented in Figure 5-7. Of these 64 triplets, 61 code for an amino acid; that is, they make amino acid sense, and are called *sense codons*. Three triplets, UAA, UAG, and UGA, do not code for an amino acid; that is, they make no amino acid sense—being *nonsense codons*. Nonsense codons are used in mRNA to terminate translation; that is, they serve as *terminator codons*. Since all three of these terminator codons have the same meaning for translation, the genetic code is said to be *degenerate*. The English language is also a degenerate code since, for example, the words DAD and POP have the same meaning.

Only two amino acids have a single codon: (f)Met has AUG and Trp has UGG. Each of the other 18 amino acids is coded by 2, 3, 4, or 6 of the remaining 59 triplets (Figures 5-5 and 5-7). Hence, the genetic code is also highly degenerate in its sense codons. The major position of

FIGURE 5-7. mRNA triplet codons for amino acids. non = nonsense or polypeptide chain-terminating codons.

|  |  | SECOND BASE |  |  |  |  |
|---|---|---|---|---|---|---|
|  |  | U | C | A | G |  |
| FIRST BASE | U | UUU ⎤ Phe<br>UUC ⎦<br>UUA ⎤ Leu<br>UUG ⎦ | UCU ⎤<br>UCC ⎥ Ser<br>UCA ⎥<br>UCG ⎦ | UAU ⎤ Tyr<br>UAC ⎦<br>UAA non<br>UAG non | UGU ⎤ Cys<br>UGC ⎦<br>UGA non<br>UGG Trp | U<br>C<br>A<br>G |
|  | C | CUU ⎤<br>CUC ⎥ Leu<br>CUA ⎥<br>CUG ⎦ | CCU ⎤<br>CCC ⎥ Pro<br>CCA ⎥<br>CCG ⎦ | CAU ⎤ His<br>CAC ⎦<br>CAA ⎤ Gln<br>CAG ⎦ | CGU ⎤<br>CGC ⎥ Arg<br>CGA ⎥<br>CGG ⎦ | U<br>C<br>A<br>G |
|  | A | AUU ⎤<br>AUC ⎥ Ile<br>AUA ⎦<br>AUG Met | ACU ⎤<br>ACC ⎥ Thr<br>ACA ⎥<br>ACG ⎦ | AAU ⎤ Asn<br>AAC ⎦<br>AAA ⎤ Lys<br>AAG ⎦ | AGU ⎤ Ser<br>AGC ⎦<br>AGA ⎤ Arg<br>AGG ⎦ | U<br>C<br>A<br>G |
|  | G | GUU ⎤<br>GUC ⎥ Val<br>GUA ⎥<br>GUG ⎦ | GCU ⎤<br>GCC ⎥ Ala<br>GCA ⎥<br>GCG ⎦ | GAU ⎤ Asp<br>GAC ⎦<br>GAA ⎤ Glu<br>GAG ⎦ | GGU ⎤<br>GGC ⎥ Gly<br>GGA ⎥<br>GGG ⎦ | U<br>C<br>A<br>G |

THIRD BASE

# TRANSLATION AND ITS CODE

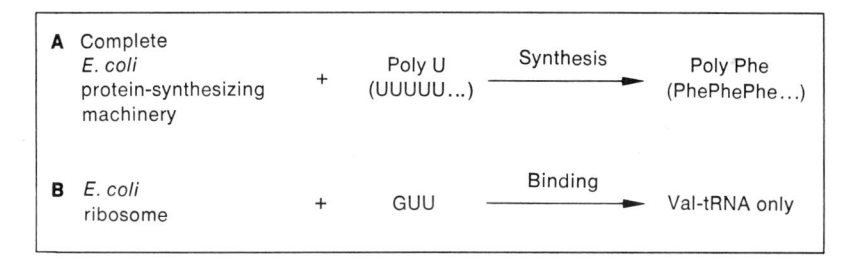

```
A  Complete
   E. coli                          Poly U        Synthesis      Poly Phe
   protein-synthesizing    +       (UUUUU...)   ──────────▶    (PhePhePhe...)
   machinery


B  E. coli                                        Binding
   ribosome                +         GUU        ──────────▶    Val-tRNA only
```

FIGURE 5-8. Sample results obtained with the two main techniques used to decipher the genetic code *in vitro*.

degeneracy is at the third base, at the end of the codon. Some degeneracy also occurs at the first and second bases of the codon.

## 5.4 The genetic code is essentially identical *in vitro* and *in vivo* in all organisms.

Two kinds of experiments helped decipher the genetic code *in vitro*. The first experiments studied protein synthesis using the protein-synthesizing machinery of *E. coli* and long artificial mRNA's of different but known base sequence. For example, a ribose polynucleotide containing only U's, poly U, was found to code for a polypeptide composed only of Phe's, poly Phe; a codon for Phe, therefore, contained only U's (Figure 5-8A). In later experiments, RNA triplets of a single kind were synthesized to order and ribosomes were exposed to them; it was then determined which charged tRNA's were bound to the ribosomes. For example, GUU acting as mRNA bound only Val-tRNA to the ribosome (Figure 5-8B). Hence, GUU was identified as a codon for Val. Such studies showed that the *in vitro* genetic code employs 61 sense and 3 nonsense triplet codons, which are read:

1. Successively (neither skipping bases nor using a single base in more than one triplet during any continuous reading).
2. Unidirectinally (GUU is a codon for Val while UUG is a codon for Leu; neither codon will yield the other amino acid).
3. Unambiguously (under appropriate conditions a codon always translates into the same amino acid).

The genetic code *in vivo* has been studied by matching a known base sequence in RNA with a known sequence of the amino acids they code. For example, the known base sequence of part of the chromosome of a phage such as $\phi$MS2 is translated by its host bacterium into protein of known amino acid sequence. It is possible to align the two sequences according to the expectations of the *in vitro* genetic code. Moreover, it is found that all 61 sense codons and 2 nonsense codons are used. Thus, the codons have the same meaning *in vivo* as they have *in vitro*. Moreover, the genetic code has been found to be essentially the same in all present-day organisms.

# WHAT THE GENETIC MATERIAL DOES

## TRANSLATION

We now consider further how the machinery for translation works on a molecular level, and some of the ways that this machinery and the translation process differ in different organisms.

**5.5** There are three general sizes of ribosomes.

All ribosomes function as the site of protein synthesis and are composed of a smaller subunit plus a larger subunit. They come in three different general sizes, however: 60S, 70S, and 80S (Figure 5-9), which differ in their ribosomal proteins and RNA's. Ribosomes belonging to the same class also differ somewhat chemically (and in S value) in different organisms.

60S ribosomes are found only in mitochondria. They are mini-ribosomes partly because they contain no 5S rRNA.

70S ribosomes are found in bacteria and chloroplasts. They consist of a 30S subunit joined to a 50S subunit. (Note that S values are not additive.)

80S ribosomes are found in the cytoplasm of eukaryotes. They consist of a 40S subunit joined to a 60S subunit.

**5.6** We are beginning to understand how ribosomes function molecularly in translation.

The ribosome performs several specific functions in translation: it binds to the mRNA, yet permits it to move a codon at a time; it is apparently active in the termination of translation; it can bind two charged tRNA's at any one time. As more becomes known about ribosomal proteins and rRNA's, we are beginning to understand more about the molecular basis for ribosomal functioning.

FIGURE 5-9. General classes of ribosomes and the protein and RNA classes of some of them.

|  | Ribosome Size Class | Ribosome Subunits | Number of Different Proteins Contained | RNA Class Contained |
|---|---|---|---|---|
| Prokaryotes | 70S | 30S<br>50S | 21<br>34 | 16S<br>23S + 5S |
| Eukaryotes |  |  |  |  |
| Chloroplast | 70S |  |  |  |
| Mitochondrial | 60S |  |  |  |
| Cytoplasmic | 80S | 40S<br>60S | 32<br>39 | 18S<br>28S + 5S |

RIBOSOMAL PROTEINS. The proteins in the smaller ribosomal subunit are given consecutive S numbers; those in the larger subunit, consecutive L numbers. It has been found that a modified S12 in the 30S subunit protects the 16S rRNA of the subunit from combining with two molecules of streptomycin. As a result, bacteria with modified S12 are resistant to streptomycin, that is, are *streptomycin-resistant*. Bacteria with normal S12, on the other hand, bind the streptomycin and, as a consequence, codons in mRNA are misread; that is, codons that normally have but one meaning now take on a new meaning—in other words, the genetic code is rendered *ambiguous*. As a result, proteins are made with wrong amino acid sequences, the bacterium is unable to metabolize properly, and dies. Such bacteria are *streptomycin-sensitive*. Because S12 affects ambiguity, it is likely that both it and the streptomycin-sensitive region(s) in the 16S rRNA are located in the ribosome near the site where translation occurs.

Other results indicate that in the 50S subunit, proteins L7 and L12 are associated with the termination of polypeptide chains, and that proteins L18 and L25 help tRNA to bind to 5S rRNA.

RIBOSOMAL RNA'S. Some details of the role that rRNA's play in translation have been deduced from the complementarity of base sequences in rRNA's with those in mRNA or tRNA. The terminal base sequence of the single piece of rRNA in the smaller (30S and 40S) ribosomal subunits is apparently used to base-pair with mRNA in order to start and to terminate translation. This can be seen in Figure 5-10, which shows that the 16S rRNA (of the 30S ribosomal subunit) has a terminal base sequence that contains the complement of part of the leader sequence needed to start translation, and the anticodon of a terminator codon. The 18S rRNA (of the 40S subunit) has a terminal base sequence that contains anticodons for the initiator codon as well as for two different terminator codons.

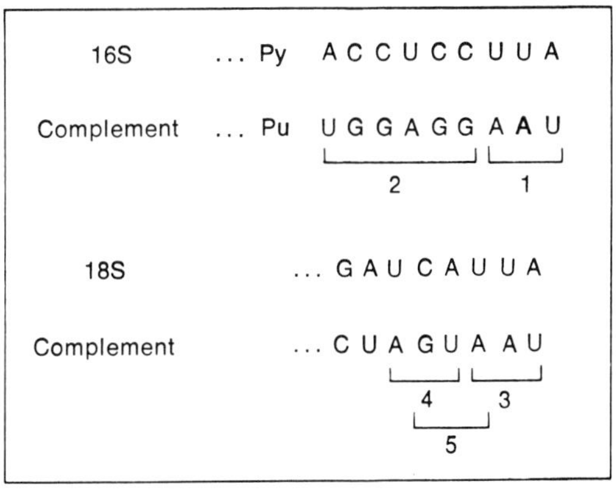

FIGURE 5-10. The terminal base sequence of 16S and 18S rRNA's. Note in the RNA complements that 1, 3, and 4 are nonsense, terminator codons; 2 is part of the mRNA leader sequence; 5 is the initiator codon AUG.

```
16S          ... Py  A C C U C C U U A

Complement   ... Pu  U G G A G G A A U
                     └─────┘ └─────┘
                        2       1


18S                  ... G A U C A U U A

Complement           ... C U A G U A A U
                         └───┘ └───┘
                           4     3
                         └─────┘
                            5
```

76  WHAT THE GENETIC MATERIAL DOES

Since part of the base sequences of 5S rRNA and of tRNA are complementary, base pairing between them may be used to hold a tRNA to the larger subunit of 70S and 80S ribosomes during translation.

RIBOSOME-BOUND ENZYMES. Ribosomes apparently have other functions related to translation. Nucleases that degrade mRNA are generally found attached to the small ribosomal subunit of prokaryotes. (In cases in eukaryotes where the mRNA survives for a relatively long time, however, no such nucleases are found on ribosomes.) In addition, polypeptide polymerase is bound to the larger ribosomal subunit in prokaryotes. We do not yet have any molecular insight as to how these enzymes are bound to ribosomes or how such binding is related to their functioning.

Although we expect that research in the near future will cast more light on the detailed molecular ways that ribosomes function in translation, it is clear that particular ribosomal proteins and particular sequences in rRNA's make specific contributions to the process of translation, to assure that it starts, continues, and stops in a correct, unambiguous manner.

**5.7 Different regions of the tRNA molecule have different functions.**

The number, type, and sequence of bases differ not only within tRNA's for different amino acids but within tRNA's for the same amino acid in different organisms (Figure 5-11). Nevertheless, all tRNA's form the same type of cloverleaf structure whose leaves base-pair to further fold the molecule into an L shape. Although the tRNA functions in the L shape (Figure 5-12), the different functions that tRNA performs are intimately related to the base sequences present in different regions of the molecule. The tRNA's that transfer different amino acids differ in base sequence in one of the leaves (the left one in Figure 5-11). Each different sequence is recognized by a different type of *aminoacyl-tRNA synthetase*. Each of these enzymes is responsible for activating a different amino acid and attaching it to the A of the terminal CCA sequence (common to all tRNA's) of the tRNA whose leaf it recognizes. Accordingly, since there are 20 different amino acids, there are 20 different such leaves, and 20 different aminoacyl-tRNA synthetases. A second leaf (the lowest one in Figure 5-11) contains an unpaired triplet that serves as the anticodon. When the ribosomes used contain 5S rRNA, the remaining leaf in tRNA contains an unpaired sequence that is common to all tRNA's (but one). This tRNA sequence apparently base-pairs with its complement in the 5S rRNA, thereby binding the charged tRNA to the ribosome. When two charged tRNA's are bound to the ribosome, their amino acids are in suitable proximity for peptide bonding.

# TRANSLATION AND ITS CODE

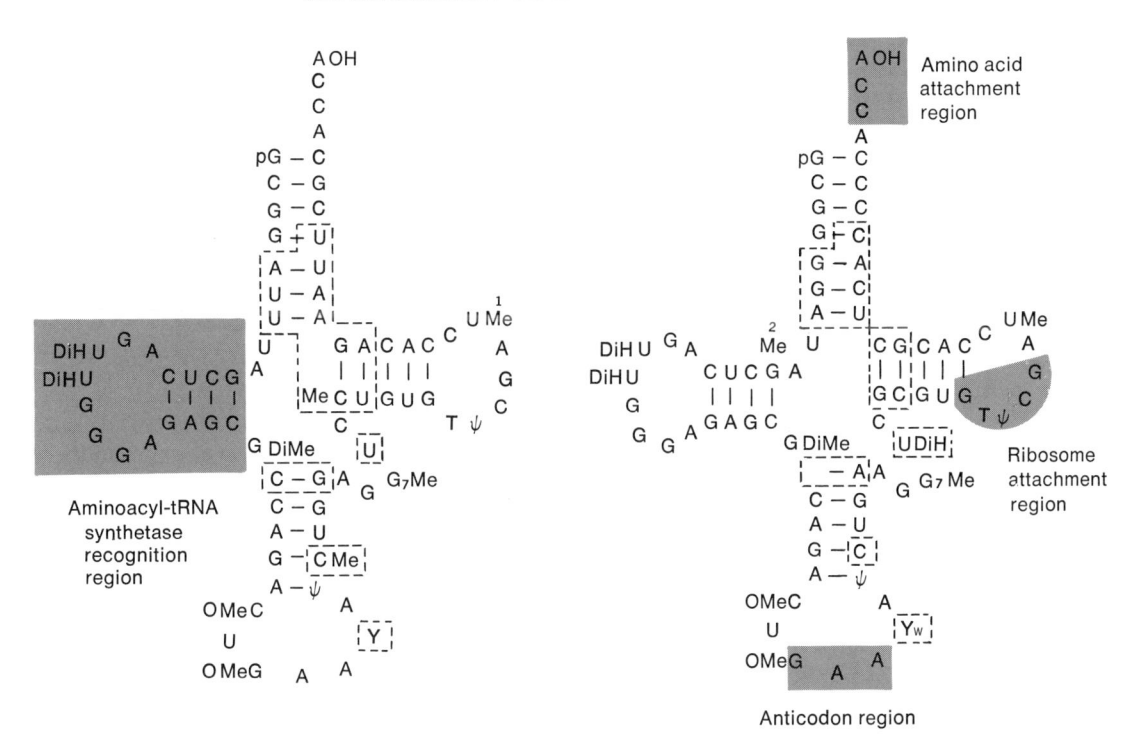

FIGURE 5-11. Two tRNA members of the same Phe-carrying family, one from yeast and one from wheat germ. The areas enclosed in dashed lines show regions of dissimilarity. (After B. S. Dudock, G. Katz, E. K. Taylor, and R. W. Holley.)

FIGURE 5-12. Schematic diagram of the three-dimensional L shape of yeast phenylalanine tRNA.

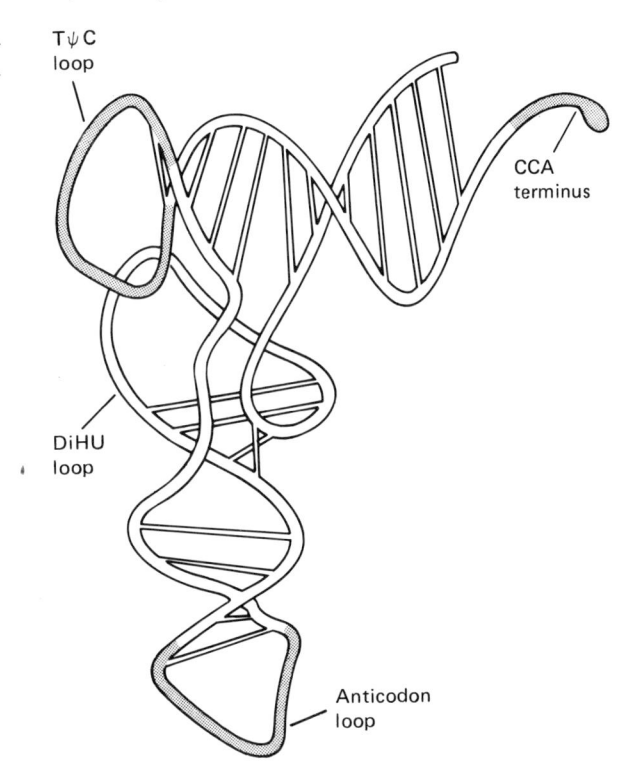

**5.8** Most degeneracy is due to one tRNA or to more than one identically charged tRNA's recognizing two or more different codons.

Most of the degeneracy in the genetic code is due to an amino acid being carried (1) by a single tRNA whose anticodon recognizes more than one codon or (2) by two or more tRNA's charged with the same amino acid having different anticodons.

1. SINGLE tRNA's. The same charged tRNA can often recognize more than one codon because of the location and nature of the first base in the anticodon (which must base-pair with the third base of the codon, for reasons of polarity). The first base in the anticodon occurs at a turn in the molecule where there is minimal constraint against making an unusual base pair (UG or GU) in addition to a normal base pair (UA or GC). Specifically, a U in the anticodon can base-pair with an A or a G in a codon (Figure 5-13), and a G in the anticodon can base-pair with a C or a U in a codon. Degeneracy also occurs when the first base in the anticodon is one of those of minor frequency that are modified or rearranged (for example, hypoxanthine, I), which can form two or three different base pairs. The flexibility in base pairing at the first base of the anticodon explains much of the degeneracy at the third base of the codon.

2. MULTIPLE tRNA's. There are 30 to 40 different tRNA's in *E. coli* that differ in their anticodons. Since there are only 20 kinds of amino acid, this must mean that some amino acids are carried by two or more tRNA's having different anticodons. Multiple tRNA's for the same amino acid can explain some of the degeneracy at the first and second bases of the codon.

FIGURE 5-13. Normal and unusual base pairs between the base at the end of the codon and the base at the start of the anticodon.

| Anticodon | Codon |
|---|---|
| U | A<br>G |
| C | G |
| A | U |
| G | C<br>U |
| I | U<br>C<br>A |

**5.9** Translation requires an energy source and additional proteins for initiating, lengthening, and terminating an amino acid chain.

Translation requires, in addition to mRNA, ribosomes, charged tRNA's, and polypeptide polymerase, a source of chemical energy for certain aspects of the process as well as a variety of other proteins. The energy supplier is guanosine triphosphate. The initiation of protein synthesis requires three or four protein initiation factors; the continuation of protein synthesis requires three protein elongation factors; and the termination of protein synthesis and the release of the last amino acid from its tRNA require three protein termination factors. We see, therefore, that protein synthesis is expedited by at least nine proteins in addition to the polypeptide polymerase. It should be noted that DNA and RNA viruses may code for certain components of the translation machinery that will result in perferential translation of viral RNA. Most of the components they use, however, are coded in host DNA.

**5.10** Depending on the length of the mRNA molecule, several to many ribosomes can be joined to the same messenger.

Any ribosome can accommodate only one mRNA and can make only one polypeptide at a time. Since most mRNA's are much longer than the diameter of a ribosome, however, several to many ribosomes can be attached to the same mRNA simultaneously. Ribosomes along a single mRNA carry successive stages in the synthesis of a given polypeptide (Figure 5-14). Such aggregates of ribosomes attached to a single mRNA are called *polyribosomes* or *polysomes* (Figure 5-15).

Since the mRNA's of the nucleus code for single polypeptides, whereas those of prokaryotes usually code for two or more, the polyribosome of a complete nuclear mRNA tends to be shorter than that of a complete prokaryotic mRNA.

**5.11** The polyribosomes of prokaryotic and nuclear mRNA's differ in cell location.

In prokaryotes, translation of an mRNA usually starts before the synthesis of the mRNA is complete. This is possible because the end of mRNA synthesized first is always the end translated first. Attachment of the first ribosome of a polyribosome can occur, therefore, just as soon as the mRNA is long enough to have a free leader sequence. Figure 5-16 shows longer and longer polyribosomes attached to mRNA's at successive stages of synthesis by a succession of transcriptase molecules. The coupled transcription and translation proceeds at about 1000 nucleotides per minute (a rate that is about 100 times slower than that for DNA replication). At this rate, the initial appearance of the first and last

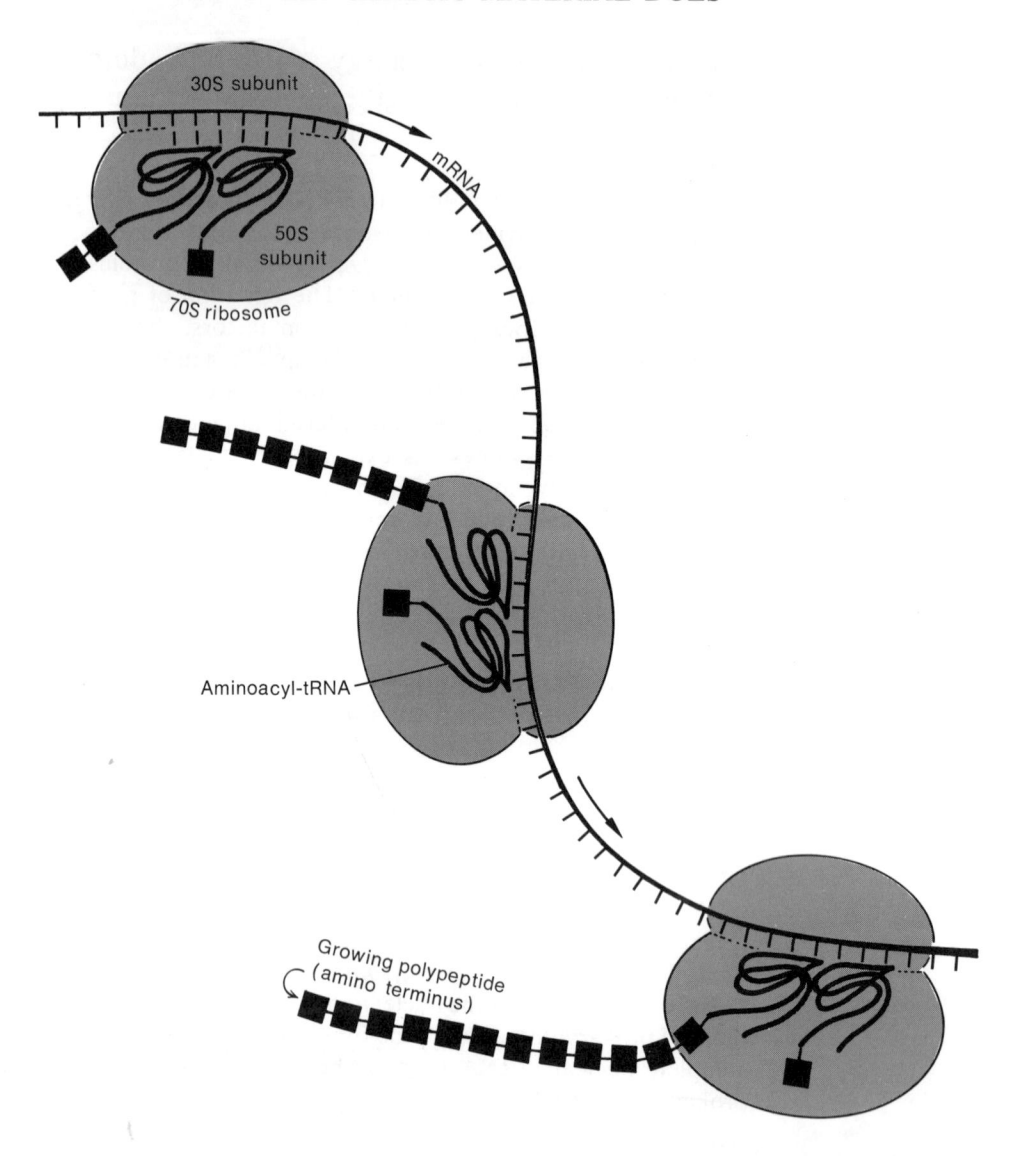

FIGURE 5-14. Schematic drawing of a polyribosome in *E. coli*. Arrows show the direction in which the ribosome moves during translation.

proteins coded in a long mRNA may be separated by as much as 5 to 10 minutes. As expected, proteins needed in metabolism sooner are coded nearer the beginning of the mRNA than those needed later.

In eukaryotes, ribosomes are present in the cytoplasm but not in the nucleus. Accordingly, transcription and translation are uncoupled; mRNA synthesis must be completed and the mRNA transported to the cytoplasm before translation and the formation of polyribosomes can occur.

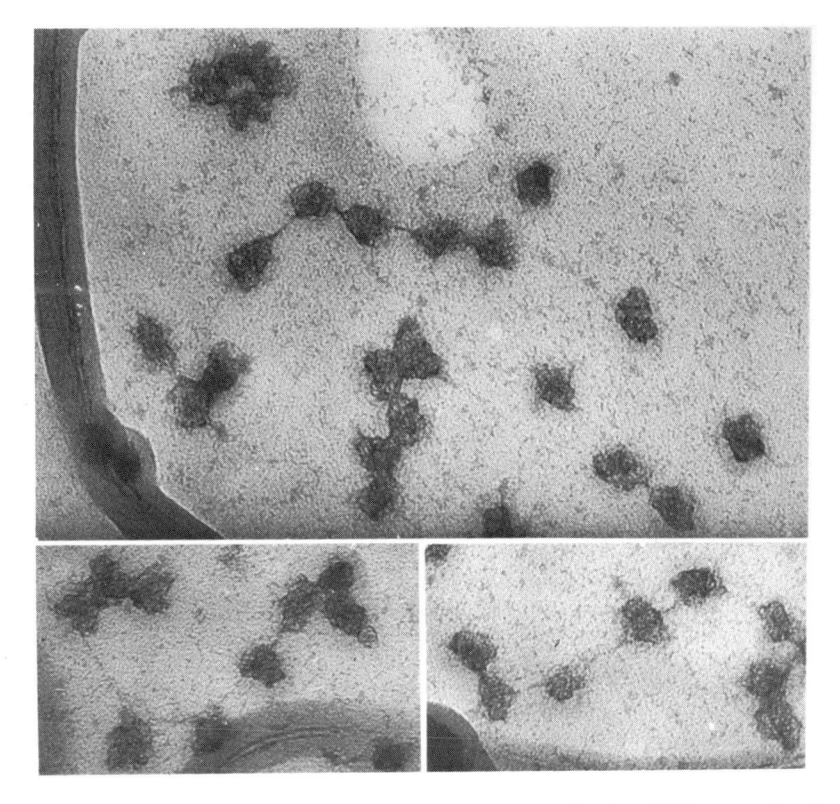

FIGURE 5-15. Electron micrographs of polyribosomes showing RNA connecting the ribosomes. (Courtesy of A. Rich.)

## 5.12 Signal polypeptides are used to attach polyribosomes to membranes.

Although the polyribosomes of prokaryotic and nuclear mRNA's differ in cell location, they are similar in that they are of two types—attached or unattached to cellular membranes. Polyribosomes that are attached to cellular membranes synthesize proteins that are either (1) incorporated as part of these membranes, or (2) pass through these membranes to leave the cell as secretions or to enter organelles such as a chloroplast or mitochondrion. The proteins synthesized by membrane-bound polyribosomes usually start with a *signal polypeptide*. This signal sequence is 15–30 amino acids long and hydrophobic. The hydrophobic nature probably helps the signal region to enter or penetrate the lipid phase of cellular membranes. After functioning in this way, the signal polypeptide is removed enzymatically. Thus, proteins being made in membrane-bound polyribosomes are more extensively tailored than proteins being made in membrane-free polyribosomes, which usually only lose their initial Met.

In prokaryotes, which have no endoplasmic reticulum, most mRNA's are translated on membrane-free polyribosomes to produce proteins

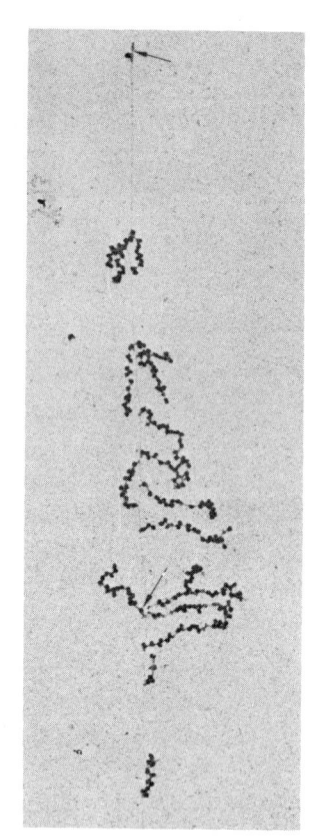

FIGURE 5-16. Electron micrograph of transcription and translation in *E. coli*. The vertical straight line visualizes bacterial DNA which is being transcribed in the top-to-bottom direction. The upper arrow indicates a transcriptase molecule at or very near the transcription initiation site; the lower arrow indicates a transcriptase molecule whose partially completed RNA transcript (seen as a thin side branch) is loaded with ribosomes (seen as black dots) that translate the mRNA behind the transcriptase, that is, as soon as the RNA is synthesized. Accordingly, the polyribosome becomes larger as the mRNA lengthens. (This preparation, like that in Figure 5-15, does not visualize the polypeptide chains being synthesized on the ribosomes.) (Courtesy of O. L. Miller, Jr., B. A. Hamkalo, and C. A. Thomas, Jr., 1970. *Science*, **169**: 392. Copyright 1970 by the American Association for the Advancement of Science.)

used for metabolism within the cell. In prokaryotes, the synthesis of proteins that are secreted into the medium occurs on ribosomes that seem to be attached to the cell membrane. In eukaryotes, which have an endoplasmic reticulum, membrane-free ribosomes seem to synthesize proteins used within the cytoplasm, whereas those that are attached to the endoplasmic reticulum seem to synthesize proteins that are secreted from the cell. The general eukaryotic mechanism for protein secretion (Figure 5-17) seems to use polyribosomes whose protein product enters

FIGURE 5-17. Secretion of proteins synthesized by membrane-bound ribosomes.

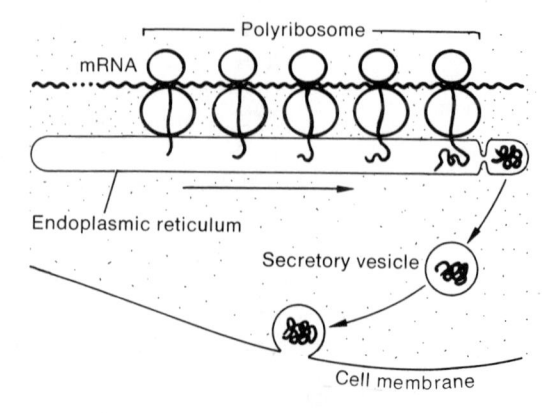

the cavity of the endoplasmic reticulum. The endoplasmic reticulum membranes pinch to form a vesicle containing the protein; this vesicle then migrates and, after certain changes, becomes a secretory vesicle whose membrane fuses with the cell membrane, emptying its contents outside the cell.

## SUMMARY AND CONCLUSIONS

The great majority of organisms contain DNA genetic material that is transcribed to mRNA that is then translated into the amino acid sequence of the various proteins that characterize each organism. Cellular DNA also contains information for the translation machinery that enables base sequence to specify amino acid sequence. Elements of this machinery include tRNA's, rRNA's and ribosomal proteins, aminoacyl-tRNA synthetases, polypeptide polymerase, a source of chemical energy, and at least nine additional proteins used or needed to produce factors for the initiation, lengthening, and termination of polypeptides. Although organisms differ in the fine structure of their machinery for translation, all of them translate mRNA by charged tRNA's at the ribosome using a generally universal genetic code.

The universal (*in vivo* as well as *in vitro*) features of the genetic code include:

1. An mRNA message read unidirectionally in successive, nonoverlapping base triplets or codons.
2. A special initiator codon (AUG) that places (f)Met at the beginning of all polypeptides, after which f or all of (f)Met may be removed enzymatically.
3. Nonsense codons that signal polypeptide termination.
4. Degeneracy that is due largely to single or multiple tRNA's charged with a given amino acid recognizing two or more codons.
5. No ambiguity under normal circumstances.
6. Codons that generally have the same translational meaning in all organisms (and *in vitro*).

## QUESTIONS AND PROBLEMS

**5.1** *Anagrams*: Unscramble and define each of the following: **a.** tartinnalos. **b.** nocandoit. **c.** omibores. **d.** buyamigit. **e.** ceedangrey.

**5.2** *Riddle*: When does a triplet in a DNA sense strand have a complement whose bases have the same sequence as its codon?

**5.3** *Riddle*: What codons are usually found at least twice in prokaryotic mRNA's but are found only once in eukaryotic mRNA's?

**5.4** What is the base sequence in the DNA that is the template for the codon whose anticodon is GAU?

**5.5** An RNA has the following base sequence:
U U U C C C A A A G G G U A A
What will it translate into? Explain.

**5.6** Give an example of an ambiguous word in the English language.

**5.7** If the ribosome is likened to the lock on a zipper, what does the zipper itself represent? In what way does zippering fail to represent translation?

**5.8** What do you suppose would happen if your cells were injected with monkey mRNA's? Explain.

**5.9** How are polypeptides tailored?

# 84  WHAT THE GENETIC MATERIAL DOES

**5.10** What is the molecular basis for streptomycin killing certain bacteria, but not: **a.** other bacteria?   **b.** human cells?

**5.11** What specific translational functions have different parts of: **a.** a ribosome?   **b.** a charged tRNA?   **c.** an mRNA?

**5.12** How many different protein-coding genes are needed to make translational machinery and perform translation?

**5.13** In what specific details do the polyribosome of a prokaryote and eukaryote differ?

**5.14** *Balloons*: Fill appropriately.

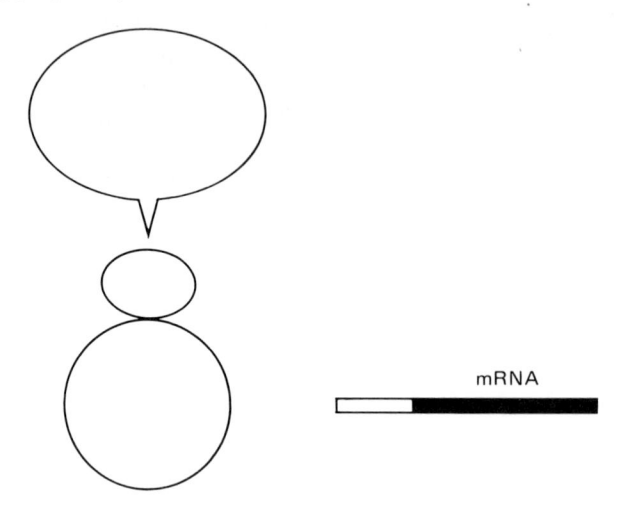

mRNA

**5.15** *Balloons*: Fill appropriately.

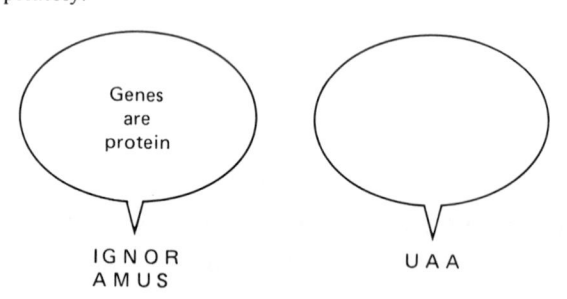

Genes are protein

IGNORAMUS                UAA

**5.16** *Balloons*: Fill appropriately.

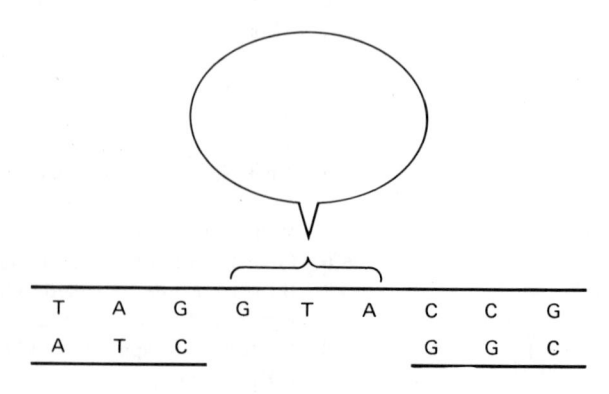

```
T  A  G   G  T  A   C  C  G
A  T  C             G  G  C
```

*PART III*

HOW GENETIC
MATERIAL IS
VARIED, PACKAGED,
AND DISTRIBUTED

# Chapter 6

## Mutation

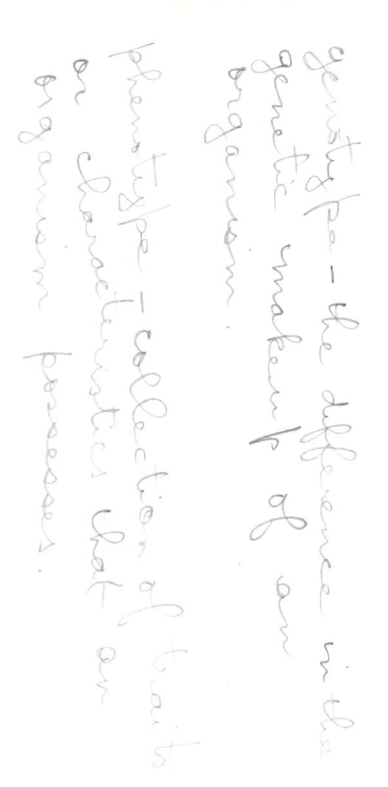

*[Handwritten marginalia: genotype - the difference in the genetic makeup of an organism. phenotype - collection of traits or characteristics that an organism possesses.]*

DIFFERENT KINDS OF present-day organisms differ in the sum total of their genetic material; that is, they have different genetic makeups, or *genotypes*. Since all organisms on earth seem to have had a common ancestor, it follows that an ancestral genotype must have undergone many changes during the course of evolution to produce the multitude of different genotypes now in existence. Such changes in genotype are called *mutations*. Mutations have the following characteristics:

1. They can *change the number, kind, or sequence of genetic nucleotides*.
2. They are more-or-less *permanent* changes. (The making and breaking of H bonds in duplex DNA is a temporary, hence nonmutational, change.)
3. They are *uncoded* or unprogrammed changes. (Methylation of *E. coli* DNA is not a mutation when it is produced by a methylase coded by *E. coli* DNA.)
4. They are *relatively rare*, unusual changes.

The product of a mutation is called a *mutant*, a term applied to a genotype, cell, or individual.

This chapter deals with mutational variation of the genetic material. We start with a general consideration of the types of mutation that occur and the effects these have on the *phenotype*, that is, the collection of traits or characteristics that an organism possesses. (The phenotype, the product of the action of the genotype and the environment, is often used to determine the usually less readily observed genotype.) We continue by discussing, at the molecular level, the production of specific types of mutation due to external and internal factors in the environment, as well as the reversal or the prevention and repair of mutations.

The remaining chapters in Part III will deal largely with the programmed ways in which organisms package and distribute their genetic material within and between generations.

## 6.1 Mutations can affect a portion of a nucleotide, or one to many whole nucleotides.

Some mutations affect only a part of a nucleotide, changing only the phosphate, sugar, or base portion. With regard to the phosphate group, for example, substitution of normal, nonradioactive phosphorus, $^{31}P$, by radioactive phosphorus, $^{32}P$, is a mutation. Moreover, when radioactive $^{32}P$ releases an electron, it decays into ordinary, nonradioactive sulfur, $^{32}S$—another mutation. In addition, when the $^{32}P$ atom releases an electron, it recoils, as does a rifle when a bullet is shot from it. This recoil often causes the phosphate to lose its connection to sugar, thus producing a break in the nucleotide sequence—another mutation. With regard to the sugar portion, mutation may substitute ribose for deoxyribose in DNA, or substitute deoxyribose for ribose in genetic RNA. Other mutational changes of sugars include substituting sugars such as mannose or arabinose for deoxyribose in DNA.

Two types of mutation can occur with regard to the base portion of a nucleotide. One type yields a normal base, the other an abnormal base, in place of the original one. In the former type of change, a substitution of one purine by another ($A \rightleftharpoons G$) or of one pyrimidine by another ($C \rightleftharpoons T$ or $U$) is called a *transition*; and a substitution of a purine by a pyrimidine, or the reverse (A or G $\rightleftharpoons$ C or T or U), is called a *transversion*. In the latter type of base mutation, a base is chemically modified to, or substituted by, an abnormal base.

Other mutations affect one or more whole nucleotides. Such mutations are often the result of breaking the backbone of genetic nucleic acid at two or more places. The result can be the addition, deletion, inversion, or transposition of one or more nucleotides (Figure 6-1). All these mutations except inversion are possible for single-stranded nucleic acids. (Note that inversion requires a double-stranded nucleic acid in order to maintain the same polarity along a strand.)

FIGURE 6-1. Whole nucleotide changes. The arrows show the polarity.

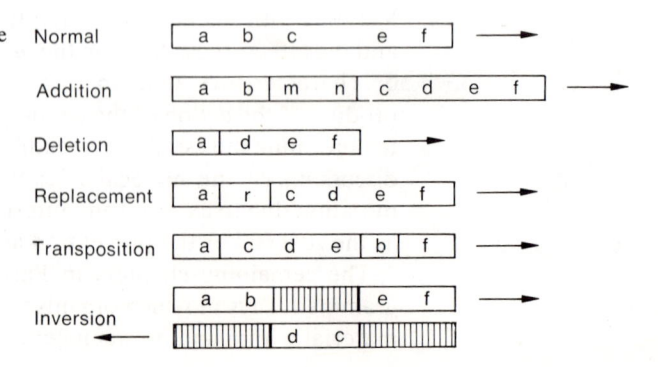

Normal: a b c e f

Addition: a b m n c d e f

Deletion: a d e f

Replacement: a r c d e f

Transposition: a c d e b f

Inversion: a b ||| e f / ||| d c |||

# MUTATION

**6.2** Although mutants can be detected directly, by their changed genetic material, most are detected indirectly, by their molecular or gross phenotypic consequences.

In theory, a mutation can initially be detected directly, from the change it produces in the number, type, or sequence of nucleotides. In practice, at present, only relatively large changes in number or sequence are readily detected. For example, the chromosome sets of newborn babies may be routinely examined in hospitals for the gain or loss of part or a whole chromosome, or for changes in the sequence of relatively large segments of chromosomes. The direct detection of mutations affecting only one or a few nucleotides is feasible at present only in viruses with a relatively small number of genetic nucleotides. It is possible, for example, to sequence the RNA of an RNA phage, grow the phage for a number of generations, and then resequence its RNA to detect the mutations that occurred (and were viable). Now that base sequencing has become relatively easy, more complete sequences of increasingly complex organisms will become available, increasing the feasibility of first detecting mutants from their changed genetic material.

Most mutations are first detected by the phenotypic effects they produce. Mutations can have a variety of effects on the phenotype at the molecular level. (1) They can modify the action of DNA or RNA polymerase by changing the genes that code them or that are needed to start or stop their action. (2) They can change the enzymes that normally modify DNA. (3) They can change genes that affect the frequency of mutation. (4) They can change the amino acid content specified by other individual genes by changing their mRNA's. (5) They can change the amino acid content of proteins specified by groups of genes by mutating other, individual genes whose transcripts are required for the translation process. The initial detection of mutations from their molecular effects on the phenotype is more common in prokaryotes and single-celled eukaryotes than in multicellular eukaryotes. In multicellular eukaryotes, such as human beings, most mutations are first detected from the changes they product in the relatively gross (supramolecular) characteristics of the organism—for example, its color, size, intelligence, growth pattern or rate, and susceptibility to infection or drugs. After a mutation is detected from its gross phenotypic change, attempts are made with increasing frequency to determine its molecular phenotypic basis as well as its primary basis in nucleotide change.

**6.3** Some mutations have sense, missense, or nonsense effects on proteins. Many mutations are presently undetected.

Consider the effect of a transition or transversion of a base pair in protein-coding DNA. Since base substitution will have occurred in the sense strand of the DNA duplex, the mRNA will contain a new triplet. This new triplet may be a *sense* codon because it translates into the same amino acid as the original triplet did; it may be a *missense* codon because it is translated into a different amino acid; or it may be a *nonsense* codon because it is a codon for the termination of translation. (If the original triplet were a nonsense codon, the triplet of the mutant might code for an amino acid, permitting readthroughs.) The addition or subtraction of a nucleotide to the sense strand of DNA, however, will cause the reading of the mRNA to be shifted out of phase or frame (Figure 6-2). As a consequence of such *phase-shift* (or *frame-shift*) *mutations*, the mRNA may contain a variety of sense, missense, and nonsense codons.

A single sense change in an mRNA produces no chemical change in the coded protein. A single missense or nonsense change, however, produces a chemically changed protein of normal or shorter lengths, respectively, whose functional effect depends upon the nature and location of the change in the protein. Thus, a nonsense codon appearing near the end of an mRNA coding one protein (and permitting the synthesis of almost the entire protein before acting as a terminator) is expected to produce less of a detrimental effect on the functioning of the protein than one appearing near the beginning of the mRNA (and permitting synthesis of only the first portion of the protein).

Mutations affecting more than one codon, such as frame-shift mutations, are, of course, more likely to produce detrimental effects on the protein involved than are mutations affecting only one codon. Mutations in a promoter or an operator gene affect the production of all the polypeptide-coding genes in the transcriptional unit. Mutations that change the enzymes or machinery for replication, transcription, or translation are potentially the most detrimental of all, since one of these mutations results in the synthesis of many defective genes and proteins.

Since many mutations are identified from their phenotypic effects at the molecular and especially at the gross level, all those with no known phenotypic effect are ordinarily undetected. Such mutations include

FIGURE 6-2. Frame-shift consequences of adding a base pair to and of deleting a base pair from mRNA-coding, duplex DNA.

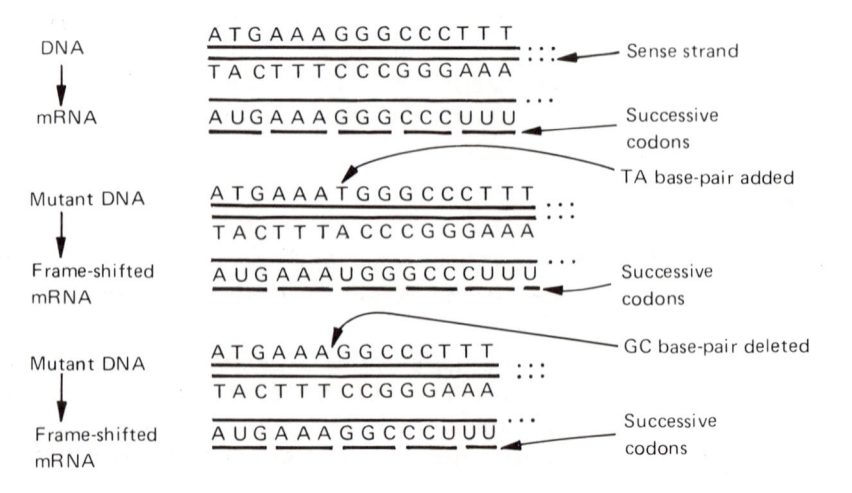

changes (1) between ribose and deoxyribose; (2) in transcription-silent regions; (3) in a transcription-active region whose transcript is discarded during the maturation of tRNA, rRNA, and nuclear mRNA; as well as (4) changes that are sense changes. Many mutants with known kinds of phenotypic effects are undetected because these effects are too small to be detected by presently available means. Organisms are often able to hide the detrimental phenotypic effects of a mutation by being redundant for the gene in question. In eukaryotes, for example, the detrimental phenotypic effect of a mutation in a single rDNA or tDNA gene is masked by the action of the other nonmutant rDNA or tDNA genes present. More (but probably not all) mutants will be detected phenotypically, as we learn more about the function of genetic material that has presently no known phenotypic effect, and as our techniques for detecting small phenotypic changes become more sensitive.

**6.4** Unusual purine–pyrimidine base pairs lead to transitions; other base pairs lead to transversions.

The usual base pairs in DNA are GC (or CG) and AT (or TA). Each of the four bases can occasionally spontaneously undergo a slight, temporary, molecular rearrangement (designated *) which permits it to form a new purine–pyrimidine (AC or GT) combination. Thus, A* can pair with ordinary C to form the A*C base pair; C* can pair with ordinary A to form the AC* base pair; G* can pair with ordinary T to form G*T; and T* can pair with ordinary G to form GT*. These new purine–pyrimidine base pairs can form during semiconservative replication in two ways: either the base in the template becomes rearranged and attracts for its complement a free nucleotide with the wrong base, or the reverse happens—that is, the normal base in the template attracts for its complement a free nucleotide with a modified, and wrong, base. In either case, the unusual base pair will become part of duplex DNA (Figure 6-3I). At the time of the next semiconservative replication, however, both members of the unusual base pair are expected to be in their normal configurations (Figure 6-3II) and to produce progeny duplexes with the usual base pairs (Figure 6-3III). One of these progeny base pairs is CG (or GC) and the other is TA (or AT). Since the ancestral DNA will sometimes have had a CG (or GC) pair at this position and other times it will have had TA (or AT), the presence of both types of pairs in its descendant DNA's means that the spontaneously rearranged bases can result in base-pair transitions in both directions (CG $\rightleftharpoons$ TA).

FIGURE 6-3. Base-pair transitions in both directions following the formation of unusual base pairs. See the text.

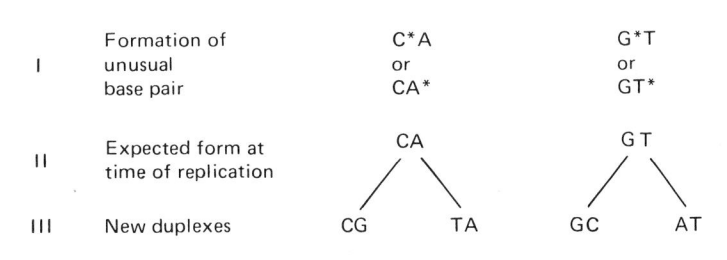

| | | | |
|---|---|---|---|
| I | Formation of unusual base pair | C*A or CA* | G*T or GT* |
| II | Expected form at time of replication | CA ╱╲ | GT ╱╲ |
| III | New duplexes | CG    TA | GC    AT |

Unusual base pairs that produce transitions may also be formed between a normal base and an abnormal complementary base. Abnormal bases that resemble one or more of the normally present bases in nucleic acid are called *base analogs*. For example, *bromouracil* (*BU*) is a base analog of T. BU's usual complement is A, but when it spontaneously rearranges (which it does much more often than T does) it can form the BU*G base pair. As a consequence, the BU analog produces base-pair transitions in both directions more often than T does. BU is a *mutagen*, a term that refers to physical or chemical agents that greatly increase the frequency of mutations. Another chemical mutagen is *aminopurine*, which is a base analog of A and G. It also produces transitions in both directions.

Other mutagenic agents act to produce changes in some of the bases already present in DNA. Some chemical mutagens act on specific bases and change them in particular ways. For example, *hydroxylamine* ($NH_2OH$) only mutates C so that it can base pair only with A. The final result is the unidirectional GC to AT transition (Figure 6-4A). Another chemical mutagen, *nitrous acid* ($HNO_2$), often attacks A and C and results in base-pair transitions in both directions. A group of chemical mutagens, *alkylating agents*, which can produce various large and small deformations of base structure, produces base-pair transversions as well as transitions (Figure 6-4B). Transversions can occur either because a purine has been so reduced in size that it can accept another purine for its complement, or because a pyrimidine has been so increased in size that it can accept another pyrimidine for its complement. In both cases, the diameter of the mutant base pair is close to that of a normal base pair.

FIGURE 6-4. Types of transitional and transversional base-pair changes produced by various mutagens.

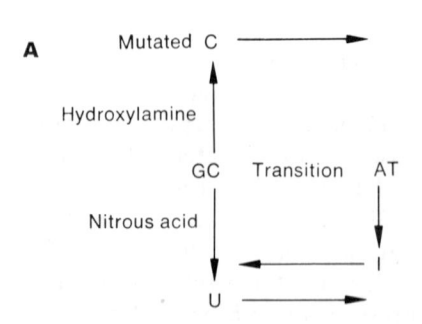

A

```
           Mutated C ─────────▶
               ▲
        Hydroxylamine
               │
              GC   Transition   AT
                                 │
        Nitrous acid             ▼
               │    ◀─────────
               ▼              I
               U ─────────▶
```

B

| Mutagen | Base-Pair Changes ||
|---|---|---|
| | Transitions | Transversions |
| Hydroxylamine | GC ──▶ AT | |
| Nitrous acid | AT ◀──▶ GC | |
| Bromouracil | AT ◀──▶ GC | |
| Aminopurine | AT ◀──▶ GC | |
| Ethyl ethanesulfonate | GC ──▶ AT | GC ──▶ TA<br>GC ◀── CG |

## 6.5 Energetic, penetrating radiations are powerful physical mutagens.

High-energy radiations can penetrate all organisms and produce large numbers of a great variety of mutations. We will first discuss X rays briefly (they will be discussed in more detail in Chapter 27) and then discuss ultraviolet light.

X RAYS. X radiation is a powerful physical mutagen. It can break chromosomes, add oxygen to deoxyribose, remove amino or hydroxyl groups, and form peroxides. As a result, all the kinds of mutations previously mentioned can be produced by X rays. This complete range of types of mutations induced by X rays is in contrast with the narrower range of types induced by many chemical mutagens, as indicated in the previous section.

ULTRAVIOLET LIGHT. Although less energetic than X rays, ultraviolet light is also a powerful mutagen, largely because the bases of nucleic acid absorb energy of its wavelengths. Although UV's energy can also break chromosomes, its two main chemical effects are on pyrimidines. One effect is the addition of a molecule of water to the pyrimidine (Figure 6-5A), which weakens the H bonding with its purine complement and permits localized strand separation. The other effect is to join two pyrimidines to make a *pyrimidine dimer* (Figure 6-5B). This dimerization can produce TT, CC, UU, and mixed pyrimidine dimers (for example, CT).

FIGURE 6-5. Effect of ultraviolet light upon pyrimidines. (The H atoms attached to ring C atoms are shown.) **A**: Hydration. **B**: Dimer formation. The arrangement of adjacent T's in a dimer is approximately that observed by folding together the right and left sides of the dimer illustrated.

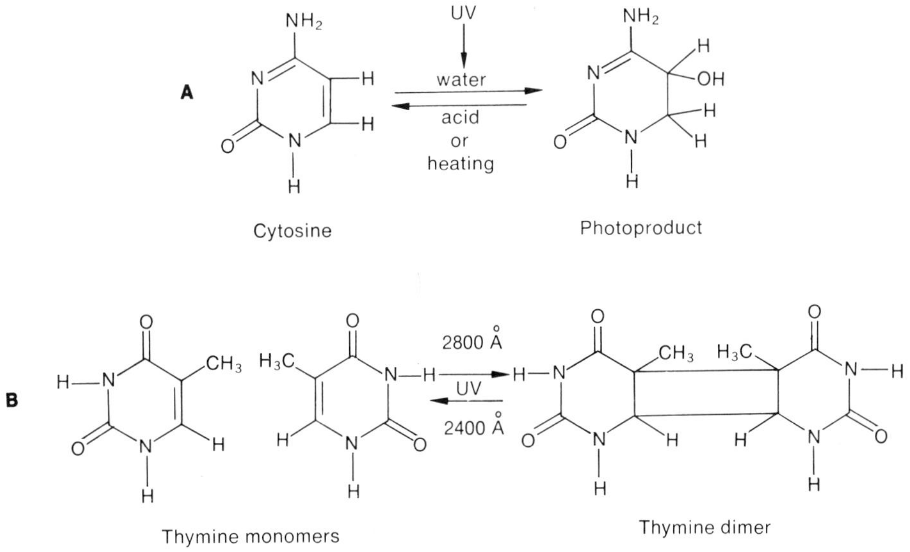

UV-induced localized strand separation gives pyrimidines greater freedom of movement, thereby increasing the chance for dimerization of bases that are adjacent in the same strand, or are widely separated in the same strand, or are in different strands. Dimerization interferes with DNA and RNA synthesis. Interstrand dimers cross-link nucleic acid chains, inhibiting strand separation and distribution. It is noteworthy that some chemical mutagens, for example *nitrogen mustard* and the antibiotic *mitomycin C*, cross-link nucleic acid chains by dimerizing purines (Figure 6-6).

## 6.6 Dimers and certain other deformities can be removed enzymatically from duplex nucleic acid and the duplex repaired.

Many mutational changes in nonduplex genetic material are so drastic they cannot be reversed; other, less drastic mutations can be reversed but only by a reverse mutation—a rare event. In the case of duplex genetic material, however, the possibility exists that if only one strand in a duplex region is mutated in any respect, the complementary strand can be used as a template to reverse or repair the mutation. This possibility is an actuality, and demonstrates the great (and perhaps the greatest) value of genetic material existing in the double-stranded state.

Dimers (and cross-links) produced by ultraviolet light and other physical or chemical mutagens are reparable by three mechanisms. All three are used by prokaryotes; the last two are used by mammals, including human beings.

1. PHOTOREPAIR. In the presence of blue light, bacteria can enzymatically split intrastrand and interstrand dimers into monomers. This repair mechanism does not require complementary information, and so

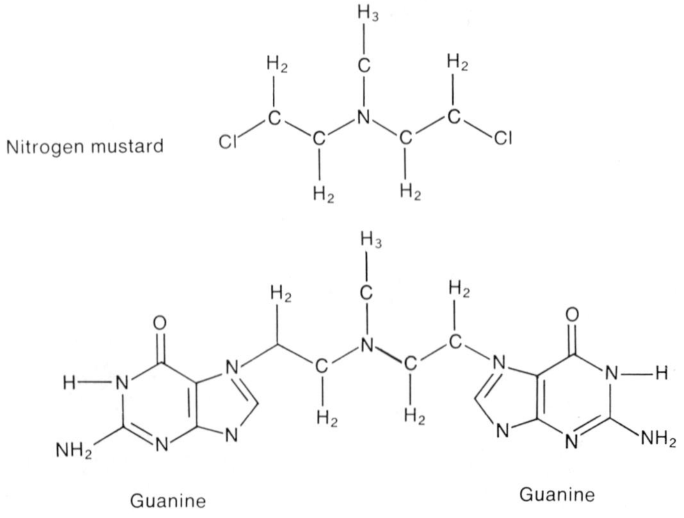

FIGURE 6-6. Nitrogen mustard is able to crosslink guanines. Such crosslinks in DNA can be repaired.

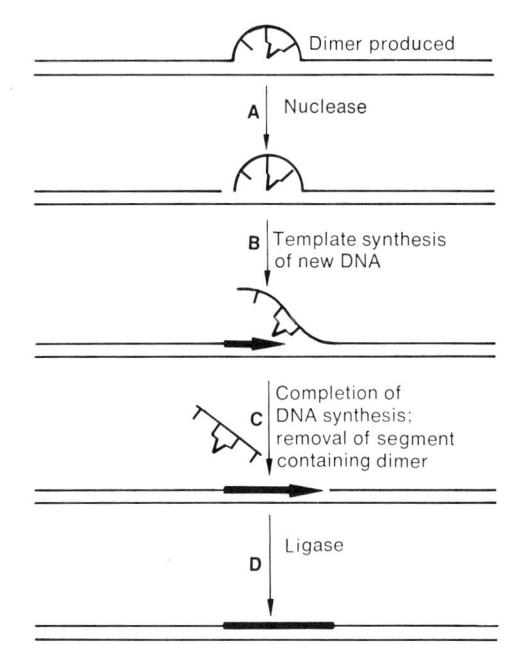

FIGURE 6-7. Repair of DNA by the excision of a dimer. (After R. B. Kelly, M. R. Atkinson, J. A. Huberman, and A. Kornberg, 1969.)

can be used to repair dimers in single-stranded as well as double-stranded nucleic acids.

2. EXCISION. This repair is accomplished by removing the dimers from the DNA (even in the dark) as shown diagrammatically in Figure 6-7. A nuclease nicks the DNA to one side of the dimer (A); synthesis of new DNA starts at the nick using the nondimerized end as a primer and the sequence on the normal complementary strand as template (B); the segment containing the dimer is removed and the synthesis of complementary DNA is completed (C); the ends of the new and old DNA are ligated, completing the repair process.

3. REPLICATION. When a region in one strand undergoes dimerization, the complementary region is usually unaffected. Accordingly, when double-stranded DNA replicates, a normal (nonmutant) copy of the mutant region will be produced when its normal complementary region is used as template. Even when dimers occur in both complements, as they often do, a completely dimer-free duplex can nevertheless be constructed by ligating dimer-free segments of the original and newly-synthesized complements (Figure 6-8).

Consider which specific mutations are not reparable in duplex genetic material. As already noted, destruction of both Watson and Crick in a given region is irreparable. Base-pair transitions and transversions, and all mutations that add or subtract one or more nucleotide pairs (Figures 6-1 and 6-2), do not deform the DNA in any way and are not reparable. Mutations that produce small deformations, such as the presence of bromine in place of the methyl group in T, and the addition of methyl,

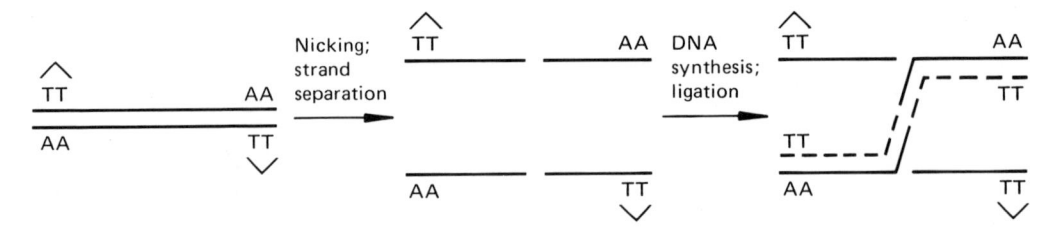

FIGURE 6-8. Formation of a dimer-free duplex by ligating segments of the original and newly synthesized complements.

hydroxyl, and so on, groups where they are not coded for, are not repaired. Only when the deformation of a duplex is sufficiently great and due to change in one of the strands is it subject to repair.

**6.7** Phase-shift mutations are produced by errors made during repair or replication and by acridine mutagens.

Although there are mechanisms for the repair of mutations in all cellular organisms, these mechanisms are subject to error. An error in repair will itself produce a mutation. For example, repair of a deformity in one of the strands may start normally (Figure 6-9)—a nick is made to one side of the deformity and nuclease removes the deformed region. Strand separation occurs, however, followed by a looping of either the broken or the unbroken strand. (Base pairing may sometimes occur within the looped portion.) When repair continues, one or more bases will be added to the broken strand if it had looped (Figure 6-9, left side), or one or more bases

FIGURE 6-9. Phase-shift mutations produced by errors in the mutation-repair mechanism.

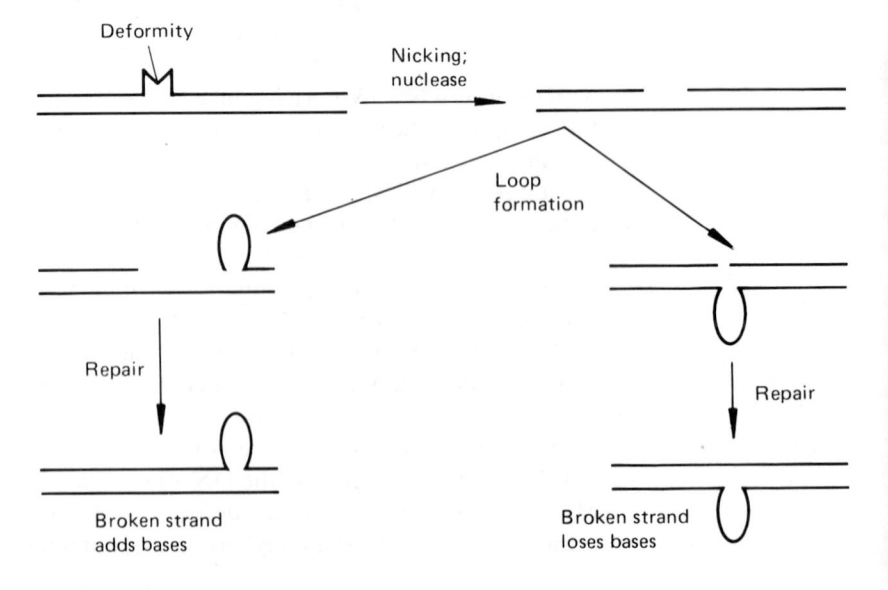

# MUTATION

Proflavin

Acridine orange

FIGURE 6-10. Two acridine dyes.

will have been lost from the broken strand if the unbroken strand had looped (Figure 6-9, right side). The next replication will produce duplexes that have gained or lost one or more nucleotide pairs and will have, therefore, undergone phase-shift mutation. Phase-shift mutations produce mRNA's that are read out of phase unless (and until) they have added or subtracted three bases or a whole multiple of three.

Phase-shift mutation can also occur as an error after a nick has been produced during normal semiconservative replication.

Phase-shift mutations can also be produced by chemical mutagens such as *acridine dyes* (Figure 6-10), and perhaps by free nucleotides or other naturally occurring substances. Acridines may bind to the outside of double-stranded nucleic acid or insert themselves between adjacent nucleotides of a strand in a duplex. The addition of one or more acridines may stretch the duplex molecule lengthwise a distance of one or more nucleotides (Figure 6-11). At the time of the next replication, an unspecified nucleotide may be inserted into the complementary strand at the position corresponding to an acridine molecule, thereby causing a frame-shift mutation by nucleotide addition. Because acridines can deform the duplex, a DNA repair process may be initiated which, through error, can result in frame-shift mutations due to nucleotide addition or subtraction in the manner already described.

**6.8** Much spontaneous mutation results from the effects of temperature and of mutation-inducing products of metabolism.

Normally occurring or *spontaneous mutants* include those induced by chemical mutagens or energetic radiations that are usual elements or

FIGURE 6-11. Acridine mutagenesis.

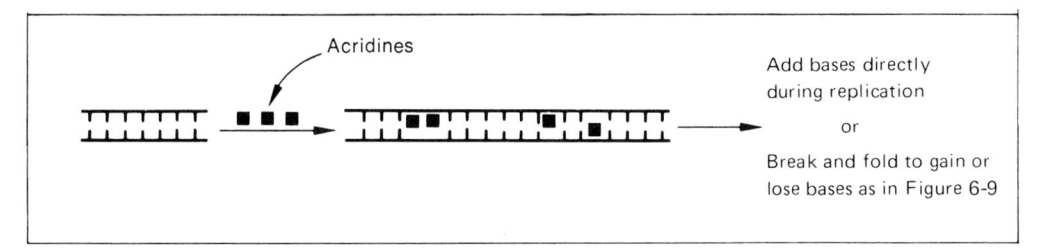

products of the external environment to which an organism is exposed. Temperature is another factor in the external environment that influences spontaneous mutation frequency. In the range of temperatures to which organisms are usually exposed, each rise of 10°C produces about a fivefold increase in mutation frequency. Violent temperature changes in either direction produce an even greater effect on mutation frequency. Actually, extreme environmental conditions of almost any kind increase the mutation frequency.

Spontaneous mutations also include those induced by factors in the internal environment of an organism. We have already seen that some spontaneous mutations result from the nature of the genetic material itself. For example, spontaneous rearrangements of bases lead to base-pair transitions. We have also seen that errors in repair or replication can lead to spontaneous mutations. Two other components of the internal environment make important contributions to the spontaneous mutation rate.

1. DNA POLYMERASES. DNA polymerases not only polymerize nucleotides using a template, but continuously proofread the growing polymer and remove nucleotides that were added in error. This proofreading activity increases the fidelity of DNA replication. Proofreading errors, however, are made, resulting in spontaneous mutations. Accordingly, normally present DNA polymerases make replication errors at a certain reproducible rate. Mutants in the genes that code for DNA polymerases sometimes generate modified DNA polymerases that consistently produce more (sometimes 10 times more) or that consistently produce fewer (sometimes 10 times fewer) mutations than normal DNA polymerases. DNA polymerases that increase or decrease the mutation rate have been identified in phages, for example. Another probable example occurs in human beings, where patients with acute lymphatic anemia have a DNA polymerase that makes more errors *in vitro* than one isolated from normal lymphocytes.

2. PURINES. In bacteria at least, many purine and purine derivatives are mutagenic. These mutagens include adenine and caffeine (which seems to interfere with ligation). When ribose is bound to one of these purines, however, the combination counteracts the effects of the purines and acts as an *antimutagen*. Even though it is not clear how these substances act to produce or prevent mutations, it is likely that a significant portion of the spontaneous mutation rate is the normal consequence of the cell's metabolism producing mutagens and antimutagens.

We see, therefore, that the spontaneous mutation frequency is the cumulative effect of a variety of mutation-causing (and -preventing) factors in the external and internal environments. As we shall appreciate in a later chapter, it is undesirable for the long-term survival of a species to have a spontaneous mutation rate that is either too high or too low.

# MUTATION

## SUMMARY AND CONCLUSIONS

Although mutations have a wide variety of phenotypic effects at the molecular level, most are detected by their phenotypic effects on the gross, supramolecular characteristics of an organism. Relatively few mutations are discovered directly by the changes they produce in the kind, number, or sequence of genetic nucleotides.

Mutations can be classified at the chemical level as changes involving only part of a nucleotide (its phosphate, sugar, or base), or changes involving one or more whole nucleotides. Temporary spontaneous changes in the internal molecular arrangement of bases can lead to base-pair transitions. Transitions occur when ordinary bases undergo such rearrangement, and occur more frequently when base analogs or mutated bases are present. Some mutated bases give rise to base-pair transversions.

Different chemical mutagens have different specificities in producing transitions and transversions. On the other hand, radiation mutagens such as X rays and ultraviolet light can produce many, if not all, types of mutation. UV is especially effective in producing pyrimidine dimers. Dimers and certain other defects that deform duplex DNA can be repaired enzymatically via photorepair, excision, or replication. Defective mutational repair, defective replication, and acridine mutagens can produce phase-shift mutations.

The spontaneous mutation rate is due to mutations produced by chemical and physical mutagens in the external environment (including temperature) and those produced by the action of mutagenic and antimutagenic compounds (including DNA polymerases) present in the internal environment as the result of the organism's own metabolism.

## QUESTIONS AND PROBLEMS

**6.1** *Anagrams*: Unscramble and define each of the following: **a.** hoppeynet. **b.** automint. **c.** revistaronns. **d.** prepairhoot. **e.** negatum.

**6.2** *Movies and TV*: **a.** "Them." **b.** "Incredible Hulk." What are the genetic applications or implications of one or both?

**6.3** Distinguish between phenotype and genotype.

**6.4** When and where do we find sulfur in DNA?

**6.5** What could we possibly learn about genetics if all the organisms under study had identical genotypes? What advantage does genotypic variation provide in this respect?

**6.6** Do sense mutants have any effect on biological fitness? Explain.

**6.7** Should the term *mutant* be restricted to genetic changes having a detectable phenotypic effect? Explain.

**6.8** The replacement of a T by a BU in DNA is a mutation. Why does this mutation lead to more mutations?

**6.9** How is cytosine affected by: **a.** hydroxylamine? **b.** ultraviolet light?

**6.10** Name three proteins that are needed to repair DNA by excising pyrimidine dimers.

**6.11** What types of mutation cannot be repaired in *E. coli*?

**6.12** What are the advantages and disadvantages of a genetic material being double-stranded versus single-stranded?

**6.13** Is the repair of a mutation itself a mutation? Explain.

**6.14** Suppose that ϕX174 undergoes dimerization while in its host. How can these dimers be repaired: **a.** in the presence of white light?   **b.** in the dark?

**6.15** Do you suppose that microwaves (such as are used in ovens) can cause mutation? Explain.

**6.16** How does the mutation frequency of *E. coli* depend on *E. coli*?

**6.17** In what respect is DNA polymerase an antimutagen?

**6.18** How are all of the following interrelated—nucleic acid folding, degeneracy, and adaptive mutations?

*Chapter 7*

# Genetic Recombination Between Viruses

DIFFERENT PORTIONS of the genetic material can change their arrangement relative to each other in several ways. These changes include the folding and unfolding of the parts of a single polynucleotide; the separation of strands in double-stranded polynucleotides and the hybridization of complementary strands; the breakage and rearrangement of segments of one or more polynucleotides; and the grouping or separation of duplexes. Such changes in the conformation of nucleic acid can affect the availability, quality, or quantity of genetic information. Of all these changes, those that modify either the sequence of genetic nucleotides or the number of polynucleotides are called *genetic recombinations* (or *recombinations*) and the resultant products are called *genetic recombinants* (or *recombinants*) (Figure 7-1). Recombination is a more-or-less permanent change in gene combination relative to some preexisting combination.

Some recombinations are unusual and uncoded and are, therefore, mutations. Mutational recombinations in eukaryotes are discussed in Chapter 12. Most recombinations result from metabolic reactions that are genetically programmed and hence are not mutational. Programmed recombination of genetic material is useful to an organism in the following ways: it distributes the genetic material into packages of suitable length, size, or amount within a generation and between generations; it combines genetic materials from the same or different individuals to produce genetic combinations that would otherwise occur much more rarely by means of mutation. It is advantageous for organisms to have an appropriate number of genetic recombinants (neither too many nor too few), for, under a given set of environmental conditions, some of the new combinations may be more successful than any of the old ones.

The present chapter deals with programmed genetic recombinations between virus chromosomes. These recombinations occur while the virus chromosomes are inside the host cell.

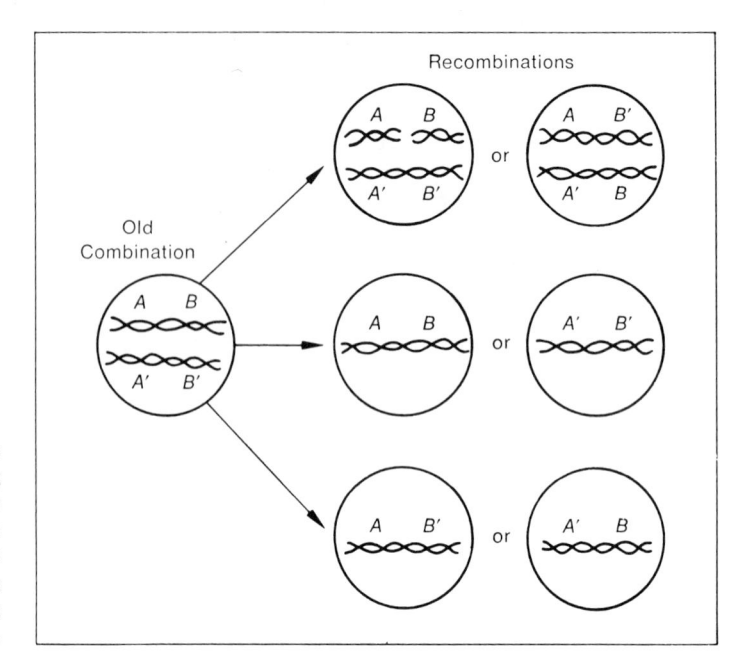

FIGURE 7-1. Some specific examples of genetic recombinations involving only nucleotide sequence (top), only duplex number (middle), or both (bottom). The old combination consists of one duplex containing genes *A* and *B* and another containing alternatives of these genes, *A'* and *B'*.

## 7.1 Each T-even virus chromosome is a molecular recombinant because it contains segments of parental and progeny DNA's.

Genetic recombination normally occurs in many DNA viruses when their chromosomes replicate and produce progeny chromosomes. Although the details are different in different viruses, we shall discuss the life and chromosome cycle of one particular group of viruses, the T-even phages ($\phi$T2, $\phi$T4, $\phi$T6) that attack *E. coli*.

A T-even phage (Figure 1-3) attaches itself tail first to its host and the DNA in the head is injected into the protoplasm of the host. During the first dozen or so minutes after injection (Figure 7-2B to D), the phage chromosome, which is a single rod-shaped duplex about 200,000 nucleotides long, is transcribed and phage mRNA is translated. This produces the proteins and enzymes needed to make virus progeny. Semiconservative replication of the virus chromosome starts at multiple points (requiring multiple nicks) in the chromosome about 5 minutes after infection. By the time empty phage heads for the progeny are produced (Figure 7-2D), the chromosome has replicated a large number of copies (Figure 7-3B). At about 9 minutes after infection, the chromosomes fall into segments due to the nicks left over from the replication process and perhaps to newly produced nicks. Base pairing then occurs between the single-stranded ends of double-stranded parental and progeny pieces to produce *joints* (Figure 7-4). Some joints are made between two parental pieces or two progeny pieces; others are made between one parental and one progeny piece. After any overlaps or gaps in the joints are repaired by nuclease and DNA polymerase, ligase seals all remaining nicks. The

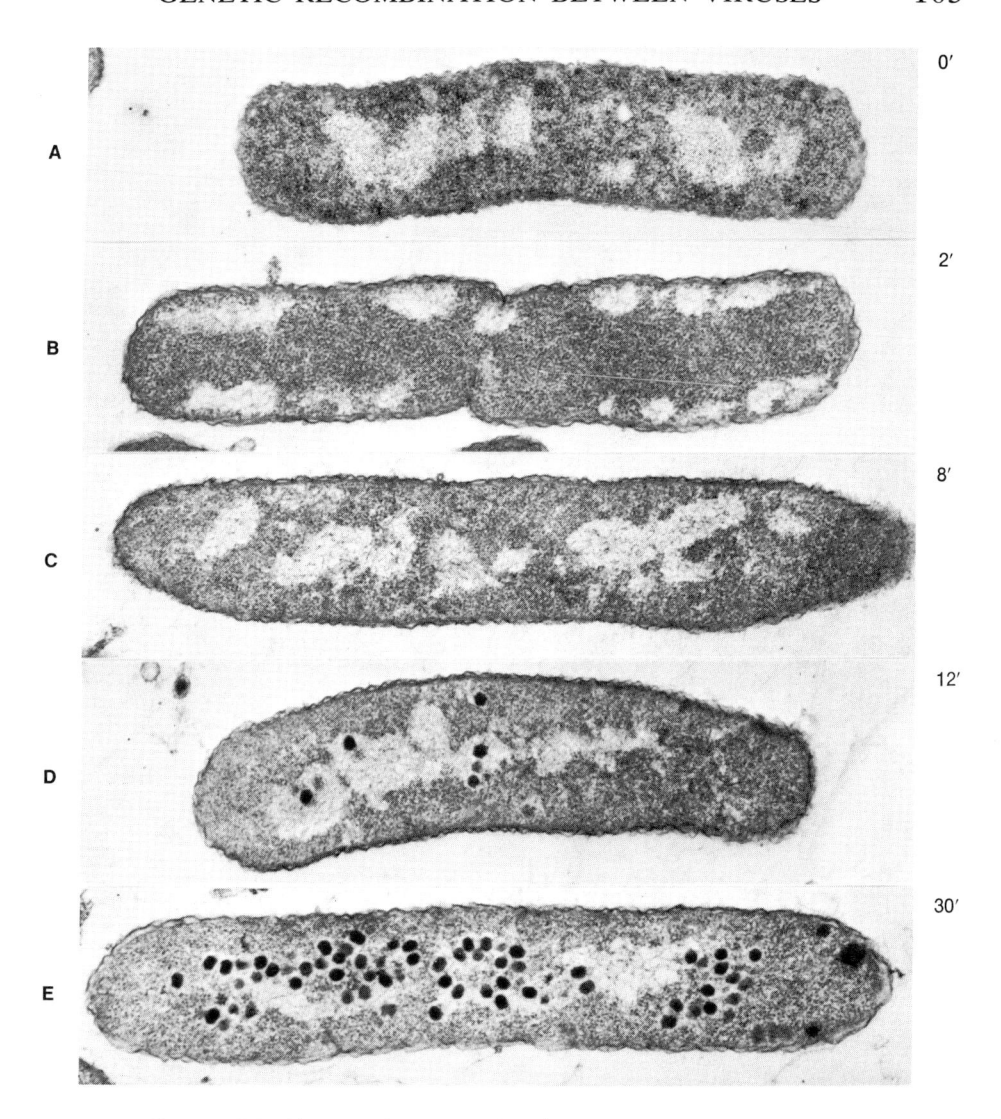

FIGURE 7-2. Electron micrographs of growth of T2 virus inside the *E. coli* host cell. **A**: Bacterium before infection. **B**: Two minutes after infection. The thin section photographed includes the protein coat of T2, which can be seen attached to the bacterial surface. **C**: Eight minutes after infection. **D**: Twelve minutes after infection. New virus particles are starting to condense. **E**: Thirty minutes after infection. More than 50 T2 particles are completely formed and the host is about ready to lyse. Original magnification, 17,500×; present magnification, 14,000×. (Courtesy of E. Kellenberger.)

result is a single giant polynucleotide that is composed of many linearly repeated phage chromosomes (Figure 7-3C). A single polynucleotide containing two or more linearly repeated chromosomes is called a *concatemer*. The phage concatemer, therefore, contains a series of chromosomes, arranged in the same linear sequence, each of which is composed of parental pieces interspersed with progeny pieces.

# 104   HOW GENETIC MATERIAL IS DISTRIBUTED

**A** Parental duplex

Semiconservative replications

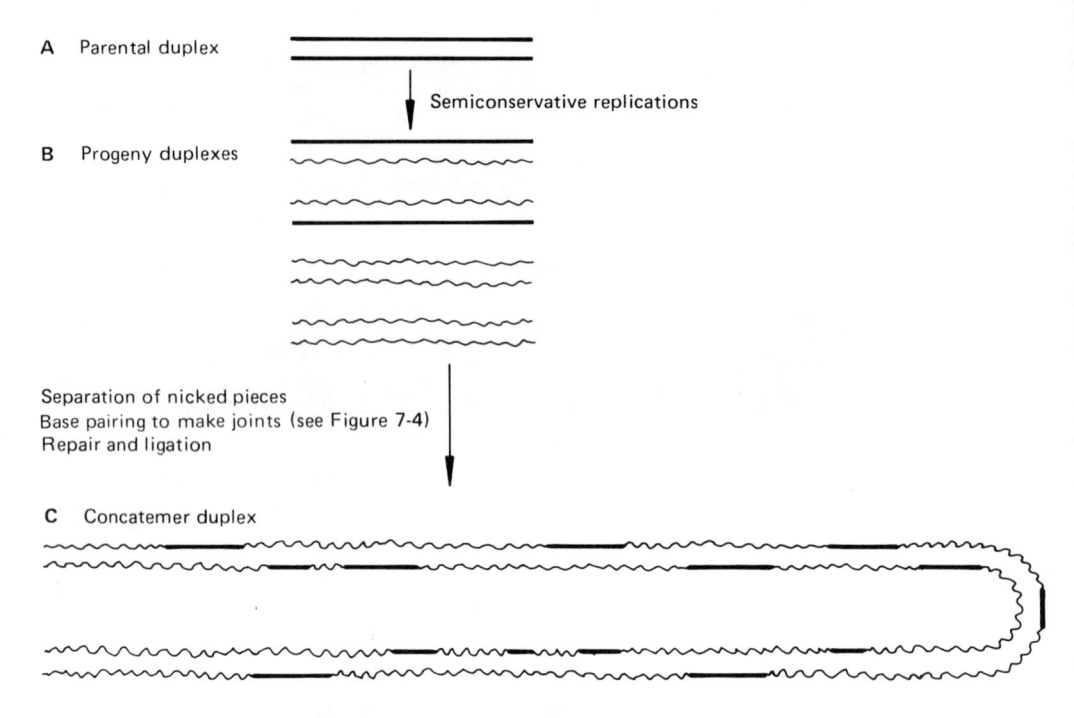

**B** Progeny duplexes

Separation of nicked pieces
Base pairing to make joints (see Figure 7-4)
Repair and ligation

**C** Concatemer duplex

FIGURE 7-3. Molecular recombination in the formation of a concatemer. Solid line = labeled parental DNA; thin wavy line = unlabeled progeny DNA.

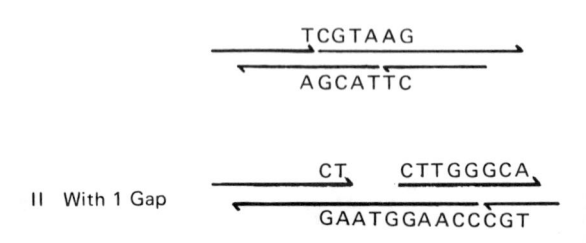

```
            TCGTAAG
            AGCATTC
```

II  With 1 Gap
```
     CT    CTTGGGCA
     GAATGGAACCCGT
```

FIGURE 7-4. Joints. I: Ready for ligation. II: Gap needs filling in before ligation. III: Overlap needs to be removed before ligation.

III  With 1 Overlap
```
         CAT
      CCG GGTGC
      GGC CCACG
```

The concatemer then becomes associated with basic proteins. With the nucleic acid thereby neutralized, one end of the concatemer can enter and fill an empty phage head. Once the head is full, a nuclease scissions the duplex, permitting the new free end of the concatemer to enter and fill another empty head. The phage head holds more than one complete sequence of genes, however. Accordingly, the first headful contains a chromosome that is terminally redundant (Figure 7-5). If the complete sequence of genes were *A B C D E F G*, this first chromosome could be *A B C D E F G A B*, which is terminally redundant for genes *A* and *B*. Successive chromosomes cut from the concatemer would be redundant for successive genes in the chromosome (Figure 7-5).

FIGURE 7-5. Scissioning (wavy line) the concatemer into successive (1, 2, 3, etc.) head-filling lengths. Each chromosome produced is terminally redundant; successive chromosomes are redundant for successive parts of the gene set (=ABCDEFG).

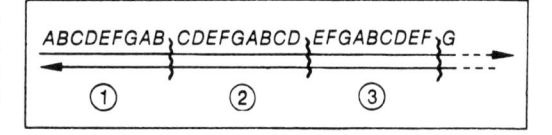

```
ABCDEFGAB CDEFGABCD EFGABCDEF G
    (1)        (2)       (3)
```

The assembly of mature phage is completed with the addition of other protein components. About 20 to 40 minutes after infection, a phage-coded enzyme ruptures or *lyses* the bacterial cell wall and a hundred or so phage progeny are liberated. T phages are said to be *virulent* (or intemperate) because infection is followed by only one course of events—the *lytic cycle* just described.

We see from the preceding that each T-even chromosome is a *molecular recombinant* since it contains a mixture of parental and progeny DNA. Since the parent and progeny DNA code for the same information, barring the rare event of mutation, such molecular recombinants are phenotypically silent: the recombinant molecule produces the same phenotype as the parent molecule. Molecular recombinants can be identified in progeny, nevertheless, if the parental DNA is labeled by being radioactive and heavy while the host is not labeled. In this case, progeny chromosomes are found to be composed of a combination of labeled and unlabeled segments.

## 7.2 Genetic recombination is expected to occur between the chromosomes of two T-even phages infecting the same host.

When two T-even phages infect a single *E. coli*, we expect that the concatemer produced will consist of fragments derived from both parents (and their progeny) interspersed among each other. This expectation is diagrammed in Figure 7-6, which supposes that the two parental phages are terminally redundant for different genes. One parent has the gene sequence *A B C A*, whereas the other parent has the gene sequence *b c a b*, where the lowercase genes are alternative forms or *alleles* of the uppercase genes. (Unless otherwise noted, alleles are assumed to differ only slightly in nucleotide sequence.) In other words, the first parent has two locations or *loci* for *A* or its alleles and only one locus for *B* or its alleles and one for *C* or its alleles, whereas the second parent has two loci for *B* or its alleles, that is, two "*B*" loci, and single loci for "*C*" and "*A*" alleles. An individual or chromosome set is said to be *haploid* for all loci present only once, and *diploid* for all loci present twice. The chromosome or gene set of a T-even phage is, therefore, partially diploid.

In the present case some joints are expected to be made between segments derived from different parents, which bear different genetic markers, to eventually form a concatemer of the type shown in Figure 7-6. When such a concatemer is cut up into mature progeny phage

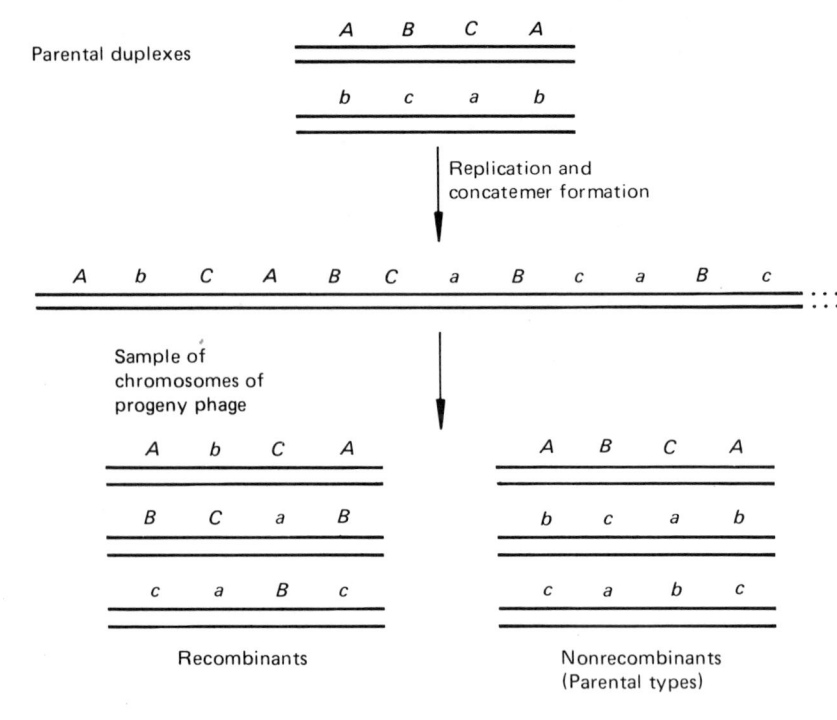

FIGURE 7-6. Genetic recombination between two T-even-phage chromosomes bearing different genetic markers.

chromosomes, some are expected to contain loci bearing only parental-type alleles (*A*, *B*, and *C*, or *a*, *b*, and *c*) and will be in this respect genetically nonrecombinant. Others, however, are expected to contain alleles derived from both parents, and are expected to be recombinants. For example, a progeny with *A*, *b*, and *C* alleles would be a recombinant for the "*B*" locus, as would a phage with *a*, *B*, and *c* alleles, whereas a phage with the genotype *a b C* or *A B c* would be recombinant for the "*C*" locus.

## 7.3 Recombinant progeny phage are detected after infecting a cell with two phages bearing different genetic markers.

To detect recombination through the use of genetic markers, *E. coli* is simultaneously infected with two strains of T-even phage that differ genetically for two markers. One strain has *A b* markers, for example, and the other strain *a B*. The progeny are recovered and their genotypes are determined from their phenotypes. (Because of their small size, phage phenotypes are almost always determined, not by direct observation, but by the phenotypic effects they and their progeny express in their hosts.) Not only do the parental genotypes (*A b* and *a B*) appear in the progeny, but also the new genotypes *A B* and *a b*. Note, however, that a rare phage may undergo a spontaneous mutation from $A$ to $a$, or the reverse. Such mutations would also give rise to progeny of $AB$ and $ab$ types. Since the observed frequency of the new types is much greater than the spontaneous mutation frequency at the "$A$" or "$B$" locus, most of the new types must be genetic recombinants. In other words, during phage reproduction nicks must sometimes occur between the "$A$" and "$B$" loci of both parental types of chromosome, so that a double-stranded segment with single-stranded ends containing a marked locus from one parent can base-pair and make a joint with a similar segment containing a marker for the other locus from the other parent; and a concatemer must be formed, which when cut up yields a recombinant phage chromosome.

Recombination involving nucleotide sequence is known or expected to occur through breakage and joint formation between DNA viruses that attack eukaryotes.

## 7.4 The sequence of loci in a polynucleotide can be determined from their frequencies of recombination.

Since a T-even phage contains a single chromosome, all the phage genes are located in one polynucleotide molecule, and all are physically bound or *linked* to each other. The recombinations we have described involve nicking and breakage that destroy linkage. Linkage between genes provides a kind of resistance that must be overcome by the recombination process. Thus, because the number of nicks is limited, the old parental combination with regard to two close-by loci will tend to remain in the majority of the progeny, whereas the new recombinational types will occur in a minority.

It is reasonable to suppose that the chance for a breakage to occur between two loci will increase as the distance between them increases. Thus, in a virus chromosome having a sequence of three relatively close loci, $DEF$, there should be more recombinants between $D$ and $F$ than between either $D$ and $E$ or $E$ and $F$. If this is so, it should be valid to use the percentage of all progeny that show recombination between two loci as an index of the actual distance between the two. This percentage of recombination, which is measurable, is taken as the number of *recombinational map units* between the two loci. For example, two loci producing 10 per cent recombinant types are linked recombinationally and are 10 recombinational map units apart. If two loci are found to produce equal numbers of recombinational and parental types (that is, they are 50 recombinational map units apart), we would conclude that they are not recombinationally linked, but are recombining independently, that is, at random to each other. These two loci can recombine independently because they are located either on different chromosomes (that is, they are physically unlinked) or on the same chromosome (physically linked) but so far apart that recombination between them is not limited by the distance between them. The latter situation would be expected when two loci are far enough apart so that at least one break occurs between them in every chromosome. In this case it would be simply a matter of chance whether a piece would make a joint with a piece from the same parent or another parent.

We can test the validity of recombinational map units as an indication of actual distance between loci in the following way. Two strains of T-even phage are obtained that differ genetically at three relatively close loci (Figure 7-7). Each of the genes $d$, $e$, and $f$ is deficient for most of the nucleotides in their polypeptide-coding alleles, $D$, $E$, and $F$, respectively. The deficiencies are, therefore, nonoverlapping. One phage strain is $dEf$, carrying two deficiencies; the other is $DeF$, carrying the third deficiency. Simultaneous infection of *E. coli* with these phages produces progeny whose genotypes are determined with respect to these three loci. From these genotypes we determine the frequencies of recombination between different pairs of loci. With regard to the "$D$" and "$E$" loci, we find that 98 per cent of the progeny are $dE$ or $De$ parental, nonrecombinant types and 2 per cent are $de$ or $DE$ recombinant types (Figure 7-7). The "$D$" and "$E$" loci are therefore 2 recombinational map units apart. One finds, with regard to the "$E$" and "$F$" loci, that 99 per cent of the progeny are parental, nonrecombinant types $Ef$ or $eF$ and 1 per cent

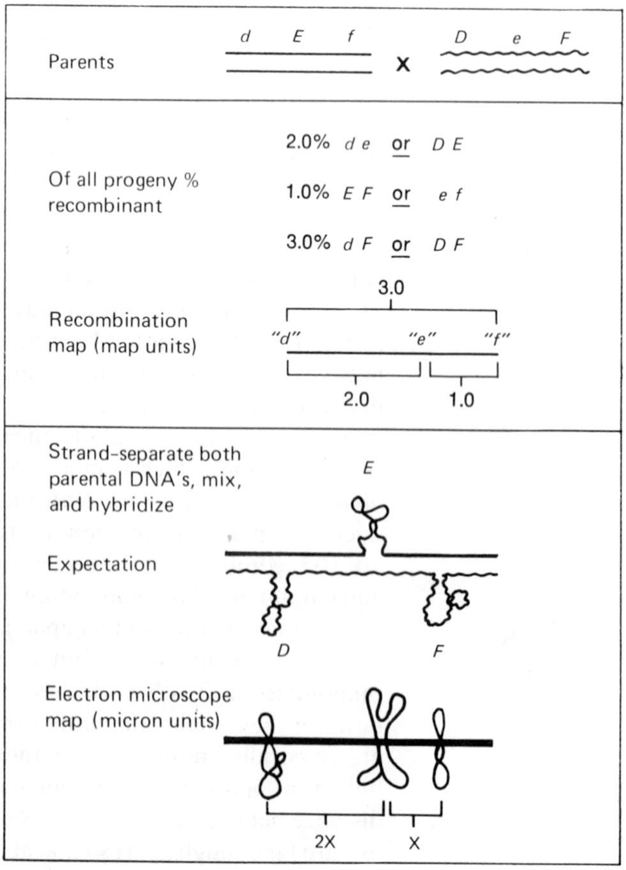

FIGURE 7-7. Congruence between maps obtained from recombination frequencies and the electron microscope. $d$, $e$, $f$ = three close, nonoverlapping deficiencies of normal genes $D$, $E$, $F$, respectively. See the text for an explanation.

are recombinant types *E F* or *e f*. The "*E*" and "*F*" loci are, therefore, 1 map unit apart. There is still one more pairing of loci for which progeny genotypes can be scored for recombination frequency. With regard to the "*D*" and "*F*" loci, 97 per cent are found to be parental types (*d f* or *D F*) and 3 per cent are recombinant (*d F* or *D f*), so these loci are 3 map units apart. A single *recombination map* can be made from the map distances determined for the three gene pairs (Figure 7-7). This map places the loci in a linear sequence, "*D*" "*E*" "*F*" (or the reverse), in which the longest distance equals the sum of the two shorter distances.

This recombinational map thus developed can now be compared with the physical positions of these loci in the chromosomes as revealed by the electron microscope. To make the loci visible, chromosomes of both strains are isolated, made single-stranded, and mixed together for hybridization. Any hybridization that reconstitutes the chromosome (or chromosome segment) of either strain, will, of course, be completely double-stranded and will appear so in the electron microscope. On the other hand, any hybridization that combines complementary strands (or segments) of the two different strains in the area of our special interest will be expected to contain three single-stranded regions (Figure 7-7) corresponding to most of the *D*, *E*, and *F* genes, and to be double-stranded elsewhere. These expectations are fulfilled; three single-stranded regions in otherwise double-stranded DNA are found in the electron microscope. When the distances between the single-stranded regions in electron micrographs are measured, the middle one is found to be just about twice as far from one of the end ones as it is from the other. The recombinational and electron microscope maps are, therefore, closely proportional or congruent. We conclude that the sequence of loci in a chromosome can be determined and placed in a recombination map from the frequencies with which these linked loci recombine.

## 7.5 The recombination map of $\phi$T4 is circular.

The recombination map obtained for many of the genetic markers in $\phi$T4 is given in Figure 7-8. The map is circular because the phage chromosome is terminally redundant, as the result of which each kind of locus is bordered by another locus on each side. In the chromosome *A B C D E F G A B*, for example, every kind of locus has an internal position where it is bordered by loci on each side, even though one locus of "*A*" and one of "*B*" kind are terminal. If each kind of locus is bordered by a locus on each side, each is recombinationally linked to a locus on each side, and the recombination map will be circular.

One would also expect a ring chromosome to produce a circular recombination map, as it does in the case of $\phi$X174. One would expect a rod chromosome, which is not terminally redundant and always has the same sequence of loci, to produce a linear recombination map—as in fact proves to be the case, for example, in human beings. One can also make a recombination map of the sites within a gene where mutation has occurred. This *intragenic mapping* can be done by crossing two different

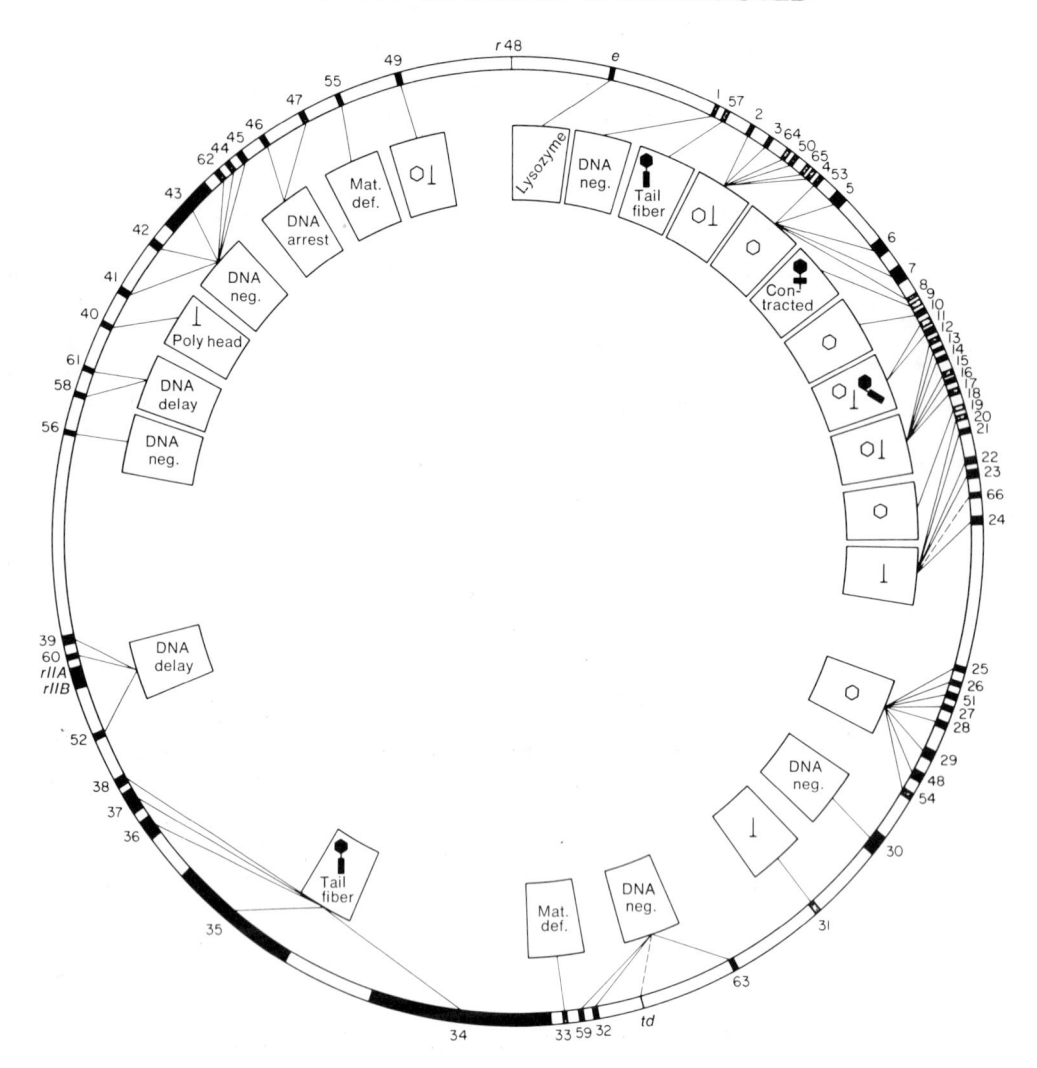

FIGURE 7-8. Recombination map of φT4D. Filled-in areas represent minimal lengths for genes. The symbols for phage components represent the typical products present in lysed cultures of mutant-infected *E. coli*. (Courtesy of R. S. Edgar and W. B. Wood, 1966. *Proc. Nat. Acad. Sci., U.S.*, **55**: 498.)

mutants of the same gene and determining the frequency of nonmutant recombinant progeny (Figure 7-9). Mutational sites that are further apart will give proportionally more nonmutant recombinants.

### 7.6 Interlocus genetic recombination also occurs between RNA viruses.

RNA viruses can be tested for their ability to undergo genetic recombination between different loci. For example, *influenza virus* is a single-stranded RNA virus that occurs in a variety of genetically different,

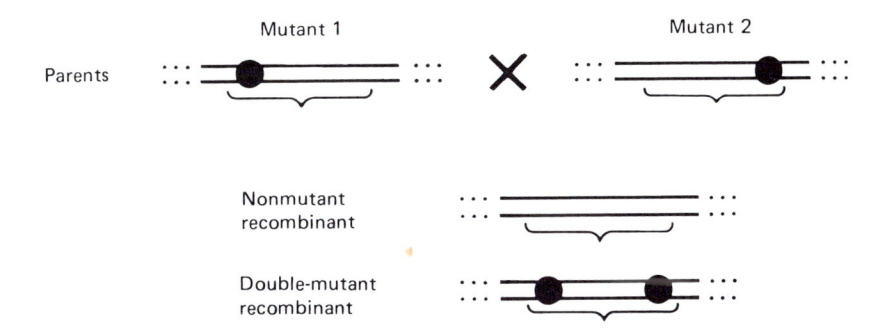

FIGURE 7-9. Mapping mutational sites within a gene by recombination frequency between different mutant alleles. Bracket = extent of gene; ● = place mutated in gene.

haploid strains. The SWE strain has genetic markers $a$ and $c$, and the MEL strain has markers $A$ and $C$. Although the usual host is the mammalian cell, the virus can also infect the cells of a developing chicken egg. When SWE and MEL viruses are allowed to infect a chicken egg simultaneously, progeny are obtained not only of the parental types ($a\,c$ and $A\,C$) but also of the recombinant types ($A\,c$ and $a\,C$). Since mutation is too rare to account for these results, we can conclude that genetic recombination also occurs between RNA viruses. The RNA of recombinant influenza particles appears to be in eight pieces.

Different strains of poliomyelitis virus (poliovirus), another RNA virus, also show genetic recombination. The resultant RNA seems to occur in one piece in these recombinants and in parainfluenza virus, just as it does in the T phages.

Because it is so difficult to infect a single plant cell with more than one TMV or other virus, we do not yet have any evidence that genetic recombination occurs between RNA viruses that attack plants.

## SUMMARY AND CONCLUSIONS

Genetic recombination changes either the sequence of genetic nucleotides or the number of polynucleotides. Since parental and recombinant T-even phages contain a single polynucleotide, genetic recombination between T phages is confined to changes in nucleotide sequence.

T-even phage genetic recombination is the normal programmed consequence of the phage's mechanism for producing progeny phage chromosomes. This mechanism includes chromosome replication, fragmentation, joint formation, and repair, thereby producing a concatemer which is then cut into phage headful lengths. When a host cell is multiply infected with two (or more) T-even phages carrying different genetic markers, joints occur using pieces derived from different parental phages, yielding phage progeny that are genetically recombinant.

Genetic recombination involving nucleotide sequence seems to have common features in all DNA organisms. These features include breakage, joint formation, and repair.

Phenotypically determined recombination frequencies permit the construction of recombination maps that have been shown to be closely proportional to the actual maps of the molecules made using the electron microscope.

Genetic recombination also occurs between RNA viruses when host cells can be multiply infected.

## QUESTIONS AND PROBLEMS

**7.1** *Anagrams*: Unscramble and define each of the following: **a.** cometracen. **b.** unlivert. **c.** nojit. **d.** eelall. **e.** poildid. **f.** boontaincrime.

**7.2** *Song*: "Everything I Have Is Yours." Does this have application in this chapter? Explain.

**7.3** *Riddle*: What joint does not move or burn?

**7.4** What are the advantages of genetic recombination? Can you state any disadvantage?

**7.5** Can a molecular recombinant ever give rise to a mutation? Explain.

**7.6** "T-even phages are partial diploids." Explain.

**7.7** A cross is made between two phages: $M n \times m N$. Give the genotypes of recombinant progeny. Which two genotypes are expected in more than 50 per cent of the progeny?

**7.8** How many map units apart are two loci if 1000 progeny include 970 of the parental types?

**7.9** Differentiate between the terms "allele" and "locus".

**7.10** State one similarity and one difference between the DNA's in a T4 phage head and in a T4 concatemer.

**7.11** "Recombinational linkage means physical linkage, and vice versa." Do you agree? Explain.

**7.12** How can linkage of three loci be demonstrated under the electron microscope?

**7.13** Design an experiment to prove that poliovirus RNA undergoes genetic recombination.

# Chapter 8

# Genetic Recombination Between Bacteria: I. Transformation and Generalized Transduction

WE HAVE ALREADY SEEN that genetic recombination preceded by breakage takes place between virus DNA's present in a bacterial host. The host provides the environment, raw materials, and most, if not all, of the machinery needed for recombining phage genes. Since phage recombinational machinery such as nuclease, DNA polymerase, and ligase are present in uninfected bacteria, breakage recombination ought to be possible between the DNA's of two bacteria—that is, between the DNA's of a bacterial donor and a bacterial recipient—if these are present in the same cell. It should, moreover, make no difference by what mechanism the donor DNA enters the recipient. This chapter will deal with the recombination of bacterial genes that enter the recipient cell (1) without the direct assistance of any organism (*genetic transformation*), or (2) with the direct assistance of a phage (*generalized transduction*). The next chapter will deal with genetic recombination of bacterial genes that enter (3) with the direct assistance of a phage (*restricted transduction*), or (4) with the direct assistance of the donor bacterium (*bacterial conjugation*).

**8.1** Bacterial genotypes are usually determined from the phenotypes of clones.

The small size of bacteria (and other cellular microorganisms) is a handicap in determining the genotype by observing the phenotype of individual bacteria. This disadvantage can be circumvented, however, by examining the phenotype of the population of cells produced from a single bacterium by successive divisions. Such a population is called a *clone*; and, barring mutation, all members of a clone are genetically identical. Genetically different clones can show phenotypic differences in the size, shape, or color they produce when grown on a semisolid medium such as agar (Figure 8-1). Genetically different clones can also respond

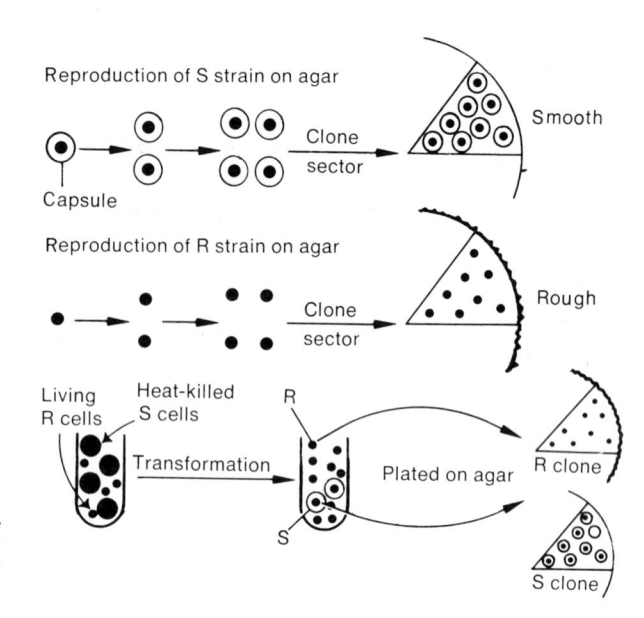

FIGURE 8-1. Cloning of S and R strains of *Pneumococcus* and the transformation of some R by heat-killed S.

differently to various dyes, drugs, and viruses. Therefore, one can readily establish the genotype of a single bacterium from the structural or physiological phenotype of the clone it produces.

A bacterium such as E. coli can grow and reproduce on a relatively simple food medium. Strains that require only such a simple medium are said to be *prototrophic*, or wild-type. Phototrophic bacteria are capable of synthesizing the numerous substances needed for metabolism and reproduction not supplied in the medium. Different mutant strains may not be able to grow and reproduce on this simple medium because they have lost the ability to synthesize one or more nutrients. For such genetically defective strains to grow and reproduce, one or more nutrients must be added to the simple medium. Such strains are said to be *auxotrophic*. One auxotrophic strain of E. coli, for example, requires the addition of threonine to the basic medium; another requires methionine.

# GENETIC TRANSFORMATION

**8.2** The genotype of one bacterium can be changed by pure DNA isolated from another bacterium. Such a genetic transformation first proved that DNA is genetic material in bacteria.

As characterized by clonal phenotype, the bacterium *Pneumococcus* occurs in several genetically different forms. One type, S, is surrounded by a capsule, and forms a clone with a smooth surface when grown on nutrient agar. Another type, R, has no capsule, and forms a clone with a rough surface when grown on nutrient agar. When heat-killed S cells are added to nutrient broth in which R cells are growing and the mixture is poured onto nutrient agar, clones of type S appear (Figure 8-1). The number of such S clones is much greater than can be accounted for by mutation from R to S. Moreover, no matter what subtype of S is heat-killed, live bacteria of that type are obtained after mixture with R cells. Thus, we see that R cells have undergone a *genetic transformation* to S cells of the specific type killed.

To determine the chemical nature of the transforming agent, we test the transforming ability of different fractions of heat-killed S bacteria. Only the fraction containing undegraded DNA is able to transform R bacteria to type S. Moreover, the purest DNA retains the full transforming ability.

Since pure DNA can transform bacteria, no contact is necessary between donor and recipient cells. Hence, genetic transformation does not require the mediation of a virus or any other transmitting organism. The DNA alone carries the specific genetic information that transforms a bacterium; and the transformed bacterium produces a clone of the same type. Genetic transformation by pure DNA was the first proof obtained that DNA is genetic material.

**8.3** Genetic transformation results from a genetic recombination in which a single-stranded piece of donor DNA replaces a corresponding segment of recipient DNA.

Studies of genetic transformation show that the process requires the following series of discrete steps.

1. BINDING AND PENETRATION OF DONOR DNA. Only at a certain stage in the cell cycle is a bacterium able to be transformed. Such competent bacteria can bind donor DNA to several receptor sites at the cell surface. This binding is at first reversible, then permanent. Permanently bound DNA then penetrates the bacterium linearly, that is, in single file. While the duplex DNA is entering the cell, or soon thereafter, one of the two complementary strands is degraded by nuclease (Figure 8-2A).

2. SYNAPSIS OF DONOR TO RECIPIENT DNA. Once inside the recipient cell, single-stranded donor fragments of DNA base pair with complementary regions of the host duplex DNA that have strand-separated (Figure 8-2B, C). Such locus-for-locus pairing is called *synapsis*. When the donor and recipient DNA's differ genetically, the base pairing between them will not be perfect. Nevertheless, the lack of complete complementarity will sometimes fail to be recognized by a nuclease. On

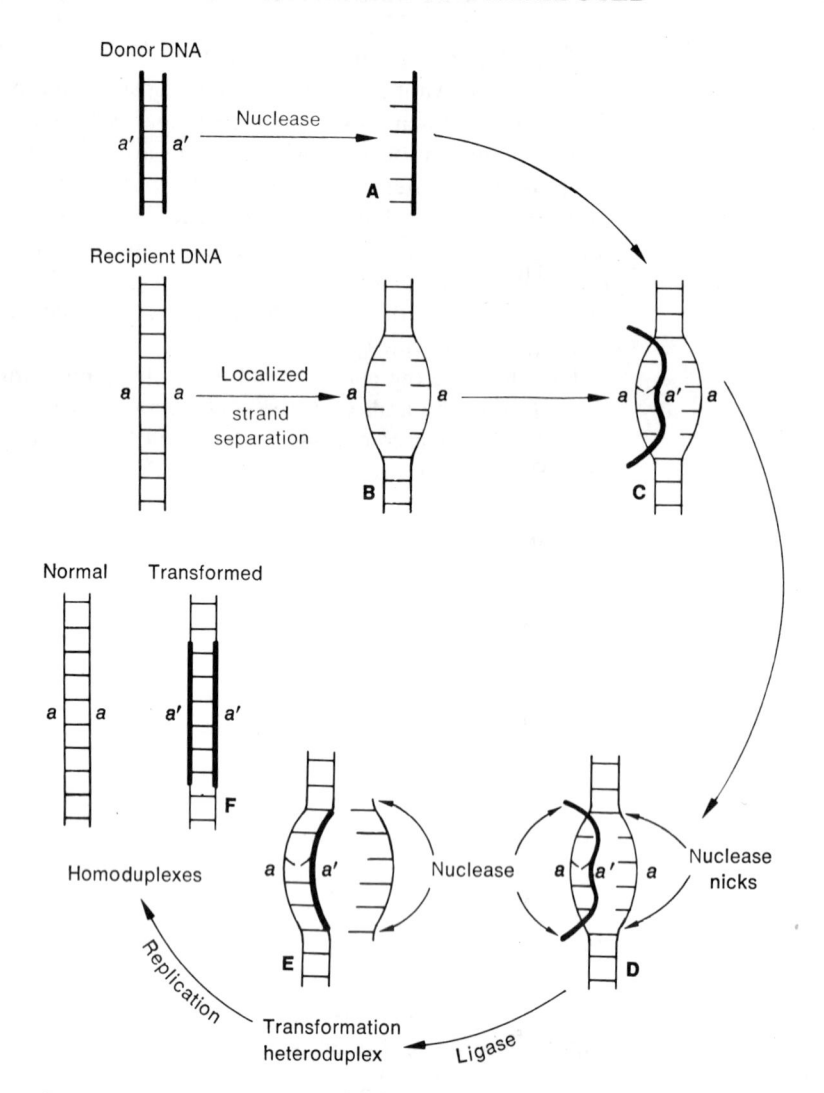

FIGURE 8-2. The mechanism of genetic transformation. See the text.

the other hand, the recipient complement that is no longer base-paired will usually be recognized by nuclease and will be excised (Figure 8-2D), as will the single-stranded ends of donor DNA that are not included in synapsis.

3. INTEGRATION OF DONOR DNA. Once the donor DNA is trimmed of non-base-paired segments, a ligase can seal the two nicks that are present, so that the donor DNA becomes an integral part of the chromosome of the recipient (Figure 8-2E). This *integration* results in the incorporation of a single-stranded segment of donor DNA in place of a corresponding segment of recipient DNA. The average length of the segment integrated is about 6000 nucleotides. When the donor and recipient DNA's differ in base sequence in the region being transformed,

as they do when R bacteria are transformed to S type, the product of integration is a region of duplex DNA in which the two strands are not exactly complementary. Such a duplex is called a *heteroduplex*. The semiconservative replication of a transformation heteroduplex produces two *homoduplexes*—duplexes whose strands are exactly complementary—one carrying the normal region of the recipient and one carrying the transformed region (Figure 8-2F). As a result, when the bacterium undergoing transformation divides, it produces one nontransformed cell and one genetically recombinant transformed cell.

**8.4** Genetic transformation occurs either experimentally or spontaneously in various bacteria as well as in eukaryotes.

Genetic transformation has been found not only in *Pneumococcus* but in other types of bacteria as well. In *Bacillus subtilis*, transformation occurs spontaneously when intact cells are mixed with recipient cells. In this case the transforming DNA is extruded from the surface of a living cell. Transformation also occurs spontaneously in *Pneumococcus*. This is demonstrated by infecting a mouse with two different *Pneumococcus* strains bearing different genetic markers, and obtaining genetically recombinant progeny. Although transformation does not occur in wild-type *E. coli*, mutant strains that do not make certain nucleases, and that are chemically treated to partially remove the cell wall to facilitate the entry of DNA, are transformable.

Good evidence exists for the occurrence of transformation in eukaryotes—including the mouse, the fruit fly *Drosophila*, and human cells cultured *in vitro*.

# GENERALIZED GENETIC TRANSDUCTION

**8.5** In genetic transduction, a phage mediates genetic recombination between bacteria by carrying bacterial genes from one host to another.

The following experiment is performed using two different auxotrophic strains of *E. coli*. One strain requires methionine (but not threonine), the other requires threonine (but not methionine). We can represent the genotypes of these strains as $met^-\ thr^+$ and $met^+\ thr^-$, respectively. The latter strain is grown in broth containing threonine and infected with $\phi$P1. After an incubation period, the culture is passed through a filter that holds back bacteria but not viruses or free DNA molecules (Figure 8-3A). To the filtered broth is added a nuclease, to destroy any free DNA that might cause transformation, and bacteria of the other strain ($met^-\ thr^+$) (Figure 8-3B). After another incubation period, the contents are poured onto an agar plate containing simple (nonenriched) culture medium. One

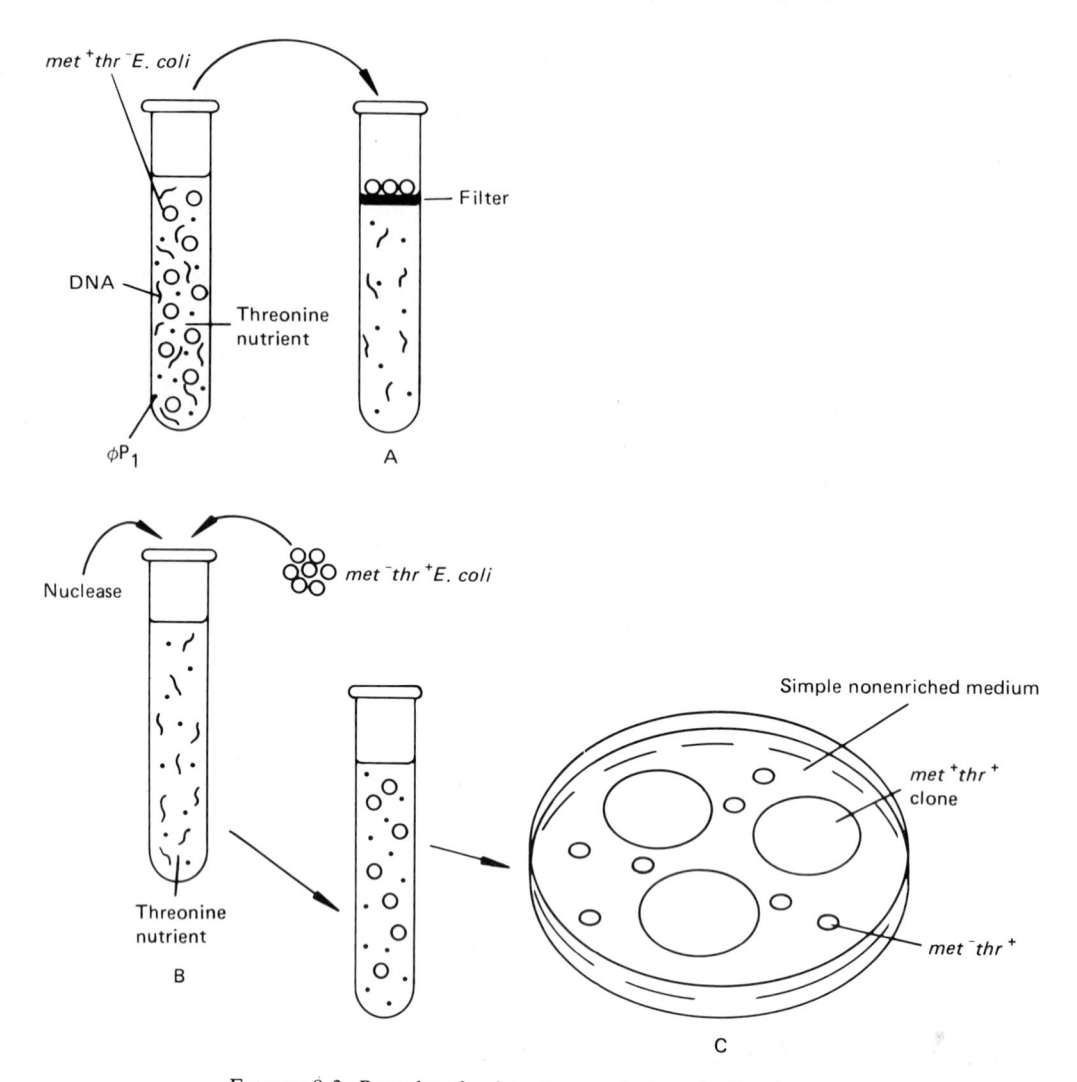

FIGURE 8-3. Procedure for detecting genetic transduction. See the text.

then observes a large number of clones growing on the agar (Figure 8-3C). These must be prototrophic clones, $met^+ thr^+$. Since mutation cannot account for all these prototrophic clones, we conclude that the $met^+$ gene must have been picked up by $\phi$P1 from its $met^+ thr^-$ host and transmitted to the $met^- thr^+$ host, where it integrated and produced a clone of prototrophs.

In support of our conclusion we find that if $\phi$P1 is excluded from the experiment, only a few prototrophic clones are found, owing to spontaneous mutation from $met^-$ to $met^+$. Moreover, if $met^- thr^+$ is the strain infected with $\phi$P1 and we reverse the procedure using the appropriate nutrient media, we get the same prototrophic result, owing to the transfer of $thr^+$ by $\phi$P1 and its integration in the $met^+ thr^-$ strain. We have proven,

GENETIC RECOMBINATION BETWEEN BACTERIA 119

therefore, that phage can mediate the genetic recombination of bacterial genes, in a process called *genetic transduction*.

## 8.6 In generalized transduction, a phage carries a headful segment of any portion of the bacterial chromosome in place of its own chromosome.

Studies of $\phi$P1-mediated transduction show that the transductions result from an error in phage reproduction. Phage P1 reproduction normally includes the formation of a concatemer of phage DNA that enters empty phage heads and is cut off in headful segments by nuclease, as occurs in the reproduction of T-even phages. On rare occasions, in $\phi$P1-infected *E. coli*, the bacterial chromosome is broken and one end enters and fills an empty phage head (Figure 8-4, column A). Once the head is full, the

FIGURE 8-4. Diagrammatic representation of generalized transduction by a phage. **A**: Formation of transducing phage. **B**: Complete transduction. **C**: Abortive transduction. Straight duplex represents bacterial DNA; wavy duplex, phage DNA. Plus represents a bacterial gene for prototrophy; minus, the allele for auxotrophy. Simple, nonenriched culture medium is used throughout.

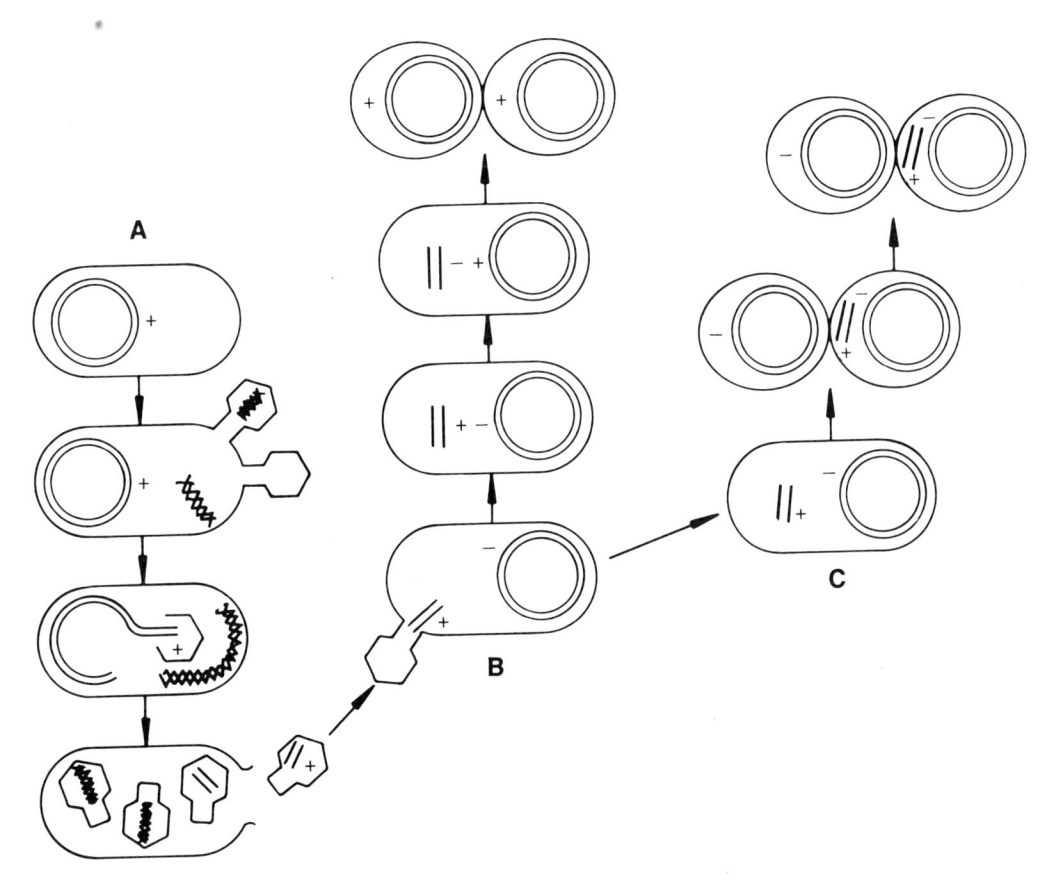

chromosome is cut by nuclease, as usual. After the rest of the phage is assembled, the result is a mature phage that is carrying a headful segment of the *E. coli* chromosome in place of the phage chromosome. Such phage function just as well as normal phage in transporting and injecting their DNA into new hosts. In other words, the mature phage is like the nose cone of an automatic interplanetary rocket—it functions regardless of the nature of its contents. Since any segment of the bacterial chromosome is subject to being included in a phage head, and hence any segment is capable of being transduced, $\phi$P1 is said to be a generalized transducing phage which causes *generalized transduction*.

In most cases of generalized transduction, the transported double-stranded segment of bacterial DNA (which is carrying a gene for prototrophy marked + in Figure 8-4) replaces a corresponding segment of the host bacterium (which is carrying the auxotrophic allele marked −), as shown in Figure 8-4, column B. The result is a transduced bacterium that can produce a clone (of prototrophs). The recombination that results in integration of the injected DNA probably requires synapsis between single-stranded regions of donated and host DNA's, breakages, and ligations. Since, in the present case, the integration of transduced DNA is complete, this result is called a *complete transduction*.

Sometimes the injected DNA cannot either recombine with the host DNA or replicate. In such cases the result is an *abortive transduction*. Sometimes, however, the injected DNA may nevertheless be transcribed. If, as in Figure 8-4, column C, the donated piece carries the prototrophic allele of a host gene that is auxotrophic, the host bacterium may be rendered prototrophic, and may be able to grow and divide. Only one of the first two daughter cells, however, receives the transduced DNA. This cell can grow and divide; the other can grow only for the period of time the necessary mRNA or polypeptide donated to it by the parent cell remains. In this case, when cultured on agar containing a simple nutrient medium, a minute clone is produced (Figure 8-5), only one of whose cells carries the prototrophic allele.

### 8.7 Genetic transduction occurs in various other bacteria and probably in eukaryotes.

Genetic transduction has been found to occur in a variety of other bacteria, including *Bacillus, Pseudomonas, Staphylococcus*, and *E. coli*'s close relative *Salmonella*.

In eukaryotes viruses known to reproduce in the nucleus sometimes incorporate host DNA during their replication and maturation. These viruses include cancer-inducing DNA viruses as well as RNA tumor viruses. Even some RNA viruses that have no nuclear stage in their life cycle, and are restricted to the cytoplasm, can incorporate segments of host DNA during virus reproduction. For example, when *vesicular stomatitis virus* (*VSV*) reproduces in human lymphocytes cultured outside the body, one progeny particle in four contains segments of DNA that are an average of 1400 base pairs long. This DNA is nuclear DNA

# GENETIC RECOMBINATION BETWEEN BACTERIA

FIGURE 8-5. Large and minute (arrows) colonies of *Salmonella*, representing complete and abortive transductions, respectively. (Courtesy of P. E. Hartman.)

that has been released from the nucleus and becomes associated with the lymphocyte's cytoplasmic membranes, including the cell membrane. When VSV progeny mature, they normally incorporate pieces of the cell membrane of their host cell. The DNA found in VSV is, therefore, probably such cell membrane DNA. When VSV reproduces in cells such as fibroblasts, which contain no cytoplasmic membrane DNA, however, progeny VSV contain no DNA. We expect, from our experience with prokaryotic viruses, that eukaryotic viruses that carry eukaryotic DNA are able to infect new hosts; that is, we expect such viruses to be transducers. We do not know whether such transduced DNA would ever be replicated or transcribed in a new host, that is, whether it would be functional genetic material. Finally, until transduced eukaryotic DNA is proven to replicate or to be integrated into eukaryotic DNA, the transduction would be classified as an abortive transduction.

## SUMMARY AND CONCLUSIONS

Alternative phenotypes are useful to determine alternative genotypes. The phenotype of the clone is especially useful in determining the genotypes of bacteria and other microorganisms.

The genetic material of bacteria can undergo breakage recombination by means of genetic transformation, the process which first proved that DNA is genetic material. For transformation to occur, donor DNA must bind to and penetrate the cell membrane of the recipient bacterium. Subsequently, a single-stranded segment of donor DNA synapses and integrates with the recipient chromosome, replacing a corresponding, if not identical, segment of the recipient DNA. When donor and recipient differ genetically in the segment replaced, the initial transformant contains a heteroduplex, which, upon replication, produces two homoduplexes, one transformed and one not. Since this replacement process requires a nuclease and a ligase, it has features in common with the processes of repair of damaged DNA and phage recombination discussed in previous chapters.

Breakage recombination between bacteria can also occur through the mediation of phages that carry duplex segments of bacterial chromosome in place of the virus chromosome. Upon injection into a bacterium, the duplex piece of donated DNA may replace a corresponding piece of the host's chromosome, resulting in a complete transduction; or the donated DNA may neither integrate in the host chromosome nor be able to replicate, resulting in an abortive transduction. Since a segment from any region of the host's chromosome can be transduced, the transduction is of a generalized type.

Transduction occurs in various bacteria and probably in eukaryotes too.

## QUESTIONS AND PROBLEMS

8.1 *Anagrams*: Unscramble and define each of the following: **a.** enloc. **b.** rutoxahop. **c.** sinpassy. **d.** gritanonite. **e.** tincandustor.

8.2 What are the advantages and disadvantages of bacteria as organisms for genetic studies?

8.3 Which type do you suppose arose first in evolution, the auxotroph or the prototroph? Explain.

8.4 After phenotypically rough bacteria are exposed to DNA from smooth bacteria, 2 per cent of the clones produced are smooth; under the same conditions, except that the smooth DNA is treated with DNA nuclease, the percentage of smooth clones is: **a.** unchanged. **b.** 1. **c.** 0.0001. **d.** 0.

8.5 Complete the following to produce a heteroduplex region that will probably not be repaired.

$$\begin{array}{c} \text{A T C G C G A} \\ \text{T A G T G C T} \end{array}$$

8.6 Suppose that transformation produces a heteroduplex at the same time that UV produces one pyrimidine dimer in the same bacterial gene. Will the dimer affect the chance to detect the transformation? Explain.

8.7 You are given as unknowns two test tubes in which clones are growing in liquid culture medium. One tube contains a $met^+$ $thr^+$ clone; the other contains a $met^-$ $thr^-$ clone. What would you do to identify which is which if no more medium is available and you cannot cross the two strains?

8.8 How would you collect and know you had a generalized transducing phage if the host's DNA and the phage's DNA could be differentially labeled by weight and/or radioactivity?

8.9 State one similarity and one difference between complete and abortive transduction.

8.10 Design two experiments: **a.** Demonstrate the transduction of globin mRNA. **b.** Demonstrate the universality of the genetic code using globin mRNA.

8.11 Two nonmotile bacteria are transduced to motility and dropped on the surface of a petri dish containing nutrient agar. How can you identify from the swimming trails produced in the agar which bacterium is the complete transductant and which one is the abortive transductant?

8.12 Is transformation or transduction a mutation? Explain.

*Chapter 9*

# Genetic Recombination Between Bacteria: II. Restricted Transduction and Conjugation

IN CHAPTER 8, we saw that breakage recombination can replace a segment of the chromosome of a recipient bacterium by a corresponding segment derived from a donor bacterium. The two processes described which transfer the DNA from one bacterium to the other were transformation and generalized transduction. In both processes the transferred DNA was simply a segment of the donor bacterial chromosome. The present chapter describes two additional processes for transferring donor bacterial DNA to a recipient bacterium for breakage recombination. In these two processes, *restricted transduction* and *bacterial conjugation*, the transferred piece of bacterial DNA is attached to another piece of DNA that is not essential for bacterial survival. In each case we will discuss the nature and behavior of the nonessential DNA before learning how it becomes connected to ordinary bacterial DNA.

## RESTRICTED TRANSDUCTION

**9.1** The normal life cycle of $\phi\lambda$ is described.

Restricted transduction occurs as an abnormality in the life cycle of certain phages. To understand how this comes about, we will first describe the normal life cycle of such phages, of which $\phi\lambda$ is an example. Phage $\lambda$ is called a *temperate phage* because it, unlike virulent phages such as T phages, does not always lyse the host cell it infects. Figure 9-1A shows the life cycle of uninfected *E. coli*. Figure 9-1B shows the life cycle of $\phi\lambda$ when it infects and lyses an uninfected cell, that is, when it undergoes a lytic life cycle. In this case the phage chromosome, which is double-stranded with complementary single-stranded ends (Figure 2-19, part J; Figure 2-21), is injected into the host, where it replicates and produces a

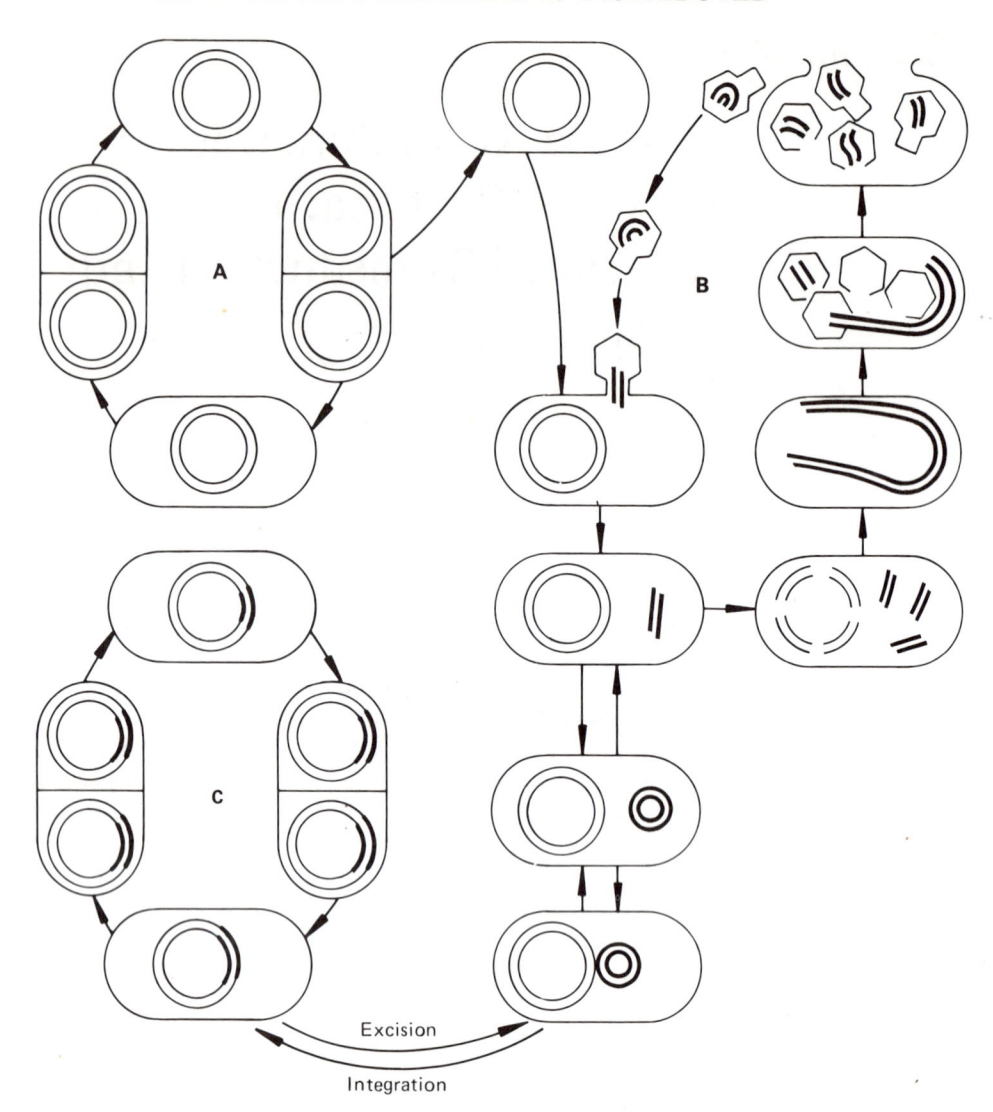

FIGURE 9-1. Diagrammatic representation of the life cycles of bacteria and a temperate phage such as λ. **A**: Cycle for uninfected bacteria. **B**: Lytic cycle of temperate phage. **C**: Cycle for lysogenic bacteria. Thin duplex represents bacterial DNA; thick duplex, phage DNA.

concatemer of phage DNA. One end of the concatemer then enters and fills an empty phage head until the concatemer is scissioned by a nuclease that recognizes a specific palindrome in the base sequence and produces two nicks. These two nicks occur in corresponding positions in the two strands, but not in the center of the palindrome (as they are in the example shown in Figure 3-11). As a result, the phage head contains a chromosome that ends with a single-stranded region, and the next chromosome in the concatemer therefore begins with the complementary single-stranded region (Figure 9-2). The newly produced single-stranded

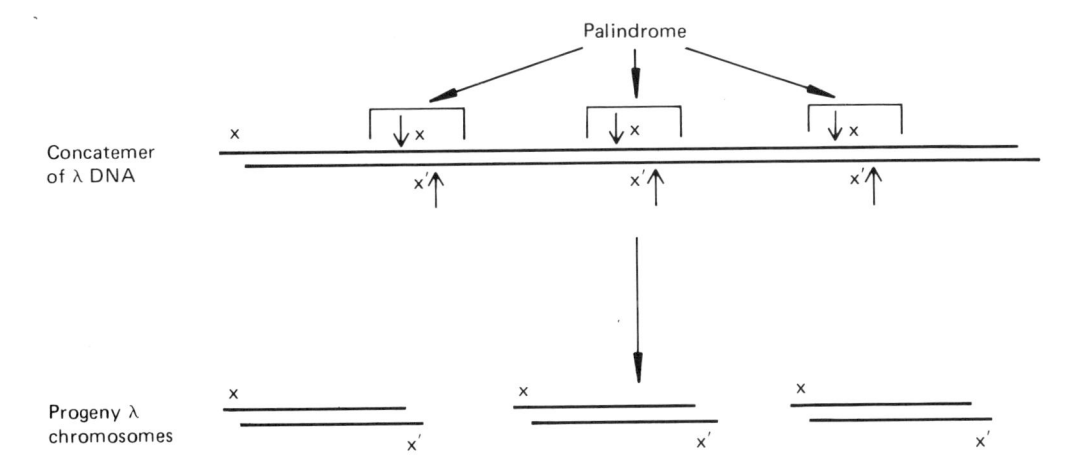

FIGURE 9-2. The production of progeny λ phage chromosomes from a concatemer cut by staggered nicks in a palindrome that occurs once per λ gene sequence. $x$ and $x'$ are complementary sequences; small arrows are sites of nicking by a nuclease.

end of the concatemer then enters a new head and the filling and cutting processes are repeated. Since this palindrome occurs only once in a complete sequence of phage genes, each head comes to contain the same chromosome with the same complementary single-stranded ends. After about 100 heads are filled and the assembly of the rest of the particles is completed, the host cell is lysed, liberating mature phage, and concluding the lytic cycle.

Upon entering its host, the phage chromosome has another possible fate, however. It may circularize by base pairing between its single-stranded ends and lose the two nicks present by the action of ligase. The resultant no-nick double-helical ring contains a different palindrome that is also present in the host chromosome. A different nuclease recognizes these two palindromes and produces off-center, staggered nicks in each. Base pairing between complementary single-stranded sequences in the *E. coli* and λ chromosomes (Figure 9-3) followed by ligation results in the integration of the λ chromosome into the *E. coli* chromosome (Figures 9-3 and 9-1). The *E. coli* cell that contains a λ chromosome within its own chromosome survives and divides to produce a clone of such individuals (Figure 9-1C). In such a clone, however, the reverse process can occur in a small fraction of the individuals. Specifically, the nuclease used in integration can recognize the two copies of its palindrome in the *E. coli* chromosome and produce two sets of staggered nicks which will excise the λ DNA from the *E. coli* chromosome. Ligation of the two ends of the λ chromosome to each other, and of the two ends of the *E. coli* chromosome to each other, re-forms the circular λ duplex and the normal *E. coli* chromosome. The nuclease that cuts up concatemers of λ DNA can then revert the λ duplex ring to its linear form, which then enters the lytic cycle.

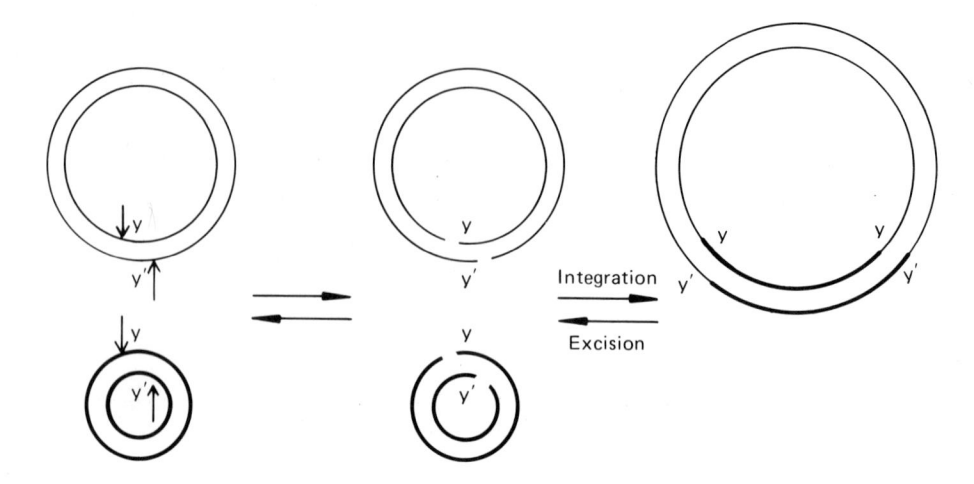

FIGURE 9-3. The integration and excision of λ DNA (thick duplex) relative to the E. coli. chromosome (thin duplex). y and y′ are complementary sequences in a palindrome; small arrows are sites of nicking by a nuclease.

We see from the foregoing that *E. coli* can harbor a λ chromosome within its own chromosome. Since each such *E. coli* has the capacity to excise the λ chromosome and generate a lytic cycle (although only a small fraction do so spontaneously), they are called *lysogenic bacteria*. Lysogenic bacteria are immune to being lysed by a subsequent phage infection, however, if the new infection is by a phage identical or closely related to a phage whose chromosome they are harboring. Different types of temperate phage have different palindromes that are also present in the *E. coli* host. As a result, *E. coli* can be lysogenic for two or more kinds of temperate phage whose chromosomes occupy different sites of integration in the *E. coli* chromosome.

The salient characteristic of λ DNA (and the DNA's of other temperate phages that can integrate in the host chromosome) is that it is optional or dispensable in the life cycle of its host, but when present can replicate either autonomously, in a lytic cycle, or as part of a host chromosome, in a lysogenic cycle. Dispensable chromosomes with these properties are called *episomes*. Dispensable chromosomes that can replicate only autonomously are called *plasmids*. The chromosome of the generalized transducing ɸP1 is a plasmid.

**9.2** Phage λ can sometimes transduce bacterial loci that are adjacent to its site of integration.

About 1 time in 1 million, excision of the λ chromosome from the *E. coli* chromosome occurs in a faulty manner. In such cases, staggered nicks are produced in a palindromic sequence that is present somewhere within the integrated λ DNA as well as in one of the bacterial loci adjacent to the site of λ DNA integration (Figure 9-4, A). Such an excision leaves a part

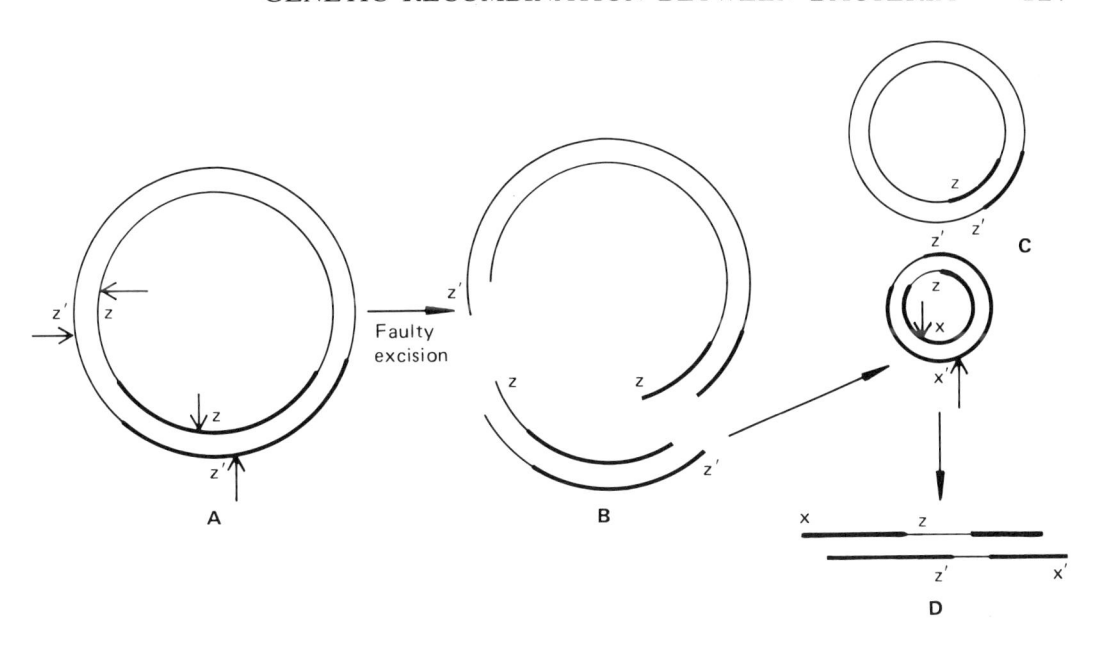

FIGURE 9-4. Formation of a restricted transducing λ chromosome by faulty excision from the *E. coli* chromosome. Thin duplex represents *E. coli* DNA; thick duplex, λ DNA. *x* and *x'* are complementary sequences in one palindrome; *z* and *z'* are complementary sequences in a different palindrome. Small arrows are sites of nicking. See the text.

of the λ chromosome within the *E. coli* chromosome, and carries a piece of *E. coli* chromosome attached to the excised part of the λ chromosome (Figure 9-4, B). Base pairing and ligation then produces two double-helical rings (Figure 9-4, C). When the defective no-nick circular λ chromosome still retains the palindrome recognized by the concatemer-cutting nuclease (Figure 9-4, C), it can be opened (Figure 9-4, D) and undergo a lytic cycle, to produce λ progeny each of which is transducing the same segment of *E. coli* chromosome attached to a defective λ chromosome. Since a λ phage head can hold no more than about 2 per cent of the length of the *E. coli* chromosome, it is clear that less than this per cent of the *E. coli* chromosome located adjacent to the λ integration site is subject to transduction by $\phi\lambda$. Phage λ is, therefore, capable of transducing only a restricted portion of the host chromosome; that is, it is capable of only *restricted transduction*.

The specific loci that border the λ integration site are *gal* loci on one side and *bio* loci on the other. The *gal* loci are more often transduced than the *bio* loci. When *gal*⁻ hosts are transduced by *gal*⁺-carrying λ phages, the host becomes diploid for this region and phenotypically gal⁺. (Phenotypes are not italicized.) Integration of the *gal*⁺ loci in place of *gal*⁻ loci can subsequently occur by breakage and ligations in the same manner that generalized transduced segments are integrated. Other temperate phages transduce other loci that are adjacent to their integration sites, and are, therefore, also restricted transducers.

128    HOW GENETIC MATERIAL IS DISTRIBUTED

# BACTERIAL CONJUGATION

**9.3**  Some bacteria contain sex particles that infect other bacteria during the process of conjugation.

Genetic transduction between bacteria can also be mediated by *sex particles*, a variety of dispensable, small, duplex ring DNA's that cause bacteria to temporarily fuse together in pairs in the process of *conjugation*. During conjugation sex particles in one partner can infect the other partner. We shall consider the nontransducing behavior of a sex particle before considering its transducing behavior in the sections that follow.

A cell that contains a sex particle has hair-like protoplasmic projections from its surface that are called *pili* (Figure 9-5). Pili enable such cells to act as *genetic donors*, that is, to be of *male* or + sex type. When a pilus of a male cell touches a cell that contains no sex particles and is, therefore, a *genetic recipient*—of *female* or − sex type—it becomes a conjugation tube that serves to connect the protoplasm of the two conjugants (Figure 9-6). During this process of conjugation, one strand of the small chromosome is nicked, and one nicked end unwinds and is sent into the conjugation tube and thence into the female (Figure 9-7). The transfer of the whole strand is accomplished in a few minutes. The sex particle chromosome also replicates during conjugation. The complement of the transferred single-stranded rod is made in the female, and the complement of the nontransferred single-stranded ring is made in the male. After ligation, the result is the synthesis of two of the small duplex rings, one in each conjugant, completing the semiconservative replication of the sex particle. Conjugants then separate, both exconjugants being male. We see, therefore, that the sex particle and hence male sexuality is infectious in prokaryotes.

FIGURE 9-5. Silhouette of a bacterium with pili projecting from its surface.

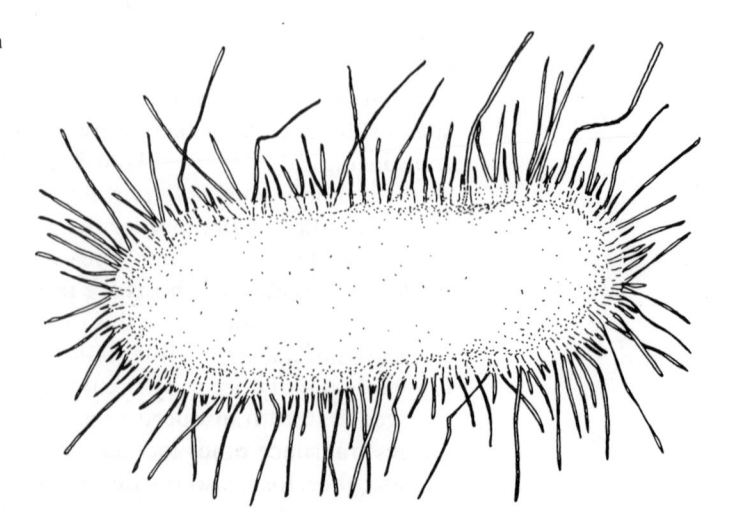

GENETIC RECOMBINATION BETWEEN BACTERIA 129

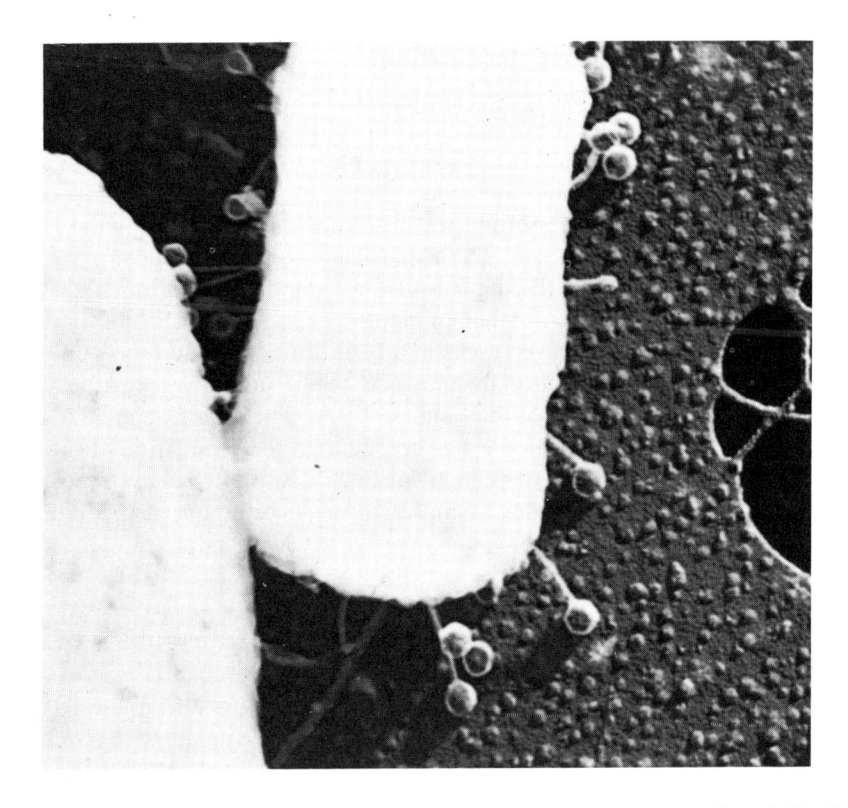

FIGURE 9-6. Electron micrograph showing conjugation between male and female *E. coli*. The female cell is labeled with tadpole-shaped bacteriophage lambda particles; the male cell is not. In the zone of contact the cell walls seem to have disappeared. (Courtesy of T. F. Anderson, E. L. Wollman, and F. Jacob, 1957. *Ann. Inst. Pasteur*, **93**: 450–455.)

FIGURE 9-7. The replication of a sex particle during conjugation. Broken-line arrows indicate the complements being synthesized.

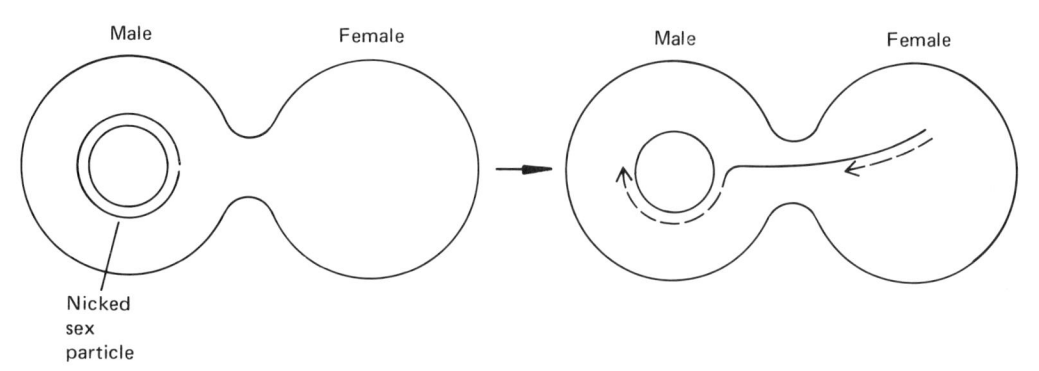

## 9.4 Some sex particles are plasmids; others, such as the F particle, are episomes.

Some sex particles are plasmids that replicate in synchrony with bacterial division to produce clones of males of the same type. Other sex particles are episomes, the best known example being the *F particle* which occurs in *E. coli*. The F particle exists and replicates autonomously in the $F^+$ male (Figure 9-8A). One or more palindromes in F also occur a relatively large number of times in the *E. coli* chromosome. On rare occasions, staggered nicks in a palindrome in F and the same palindrome in the *E. coli* chromosome can result in the integration of F into the *E. coli* chromosome (Figure 9-8A). The result is a male of the *Hfr sex type*. Since integrated F has no effect on the replication of the *E. coli* chromosome, Hfr males can produce clones of Hfr males. On other rare occasions, an Hfr male can undergo the reverse genetic recombination to produce an $F^+$ male.

## 9.5 Hfr males are generalized transducers of bacterial genes.

Although integrated, the F in an Hfr male acts like a sex particle. Accordingly, an Hfr male can conjugate with a female, called $F^-$, which carries no F particles (Figure 9-8B). A nick occurs in one of the strands of the *E. coli* chromosomes adjacent to, or within, F (we are not sure just where it occurs), and the free end not bearing F is sent into the $F^-$ conjugant. If conjugation lasts long enough and the strand being trans-

FIGURE 9-8. **A**: F as an episome. **B**: Transduction of bacterial markers by integrated F during conjugation. Thick duplex represents F DNA; thin duplex, *E. coli* DNA.

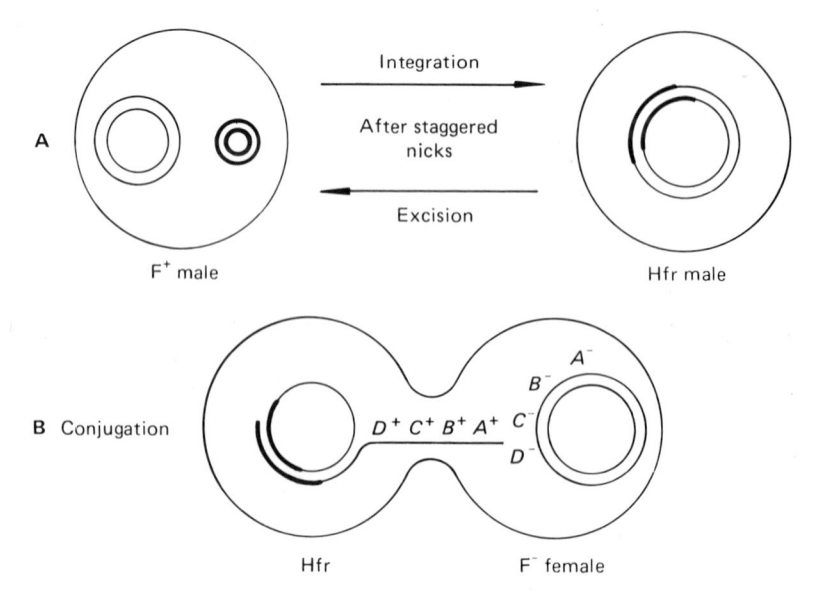

ferred is not broken, a complete complement is transferred to the F⁻ conjugant. If conjugation lasts long enough and the strand being trans-semiconservative replication proceeds during the transfer of the Hfr strand. Once any segment of the Hfr chromosome is transferred, it may undergo synapsis and recombination in the corresponding region of the F⁻ chromosome. Recombination requires the integration of a single-stranded segment in place of the equivalent one in the F⁻ chromosome—the process being the same as in transformation. In this way, if conjugation lasts long enough to deliver all the bacterial loci as well as F, F⁻ may integrate segments containing any of the bacterial loci as well as the locus of F. In the latter case, the F⁻ cell becomes an Hfr cell.

Usually, however, the conjugation tube breaks before the entire Hfr complement is transferred, so that the F⁻ exconjugant is only partially diploid and is still F⁻. Any single strain of Hfr, therefore, is usually a restricted transducer of bacterial loci, a High frequency of recombination (hence the symbol Hfr) being obtained only for the loci transferred first. Since, however, different Hfr strains have F integrated at any one of a score of widely separated regions in the *E. coli* chromosome (Figure 9-9), and since all loci are occasionally transduced by a given Hfr, Hfr strains are considered to be generalized transducers of bacterial loci.

Conjugation between Hfr and F⁻ strains can be used to sequence the loci in the *E. coli* chromosome. This sequence can be determined by artificially interrupting the conjugation of Hfr and F⁻ at different times and scoring the earliest times that exconjugant F⁻ cells are recombinant

FIGURE 9-9. Linkage map showing the point of origin of chromosome transfer for several Hfr strains of *E. coli*. Arrowheads on the inner circle indicate the direction of transfer. The first and last markers known to be transferred by each Hfr are displayed on the outer circle. Genetic markers are shown at their approximate positions only; precise map locations are given in Figure 9-10. (From A. L. Taylor and C. D. Trotter, 1967. *Bact. Rev.*, **31**:332–353.)

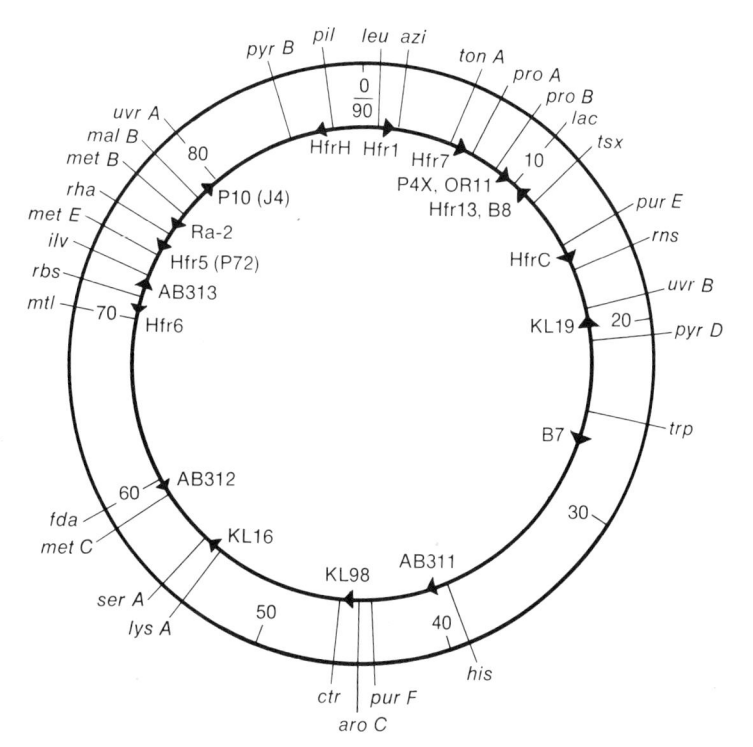

for different marker genes. For example, if an F⁻ is marked by genes $A^- B^- C^- D^-$ and Hfr is marked by alleles $A^+ B^+ C^+ D^+$, as in Figure 9-8B, the earliest F⁻ exconjugants may be $A^+$ recombinants; later ones may be recombinants containing $A^+$, $B^+$, or $A^+ B^+$; still later ones may

FIGURE 9-10. Scale drawing of the linkage map of *E. coli*. The inner circle, which bears the time scale from 0 through 90 minutes, depicts the intact circular linkage map. The map is graduated in 1-minute intervals beginning arbitrarily with zero at the *thr A* locus. Selected portions of the map (for example, the 10- to 12-minute segment) are displayed on arcs of the outer circle with a 4,5-times expanded time scale to accommodate all the markers in crowded regions. Markers in parentheses are only approximately mapped at the positions shown. A gene identified by an asterisk has been mapped more precisely than the markers in parentheses, but its orientation relative to adjacent markers is not yet known. (From A. L. Taylor, 1970. *Bact. Rev.*, **34**: 155–175, Fig. 1.)

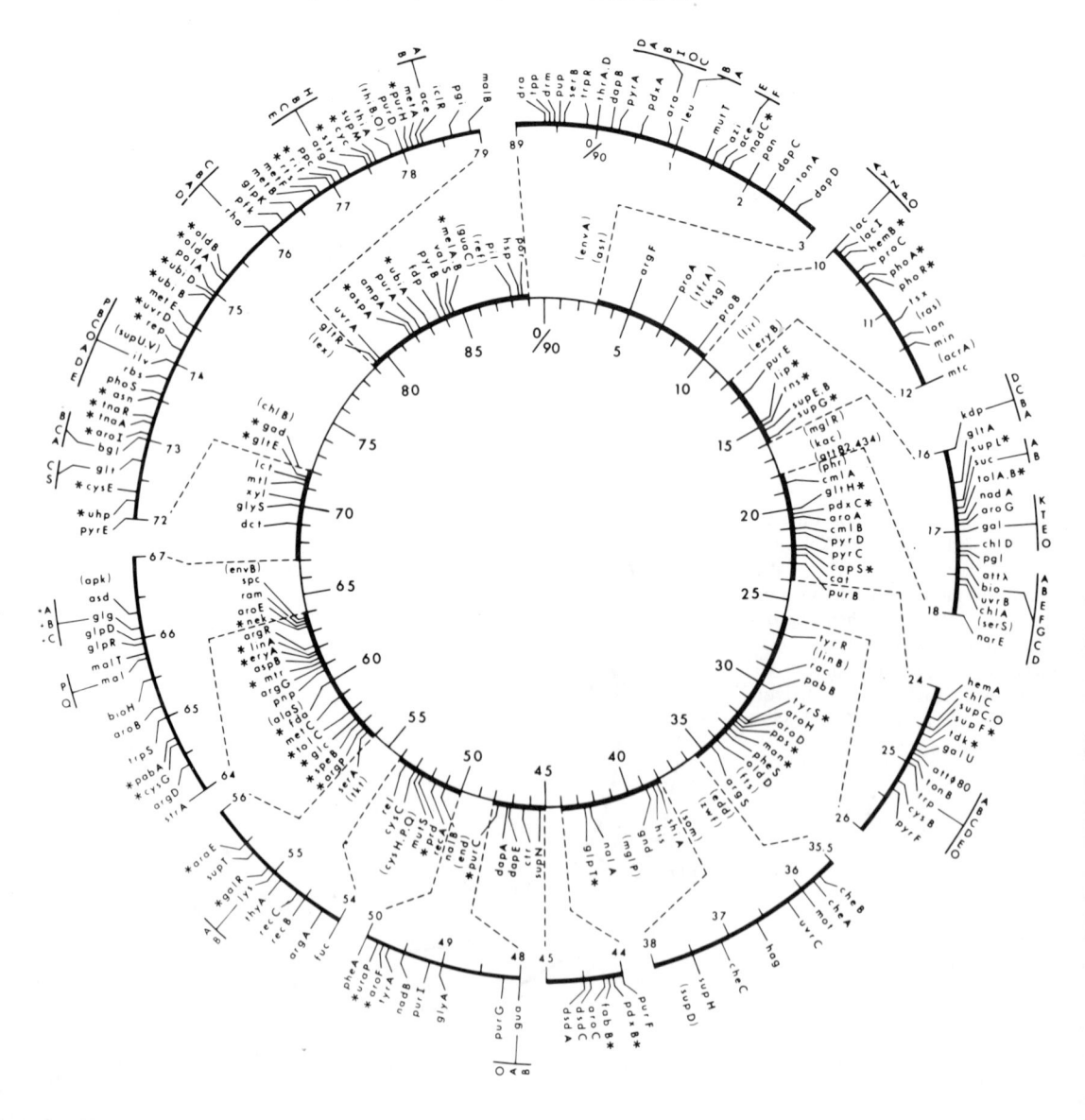

be recombinants containing $A^+$, $B^+$, and/or $C^+$; and so forth, proving that the sequence of loci in this Hfr strain is "$A$" "$B$" "$C$" "$D$". Maps of different Hfr strains can be combined to yield a circular map of the *E. coli* chromosome (Figure 9-10).

## 9.6 Faulty excision of integrated F can produce free, restricted-transducing derivatives of F.

We have already seen how faulty excision of the λ episome produces restricted-transducing λ phages. The episome F can have a similar fate. On rare occasions, staggered nicks occur in a palindrome located both within an integrated F particle and a region of the *E. coli* chromosome. This leads to the excision of an F particle carrying some bacterial loci (as well as a bacterial chromosome carrying a segment of F) (Figure 9-11A). Some such derivatives of F can still function as sex particles. In such cases, conjugation can occur with an $F^-$ individual; and transfer of the *F derivative* to $F^-$ takes place with the bacterial loci delivered first, as in Hfr conjugations (Figure 9-11B). Because the bacterial segment is often short, the $F^-$ cell often receives the entire F derivative sequence and thereby also becomes a male with a derived F particle. Such derived F particles may also be episomes. They may replicate autonomously not only during conjugation but before each ordinary bacterial division; they may also integrate into the *E. coli* chromosome, usually at a locus corresponding to one of the bacterial loci they are transducing, where they are replicated as a part of the *E. coli* chromosome. Since derived F

FIGURE 9-11. **A**: Production of a restricted-transducing F derivative. **B**: Transduction by a derived F particle during conjugation. Thin duplex represents *E. coli* DNA; thick duplex, F DNA.

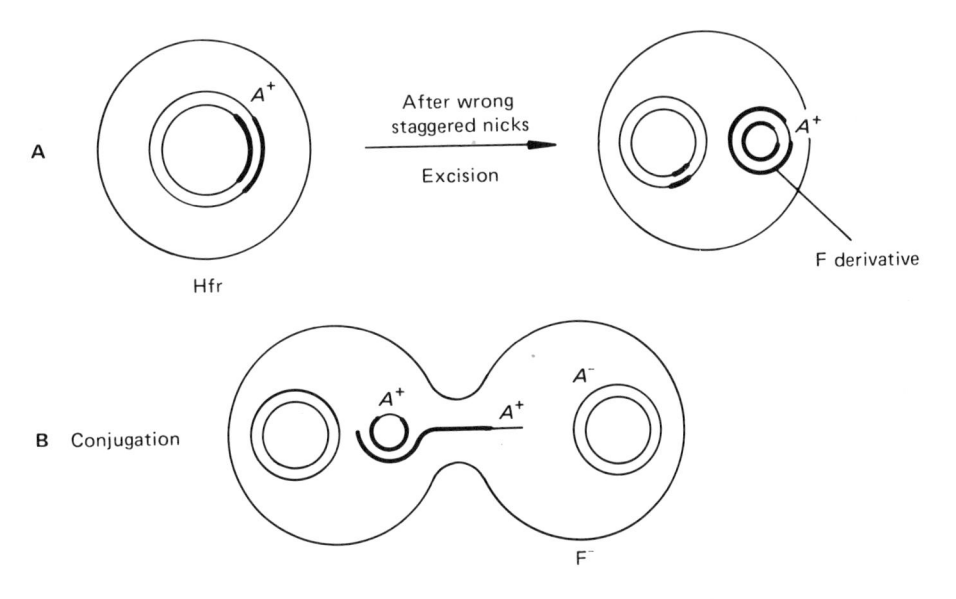

particles can originate from only the limited number of different sites in the *E. coli* chromosome where F can integrate, and since the transduced segment is usually short, only about 10 per cent of the loci in the *E. coli* chromosome are transduceable by F derivatives. Therefore, free F derivatives are only restricted transducers, although integrated F derivatives may be, like integrated F, generalized transducers.

**9.7** A bacterium may contain a variety of plasmids and sex particles that can undergo a variety of breakage recombinations.

*E. coli* and other bacteria may contain several different kinds of dispensable, duplex ring DNA's. Some of these duplex rings are other kinds of plasmid or episome-like sex particles. They cause conjugation and are transferred to females along with any bacterial genes they may have incorporated by recombination. Other duplex rings are plasmids that are not sex particles. These plasmids may code for enzymes that break down unusual compounds (such as camphor or salicylic acid) to liberate energy; or they may code for substances that can kill bacteria that do not carry such plasmids; or they may bind to a class of antibiotics so that the bacterial host is rendered resistant to the antibiotics. Still other duplex rings are recombinants between sex particles and non-sex-particle plasmids, or between two kinds of sex particles, or between a sex particle or non-sex-particle plasmid and a phage episome. A great variety of genetic recombinations are possible.

The group of duplex ring DNA's that confer resistance to antibiotics has been well investigated. Such *drug-resistance particles* may confer upon their bacterial host resistance to as many as eight different antibiotics (including sulfonamide, chloramphenicol, tetracycline, penicillin, furazolidone, and four different mycins). Some drug-resistance particles are non-sex-particle plasmids; others include a sex particle that can transduce the drug-resistance loci and sometimes bacterial loci.

Many of the recombinants mentioned in this chapter result from the action of the palindrome-recognizing nucleases, coded by plasmids or episomes, that produce staggered nicks (Figure 9-12). Any two separate DNA sequences with the same palindrome can be recombined by the sequential use of an appropriate nuclease to produce staggered nicks and a ligase, thereby producing a hybrid molecule.

SUMMARY AND CONCLUSIONS

Genetic recombination between bacteria is facilitated by episomes, through the process of episome-mediated transduction. This process requires the transfer of bacterial genes attached to an episome.

The two types of episome described differ in the mechanism employed for introducing a chromosome segment of the donor bacterium into the recipient. The episomes of temperate phages such as λ use phage infection to inject double-stranded transducing DNA into the recipient

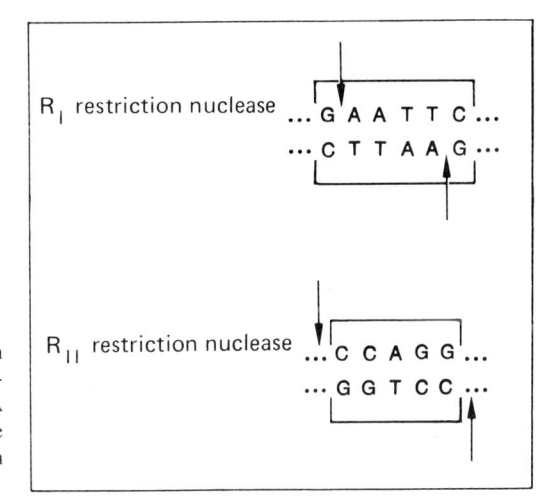

FIGURE 9-12. Staggered nicks produced by certain restriction enzymes result in complementary single-stranded (cohesive) ends. Note that any two separate DNA sequences with sites recognized by the $R_I$ (or $R_{II}$) enzyme can be joined by treating first with the nuclease and then with ligase. Arrows show the cleavage sites.

$R_I$ restriction nuclease ...G A A T T C...
...C T T A A G...

$R_{II}$ restriction nuclease ...C C A G G...
...G G T C C...

bacterium. A sex-particle episome such as F employs conjugation to send single-stranded transducing DNA into the recipient bacterium. Both types of episome become transducing after a faulty excision from the bacterial chromosome replaces part of the episomal chromosome by a fragment of the bacterial chromosome. Since excised transducing episomes usually carry only small pieces of bacterial chromosome adjacent to their sites of integration, such free episomes are capable only of restricted transduction. When integrated, however, F and F derivatives act as sex particles which mobilize the entire bacterial chromosome, so that it is possible (although rare) for them to transduce an entire bacterial chromosome. For this reason, and because they can integrate at any one of a variety of sites in the bacterial chromosome, integrated sex particles are capable of generalized transduction.

*E. coli* and other bacteria can harbor a variety of dispensable chromosomes—those of episomal phages, of plasmid or episomal sex particles, or of plasmids of non-sex-particles having various phenotypic effects. These chromosomes can undergo breakage recombination with each other or with bacterial chromosomes to produce a great variety of genetic combinations. These and most of the other breakage combinations described in this chapter seem to use ligase after staggered nicks are produced by palindrome-recognizing nucleases.

## QUESTIONS AND PROBLEMS

9.1 *Anagrams*: Unscramble and define each of the following: **a.** apterteem. **b.** sipeemo. **c.** dimpals. **d.** singleocy. **e.** sinexico.
9.2 *Riddle*: How can an $F^-$ *E. coli* become $F^+$ without conjugating?
9.3 How would you collect and know you had a specialized transducing phage if the host's DNA and the phage's DNA could be differentially labeled by weight and/or radioactivity?
9.4 State one difference between a temperate and a virulent phage.
9.5 The chromosome of $\phi\lambda$ has a different linear sequence of its loci at different times. Explain.

**9.6** The chromosome of $\phi\lambda$ can become a double-stranded ring in its host. Why is this chromosome not already circular in the phage particle?

**9.7** What procedure would you follow to lengthen the *E. coli* chromosome a total of about 2 per cent, about 4 per cent, and finally about 6 per cent?

**9.8** Is either integration or excision of $\phi\lambda$ DNA a mutation? Explain.

**9.9** When ordinary *gal*⁻ *E. coli* are exposed to *gal*⁺-carrying DNA isolated from transducing $\phi\lambda$, *gal*⁺ *E. coli* are not produced. When ordinary $\phi\lambda$ are added to the mixture, however, *gal*⁺ *E. coli* are obtained. What is the molecular explanation of the second result?

**9.10** How do male and female *E. coli* differ phenotypically?

**9.11** How could *E. coli* be obtained that had two *lac* loci in tandem?

**9.12** Name four different kinds of dispensable circular DNA's found in *E. coli*.

**9.13** Name one female and three male sex types in *E. coli*. State one way in which each is unique.

**9.14** How could you integrate a double-stranded segment of frog DNA into the DNA of a drug-resistance particle?

**9.15** Using the information in this chapter, state a circumstance when a hospital may not be the best place to get well.

**9.16** *Balloons*: Fill appropriately.

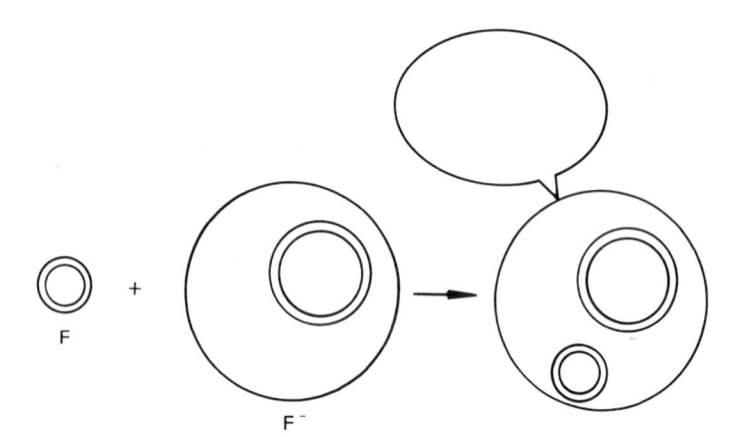

# Chapter 10

# Genetic Recombination in Eukaryotes: I. Mitosis, Meiosis, and Segregation

THE REPLICATION of the bacterial chromosome is followed by a bacterial division which (barring mutation) results in each of the progeny cells receiving an exact copy of the parent chromosome. The orderliness of this distribution of chromosomes is accomplished by the attachment of the replicating bacterial chromosome to an infolding of the cell membrane (Figure 2-26), so that when this region of the cell membrane divides, it automatically separates the halves of the replicated chromosome and places them in different progeny cells. This process for chromosome distribution, therefore, assures that each parent and each progeny cell has the same chromosomal constitution at the start of its life cycle, and is a genetically programmed mechanism to prevent genetic recombination between bacterial generations. In eukaryotes, each of which contains much larger amounts of DNA and several chromosomes, two special programmed mechanisms have evolved for the orderly distribution of replicated nuclear chromosomes to progeny nuclei. One is a mechanism for the nonrecombinational distribution of replicated nuclear chromosomes to progeny nuclei (*mitosis*) and is described in the first part of this chapter. The second part of the chapter describes the second mechanism (*meiosis*), which programs recombination between and within DNA duplexes.

## MITOSIS

**10.1** Nuclear DNA replicates before nuclear division.

Unlike a bacterial cell, a eukaryote's (nuclear) DNA is bounded by a nuclear membrane during most of the life cycle of the cell. During a cell cycle there are four successive periods (Figure 10-1): a period of growth (G1), a period in which the DNA replicates (S), a period of more growth

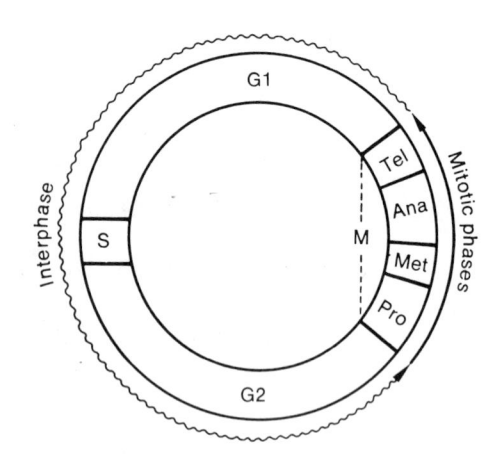

FIGURE 10-1. Periods in a mitotic cell cycle.

(G2), and a period of nuclear division (M). The first three periods (G1, S, and G2) combined make up what is called *interphase*. The fourth period (M) separates the replicated DNA nonrecombinationally so that each of the two progeny (daughter) nuclei have the same chromosome composition as the parent nucleus did, and is called *mitosis*.

## 10.2 Mitosis uses a spindle to produce nonrecombinant daughter nuclei.

We will describe the process of mitosis as it is seen under the ordinary light microscope. During interphase (Figure 10-2A) the nuclear DNA-containing chromosomes are partly unwound and hence too long and thin to be seen as discrete bodies. Among the first indications that the nucleus is preparing to divide is the appearance (due largely to supercoiling) of a mass of thin, separate chromosomes (Figure 10-2B), some of which are associated with nucleoli. Each such chromosome is composed of two visible threads, or *chromatids*, irregularly twisted about each other; each chromatid contains a DNA duplex as well as basic and acidic proteins, and RNA. The appearance of these chromosomes marks the start of *prophase*, the first stage of mitosis (Figure 10-1). By continuing to supercoil and by folding as prophase continues, the chromatids of each chromosome become shorter and thicker and untwist from one another (Figure 10-2C); the nucleoli become smaller. By the end of prophase (Figure 10-2D), the nucleoli and nuclear membrane have disappeared, and the chromosomes are seen as thick rods that move nonrandomly for the first time. This directed movement is restricted to a part of the chromosome called the *centromere*.

Each centromere becomes joined to small tubes of the *spindle*, which has been forming throughout prophase. The spindle is a symmetrical structure, each half of which is made up of numerous small tubes gathered together at one end, the pole. The two halves are joined at the other, unbound tips of the tubes in a form like that of a pair of hands with the spread fingertips touching. Chromosomes are moved by the tubes until

A Interphase

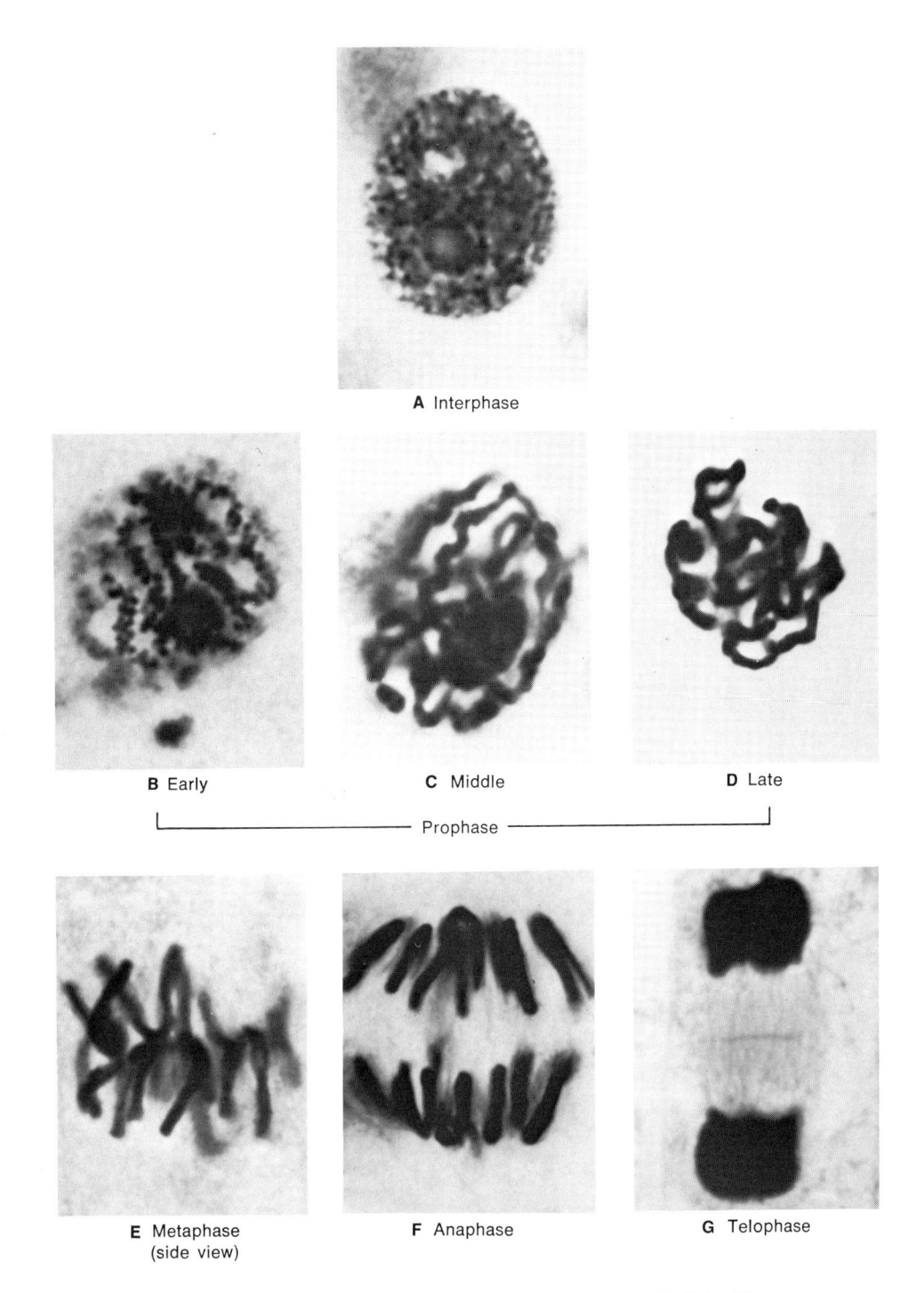

B Early  C Middle  D Late
└──────────── Prophase ────────────┘

E Metaphase  F Anaphase  G Telophase
(side view)

FIGURE 10-2. Mitosis in the onion root tip. (Courtesy of R. E. Cleland.)

each centromere lies in a single plane passing between the two symmetrical halves of the spindle (corresponding to the plane determined by the points at which fingertips touch). The rest of each chromosome can assume essentially any position. When all the centromeres have assumed their position in the central plane of the spindle, mitosis has reached its middle stage, or *metaphase* (Figure 10-2E).

At metaphase the two chromatids of a chromosome are attached to each other at the centromere, although elsewhere they are largely free (Figure 10-3). After metaphase they separate at the centromere and the two daughter centromeres, which are attached to spindle tubes that pull them toward opposite poles of the spindle, suddenly move apart, one going toward one pole, the other toward the other pole. Once a separate entity, each chromatid becomes, by definition, a chromosome. The stage at which the chromatids separate and move to opposite sides as chromosomes is called *anaphase* (Figure 10-2F).

FIGURE 10-3. Nuclear chromosomal complement of a normal human female. The cell was in the mitotic metaphase (hence chromosomes appear double except at the centromere) when squashed and photographed. Chromosomes can be cut out and paired, as can be seen in Figure 10-4. (Courtesy of K. Hirschhorn.)

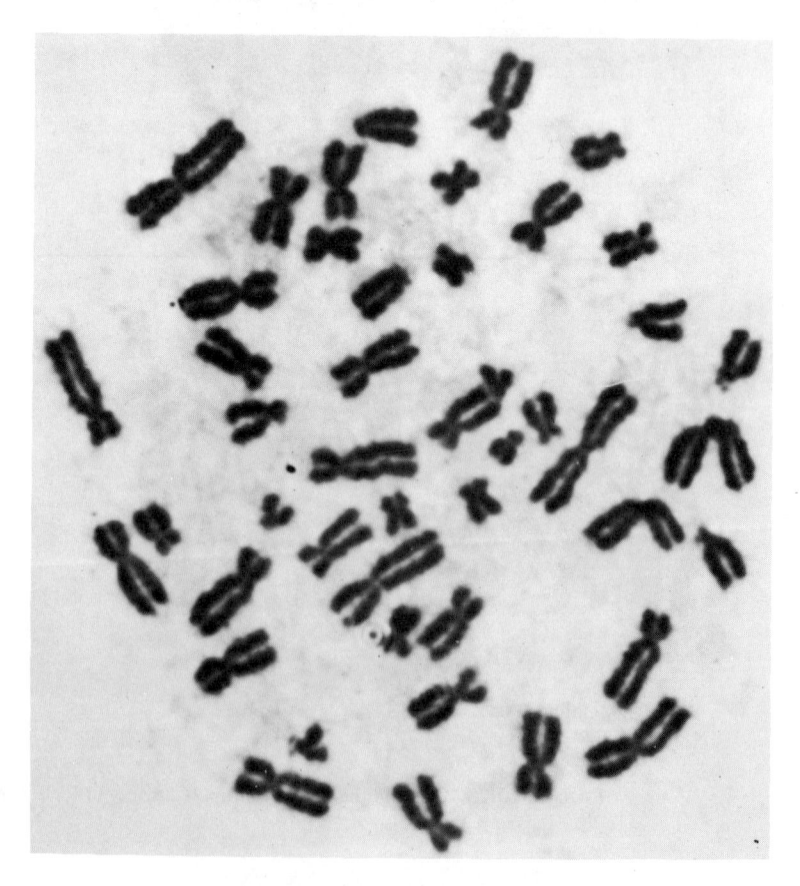

# GENETIC RECOMBINATION IN EUKARYOTES

When the chromosomes reach the poles, the last stage, *telophase*, begins (Figure 10-2G). The subsequent events are the reverse of those of prophase: the spindle disintegrates, a new nuclear membrane envelops the chromosomes, and nucleoli reappear. The chromosomes once more become thinner and longer by uncoiling and unfolding and can be seen to consist of single threads. Finally, as the chromosomes become indiscrete under the light microscope, the nucleus enters interphase (Figure 10-2A). The chromosome comes to contain two chromatids again when the DNA replicates during the S stage.

From this description we see how mitosis results in the exact distribution of replicated nuclear DNA chromosomes, so that daughter nuclei have the same chromosomal constitution at their time of formation as their parent did when it was formed. In other words, mitosis produces daughter nuclei which are nonrecombinants.

Mitotic division of the nucleus is usually followed by a cytoplasmic division, in which the cytoplasmic components are less equally divided, to produce two daughter cells.

## 10.3 Various features of mitosis are known to be genetically programmed.

Nuclei almost always divide mitotically. Since eukaryotic cells in a more-or-less constant environment are sometimes undergoing mitosis and sometimes not, the onset of mitosis must be genetically programmed. The genetic program for mitosis is apparently separate from that for DNA replication, since certain special nuclei routinely replicate their DNA's without subsequently undergoing mitosis, and other special nuclei undergo successive mitotic divisions without intervening DNA replications. The genetic program for mitosis is apparently also separate from that for cell division, since certain mitotic divisions in an organism are routinely followed by cell division while others are not. For example, cell division follows mitosis in most cells of a human being but not in striated muscle.

The total number of successive mitotic divisions that a somatic cell can undergo is also genetically programmed, although we do not yet know the genetic mechanism. A somatic cell of an embryo can undergo a larger number of mitotic divisions before it fails to divide than can a somatic cell of an adult (see Section 27.4). On the other hand, the number of mitotic (plus meiotic) divisions of cells that are in the reproductive or *germ line* is apparently unlimited. The genetic program that limits the number of somatic mitoses is apparently lost or nonfunctional in somatic cells that becomes cancerous, cancer cells being unlimited in their ability to divide mitotically.

Even the position of the mitotic spindle in the cell is genetically programmed. In the fertilized egg of the snail *Limnea*, for example, the spindle assumes one of two tilted positions relative to the yolk, which lies at the bottom of the cell. The direction of tilt is determined by a single locus. When the diploid mother (who synthesized the egg) carries one particular allele at this locus, the spindle tilts in one direction, and the snail shell subsequently synthesized is coiled right-handedly. When the mother does not carry this allele, the spindle tilts in the other direction, and the shell made is coiled left-handedly. (Note that the spindle orientation is determined by the genotpye of the mother who synthesized the egg and not by the genotype of the fertilized egg.)

## 10.4 Different eukaryotes have characteristic nuclear chromosomes.

The nuclear chromosomes of different kinds of eukaryotes are characteristic in number, length, and position of the centromere. Most chromosomes are rod-shaped and have a single centromere that is located subterminally—in the middle, off-center, or near one tip—and which therefore separates the chromosome into two *arms*. For each chromosome there is usually another chromosome very similar or identical in appearance located somewhere in the nucleus of a somatic cell. Thus, chromosomes in a somatic nucleus can be paired (Figures 10-3 and 10-4), the members of a pair appearing similar because they contain the same loci (although they may differ in the alleles occupying these loci). The members of a pair of chromosomes are therefore corresponding or *homologous chromosomes*. Somatic nuclei therefore contain two sets of chromosomes and are diploid. The paired or diploid number of chromosomes in a typical somatic cell of a human being is 46 or 23 pairs (Figures 10-3 and 10-4). The chromosome number of different eukaryotes is characteristic (but not necessarily different)—being 7 pairs in the garden pea, 10 pairs in Indian corn (maize), and 28 pairs in the domesticated silkworm.

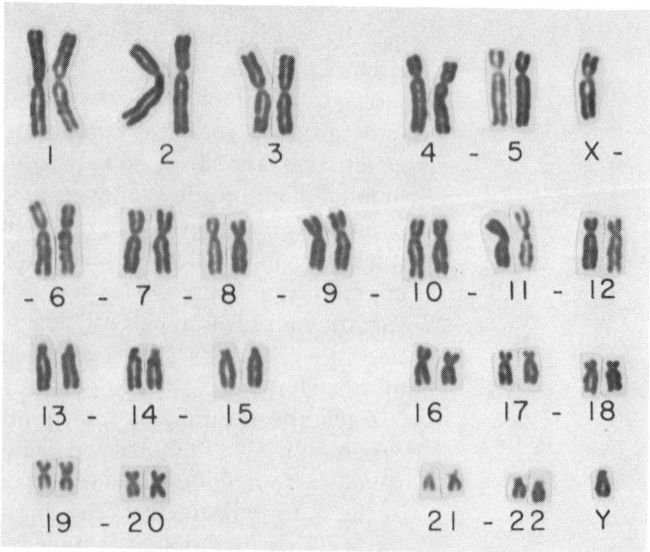

FIGURE 10-4. Metaphase chromosomes of the human male arranged in pairs. (Courtesy of K. Hirschhorn.)

## 10.5 A nuclear chromosome contains many repeats of the same or similar base sequences.

Only a relatively small fraction of the DNA of a prokaryote contains repeats of sequences more than several nucleotides long (including promoter, terminator, and rDNA sequences). In a nuclear chromosome, on the other hand, 10 to 40 per cent or more of the DNA is composed of sequences that are repeated exactly, most of these sequences being clustered on either side of the centromere. Other, nearly repetitive sequences, which can be as much as 15 to 80 per cent of the nuclear DNA's in different eukaryotes, are distributed along the arms of the chromosomes. The repetitiveness of sequences can be determined by the rapidity with which a given amount of fragmented, strand-separated DNA will re-form duplexes *in vitro*. For example, duplex formation will occur rapidly for completely repetitive DNA's, less rapidly for partially repetitive DNA's, and slowest for single-copy sequences. Thus, whereas nuclear DNA has all three rates of duplex formation, bacterial DNA has essentially only the slow component. The function of the completely repetitive DNA is completely unknown. The nearly repetitive sequences may be used to control the transcription of adjacent single-copy sequences (most of which code for proteins).

The completely repetitive sequences can be recognized under the light microscope at any one of three stages. (1) *In interphase*. The tangled mass of chromosomal fibers seen in interphase comprises *chromatin*. Most of the chromatin is relatively diffuse in appearance, being relatively uncoiled and unclumped, and is called *euchromatin*. Another portion is relatively dense or clumped, because it is supercoiled and folded, and is called *heterochromatin*. Most heterochromatin is composed of completely repetitive DNA, being heterochromatic because of its innate or intrinsic base composition. This is called *constitutive heterochromatin*. In some organisms, including the human female, some heterochromatin is due to the clumping of chromosome segments containing partially repeated and nonrepeated sequences, the clumping serving to prevent transcription (Chapter 16). Chromatin that has the facility or faculty to become heterochromatin according to the organism's needs is called *facultative heterochromatin*. (2) *In prophase*. During prophase, constitutive heterochromatin reaches the even more supercoiled and folded condition of metaphase earlier than euchromatin does. In this way one can identify the completely repetitive DNA that is located adjacent to the centromere. (3) *At metaphase*. Because metaphase chromosomes are so highly compacted, it is ordinarily impossible to distinguish heterochromatin from euchromatin at that stage (Figure 10-3). When metaphase chromosomes are treated in a particular manner before being stained, however, the stain is taken up preferentially by regions containing constitutive heterochromatin, producing the so-called *C bands* (Figure 10-5) that border the centromeres.

When the relative time of synthesis of DNA is determined by exposure of cells to radioactive raw materials of DNA at different times during the DNA synthesis (S) period of interphase, it is found as a general rule that

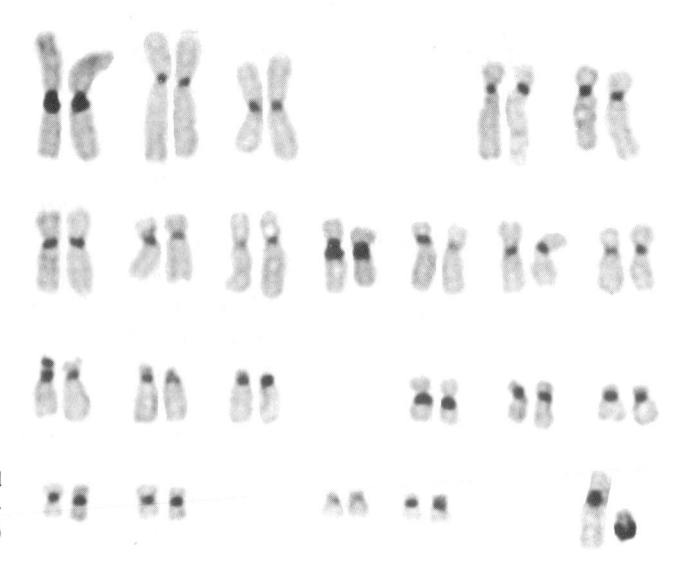

FIGURE 10-5. Human chromosomes stained to show the location of constitutive heterochromatin (C bands). (Courtesy of T. C. Hsu.)

heterochromatic regions replicate after euchromatic ones. We believe that the clumped condition inhibits or delays replication—perhaps by making the DNA template relatively unavailable to DNA polymerase and DNA raw materials.

Finally, one can identify the location of constitutive heterochromatin by base-pairing radioactive repetitive DNA with nonradioactive chromosomes that have been strand-separated, yet held in place on a slide (Section 2.9).

# MEIOSIS

Whatever the chromosome content is of a *fertilized egg*, or *zygote*, the same content is found in every cell descended from the zygote by mitotic cell divisions. If all nuclei divided by mitosis, however, a *sex cell* or *gamete* would contain the same number and kind of chromosomes as every other cell of a many-celled organism, all of them derived from the same zygote. Consequently, since the zygote results from the combination of two gametes, the number of chromosomes would increase in successive generations. Chromosome number does not increase from one generation to the next, however. This constancy is possible because gametes contain only one member of each pair of homologous chromosomes. For example, human gametes—eggs and sperms—normally contain not the diploid chromosome number—23 pairs (Figure 10-5)—but 23 chromosomes, which are *nonhomologous chromosomes*, in the unpaired or haploid condition. Since each gamete provides a haploid set of chromosomes, fertilization consequently restores the diploid chromosome constitution. Clearly, then, all cell divisions cannot be mitotic; certain cells must have a way of reducing the number of chromosomes from diploid to haploid. This reduction, which is a genetic recombination of duplex DNA's, is genetically programmed in the process called *meiosis*, which occurs only in cells of the germ line.

## 10.6 Meiosis uses the spindle in two successive divisions to reduce chromosome number.

Knowing mitosis, it is easy to invent a relatively simple mechanism to reduce chromosome number from diploid to haploid. Suppose that we were dealing with a diploid cell containing one pair of long homologous chromosomes and one pair of short homologous chromosomes (Figure 10-6A). We could have them strand-separate here and there so homologous chromosomes could undergo synapsis as in prokaryotes (Figure 10-6B). The interlocked chromosomes would then proceed to metaphase as pairs (Figure 10-6C). To complete an otherwise regular mitotic division, the members of each pair would be pulled by their centromeres to opposite poles of the spindle, producing two haploid telophase nuclei (Figure 10-6D).

Such an imaginary meiosis is impossible, however, due to a single biological complication. The centromeres of synapsed homologous chromosomes repel each other toward the end of prophase. As a consequence, synapsis between homologous chromosomes is destroyed, causing the homologous chromosomes in our model to separate and

FIGURE 10-6. A simple (but impossible) mechanism for meiosis. See the text.

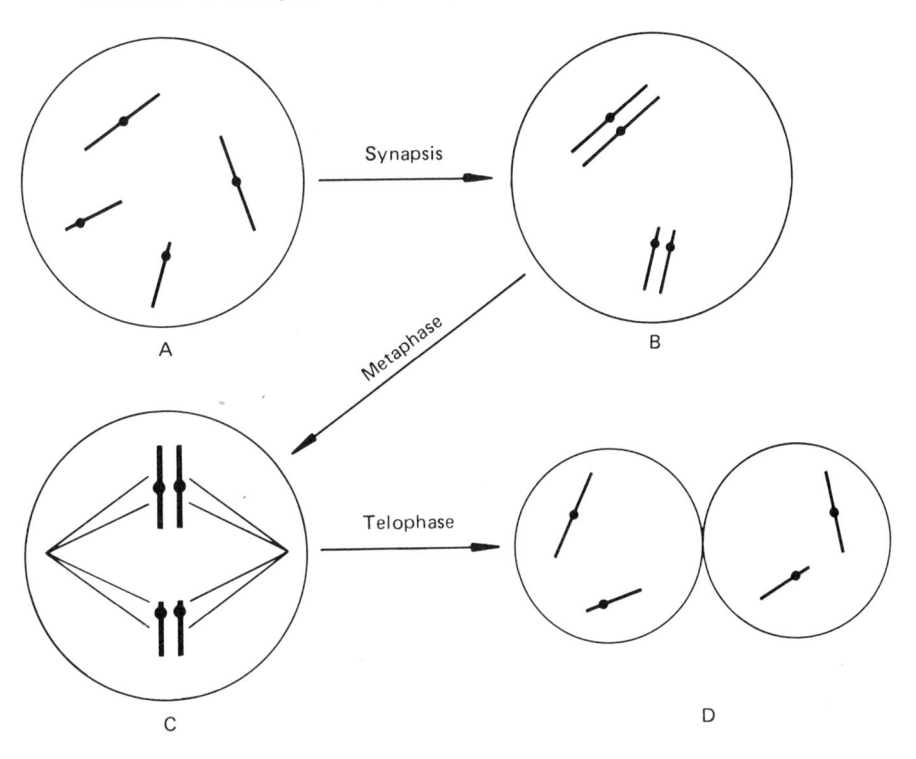

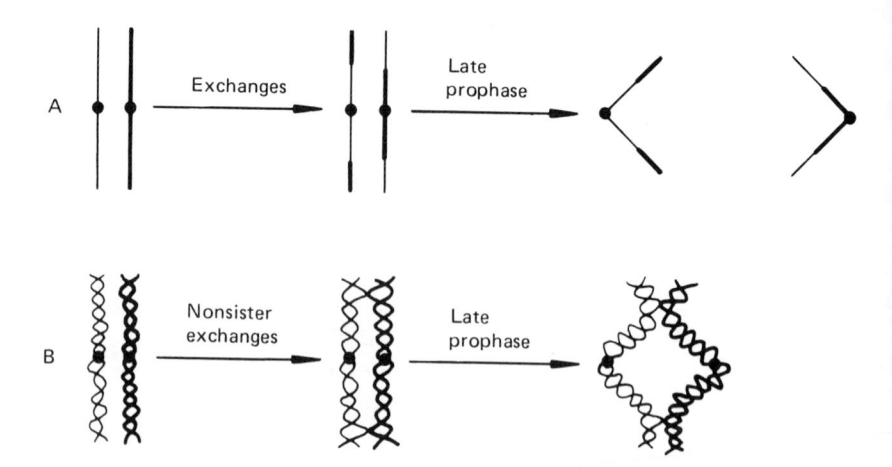

FIGURE 10-7. Exchanges that do not hold homologous chromosomes together until metaphase I (**A**) and those that do (**B**). Thick line represents DNA duplex of one chromosome; thin line, DNA duplex of homologous chromosome.

proceed to metaphase independently rather than as pairs. With all four chromosomes aligned and moving on the spindle independently, daughter nuclei could contain anywhere from one to four chromosomes. It is, moreover, no help to our simple scheme to break homologous chromosomes at corresponding loci and exchange corresponding segments, since this also would not hold the recombinant chromosomes together until metaphase (Figure 10-7A).

The most common biological solution to this dilemma has been to have a regular S stage precede meoisis, so each chromosome has two chromatids; and to permit breakage and exchange of corresponding segments between *nonsister chromatids*, that is, chromatids belonging to different members of a pair of homologous chromosomes (Figure 10-7B). This biological solution of meiosis is diagrammed in Figure 10-8. The two pairs of homologous chromosomes (A) replicate their DNA's to produce chromosomes with two chromatids (B), synapsis occurs (C), followed by breakage recombination between two nonsister chromatids (D). Since *sister chromatids*, that is, chromatids belonging to the same chromosome, are intertwined, and the repulsion between homologous centromeres is insufficient to completely unravel all intertwining, homologous chromosomes are held together and proceed as a pair to metaphase I (Figures 10-7B and 10-8). Telophase I produces two haploid nuclei (F) in which each chromosome contains two chromatids. Accordingly, each daughter nucleus undergoes a second meiotic division. This is essentially a mitotic division, and produces four haploid nuclei each of whose chromosomes contains but a single DNA duplex (G), as occurs after mitosis.

We see from the preceding that homologous chromosomes will go to metaphase I as a pair if they are held together by exchanges between two nonsister chromatids (but not by the other two nonsister chromatids at the same locus). The reduction from the diploid to the haploid chromosome number is accomplished in a single nuclear division (*meiosis*

GENETIC RECOMBINATION IN EUKARYOTES 147

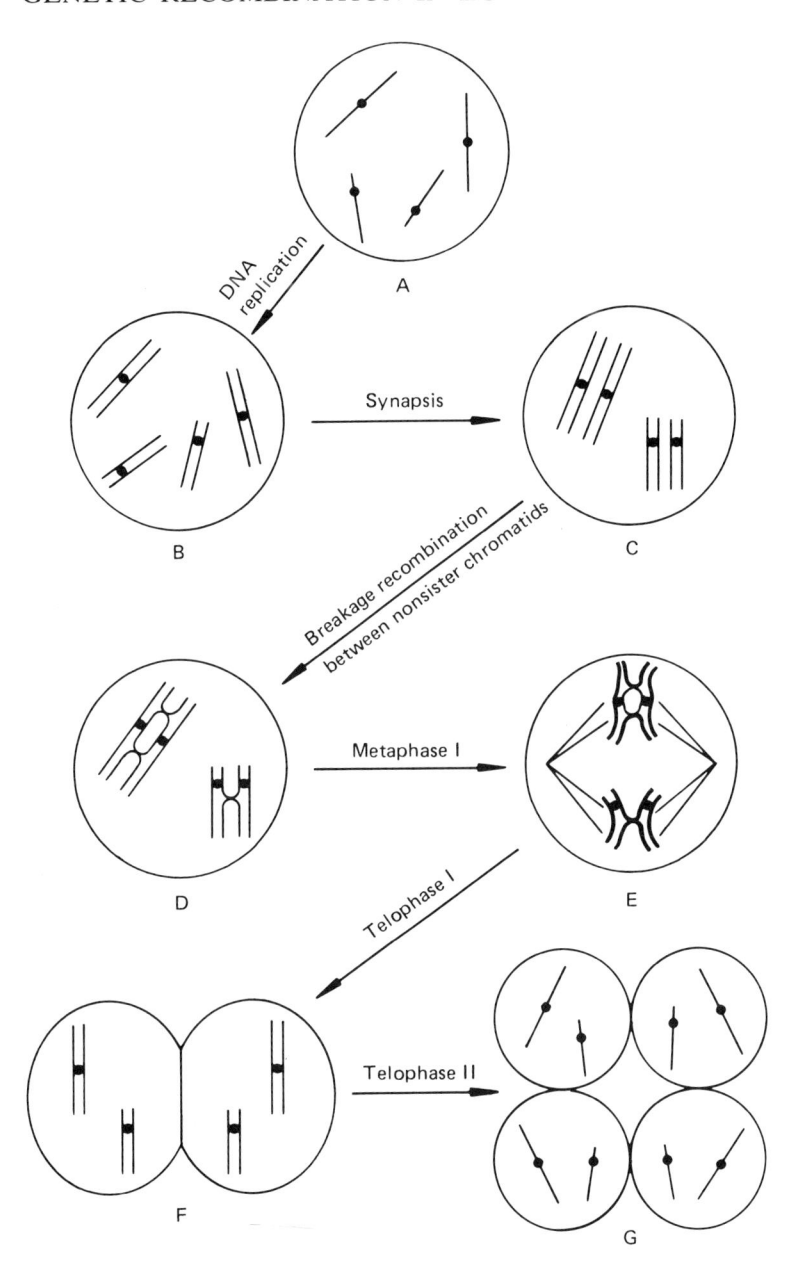

FIGURE 10-8. Meiosis, starting with two pairs of chromosomes. See the text.

*I*). To reduce the number of chromatids per chromosome, however, as occurs in mitosis, a second nuclear division (*meiosis II*) is required. Meiosis II differs from mitosis in two respects: (1) It is preceded by a short interphase (interphase I) having no S period. (S is not needed since each chromosome entering interphase I already contains two chromatids.) (2) The two chromatids in any chromosome are not sisters throughout; that

148 HOW GENETIC MATERIAL IS DISTRIBUTED

is, some chromatids are composed of alternating pieces of nonsisters, and are, therefore, in this respect molecular recombinants. We see, therefore, that to program the recombination *between* chromosomes that reduces chromosome number from diploid to haploid, the meiotic process also programs recombination *within* the chromosome.

## 10.7 Meiosis usually has special features in the female sex.

The general relation of the two meiotic divisions to each other and to the mitotic divisions that precede them is diagrammed in Figure 10-9. Note that prophase of the first meiotic division, *prophase I*, lasts longer than mitotic prophase. The meiotic process often differs in detail in the two different sexes. In the human male, for example, meiosis is as expected. For each cell entering meiosis, there are four meiotic cell products. Each of these cells is a haploid spermatid, which through various changes and without further division becomes a haploid *sperm cell*. Meiosis in the human male occurs only after puberty. In the human female, however, meiosis starts before birth, in the fetus. At the time of birth all the potential female sex cells have entered the first meiotic division but have stopped at late prophase I (Figure 10-9). These cells are called *oocytes*, and it is at this stage that they enlarge through the accumulation of nutrients as well as gene products. Upon sexual maturity, one oocyte a month ordinarily continues meiosis; after the first meiotic division (meiosis I) the cytoplasm divides unevenly to produce one small cell—a *pole cell*—and a large oocyte. The pole cell may or may not undergo meiosis II to produce two haploid pole cells. The oocyte enters meiosis II and proceeds until about metaphase II, where it stops until it is penetrated by a sperm. After a sperm penetrates, the developing oocyte

FIGURE 10-9. Stages of meiosis (heavy lines) and their relation to the preceding mitotic cell cycle (light lines).

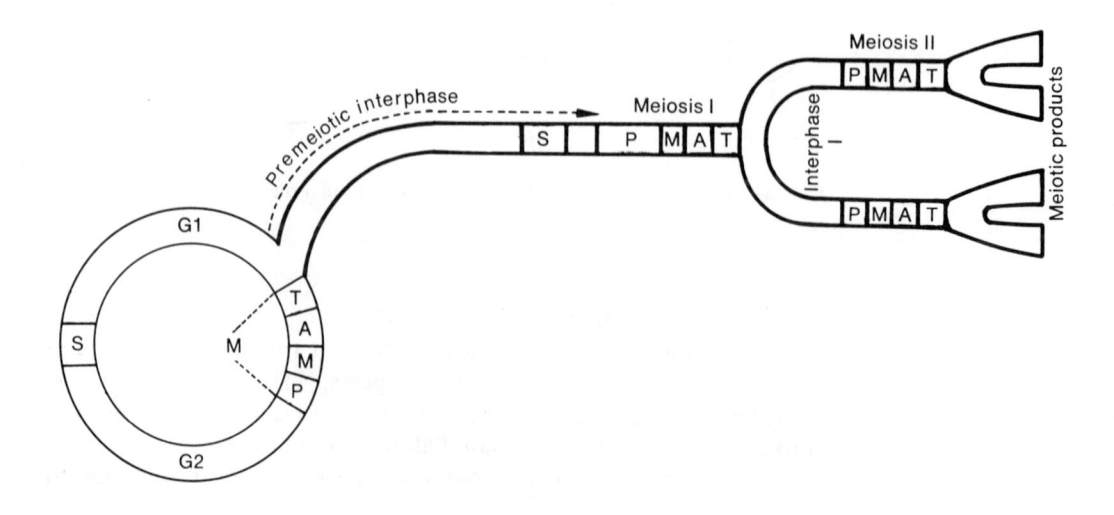

completes meiosis II, and the cell divides into a small cell—a pole cell—and the mature haploid *egg*, in which the haploid sperm nucleus has been waiting. The two gametic nuclei then fuse to produce the single, diploid nucleus of a fertilized egg. In the human female, therefore, each cell starting meiosis can produce only one functional mature egg (besides two or three small pole cells that are ordinarily discarded).

In other organisms, meiosis in the female may stop at other stages until fertilization occurs (for example, at metaphase I in *Drosophila*); and the oocyte at prophase I may grow more or less in a longer or shorter time. Other modifications of meiosis may occur in the male. In *Drosophila* males, for example, homologous chromosomes are held together in prophase I not by means of nonsister exchanges (as they are in the *Drosophila* female) but by special localized regions of stickiness.

**10.8** **Meiosis causes the segregation of the members of a pair of homologous chromosomes and, therefore, the segregation of the paired genes they contain.**

Since meiosis starts with pairs of homologous chromosomes and ends with single chromosomes, meiosis causes the separation of the members of a pair of homologous chromosomes, or *chromosome segregation*, so that a gamete contains only one member of any pair. It follows, therefore, that homologous segments of a pair of homologous chromosomes must also segregate; and that the paired loci these segments contain also segregate. In other words, somatic cell nuclei contain pairs of genes which in germ cell nuclei undergo *gene segregation* during meiosis. Thus, if the alleles of a gene pair are different, a single haploid meiotic cell will contain either one allele or the other.

For example, when a diploid of *A a* genotype undergoes meiosis, the chance is 50 per cent that its haploid gamete will contain *A* and 50 per cent that it will contain *a*. Consequently, we will obtain approximately equal numbers of both kinds of gamete if the sample of gametes is sufficiently large. Such an *A a* parent (or cell) that contains two different alleles for a locus is said to be *heterozygous* or *hybrid* in this respect; an *A A* or *a a* parent (or cell) in which both alleles are the same and which produces only one kind of gamete is said to be *homozygous* or *pure* in this respect. A heterozygous individual is called a *heterozygote* or hybrid; a homozygous individual is called a *homozygote*.

If a heterozygous parent, *A a*, is mated with a homozygous parent, *a a*, half of the resulting zygotes will be *A a* and half will be *a a*, since the homozygous parent produces only *a* gametes. If, however, both parents are *A a* heterozygotes (Figure 10-10), random fertilizations between the

FIGURE 10-10. Zygotes produced from a cross between identical hybrids.

| Parents | *A a* X *A a* |
|---|---|
| Gametes | ½ *A*, ½ *a*    ½ *A*, ½ *a* |
| Zygotes | ¼ *A A*, ½ *A a*, ¼ *a a* |

$A$ and $a$ gametes of each parent will result in zygotes of three different diploid genotypes: $A A$, $A a$, and $a a$ in the relative proportion $1:2:1$.

Many human traits are determined by single pairs of segregating genes. Such traits include *albinism*, a lack of melanin pigment; *polydactyly*, the occurrence of extra digits; *sickle cell anemia*; and blood types such as *MN*, *ABO*, and *Rh*. All of these gene-specified traits will be discussed in later chapters. Since any gene can be chemically modified in a large number of ways, nuclear genes, too, can have many alleles.

**10.9** Since pairs of nonhomologous chromosomes segregate independently, so do their genes.

Each pair of chromosomes that aligns on the spindle at metaphase I is equally likely to face the poles in one of two different ways. This is illustrated diagrammatically in Figure 10-11. One pair of homologous chromosomes is marked differently by alleles $A$ and $a$; a second pair of

FIGURE 10-11. Different pairs of homologous chromosomes are equally likely to align at metaphase I in positions I and II. $A$, $a$ = alleles marking one pair of chromosomes; $B$, $b$ = alleles marking the other pair of chromosomes; • = centromere.

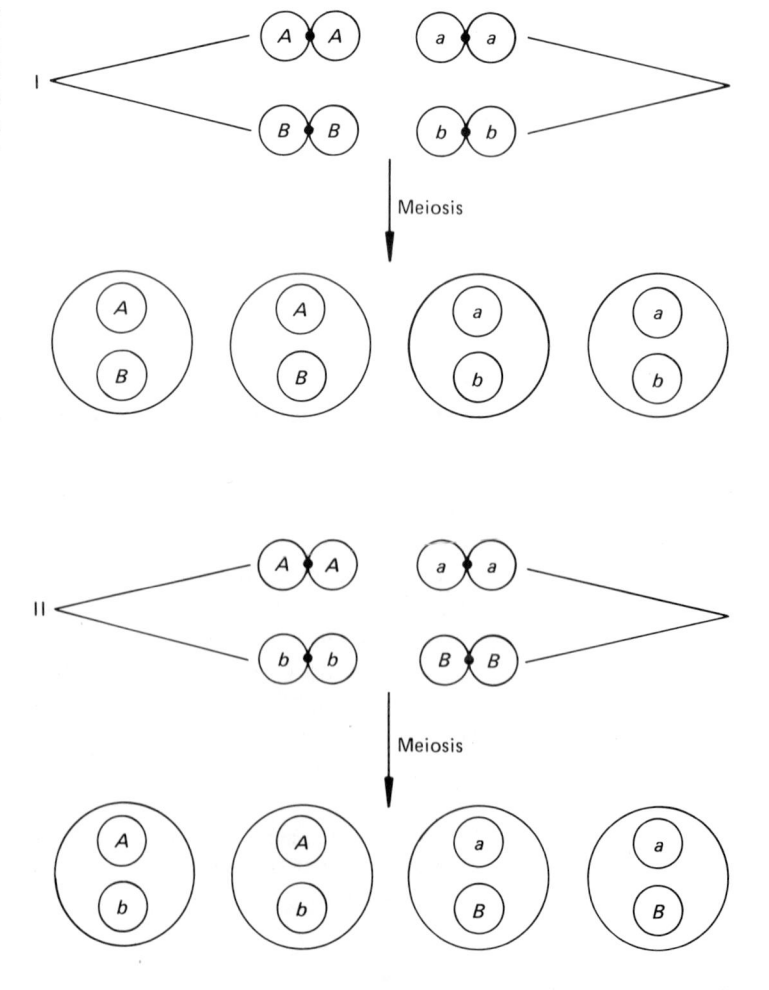

homologous chromosomes is marked differently by alleles $B$ and $b$. Since the poles of the spindle are identical, the two pairs of homologous chromosomes align at metaphase I half the time as in alternative I, where the two chromosomes containing $A$ and $B$ (and $a$ and $b$) face the same pole, and half the time as in alternative II, where the chromosomes containing $A$ and $b$ (and $a$ and $B$) face the same pole. Either alternative results in the segregation of $A$ from $a$ and $B$ from $b$; the alignment of different pairs of chromosomes at random to each other results in the segregation of $A\,a$ independently of $B\,b$. In other words, genes on different pairs of chromosomes segregate independently.

The result of *independent segregation* in our example ($A\,a\,B\,b$) is that gametes of four different haploid genotypes, $A\,B$, $A\,b$, $a\,B$, $a\,b$, have an equal likelihood of occurrence, so that if a large enough example is examined, each genotype is expected in about 25 per cent of the gametes. Any hybrid for two pairs of independently segregating genes, therefore, produces four genotypic kinds of gametes in a 1:1:1:1 ratio. From the viewpoint of the genotypes of gametes, two of the four types of gametes produced by such an individual must be the same as the gametes that formed the individual—these are the parental combinations—and the other two must be new combinations, genetic recombinants. Independent segregation in a hybrid for two pairs of genes produces equal numbers of old and new genetic combinations. To identify the old combinations in cases of independent segregation, one must know the genotypes of the parental gametes. If in Figure 10-11, the diploid undergoing meiosis had $A\,B$ for one parental gamete (say, from a pure $A\,A\,B\,B$ parent) and $a\,b$ for the other (say, from a pure $a\,a\,b\,b$ parent), the $A\,B$ and $a\,b$ gametic genotypes it produces would be parental combinations and $a\,B$ and $A\,b$ would be the recombinations. If, contrariwise, the diploid undergoing meiosis in the figure was the product of fertilization of gametes containing $a\,B$ and $A\,b$ genotypes, these genotypes in its gametes would be the parental types, and $A\,B$ and $a\,b$ would be the recombinant types.

## 10.10 Various features of meiosis are known to be genetically programmed.

The meiotic process is genetically programmed. This program is at least partially independent of DNA replication, which occurs before prophase I and does not occur in interphase I. It is also independent of cell division, which in the case of human beings always follows meiosis I and II in the male but does not always follow meiosis II if this occurs in a pole cell of the female. We consider next specific features of meiosis that are genetically programmed.

SYNAPSIS AND EXCHANGE. Synapsis between homologous chromosomes may depend on the presence of a specific allele. When this allele is absent in maize, for example, synapsis is prevented or destroyed between all homologous loci. In the absence of synapsis, no exchange occurs between the homologous chromosomes, and chromosome distribution is irregular during anaphase I. It is generally true that the sex that contains both X and Y chromosomes (see Section 11.1) is genetically programmed to have a reduced number or no exchanges. In *Drosophila*, for example, meiosis in the female (XX) is typical—it has synapsis and exchanges. In the male (XY), however, as noted in Section 10.7, homologous chromosomes stick to each other only in a localized heterochromatic region near the centromere during prophase I; although this arrangement serves to get pairs of homologous chromosomes to metaphase I, the lack of synapsis in euchromatic regions prevents exchange. (Since there is no exchange in the *Drosophila* male, all genes in a chromosome keep the original sequence in all sperm.)

CENTROMERE BEHAVIOR. The specific behavior of centromeres during meiosis is genetically programmed. At metaphase II, sister centromeres normally lie so close together as to appear fused, so that when one of them faces one pole of the spindle, the other one automatically faces the opposite pole (Figure 10-12A). In the presence of a particular allele in *Drosophila*, however, sister centromeres at metaphase II separate early, so that each is able to orient to the poles independently. As a consequence, both sister centromeres sometimes orient to the same pole (Figure 10-12B), and hence are not distributed to daughter nuclei properly. This allele has no effect on mitosis or meiosis I.

SPINDLE SHAPE. Abnormal alleles are also known in maize and *Drosophila* that change the shape of the spindle during meiosis but not mitosis. The ends of normal spindles converge (Figure 10-13A), so that all the chromosomes arriving at a pole form a single group and are readily included in a telophase nucleus. An abnormal allele causes the meiotic spindle to have divergent ends (Figure 10-13B). As a result, the

FIGURE 10-12. Arrangement of sister centromeres on a spindle at metaphase. **A**: Normal orientation. **B**: Abnormal orientation made possible by the precocious separation of sister centromeres. **A** leads to normal separation, **B** to abnormal separation, at anaphase.

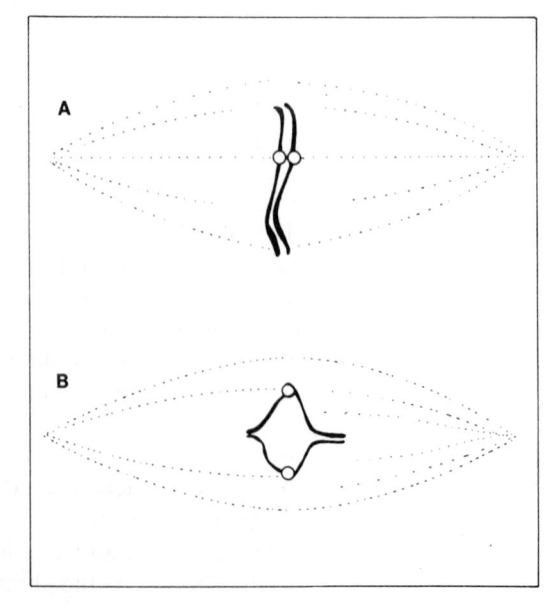

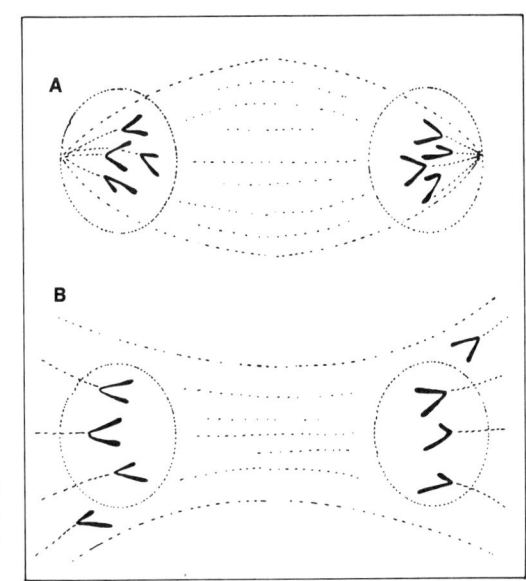

FIGURE 10-13. **A**: Normal spindle with convergent ends. **B**: Abnormal spindle with divergent ends. Dotted ovals indicate which chromosomes are likely to be included in the same telophase nucleus. Chromosomes left out in **B** are degraded.

chromosomes are so spread out at the end of anaphase that some are left out of the telophase nuclei.

SPINDLE ORIENTATION. Both spindles in meiosis II are usually oriented in the same direction. When this direction is the same as that of meiosis I, the result is a row of four nuclei or cells (Figure 10-14A). When this direction is opposite to that of meiosis I, however, the result is a cluster of four cells or nuclei (Figure 10-14B). Different species or sexes orient the spindle in meiosis II in these two different ways. Although this spindle orientation is doubtless programmed genetically, we know none of the genetic details.

## SUMMARY AND CONCLUSIONS

The orderly separation of replicated nuclear DNA is accomplished by mitosis, a genetically programmed mechanism that uses a spindle to produce two daughter nuclei that are chromosomally (and thus genetically) identical to the parent nucleus at the same stage. Mitosis, therefore, is a mechanism of nuclear division that prevents recombination between nuclear generations.

Nuclear chromosomes are species-characteristic in number and form. They contain highly repetitive DNA sequences that are usually found on either side of the centromere and that can be identified cytologically because they are constitutively heterochromatic; they also contain nearly repetitive and single-copy DNA's that can be identified cytologically because they are euchromatic.

In eukaryotic organisms that reproduce sexually, the problem of maintaining a constant number and kind of nuclear chromosomes is solved by meiosis. This genetically programmed process reduces the diploid chromosome number of the zygote to the haploid number of the

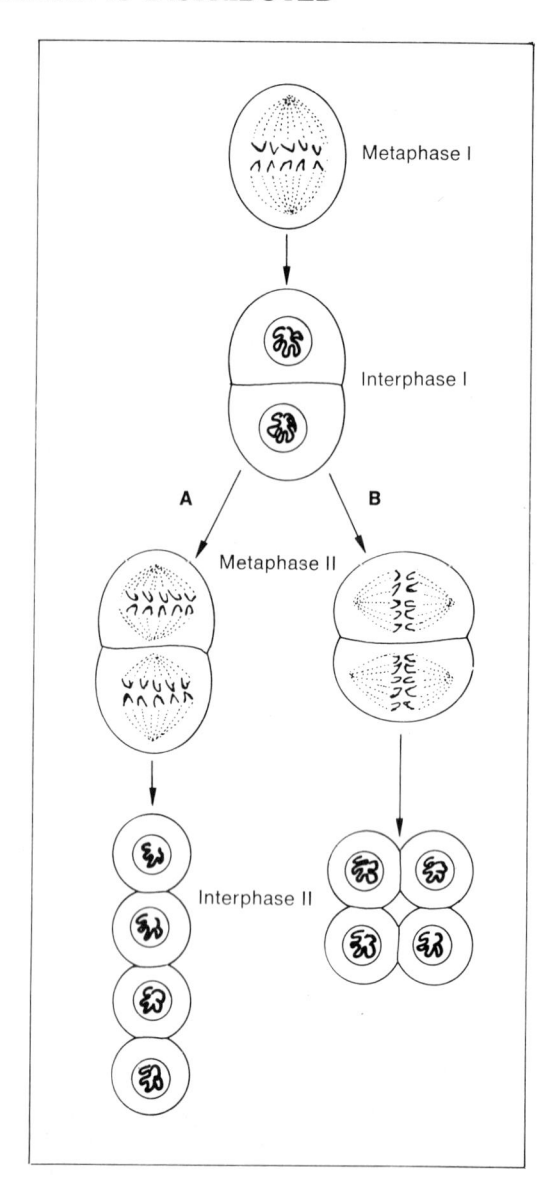

FIGURE 10-14. Effect of spindle orientation during meiosis. **A**: Metaphase II orientation is the same as metaphase I orientation. **B**: Metaphase II orientation is opposite to metaphase I orientation.

gamete by the following means: replicated homologous chromosomes that synapse and are held together by exchanges between nonsister chromatids undergo two successive spindle-using divisions without replicating again. Meiosis, therefore, programs nonbreakage as well as breakage recombination between homologous chromosomes.

The result of the two meiotic divisions is chromosome (and gene) segregation—only one member of each pair of homologous chromosomes (or genes) is present in a gametic nucleus. Since the centromeres of pairs of homologous chromosomes (and of single chromosomes) face either pole of a metaphase spindle at random, the segregation of one pair of homologous chromosomes (and of the gene

# GENETIC RECOMBINATION IN EUKARYOTES   155

pairs it contains) occurs independently of the segregation of other pairs of homologous chromosomes (and of the gene pairs they contain). When a hybrid for two pairs of genes on different pairs of chromosomes undergoes meiosis, independent segregation is expected to produce, among a large number of gametes, four genotypes in equal frequency, two of them parental gametic types, and two recombinational.

The genetic programming of mitosis and meiosis is revealed by alleles that change the program and result in mitotic or meiotic abnormalities.

## QUESTIONS AND PROBLEMS

**10.1** *Anagrams*: Unscramble and define each of the following: **a.** isomist. **b.** omissie. **c.** charmtoid. **d.** slednip. **e.** poperash. **f.** thinocram.

**10.2** *Riddle*: What has arms that are not used for fighting, waving, or embracing?

**10.3** *Riddle*: Barring mutations, what sisters are always genetically identical?

**10.4** *Riddle*: When is euchromatin not euchromatin?

**10.5** *Songs*: **a.** "There Will Never Ever Be Another You."  **b.** "Brand New Me." Do you agree? Explain.

**10.6** *Song*: "Come Together." Give three different examples described in this book.

**10.7** *Song*: "A Little Bit Me, a Little Bit You." What genetic application has this song?

**10.8** What evidence have you that nuclear DNA is genetic material?

**10.9** If the maximum distance between the two poles of a spindle is 20 micrometers, how long can the longest chromosome be when maximally coiled and supercoiled?

**10.10** In what respect does the mitotic prophase nucleus differ from a daughter telophase nucleus?

**10.11** Distinguish between cells of the somatic and germ lines.

**10.12** State one evidence that normal mitosis is under genetic control.

**10.13** How are mitosis and meiosis similar? Different?

**10.14** In what male human cells are there only nonhomologous chromosomes?

**10.15** How could you tell whether a human cell was at mitotic metaphase, metaphase I, or metaphase II?

**10.16** Are all interphases in the same diploid organism identical? Explain.

**10.17** Does the human female ever produce a metaphase that has, relative to another metaphase, half the amount of DNA: **a.** per chromosome? **b.** in toto?    Explain.

**10.18** Distinguish between segregation and independent segregation.

**10.19** In a given human being, is the spindle at mitotic metaphase identical to that at metaphase II? Why?

**10.20** If all gene pairs are segregating independently, how many types of gametes are possible when the number of heterozygous gene pairs is: **a.** 1?  **b.** 2?  **c.** 3?  **d.** 4?  **e.** $n$?

**10.21** If all gene pairs are on different chromosome pairs, how many different gametes are possible for the following diploid genotypes? **a.** $R r\ S s$. **b.** $M m\ N n\ o o\ P p$. **c.** $A a\ B b\ c c\ D D\ E e\ F f$.

**10.22** What kind of evidence do we have that normal meiosis is genetically controlled?

**10.23** Which one of the following statements about mitosis is false? **a.** It produces two chromosomally identical daughter nuclei.  **b.** It does not include interphase. **c.** It is preceded by an interphase containing an S period. **d.** It always uses a spindle. **e.** It occurs only in bacteria.

**10.24** Which one of the following statements about nuclear chromosomes in human beings is false? **a.** Except for ring chromosomes, all have two ends. **b.** Before they replicate they usually contain one DNA duplex. **c.** Their DNA doubles during the S stage. **d.** They go to mitotic metaphase as pairs. **e.** Each contains two chromatids during prophase.

**10.25** Which of the following is in a correct sequence for a cell cycle? **a.** Prophase, metaphase, anaphase, telophase, mitosis, G1, S, G2. **b.** Telophase, interphase, prophase, metaphase, anaphase. **c.** G1, G2, S, mitosis. **d.** G1, S, G2, interphase, mitosis. **e.** G1, S, mitosis, G2, interphase.

**10.26** What process is diagrammed? Draw a circle and an arrowhead and place the letters in their correct sequence.

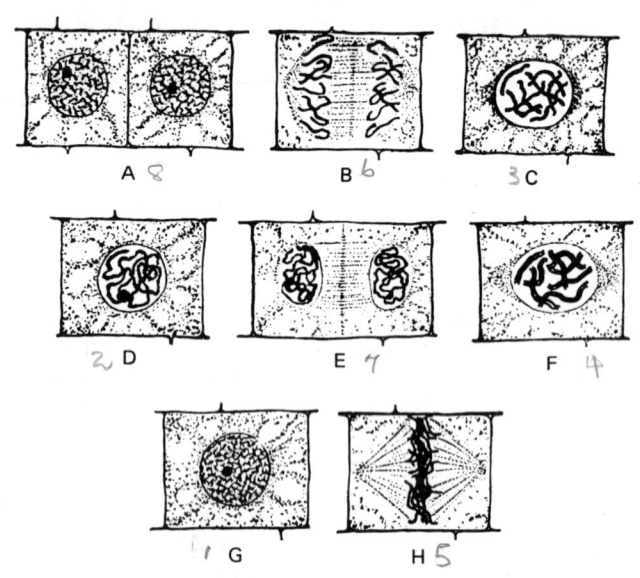

**10.27** Examination of the female reproductive tract in the roundworm *Ascaris* revealed the stages shown below. List the correct order of these stages.

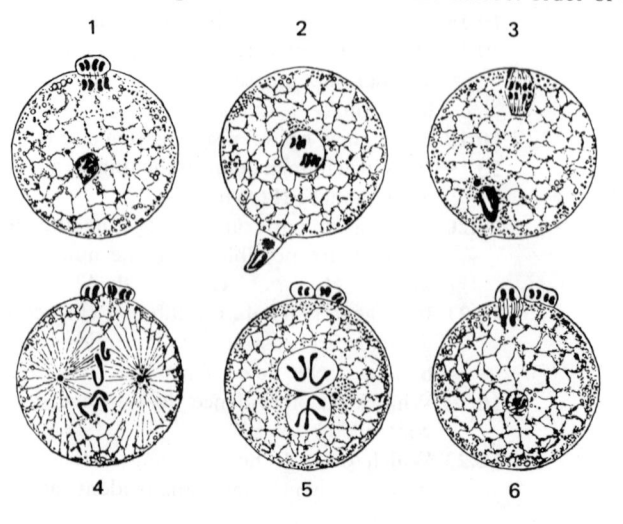

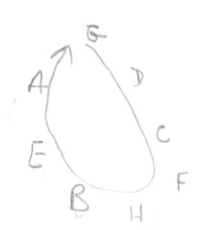

# GENETIC RECOMBINATION IN EUKARYOTES

**10.28 a.** What processes are shown in the diagrams below? **b.** Considering diagram 5, how many arrangements are possible in diagram 6? **c.** From what is shown in 6, in what respect is 7 inaccurate?

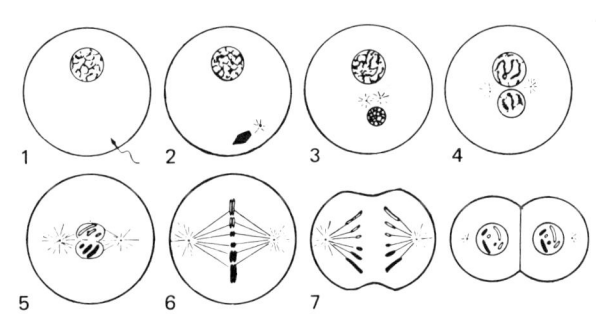

**10.29** What process is diagrammed? In what respects is the series of diagrams incorrect? How many different combinations of maternally derived and paternally derived centromeres are possible in the haploid eggs of this species?

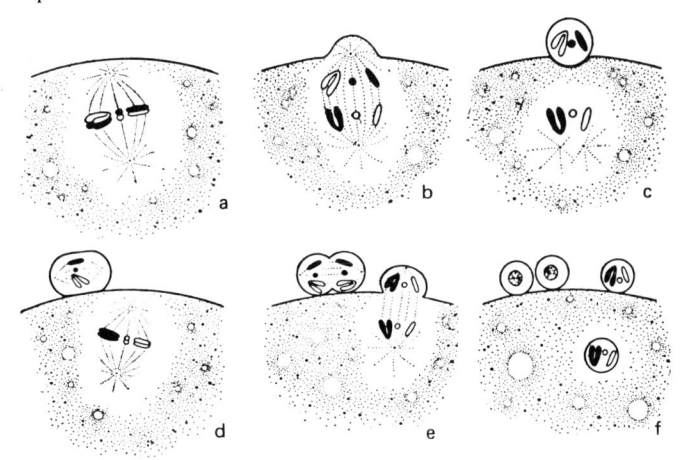

*Chapter 11*

# Genetic Recombination in Eukaryotes: II. Sex Linkage and Crossing Over

GENES LOCATED in different pairs of homologous chromosomes are *nonhomologous genes* located in nonhomologous chromosomes. We found in the last chapter that nonhomologous nuclear genes segregate independently. (All the gene pairs in a pair of homologous chromosomes are homologous gene pairs.) All the genes located in one homologous chromosome are physically linked to each other, as are all the genes in the other homologous chromosome of the pair. This chapter describes the linkage of genes to particular chromosomes (*sex linkage*) and the breakup of linkage due to exchanges between nonsister chromatids during meiosis (*crossing over*).

## SEX LINKAGE

**11.1** In many eukaryotes, different sexes have different sex chromosome compositions.

Different eukaryotes genetically program different sexes in different ways. The main different ways are discussed in Chapter 19. Of present interest are those eukaryotes whose different sexes have different compositions with regard to certain homologous chromosomes. In human beings, for example, of the 23 pairs of chromosomes seen at mitotic metaphase, 22 pairs are the same in both males and females, and are called *autosomes*. Since the remaining pair differs in males and females, it can be used to distinguish between sexes; its members are called *sex chromosomes*. The human female has two sex chromosomes, called *X chromosomes*, that look identical under the microscope. The human male has one X chromosome and a *Y chromosome* that has a

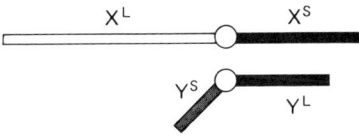

FIGURE 11-1. Diagrammatic representation of the human X and Y chromosomes showing their unique (differently shaded) and homologous (same shaded) regions.

different and unique appearance. The XX = female (♀),* XY = male (♂) sex chromosomes situation is true for all mammals and many other animals.

In certain animals, however, the sex chromosome situation is reversed. For example, in birds, some reptiles and amphibians, as well as in moths and butterflies, the male is XX and the female is XY. Regardless of the sex involved, all gametes of XX individuals carry an X, whereas half the gametes of an XY individual carry an X and half carry a Y. Random fertilization of such gametes produces equal numbers of XX and XY zygotes that have the genetic potential to develop into equal numbers of males and females.

**11.2 The X and Y chromosomes have loci in common as well as unique loci.**

During meiosis in the sex type containing X and Y chromosomes, segments of the X and Y that contain homologous loci undergo synapsis. In the case of human beings, the shorter arm of the X chromosome ($X^S$) seems to carry many of the loci present in the longer arm of the Y chromosome ($Y^L$) (Figure 11-1). It is the loci held in common that make X and Y homologous chromosomes. The other parts of the sex chromosomes are unique, and exist in only one copy in the XY individual. In human beings, the long arm of the X chromosome ($X^L$) is unique and contains many loci not found in the Y; such loci are said to be *X-limited*. In the Y chromosome, the smallest human chromosome, the loci in the short arm, $Y^S$, are unique—these being primarily concerned with determining maleness; such loci are said to be *Y-limited*. Thus, the loci held in common between X and Y chromosomes make them homologous chromosomes, whereas the X-limited and Y-limited loci distinguish X from Y genetically and cytologically (under the microscope).

**11.3 Albinism is usually due to an autosomal gene; red-green colorblindness to an X-limited gene.**

Consider how autosomal and X-limited loci in human beings are distributed to future generations relative to sex type.

ALBINISM. The lack of melanin pigment that characterizes *albinism* is due to the absence of an autosomal allele, *A*, that codes for a specific

---

* The female symbol, ♀, is the symbol of Venus and resembles a looking glass; the male symbol, ♂, is the symbol of Mars and has an arrow.

protein. Thus, whereas $AA$ and $Aa$ individuals are normally pigmented, $aa$ individuals are albinos. Since the "$A$" locus is autosomal, it undergoes segregation independently of the sex chromosomes. So, for example, when two normally pigmented heterozygous persons are the parents, or $P_1$ (Figure 11-2), their gametes ($G_1$) occur in four equally frequent genotypes in the father and in two equally frequent types in the mother. The figure develops these gametic types by the following reasoning. In the father, $\frac{1}{2}$ of the sperm will get $A$ and $\frac{1}{2}$ will get $a$, as a result of the segregation of $Aa$; of the $\frac{1}{2}$ of the sperm that get $A$, $\frac{1}{2}$ will also get an X chromosome and the other half will get a Y chromosome, as a result of the independent segregation of the XY chromosome pair; of the $\frac{1}{2}$ of the sperm that get $a$, $\frac{1}{2}$ will also get an X chromosome and the other $\frac{1}{2}$ will get a Y chromosome, as a result of the independent segregation of XY. The result is four different equally frequent types of sperm (since segregation is occurring for two nonhomologous hybrid pairs—one genic, the other chromosomal).

In the mother, $\frac{1}{2}$ of the eggs will get $A$ and $\frac{1}{2}$ will get $a$, as a result of the segregation of $Aa$; each of these will also get an X, owing to the independent segregation of XX; as a result, eggs occur in two equally frequent genotypes.

FIGURE 11-2. Autosomal linkage. The expected genotypic and phenotypic results of a cross between two heterozygotes for albinism in human beings.

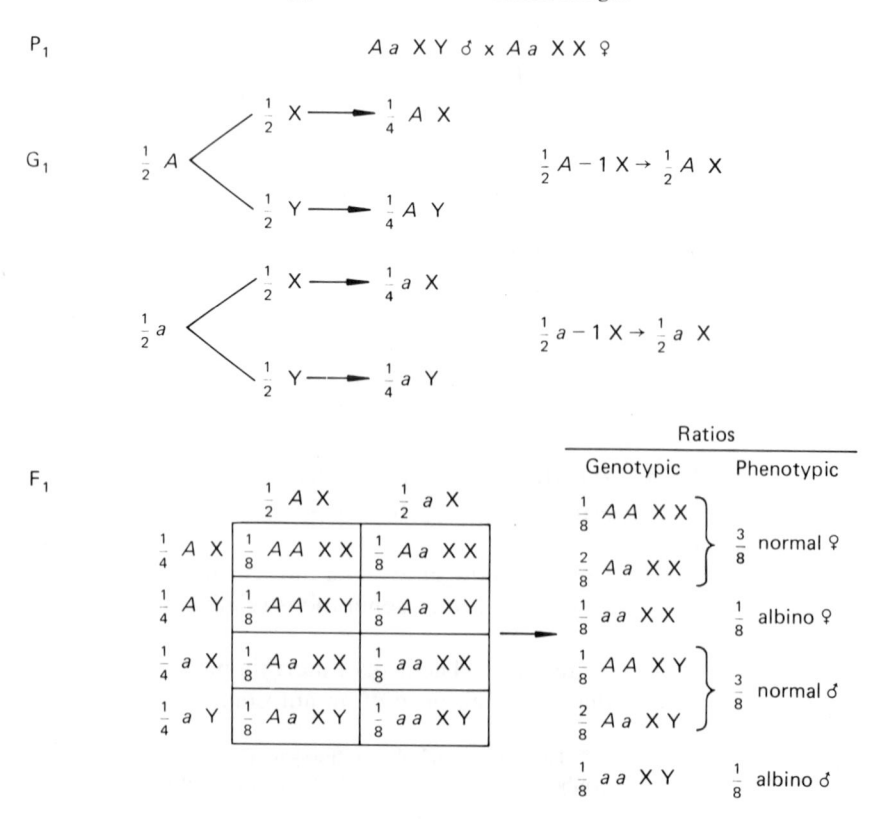

Since eggs and sperms ordinarily combine with each other at random, without regard to the genotypes under consideration, we can combine them at random, and obtain the genotypes expected in the progeny, or $F_1$, by using the checkerboard method. In this method, the types and frequencies of male and female gametes are placed along two sides of a grid or checkerboard (as shown for the $F_1$ in Figure 11-2); the grid positions are filled in by the genotype of the zygote resulting from the union of the male and female gametes in that row and column. The frequency of fertilization shown in each box is the product of the frequencies of the parent gametes. When like classes are summed, we find the genotypic results with regard to albinism among daughters (XX's) and sons (XY's). Specifically, the genotypic expectation is $1:2:1$ as $AA:Aa:aa$ among both daughters and sons; the phenotypic expectation is $3:1$ as normal:albino among both daughters and sons.

Had the parents been $Aa$ and $aa$, the expected genotypic ratio in the $F_1$ would be $1\ Aa:1\ aa$ and the expected phenotypic ratio 1 normal:1 albino; these ratios would apply to both daughters and sons. We see, therefore, that any genotypic or phenotypic ratio for segregating autosomal genes will be independent of the sex of the progeny.

RED-GREEN COLORBLINDNESS. Most persons who cannot distinguish between red and green are red-green colorblind for genetic reasons. The defect is due to the absence of allele $C$, which apparently codes for a protein needed to distinguish between the two colors. Since the locus for $C$ is X-limited, females have two such loci in their two X's and males one such locus in their single X. (Notice that sex chromosomes contain loci that have no phenotypic effect on sexuality.) Colorblind females are, therefore, $X^c X^c$ and colorblind males $X^c Y$. Normal females have one of two genotypes: $X^C X^C$ or $X^C X^c$.

Consider the results expected from marriages between colorblind women and normal men (Figure 11-3). The expected genotypic ratio in the $F_1$ is $1 X^C X^c : 1 X^c Y$; the expected phenotypic ratio in the $F_1$ is, therefore, 1 normal daughter:1 colorblind son. Note that the genotypic and phenotypic expectations with regard to the colorblind locus are different for the two sexes, contrary to the situation for an autosomal locus. In the present case, all daughters will be normal because each

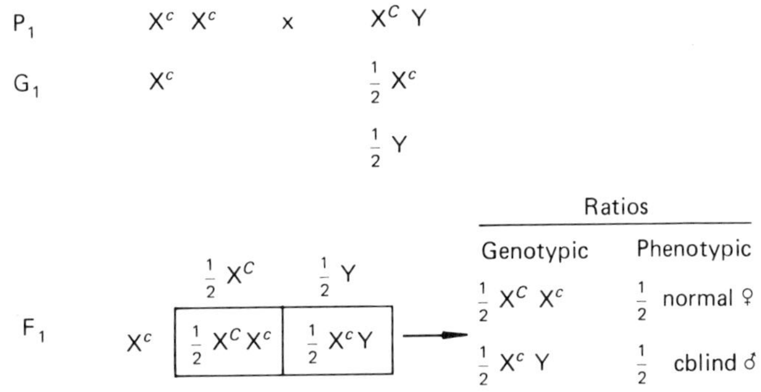

FIGURE 11-3. X-linkage. The expected genotypic and phenotypic results of a cross between a red-green colorblind woman and a normal man.

$P_1$ : $X^c X^c$ × $X^C Y$

$G_1$ : $X^c$ ; $\frac{1}{2} X^C$, $\frac{1}{2} Y$

$F_1$:

|  | $\frac{1}{2} X^C$ | $\frac{1}{2} Y$ |
|---|---|---|
| $X^c$ | $\frac{1}{2} X^C X^c$ | $\frac{1}{2} X^c Y$ |

Ratios

Genotypic: $\frac{1}{2} X^C X^c$, $\frac{1}{2} X^c Y$

Phenotypic: $\frac{1}{2}$ normal ♀, $\frac{1}{2}$ cblind ♂

$P_1$     $X^C X^c$     ×     $X^C Y$

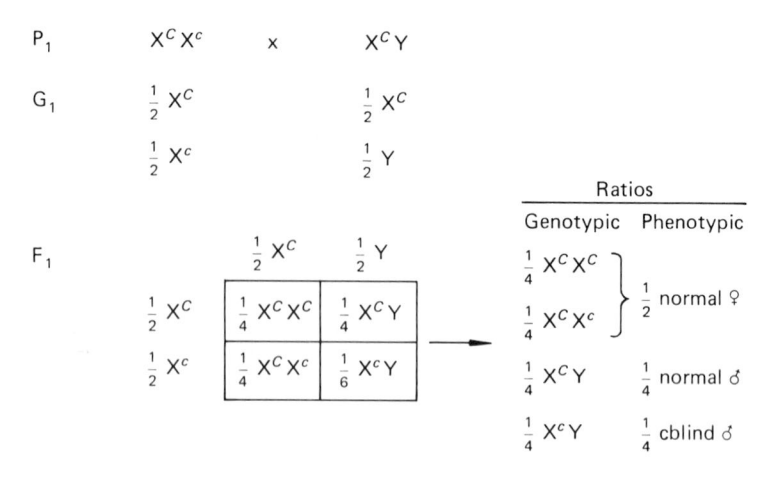

$G_1$     $\frac{1}{2} X^C$          $\frac{1}{2} X^C$

      $\frac{1}{2} X^c$          $\frac{1}{2} Y$

|  | | Genotypic | Phenotypic |
|---|---|---|---|
| $F_1$ Punnett square: $\frac{1}{2}X^C$, $\frac{1}{2}Y$ across; $\frac{1}{2}X^C$, $\frac{1}{2}X^c$ down; cells: $\frac{1}{4}X^CX^C$, $\frac{1}{4}X^CY$, $\frac{1}{4}X^CX^c$, $\frac{1}{6}X^cY$ | | $\frac{1}{4} X^C X^C$ <br> $\frac{1}{4} X^C X^c$ | $\frac{1}{2}$ normal ♀ |
| | | $\frac{1}{4} X^C Y$ | $\frac{1}{4}$ normal ♂ |
| | | $\frac{1}{4} X^c Y$ | $\frac{1}{4}$ cblind ♂ |

FIGURE 11-4. X-linkage. The expected genotypic and phenotypic results of a cross between a woman heterozygous for red-green colorblindness and a normal man.

received $X^C$ from their father; all sons will be colorblind, because each receives their single X, in this case $X^c$, from their mother.

Consider next the results expected for the colorblind locus from marriages between heterozygous women and normal men (Figure 11-4). Note that the genotypic and phenotypic expectations with regard to the colorblind locus are again different for the two sexes. Specifically, whereas all daughters are expected to be phenotypically normal, half the sons are expected to be normal and half colorblind. In this case, every daughter is phenotypically normal because she receives $X^C$ from the father; half the sons are normal and half are colorblind because half receive $X^C$ and half $X^c$ from the mother.

Since the red-green colorblindness locus is X-limited, sons and daughters will always differ genetically (at least in the number of loci present); they will differ phenotypically, however, only in the progeny of the two crosses described last. They will not differ phenotypically in any of the following crosses: $X^C X^C \times X^C Y$; $X^c X^c \times X^c Y$; and $X^C X^c \times X^c Y$. In the last cross, the genotype of the mother's egg alone determines whether the $F_1$ will be normal or colorblind (the sperm genotype does not contain the allele for normal color vision) and the genotype of the father's sperm alone determines the sex of the $F_1$ individual (the egg genotype has no effect on $F_1$ sex type). Random fertilization therefore gives a 50 per cent chance for normal vision and a 50 per cent chance for colorblindness to both sons and daughters.

ALBINISM AND COLORBLINDNESS. We can follow the independent segregation of an autosomal locus and an X-limited locus simultaneously in the mating of an albino man and a red-green colorblind woman who had an albino parent (Figure 11-5). (By custom, only the abnormal aspects of the phenotype are specified.) Thus, the man's genotype is $a\,a$ (due to albinism) and $X^C Y$ (normal for the colorblindness locus). The

# GENETIC RECOMBINATION IN EUKARYOTES 163

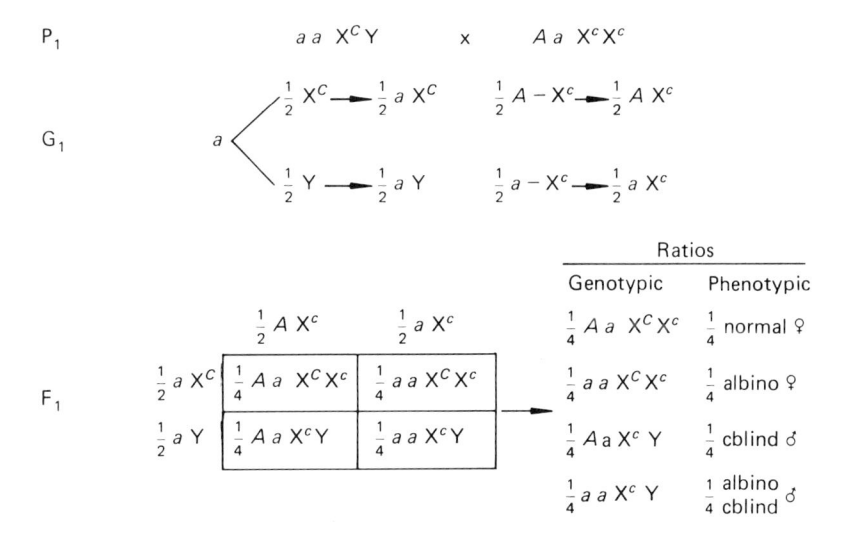

P₁      $aa\ X^c Y$    ×    $Aa\ X^c X^c$

G₁: gametes combining to give the F₁ as shown.

FIGURE 11-5. Autosomal and X-linkage. The expected genotypic and phenotypic results of a cross between an albino man and a red-green colorblind woman who is heterozygous for albinism.

woman's genotype is $Aa$ (her normal pigmentation reveals that $A$ is present) and $X^c X^c$ (due to colorblindness).

Four different phenotypes are expected to occur with equal frequency among the F₁. Note that the phenotypic expectation for albinism (1 albino : 1 normal) is the same for both sexes, as expected for an autosomal locus; and that the phenotypic expectation for colorblindness (1 normal : 1 colorblind) differs for different sexes, as expected as a result of X-limitation.

## 11.4 Phenotypic ratios may reveal if and how a locus is sex-linked.

As long as both sexes are equal in their ability to express different alternatives for a given trait (such as normal color vision versus colorblindness), any mating that gives different phenotypic results for sons and daughters is due to a locus that is located in a sex chromosome, that is, to a locus showing *sex linkage*. We have already seen how X-limited loci will give such results.

Loci present both in the X and Y chromosomes can also give such results; this can be illustrated hypothetically with a pair of homologous loci that carry either the normal protein-coding gene $D$ or its abnormal allele $d$. In a cross between an abnormal woman ($X^d X^d$) and a normal man whose mother was abnormal ($X^d Y^D$), all sons are expected to be normal and all daughters are expected to be abnormal since the $d$ and $D$ alleles in the male parent remain in X and Y chromosomes in the progeny. Alleles in X and Y chromosomes cannot switch positions due to the prevention of exchanges between the X and Y (Section 10.10). Since, as

FIGURE 11-6. Identification of types of sex-linkage from phenotypic results.

| | |
|---|---|
| X-limited | All daughters are normal whereas half or all sons are abnormal. |
| X- and Y-linked | All sons are normal whereas half or all daughters are abnormal |
| Y-limited | All sons are the same type as their father. |

specified, the sex-chromosome linkages in the male parent are retained in the progeny, the cross of a heterozygous female ($X^D X^d$) by the same type of male ($X^d Y^D$) would also produce all normal sons, but half the daughters would be normal and half would be abnormal.

Different phenotypic results for different sexes can also result from Y-limited loci. In this case, whatever allele is carried by the father will be transmitted and expressed in all his sons and in none of his daughters. (As noted previously, the loci for male sexuality are Y-limited, being restricted to loci on $Y^S$.)

From the phenotypic results of crosses described in this and the preceding section, we can state criteria that enable us to identify three different genetic bases for sex linkage (Figure 11-6), provided the abnormal phenotype results from the absence of the normal allele. The locus is X-limited when all daughters are normal and half the sons (Figure 11-4) or all the sons (Figure 11-3) are abnormal. The locus is present both in the X and Y chromosomes when all sons are normal and half or all the daughters are abnormal. The locus is Y-limited if all sons express the same phenotype as their father. When a cross involving an unknown nuclear locus yields the same phenotypic result in sons and daughters, one cannot determine whether the locus is autosomal or sex-linked because the parental genotypes were unsuitable.

**11.5** On rare occasions, homologous chromosomes or daughter chromosomes, and hence the genes they carry, fail to separate during meiosis.

In the human female, the behavior of the X chromosomes is usually normal during meiosis (Figure 11-7A). By metaphase I the two replicated X's have aligned on the spindle as a pair, and after the two meiotic divisions are completed, four nuclei are (sometimes) produced, each of which contains one X. One of these nuclei serves as the egg nucleus. On rare occasions, however, segregation of the X chromosome is aberrant, as follows:

1. At anaphase I both chromosomes go to the same pole (Figure 11-7B); hence, none go to the other pole. The latter nucleus, containing no sex chromosomes, then undergoes the second meiotic division normally to produce two nuclei, neither one having an X. The other nucleus, with two replicated X's, undergoes the second meiotic division normally (each replicated X goes to metaphase independently) to produce a pair of daughter nuclei that each contain two X's. The failure of the X's to separate at

# GENETIC RECOMBINATION IN EUKARYOTES 165

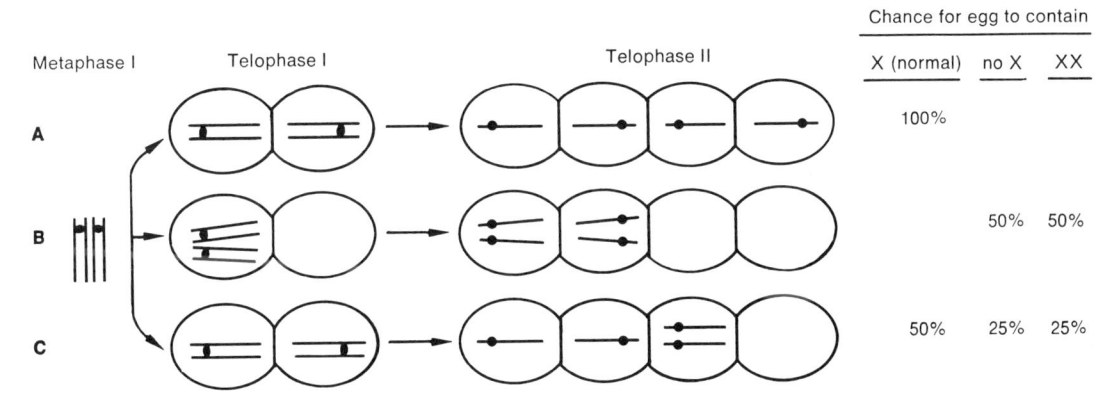

FIGURE 11-7. Consequences of normal separation of X chromosomes (row A) and of its failure to occur (rows B and C).

anaphase I thus results in a gamete with a 50 per cent chance of carrying two X chromosomes and a 50 per cent chance of carrying none.

2. The first meiotic division may be normal, producing two daughter nuclei with one replicated X in each. The second meiotic division, however, may proceed abnormally in one of these nuclei (Figure 11-7C): the chromatids of one X may fail to separate at anaphase II, both going instead to the same pole. Consequently, one daughter nucleus will contain two X's and the other will have none. Overall, a meiosis in which the chromatids in one of the two metaphase II nuclei fail to separate at anaphase II will result in a gamete with a 25 per cent chance of carrying two X's, a 25 per cent chance it will carry none, and a 50 per cent chance it will carry one.

By either mechanism, the failure of X chromosomes to separate or disjoin, *chromosomal nondisjunction*, results in rare gametes that have either two X's or none. Consider the consequences of fertilizations of such exceptional gametes by regular sperms (Figure 11-8). Of the four different exceptional zygotes possible, three are sexually abnormal and one dies. These abnormal sex types are described in detail in Chapter 19.

The usual rarity of chromosomal nondisjunction indicates that it is not programmed and is, therefore, a mutation. (Section 10.10 and Figures 10-12 and 10-13, however, describe three examples of genetically programmed nondisjunction during meiosis.) Nondisjunction occurs for autosomes as well as sex chromosomes; it can occur during mitosis as well as meiosis; and it occurs in males as well as in females.

FIGURE 11-8. Genotypic and phenotypic results in human beings when normal sperm fertilize exceptional eggs produced after X chromosome nondisjunction.

|  |  | Zygote |  |
| --- | --- | --- | --- |
| Exceptional Egg | Fertilized by Regular Sperm | Genotype | Phenotype |
| X X | X | X X X | abnormal ♀ |
| X X | Y | X X Y | abnormal ♂ |
| 0 | X | X 0 | abnormal ♀ |
| 0 | Y | Y 0 | dies |

# HOW GENETIC MATERIAL IS DISTRIBUTED

$$P_1 \quad X^c X^c \quad \times \quad X^C Y$$

$$G_1 \quad X^c \quad\quad \tfrac{1}{2} X^C \text{ (regular)}$$

$$\tfrac{1}{2} Y \text{ (regular)}$$

$$X^C Y \text{ (exceptional)}$$

$$0 \text{ (exceptional)}$$

F$_1$

|  | $X^c$ | Corresponding Phenotypes |
|---|---|---|
| $\tfrac{1}{2} X^C$ | $\tfrac{1}{2} X^C X^c$ | 1 normal vision<br>2 normal ♀ |
| $\tfrac{1}{2} Y$ | $\tfrac{1}{2} X^c Y$ | 1 colorblind<br>2 normal ♂ |
| $X^C Y$ | $X^C X^c Y$ | normal vision abnormal ♂ |
| 0 | $X^c 0$ | colorblind abnormal ♀ |

FIGURE 11-9. Genotypic and phenotypic results expected when regular sperm and exceptional sperm (produced after nondisjunction at meiosis I) of normal men fertilize regular eggs from red-green colorblind women.

Consider the genetic consequences of chromosomal nondisjunction during meiosis I in the human male (where it occurs less often than in the human female). Suppose that the mating under consideration is between red-green colorblind women and normal men (Figure 11-9). Nondisjunction at meiosis I will produce rare sperm carrying $X^C$ Y or no sex chromosomes. Fertilizations that use normal gametes will produce the usual normal daughters and colorblind sons. Fertilizations that use nondisjunctional sperm, however, will produce rare colorblind daughters (having sexual abnormalities) and rare normal vision sons (having sexual abnormalities). We see, therefore, that when chromosomes fail to separate or segregate, so do the loci they contain. In other words, gene and chromosomal nondisjunction always go together.

## CROSSING OVER

**11.6** The process of exchange between nonsister chromatids during meiosis is called *crossing over* and the products are called *crossovers*.

While a pair of homologous chromosomes are synapsed during prophase of the first meiotic division, breakage recombinations occur between nonsister chromatids that serve to keep the homologous chromosomes together until metaphase I (Section 10.5). Such recombinations also break the linkage between different genes located in the same chromosome. We can detect these breakage recombinations genetically when the two homologous chromosomes differ in the alleles they carry at two loci, and when nonsister exchange occurs between these two loci. In Figure 11-10, for example, one homologous chromosome is marked by *A*

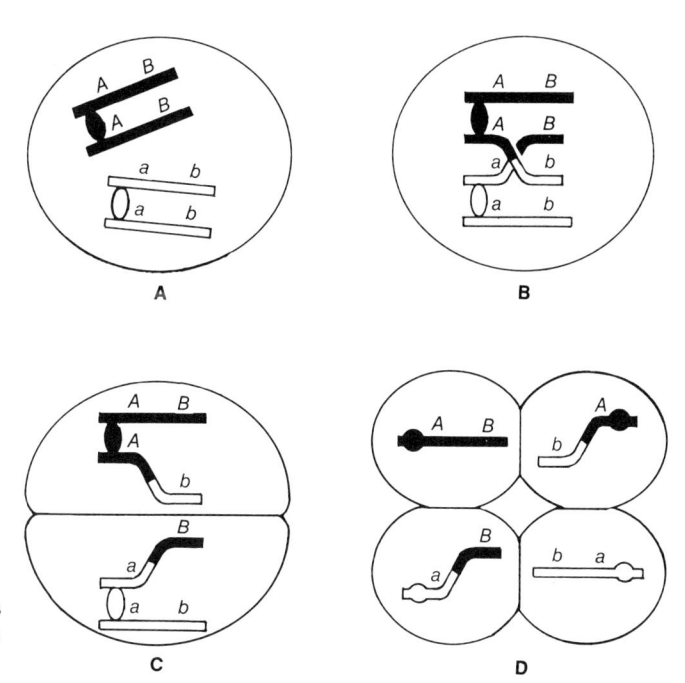

FIGURE 11-10. Genetic consequences expected after a crossing over between linked genes.

and $B$ and the other homologous chromosome is marked by alleles $a$ and $b$. The nonsister exchange shown between the "$A$" and "$B$" loci in Figure 11-10B required the breakage of two nonsisters and the cross union of corresponding segments in a process called *crossing over*. The results of such a crossing over are two recombinant chromatids, called *crossovers*, which in our example are $A\ b$ and $a\ B$, and two nonrecombinant parental chromatids, $A\ B$ and $a\ b$ (Figure 11-10C). Note that the two crossovers from any crossing-over event are reciprocal for the marker genes; that is, in our example, for every $A\ b$ crossover chromosome there will be an $a\ B$ crossover chromosome. When cells have a single crossing over between two marked loci, therefore, half the gametes they produce will contain crossovers and half will not.

## 11.7 Linked genes tend to stay in their original sequence.

Crossing over occurs in both arms of each chromosome pair, provided that the arms are sufficiently long. The longer the chromosome, the more crossing over occurs—as many as seven crossing overs have been detected in certain long chromosomes. Since two or more crossing overs occur in a synapsed pair of homologous chromosomes, and since the nonsisters that undergo crossing over at one position may be the two that have not undergone crossing over at another position, it is likely that every chromosome entering a gamete is a crossover. The chance for crossing over is not equal in different parts of a chromosome—being less in regions that are constitutively heterochromatic, such as regions adjacent to centromeres and the tips of chromosomes. In the main euchromatic region of a chromosome, different regions have a roughly equal chance of undergoing crossing over.

Although every chromosome is likely to be a crossover for some genes in euchromatin, it is more and more unlikely for a chromosome to be a crossover for two particular genes as their loci get closer and closer to each other. When two loci are relatively far apart it may be possible for a crossing over to occur between them in every cell undergoing meiosis. Even so, because only two of the four chromatids become crossovers, the maximum frequency of crossovers with respect to two loci is 50 per cent. In this case, the old, parental arrangement of alleles at two loci is retained in the progeny 50 per cent of the time. As the distance between two loci decreases, so does the fraction of all cells undergoing meiosis that have a crossing over between these loci. As a result, more and more cells undergoing meiosis yield only parental combinations; and in the fewer and fewer cells that have a crossing over between the loci under test, half the meiotic products are also parental combinations, and only half are crossovers. In other words, the specific sequence of alleles in a chromosome tends to be maintained generation after generation, although it can be changed by crossing over with a homologous chromosome bearing different alleles.

## 11.8 Crossover frequencies can be used to construct recombination maps for linked loci.

In accordance with the discussion in Section 11.7, we expect that as the distance between two loci increases, the chance that a crossing over will occur between them will increase and thus the frequency of crossovers will increase. Accordingly, we shall use the relative frequencies of crossovers to indicate relative distances between linked loci (just as we used relative frequencies of phage recombinants to indicate relative distances between phage loci). By definition, a *crossover map unit* is that distance between linked genes that results in 1 crossover per 100 meiotic products.

If 10 per cent of cells at prophase I have a crossing over between the linked loci "*a*" and "*b*," at the completion of meiosis, 5 per cent of the haploid products will contain crossover chromosomes with respect to these loci. (Recall that only 5 per cent are crossovers because only two of the four chromosomes produced from each crossing over are crossovers.) Thus, the distance between the "*a*" and "*b*" loci is 5 crossover map units.

Since crossover percentages equal crossover map units, it is possible to determine the sequence of three or more linked loci from the percentages of crossovers between them; just as we were able to sequence phage loci from the percentages of recombinants for these loci. For example, if three loci "*M*," "*N*," and "*O*" are present in the same autosome, their sequence can be determined from the results of three crosses. One cross determines the percentage of crossover gametes with respect to the "*M*" and "*N*" loci in an individual who, let us say, received $M$ and $n$ from one parent and $m$ and $N$ from the other parent (Figure 11-11). In this case,

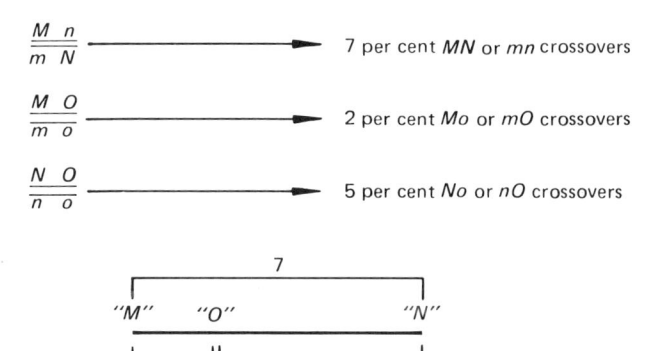

FIGURE 11-11. Mapping the sequence of three linked autosomal loci from the percentages of crossovers produced by individuals hybrid for two gene pairs.

$\frac{M\ n}{m\ N}$ ⟶ 7 per cent *MN* or *mn* crossovers

$\frac{M\ O}{m\ o}$ ⟶ 2 per cent *Mo* or *mO* crossovers

$\frac{N\ O}{n\ o}$ ⟶ 5 per cent *No* or *nO* crossovers

```
            7
      ┌──────────────┐
  "M"    "O"         "N"
      └──┴───────────┘
         2     5
```

the crossovers are either $MN$ or $m\ n$—and suppose that we find 7 per cent crossovers. A second cross determines the percentage of crossover gametes with respect to the "$M$" and "$O$" loci, using an individual who, let us say, received $M$ and $O$ from one parent and $m$ and $o$ from the other parent (Figure 11-11). In this case, the crossovers are either $Mo$ or $m\ O$—and suppose we find that 2 per cent are crossovers. The last cross determines the percentage of crossover gametes with respect to the "$N$" and "$O$" loci, using an individual who, let us say, received $N$ and $O$ from one parent and $n$ and $o$ from the other parent (Figure 11-11). In this case, the crossovers are either $N\ o$ or $n\ O$—and suppose that we find 5 per cent crossovers. The largest map distance, 7 map units, is between the "$M$" and "$N$" loci, so these loci are farthest apart. The "$O$" locus lies between them, 2 map units from the "$M$" locus and 5 map units from the "$N$" locus. Therefore, these loci are actually in the relative linear order "$M$" "$O$" "$N$" (or "$N$" "$O$" "$M$") in the chromosome.

The crossover map distances between loci are not always exactly congruent with the actual distances between them in the chromosome, however, because of the reduced likelihood of crossing over near the centromere and at the ends of the chromosome. In other words, the crossover map distances are closely congruent with actual chromosome distances only for loci located in euchromatin. It should be noted, however, that crossover maps always correctly sequence linked loci, no matter where they may be in the chromosome. Figure 11-12 is a recombinational linkage map of the human X chromosome determined from crossover and other recombinational data.

## SUMMARY AND CONCLUSIONS

Many eukaryotes have autosomes and a pair of sex chromosomes that are usually XX in one sex and XY in the other. Loci occur in pairs, of course, in paired autosomes and in paired X chromosomes. Although some sex-linked loci are paired in XY individuals (these make X and Y homologous chromosomes), others are restricted to one type of sex chromosome and are, therefore, X-limited or Y-limited. Thus, the differences in cytological appearance of the X and Y chromosomes are paralleled by differences in the loci they carry.

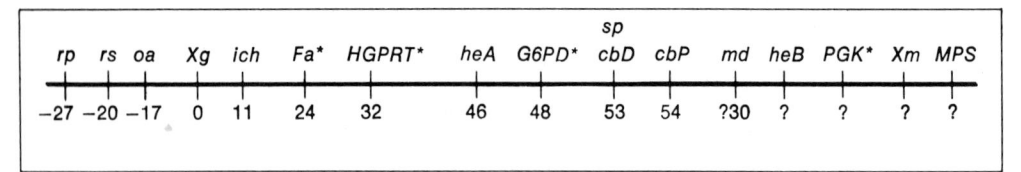

| | | | | | | | | | sp | | | | | |
|---|---|---|---|---|---|---|---|---|---|---|---|---|---|---|
| rp | rs | oa | Xg | ich | Fa* | HGPRT* | heA | G6PD* | cbD | cbP | md | heB | PGK* | Xm | MPS |
| −27 | −20 | −17 | 0 | 11 | 24 | 32 | 46 | 48 | 53 | 54 | ?30 | ? | ? | ? | ? |

* Established by cell hybridization.

FIGURE 11-12. Genetic recombinational linkage map of the human X chromosome. Key to symbols: *rp*, retinitis pigmentosa; *rs*, retinoschisis; *oa*, ocular albinism; *Xg*, Xg blood group; *ich*, ichthyosis; *Fa*, Fabry's disease, angiokeratoma, or β-galactosidase deficiency; *HGPRT*, hypoxanthine–guanine phosphoribosyltransferase; *heA*, hemophilia A; *G6PD*, glucose 6-phosphate dehydrogenase; *cbD*, deutan color blindness; *cbP*, protan color blindness; *sp*, scapuloperoneal syndrome; *md*, Duchenne muscular dystrophy; *heB*, hemophilia B; *PGK*, phosphoglycerate kinase; *Xm*, Xm serum protein type; *MPS*, Hunter syndrome. (After V. A. McKusick, *Human Chromosome Mapping Newsletter*, December 1972.)

When parents have appropriate genotypes and a sufficient number of progeny are examined, it is possible to identify from the phenotypic results in the $F_1$ whether the loci under test are autosomal, X-limited, X- and Y-linked, or Y-limited.

On rare occasions, homologous or daughter chromosomes undergo nondisjunction, and so do the loci they carry. In human beings, nondisjunction of sex chromosomes during meiosis yields offspring who either are sexually abnormal or die; as expected, the abnormal distribution of sex chromosomes also abnormally distributes sex-linked loci, such as that for colorblindness, that have no phenotypic effect on sexuality.

The nonallelic genes in a given chromosome are linked to each other and tend to be transmitted together from one generation to another. Crossing over prevents such linkage from being permanent or complete. During meiosis, crossing over is programmed to occur between nonsister chromatids of a pair of homologous chromosomes, and to produce crossover chromosomes that are reciprocal for marker genes.

Crossover frequency is directly (but not always simply) related to the distance between loci in a chromosome, 1 unit of crossover distance being defined as one crossover per hundred haploid meiotic products. Crossover distances can be used to sequence linked loci in a linear crossover map.

## QUESTIONS AND PROBLEMS

**11.1** *Anagrams*: Unscramble and define each of the following: **a.** umatoose. **b.** labminis. **c.** sidojin. **d.** scoverros.

**11.2** What are two important consequences of exchanges between nonsister chromatids?

**11.3** A woman heterozygous at the red-green colorblindness locus ($X^C X^c$) who marries a normal man ($X^C Y$) usually produces: **a.** sons who are all colorblind. **b.** daughters who are all colorblind. **c.** sons who are all normal. **d.** daughters who are all normal. **e.** normal as well as colorblind sons and daughters.

# GENETIC RECOMBINATION IN EUKARYOTES

**11.4** A red-green colorblind, albino woman has a daughter. What is the most likely genotype and phenotype of the daughter? $X^c X^c\ Aa$; normal

**11.5** Two phenotypically normal parents have a red-green colorblind, albino son. Give the genotypes of parents and child. $X^C X^c\ Aa,\ X^C Y\ Aa;\ X^c Y\ aa$

**11.6** A colorblind woman marries an albino man. Their first child is albino. What is the chance that their second child will be a phenotypically normal daughter? 25%

**11.7** Nondisjunction of the members of a pair of genes: **a.** rarely occurs in meiosis and never in mitosis. **b.** does not require chromosomal nondisjunction. **(c.)** during meiosis is a violation of segregation. **d.** violates linkage. **e.** often occurs in mitosis but is rare in meiosis.

**11.8** Which one of the following statements about crossing over is false? **a.** It always produces crossovers. **b.** It occurs in males and females. **(c.)** It always occurs between sister chromatids. **d.** It is a programmed recombination. **e.** It only occurs between linked loci.

**11.9** Suppose that 2 of each 10 cells entering meiosis have a crossing over between two linked and appropriately marked loci. What is the map distance between these two loci?

**11.10** In *Drosophila*, the locus for miniature wing ($m$ = miniature, $M$ = normal, long wing) is X-limited. A miniature male is mated to a normal female whose father was miniature. **a.** Give the genotypes of this mating. **b.** Give the usual types and frequencies of the gametes produced by these parents. **c.** What are the frequencies of the genotypes and phenotypes expected in the progeny produced from these gametes? **d.** What rare genetic constitutions could unfertilized eggs have as the result of maternal nondisjunction?

**11.11** What can you conclude about the precise location of the genes involved in each of the two cases when an individual produces the gametes indicated?

| Case A | | Case B | |
|---|---|---|---|
| B d | 450 | R S | 417 |
| b D | 450 | r S | 391 |
| B D | 45 | R s | 400 |
| b d | 35 | r s | 412 |
| | 1000 | | 1620 |

**11.12** What is the relation between crossing over and constitutive heterochromatin?

**11.13** What are the usual types and frequencies of gametes expected from $B\ d/b\ D$ when the loci are 8 crossover map units apart in a *Drosophila*: **a.** female? **b.** male?

**11.14** Which alternatives are correct? Characters due to rare recessive genes confined to the X chromosome: **a.** appear in men only. **b.** may appear in some of the grandsons produced by a daughter of a man with the trait. **c.** appear more often in men than in women. **d.** appear in the daughters, but not the sons, of an affected man. **e.** have been studied, and the alleles for these characters are found mostly in women.

# Chapter 12

## Gross Changes in Nuclear Chromosomes

WE HAVE ALREADY CONSIDERED the changes in nuclear chromosomes associated with mitosis, meiosis, and fertilization. Nuclear chromosomes can undergo other programmed or mutational changes that are gross enough to be detectable cytologically. These changes include the gain or loss of individual chromosomes or entire chromosome sets as well as the rearrangement of large chromosome segments produced by chromosome breakage. This chapter considers the origin and some of the consequences of such gross chromosomal changes. Changes involving unbroken chromosomes are taken up before those involving broken chromosomes.

**12.1** **Most gross chromosomal changes in human beings are detected cytologically at metaphase.**

Ordinary chromosomes in interphase are poor material for cytological study since they are too clumped, or thin and intertwined, to permit a distinction between different chromosomes and parts. Metaphase chromosomes, however, are readily studied unstained, uniformly stained, or differentially stained. Differential staining gives the metaphase chromosome a banded appearance. One staining procedure produces C bands (Figure 10-5), another (with quinacrine mustard) produces fluorescent Q bands, and two others (with Giemsa stain) produce G bands (Figure 12-1) or R bands; the banding pattern for a given chromosome is constant and different with different procedures. Since each of the banding patterns is unique for practically every type of chromosome, almost all rearrangements that involve one or more bands are detectable at metaphase through differential staining. The grosser the changes are, of course, the more readily they are detected in unstained or uniformly stained metaphases.

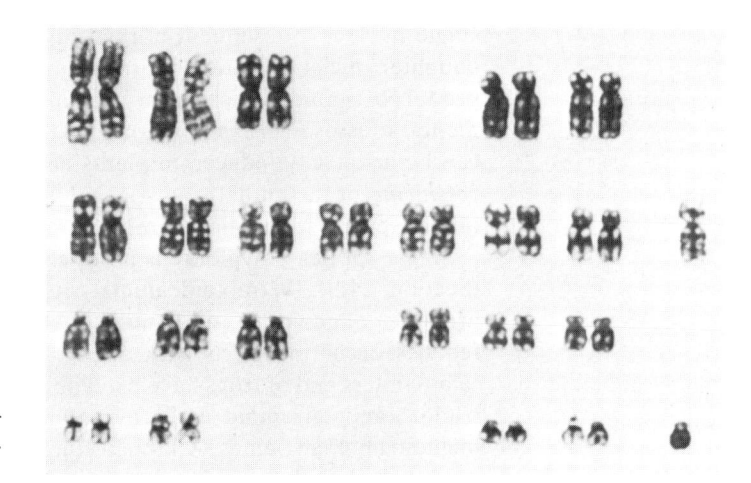

FIGURE 12-1. Human chromosomes treated to exhibit G bands. (Courtesy of T. C. Hsu.)

Although the metaphase stage is the favorite stage for cytological studies in human beings, in other organisms prophase I chromosomes or the giant interphase chromosomes that contain many chromatids make excellent cytological material.

# UNBROKEN CHROMOSOME CHANGES

**12.2** A nucleus may contain one or more extra sets of chromosomes.

Some species are haploid; most sexually reproducing eukaryotes are diploid. Whenever an individual or cell contains one or more extra sets of chromosomes, it is said to be *polyploid*. A polyploid with three sets of chromosomes is *triploid*; one with four chromosome sets is *tetraploid*. Some polyploid plants have as many as six or eight chromosome sets. The sets of chromosomes in polyploids are either all derived from a single species or they are derived from two or more species. These two kinds of polyploids are discussed separately.

SINGLE-SPECIES POLYPLOIDY. This kind of polyploidy may occur as a normal event, because it is genetically programmed, or as a nonprogrammed mutational event. It is a normal event in some of the liver cells in a human being. In these cases tetraploid cells are produced because cell division did not follow mitotic chromosome duplication in a diploid cell. Polyploidy is the result of a variety of mutational events: (1) A haploid egg may be fertilized by two or more sperm, producing a zygote with three or more sets of chromosomes. (2) Two haploid meiotic nuclei may fuse to produce a diploid gamete, so that after union with a haploid gamete, a triploid zygote is formed. (3) Mitotic anaphase may be abnormal, so that the doubled number of chromosomes may be included in a single nucleus. Subsequent normal divisions will give rise to polyploid daughter nuclei. Mitotic abnormalities that give rise to polyploidy are produced by mechanical injury, high-energy radiations, and environmental stresses such as starvation or extremes of temperature. Polyploidy can be artificially induced in plants and animals by treating them with *colchicine* or its synthetic analog, colcemide, which prevents the formation of spindle tubes and, thereby, chromosome separation in anaphase.

Single-species polyploidy is more common in plants than in animals (Section 24.9). Tetraploid animal species do occur, however, in sea urchins, roundworms, and moths. Abnormal triploid and tetraploid females occur in insects such as *Drosophila*, and polyploid larvae of salamanders and of frogs have been obtained experimentally. Abnormal triploid and tetraploid embryos are found in a variety of mammals, including human beings. Although complete triploidy in human beings is ultimately lethal (some such individuals are born alive but die soon thereafter), individuals who are diploid in some cells and triploid in a substantial fraction of others may be viable but defective. The triploid cells of such individuals may contain XXX, XXY, or XYY, besides three sets of autosomes. Specific phenotypic effects of single-species polyploidy are discussed further in later chapters.

MULTIPLE-SPECIES POLYPLOIDY. Cultivated wheat, rye, tobacco, and cotton all originated as polyploids between two or more species. We can appreciate how such polyploidy can come about by outlining how it was accomplished experimentally in a specific case. The goal of the experiment was to produce a new species which, by containing the genotypes of two species, might demonstrate certain desirable phenotypic features of both. One species was the radish, whose root is desirable, and the other was the cabbage, whose shoot is desirable. It is possible to cross them in the greenhouse and obtain an $F_1$ hybrid (Figure 12-2). The $F_1$ hybrid is sterile, however, since none of its chromosomes has a homologous chromosome to pair with at meiosis, and its gametes do not contain a complete set of chromosomes. To circumvent this sterility, the $F_1$ hybrid was treated with colchicine, which doubled the number of chromosomes in a cell, so that at the time of meiosis each chromosome had a homologous partner. The result was a new fertile species, containing the diploid chromosome number of both the radish and cabbage, that was genetically isolated from both the radish and cabbage. Unfortunately, the polyploid had a shoot like a radish and a root like a cabbage.

## 12.3 Individual chromosomes in a nuclear chromosome set may be added or lost.

Nuclei may contain chromosome sets that have one or more chromosomes too many or too few. For example, a zygote (and hence the entire organism it develops into) may contain one chromosome too many or too few because nondisjunction produced a gamete with one chromosome too many or too few (Sections 10.10 and 11.5). Since single

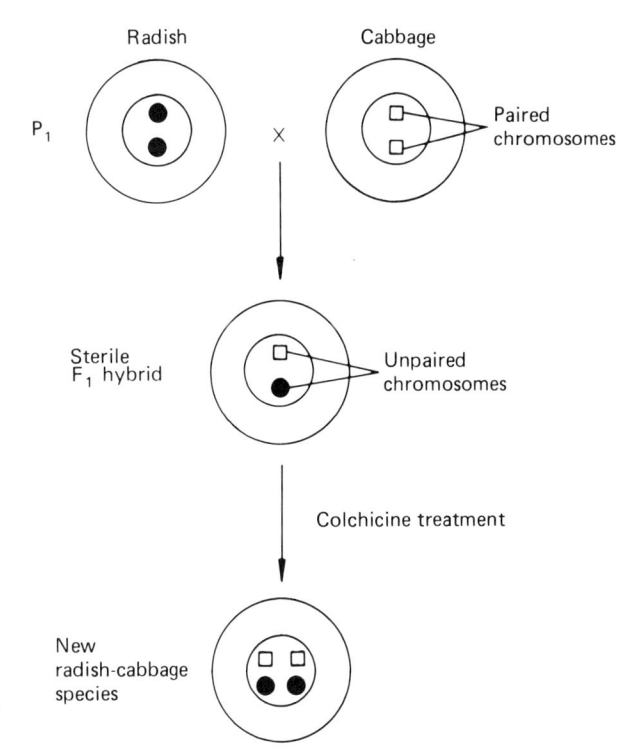

FIGURE 12-2. Multiple-species polyploidy between radish and cabbage.

chromosomes can be destroyed through mutagenic treatment, gametes and zygotes can be missing one or more chromosomes by this mechanism (to be discussed in Section 12.5). Since nondisjunction, chromosome destruction, or polyploidy can occur during mitotic divisions in the somatic or germ lines, multicellular organisms can be *mosaics* in their chromosomal constitution; that is, some parts may be chromosomally normal and other parts chromosomally abnormal.

A chromosome that fails to disjoin properly during nuclear division and is left out of the nucleus (Figure 10-13) is usually degraded by nuclease, and hence is lost. Such losses are common after two somatic cells (from the same or different species) and their nuclei fuse spontaneously or experimentally. *Somatic cell fusion* is used experimentally to locate specific genes on specific chromosomes. This localization is accomplished by correlating the loss of a particular cellular function or protein with the loss of a particular chromosome.

## 12.4 The loss of a chromosome is more detrimental than its addition.

A diploid individual contains, in its two sets of chromosomes, a balance of genes that is responsible for the proper functioning of that organism. The gain of a whole chromosome set does not generally upset this genetic balance (in XX individuals, for example). Since the rest of the cell is adapted to the presence and functioning of a diploid genotype, however, the gain or loss of whole chromosome sets is not without some phenotypic detriment. The loss of a chromosome set from a diploid has an additional disadvantage. Since a diploid is usually heterozygous for at least several loci, the chromosome set lost will usually contain some normal alleles, while the set remaining retains their abnormal alleles. The resultant haploid is therefore phenotypically abnormal with respect to the functions these genes perform.

When homologous chromosomes are heterozygous at one or more loci, the loss of a single chromosome is likely to be more phenotypically detrimental than the addition of that chromosome, for the same reason. The loss of a single chromosome is also more detrimental than its addition because it brings about more of a genetic imbalance relative to the rest of a diploid genotype. We can see how loss of a chromosome produces a greater genetic imbalance than its gain by the fact that when a homologous chromosome is present only once in a diploid, being *monosomic* (rather than *disomic*), the dosage of its genes is 100 per cent too small relative to the rest of the genes, whereas when it is present three times, being *trisomic*, the dosage of its genes is only 33 per cent too large. Thus, if a particular trisomic is phenotypically detrimental, its monosomic is expected to be even more detrimental. When a single chromosome is added or lost in a single-species tetraploid, however, we expect relatively less phenotypic detriment than in a diploid since the shift in gene balance is relatively less in the former than in the latter.

## 12.5 Monosomy and trisomy occur in human beings.

In human beings, the largest autosome is numbered 1; the smallest, 22. The sex chromosomes are not numbered—the X being middle-sized, and the Y the smallest of all the chromosomes (Figure 10-4). Each of these 24 chromosomes can undergo nondisjunction during meiosis (or mitosis) in men and in women, resulting in a variety of monosomic or trisomic zygotes; most of these are lethal in the embryonic stage.

*Down's syndrome* (*mongolism*) in human beings is usually the result of trisomy for chromosome 21 (Figure 12-3). Affected individuals, who usually have a happy disposition, are handicapped by severe mental and physical retardation. They have small brains, containing disproportionately small frontal lobes, brain stem, and cerebellum; they usually have a mental age of three to seven years. They are characteristically obese and have a thick tongue, sagging mouth, unusual palm and sole prints, and an eyelid fold, which is different, however, from that of members of the Mongoloid race. Although females are fertile, males are sterile. The variability that exists among Down's syndrome individuals is due partly to environment and partly to differences in the alleles they carry in their 21's and other chromosomes.

The overall frequency among live births of Down's syndrome due to trisomy is approximately 0.2 per cent. A plot of the proportion of all trisomic Down's children born to mothers of different ages (Figure 12-4)

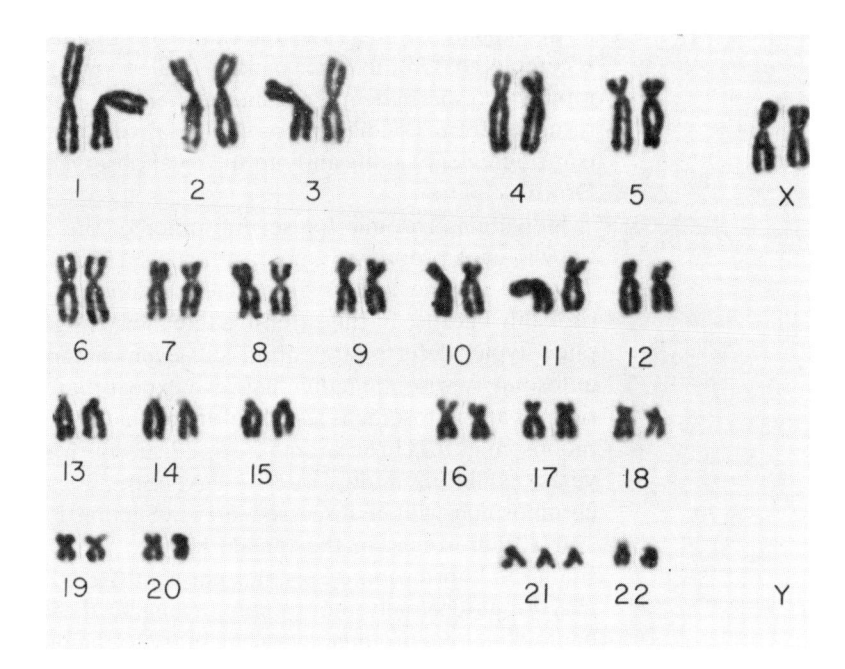

FIGURE 12-3. Chromosomal constitution of a female showing Down's syndrome. (Courtesy of K. Hirschhorn.)

shows that the defect occurs more frequently as mothers age. More extensive data indicate that such children are 50 times more likely to be born to older than to younger mothers (2 per cent compared to 0.04 per cent). The defect is due primarily to the production of disomic 21 eggs formed as the result of nondisjunction during meiosis. A defect associated with the age of the oocyte—which has been aging since before

FIGURE 12-4. The percentages of all normal children (or of certain abnormal children) born to mothers of different ages. (Modified from L. S. Penrose, 1964. *Ann. Human Genet.*, **28**: 199–200.)

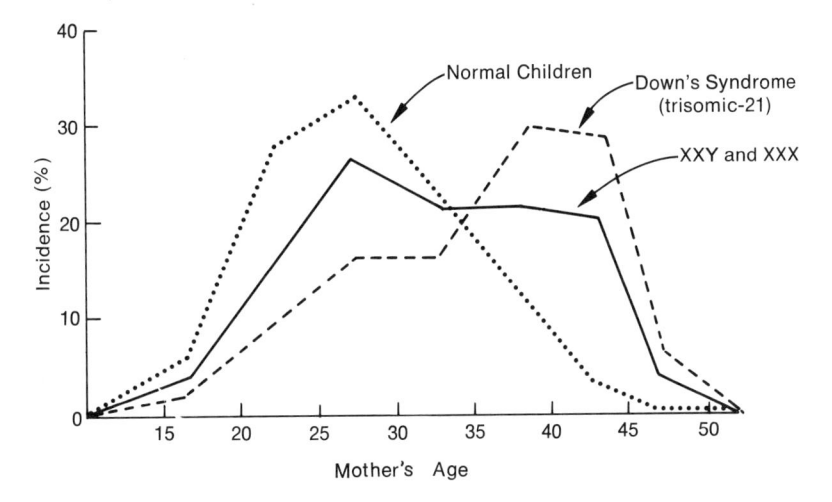

birth—apparently increases the chance for nondisjunction when meiosis is completed. On the other hand, in males—where meiosis is completed rapidly and sperm do not age more than a few weeks—disomic 21 sperm resulting from nondisjunction are relatively rare and do not seem to contribute very significantly to the total observed frequency of trisomic Down's children.

Individuals trisomic for several other of the smaller autosomes are known, each producing a characteristic set of congenital abnormalities. Trisomy for the largest autosomes is apparently lethal before birth, probably because of the extensive imbalance of genes. Since very severe phenotypic defects are observed even among the least affected autosomally trisomic individuals, we expect that the monosomic condition for any autosome is lethal before birth. On rare occasions, however, monosomics for chromosome 21 or 22 survive for a period of months to years, exhibiting multiple defects. Abnormalities in the number of sex chromosomes will be discussed later in Section 19.6.

If all autosomes in the female have a frequency of nondisjunction similar to chromosome 21's, 4.4 per cent (22 × 0.2 per cent) of zygotes may be autosomally monosomic, owing to the equal chance that the haploid meiotic product complementary to the one that is disomic—the *nullosomic* one—will become the egg. Actually, more nullosomic than disomic eggs are expected, since the chromosome left out of one daughter nucleus due to nondisjunction may not be included in the sister nucleus and, therefore, is lost. Supporting the expectation of a high normal frequency of monosomy and trisomy is the observation that about one fourth of spontaneously aborted human fetuses have abnormal chromosome numbers. It is expected, moreover, that many conceptions involving monosomy are lost so early that they go unnoticed.

# BROKEN CHROMOSOME CHANGES

**12.6** Broken ends of chromosomes can be ligated in pairs.

As implied in Section 6.1 and Figure 6-1, chromosomes can be broken and pairs of broken ends can be ligated. Whereas the normal ends of a nuclear rod chromosome cannot be ligated to any other normal or broken end, any broken end may be ligated to any other broken end provided that the polarity of the ligating ends is correct. Therefore, in the case of double-stranded DNA chromosomes, to which our attention will be confined, this ligation occurs only when the two required unions produce a duplex whose polarity is continuous and antiparallel in complementary strands. We do not know whether the broken pieces that join find each other because they have complementary single-stranded ends that base-pair and are repaired as needed before being ligated; or whether they have double-stranded ends that meet by accident and are ligated; or both. In any event, our present concern is the gross cytological and general phenotypic consequences of one or more *chromosomal breaks*, that is,

## 12.7 The consequences of a single chromosome break are described.

Let us consider the consequences of a single chromosome break (Figure 12-5). Part 1 shows a normal chromosome whose centromere is indicated by a black dot. In part 2 the chromosome is broken. The most common event is for the two broken ends to reunite. Because proximity of broken ends favors their union, the two ends from a single break usually join together even when other broken ends occur in the same nucleus. Sometimes, however, the ends do not join before the chromosome replicates. Accordingly, two daughter chromosomes are produced like the parent chromosome, each with a single break (part 3). Even so, unions can occur that restitute the original chromosomal arrangement. The unions of a with b and a' with b', or a' with b and a with b' would be *restitutional unions*. In part 4 we see the results of *nonrestitutional unions*: one union between sister ends (a with a') produces a chromosome without a centromere, an *acentric chromosome*; and one union between the other sisters (b with b') produces a chromosome with two centromeres, a *dicentric chromosome*. Part 5 shows the next anaphase in which the acentric chromosome cannot be pulled to either pole, and so is left out of both daughter telophase nuclei, and is lost. The sister centromeres in the dicentric, however, are pulled to opposite poles to produce a *bridge* that, by hindering migration to the poles, can prevent both daughter nuclei from receiving any of its chromosomal material. Thus, the dicentric, too, may be lost. Sometimes, however, the bridge snaps (usually into unequal pieces) so that a broken piece with one centromere goes to each pole. After replication, a new dicentric chromosome can form, by union between sister broken ends, in one or both daughter nuclei, and once

FIGURE 12-5. Consequences of a single nonrestituting chromosome break. In part 4, the chromosomes are contracted prior to metaphase.

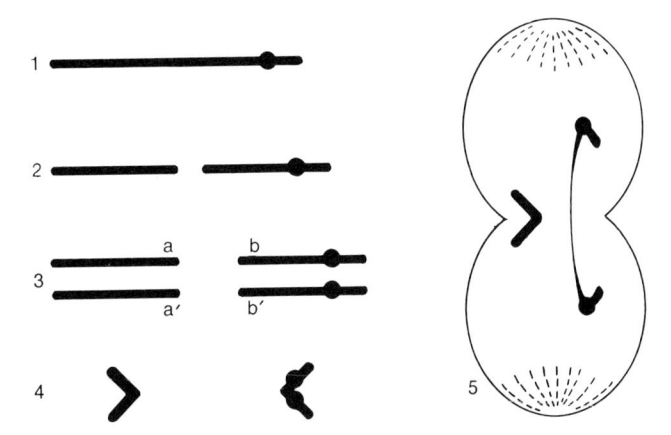

again make a bridge at the next anaphase. In this manner, a *bridge-breakage-fusion-bridge cycle* can occur at successive nuclear generations. (Under similar conditions, shorter dicentrics break more often than longer ones.) A bridge that fails to break can tie two daughter nuclei together and interfere with subsequent nuclear divisions. Such inteference may have a much greater effect than the unequal distribution of the genes located in the bridge. Nevertheless, unequal distribution of genes located in the bridge can have detrimental phenotypic effects; and the loss of chromosomal material, after an acentric or dicentric fragment is left out of a daughter nucleus, can be detrimental or lethal.

Chromosome breaks can occur in either the somatic or the germ line; those in the germ line of either sex can give rise to gametes having incorrect multiples of some genes. Since little, if any, transcription or translation occurs in the mature gametes of animals, a genetically imbalanced gamete can ordinarily form a zygote with a normal gamete. Harmful or lethal effects may occur, however, in early or later development of the zygote. In many plants, on the other hand, the meiotic products form a gamete-producing generation in which numerous genes are active, so that detrimental effects of genetic imbalance usually occur before fertilization.

**12.8** After replication, a nonrestituted chromatid becomes a nonrestituted chromosome.

Sometimes a break is seen in only one chromatid of a replicated chromosome (Figure 12-6A). Restitution is more likely for such *chromatid breaks* than for chromosome breaks, since the unbroken chromatid is coiled around its broken sister and serves as a splint to hold the newly produced ends close to each other. One should note, however, that a chromatid break seen under the microscope may have two origins: a break in one of the two chromatids of a replicated chromosome; or a break in an unreplicated chromosome, one of whose daughter chromatids restituted after chromosome replication.

As might be expected, the consequences of single nonrestituted chromatid breaks are similar to those of single, nonrestituted chromosome breaks. Hence, the following discussion will be restricted to chromosome breaks that fail to restitute. We should not think, however, that chromatid breaks are less frequent or less important than chromosome breaks.

**12.9** Two nonrestituted breaks in one chromosome can lead to deficiency, inversion, or duplication.

When two breakages occur in the same chromosome, these may occur within the same arm or in different arms, as shown in Figure 12-7. When these breakages do not restitute, the result can be a deficiency, inversion, or duplication.

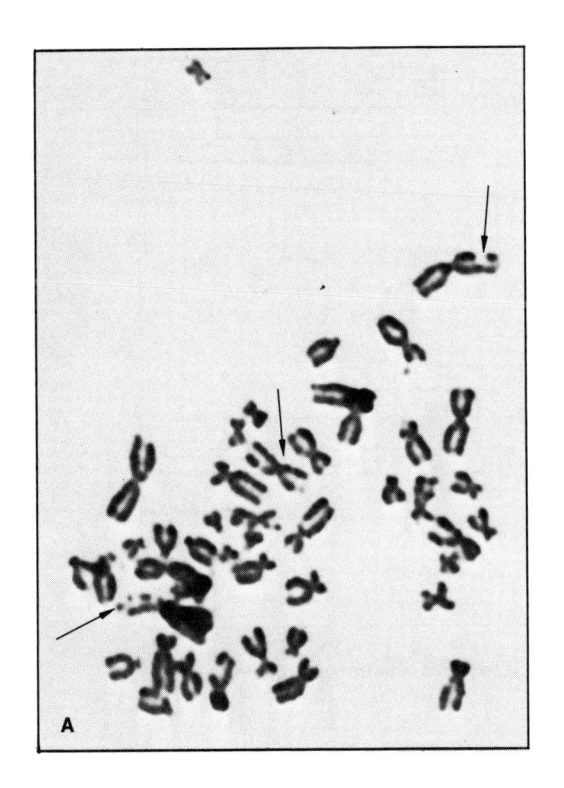

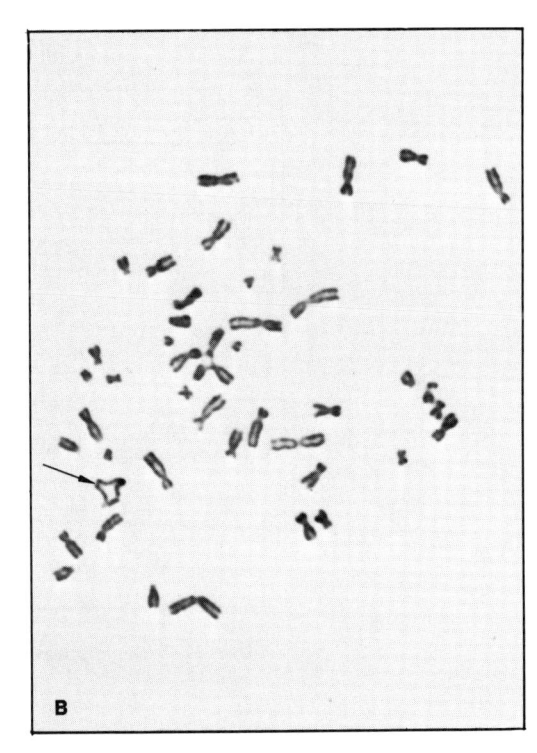

FIGURE 12-6. Structural changes X-ray-induced in normal human male fibroblast-like cells *in vitro*. The arrows show broken chromatids and chromosomes in **A**, dicentric chromosomes in **B**, and ring chromosomes in **C**. **A** and **B** are in metaphase (see Figure 10-3); **C**, is late prophase. (Courtesy of T. T. Puck, 1958. *Proc. Nat. Acad. Sci., U.S.*, **44**: 776–778.)

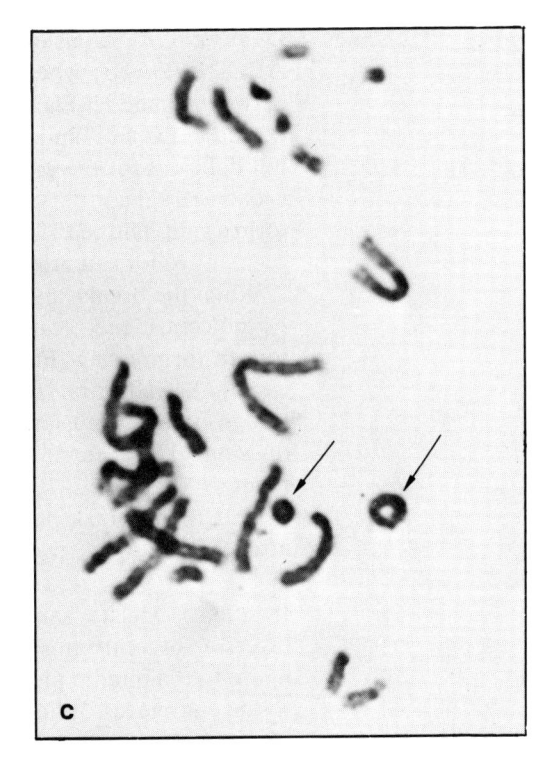

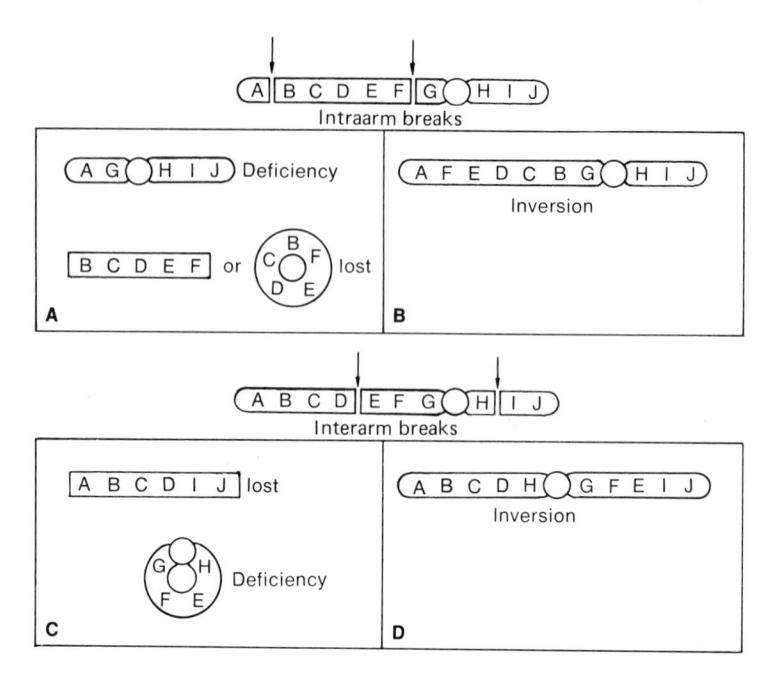

FIGURE 12-7. Some consequences of two breaks in the same chromosome.

DEFICIENCY. Consider a chromosome with segments in the order ABCDEFG.HIJ, where the centromere is represented by the period between G and H (Figure 12-7). Intraarm breakages (say between A and B, and between F and G) can give rise to chromosome AG.HIJ, centric but deficient for the piece BCDEF, after the ends at A and G join. The acentric fragment is subsequently lost, whether or not its ends join to form a ring. The centric, deficient chromosome may survive if the missing segment is not essential phenotypically.

When the breaks are interarm (for instance, between D and E, H and I), the centric piece can survive a subsequent nuclear division if its ends join to form a ring (Figure 12-6C) and if the deficient sections are not essential. (If the ends do not join before chromosome replication, a bridge-breakage-fusion-bridge cycle will be initiated at the next nuclear division.) The acentric pieces are eventually lost whether or not they join together (Figure 12-7C).

Note that acentric pieces present in the interphase nucleus are lost only after the next nuclear division.

INVERSION. The same two nonrestituted breaks can lead to the inversion of a chromosome segment (Figure 12-7B and D). We see in the figure that a middle piece of the chromosome becomes inverted relative to the end pieces. Whereas intraarm inversions do not change the shape of the chromosome, interarm inversions do whenever the two breakages are different distances from the centromere.

FIGURE 12-8. Duplication.

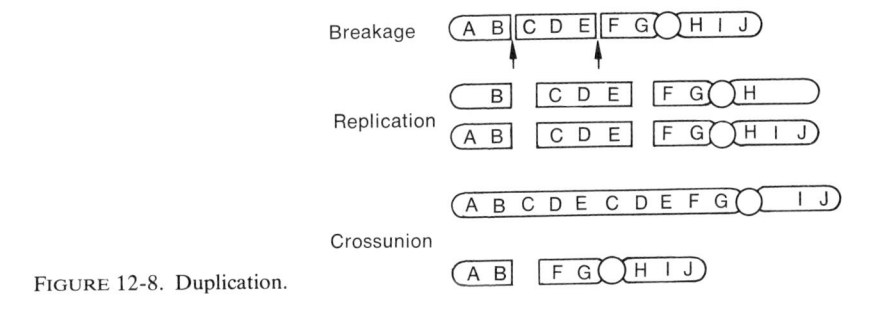

DUPLICATION. If the joining of ends made by two breaks in the same chromosome is delayed until after the chromosome has replicated, the pieces can join to form a chromosome with an internal region repeated, or duplicated (Figure 12-8). Neither, either, or both of the regions making up the duplication may be inverted with respect to the original arrangement. The remaining pieces may join to form a deficient chromosome. If the duplicated region is small and does not include the centromere, the chromosome may survive in successive generations.

**12.10  A nonrestituted break in each of two chromosomes can lead to either a reciprocal translocation or a half-translocation.**

When a single chromosome break occurs in each of two nonhomologous chromosomes, the four broken ends may sometimes undergo two nonrestitutional unions (Figure 12-9). The mutual exchange of chromosome segments between nonhomologous chromosomes produces a *reciprocal translocation*. When the reciprocal translocation is regular, the centric piece of one chromosome is joined to the acentric piece of the nonhomologous chromosome, and vice versa. When it is irregular, the two centric pieces are joined to form a dicentric chromosome, and the two acentric pieces are joined to form an acentric chromosome. Since, in the next nuclear division, the acentric will be lost and the dicentric will form a bridge if its centromeres proceed to opposite poles, the irregular reciprocal translocation is often lethal. Both types of reciprocal translocation are just about equally likely to occur. The regular type, however, may also be lethal to progeny when gametes are produced containing one

FIGURE 12-9. Reciprocal translocation between nonhomologous chromosomes.

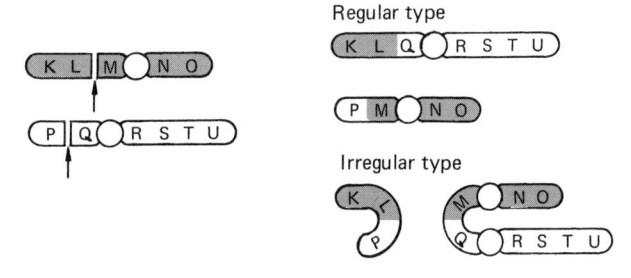

but not both members of the reciprocal translocation, that is, when they contain a *half-translocation*.

Figure 12-10 shows how segregation occurs in a diploid individual heterozygous for a regular reciprocal translocation, that is, that has one regular reciprocal translocation between two pairs of homologous chromosomes. Since the homologous black centromeres segregate independently of the homologous white centromeres, there are two alternative segregations. In segregation alternative 1, the unbroken chromosomes go to one pole and both rearranged ones go to the other. Gametes that contain either combination are genetically balanced. In segregation alternative 2, however, an unbroken chromosome and a half-translocation go to each pole. Gametes that contain either of these combinations are genetically unbalanced, each with a duplication and a

FIGURE 12-10. Diagrammatic representation of the results of segregation in a regular reciprocal translocation heterozygote. Chromatids are not shown. In both segregation alternatives, the black centromeres segregate from each other, as do the white centromeres. In segregation alternative 1, the unbroken chromosomes go to one pole and the rearranged ones go to the other. In segregation alternative 2, an unbroken and a rearranged chromosome go to each pole.

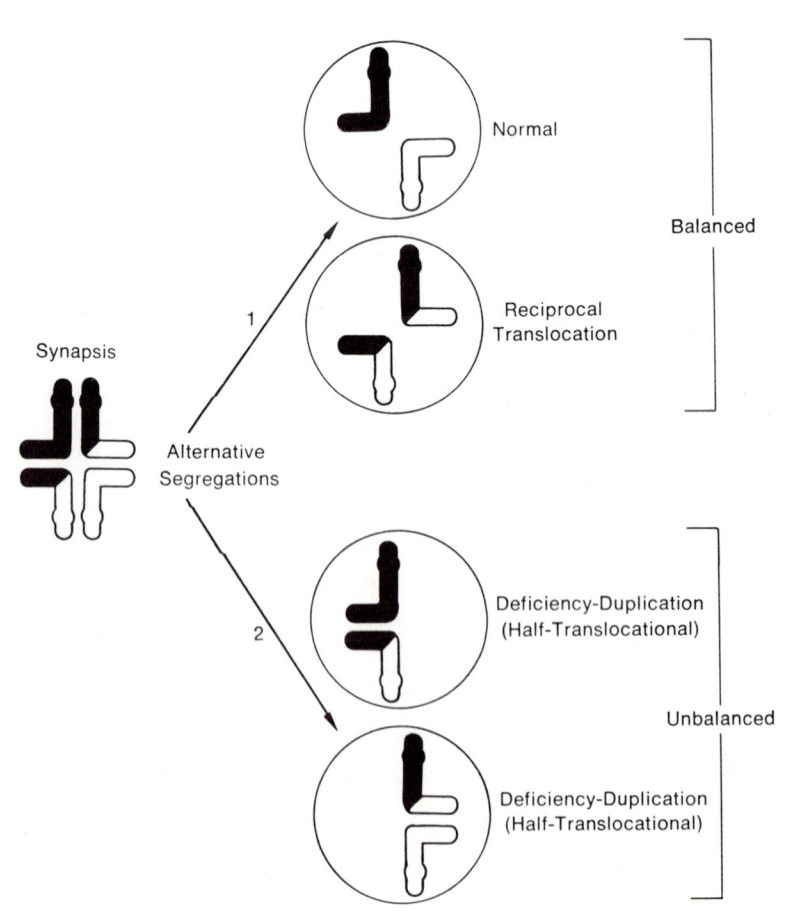

deficiency. In animals, such gametes are likely to be lethal to the developing zygote.

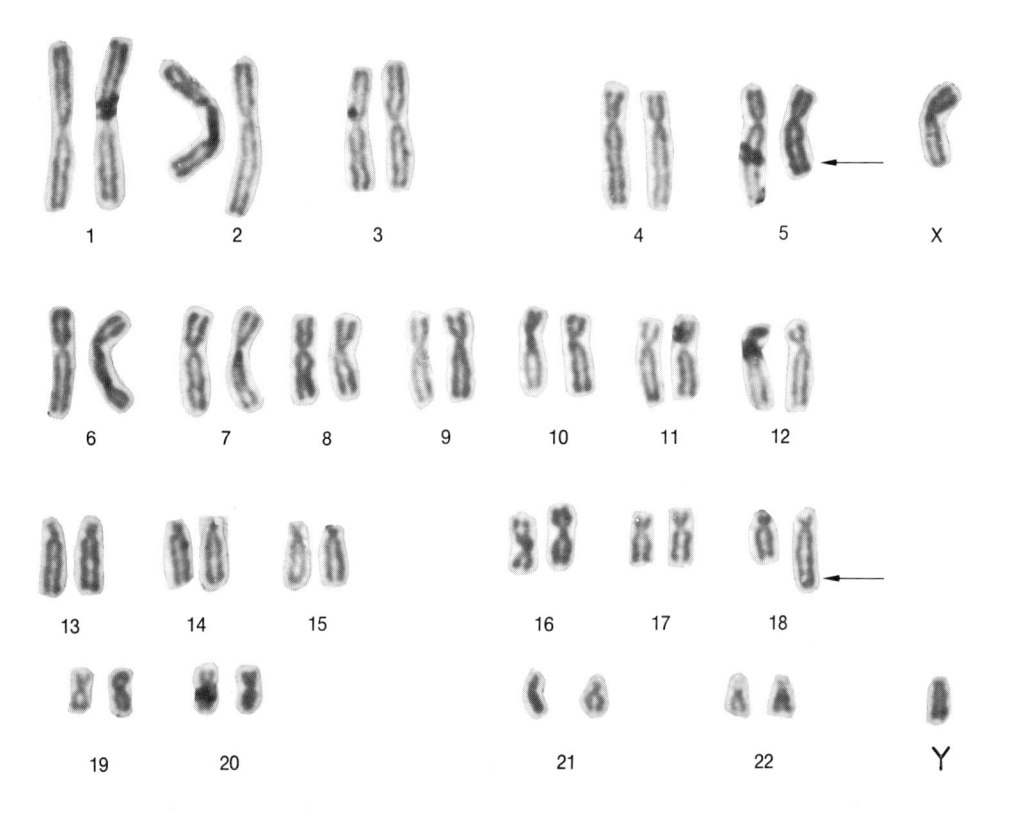

A gamete may also receive a half-translocation in cases where, after breakage, only one-half of a reciprocal translocation is formed, the two unjoined pieces being included in another gamete or in a polar nucleus.

## 12.11 Regular reciprocal and half-translocations occur in human beings.

Human beings contain a variety of gross chromosomal changes involving breakage. Some of these changes are described below; others, involving the sex chromosomes, are described in Chapter 19.

Individuals who are heterozygotes for a regular reciprocal translocation are genetically balanced. Accordingly, they usually show little or no detrimental phenotypic effect of the rearrangement, and are usually detected either accidentally or because they produce abnormal half-translocation-carrying children. Figure 12-11 shows the paired chromosomes of a normal man heterozygous for a regular reciprocal translocation between autosomes 5 and 18.

FIGURE 12-11. Male heterozygote for a regular reciprocal translocation between autosomes 5 and 18. (Courtesy of K. Hirschhorn.)

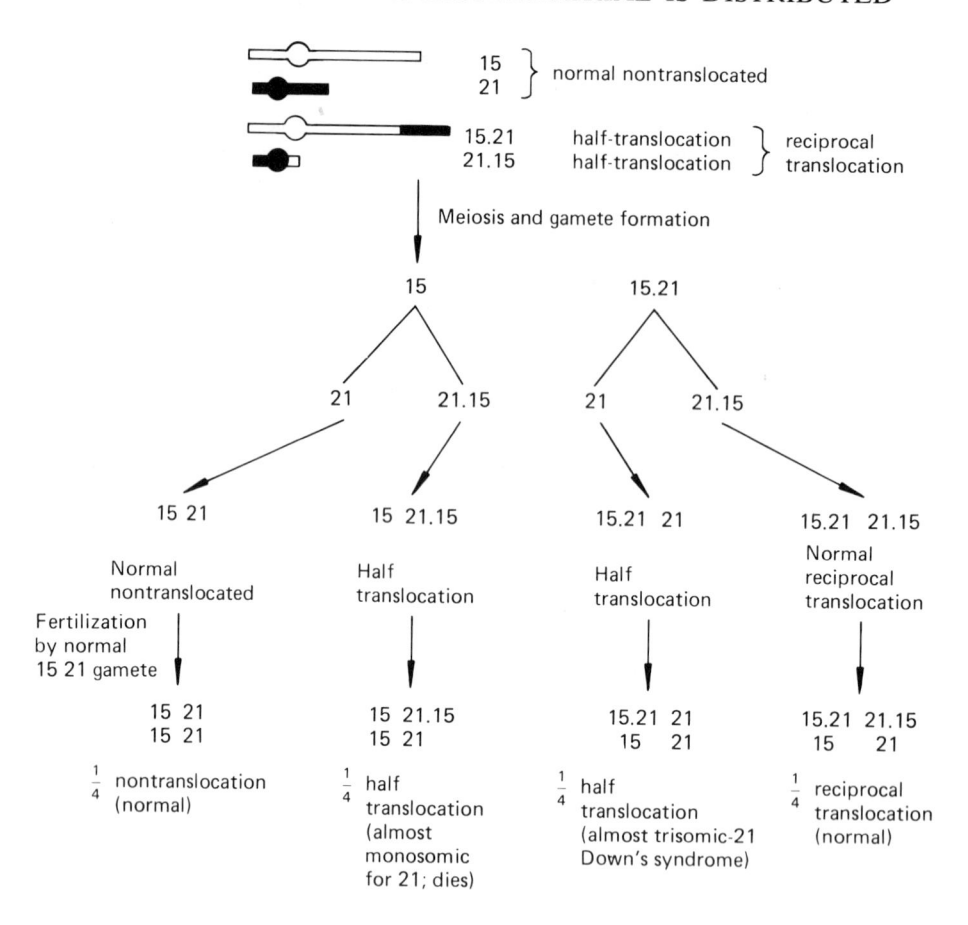

FIGURE 12-12. Down's syndrome due to a half-translocation.

Other phenotypically normal parents are heterozygotes for a regular reciprocal translocation between autosomes 15 and 21 (Figure 12-12). The break in 15 was in the heterochromatin near a tip, whereas that in 21 was in the heterochromatin of the long arm close to the centromere. The chromosome constitution can be represented by 15, 21, 15.21 (centromere of 15), and 21.15 (centromere of 21). When such a parent undergoes meiosis and gamete formation, as in Figure 12-10, the two 15 centromeres segregate independently of the two 21 centromeres. As a result, four types of gametes are produced in equal frequency that are fertilized by normal gametes, each containing 15 and 21. Of the resultant zygotes, $\frac{1}{4}$ are nontranslocational and develop normally, as do $\frac{1}{4}$ that contain the reciprocal translocation. Another $\frac{1}{4}$ of the zygotes have a half-translocation that makes them almost monosomic with regard to chromosome 21; these individuals usually die before birth. The last $\frac{1}{4}$ have a half-translocation that gives them an extra long arm of 21. These zygotes are almost trisomic for 21, and develop into individuals with Down's syndrome. Note that up to one third of the children of such a parent will have half-translocation Down's syndrome (Figure 12-12); and that no correlation exists between sex or age and these half-translocational children. The chromosomes of a male with such a half-translocational basis for Down's syndrome are shown in Figure 12-13.

Some half-translocational Down's children have a parent, apparently normal phenotypically, who contains the 15.21 half-translocation in addition to one normal 21 and one normal 15. Such a parent (with only 45 chromosomes) can apparently tolerate being haploid for the genes in the short arm of 21 and near a tip of 15. In this case, too, approximately one third of the children produced will have half-translocational Down's syndrome. About 4 per cent of all Down's children are of the half-translocational type.

Persons with the *cri-du-chat* ("cat-cry") syndrome have a similar origin. This syndrome is characterized by a mewing cat-like cry during the first year of life, numerous head defects, including widely spaced eyes and eye folds, and mental deficiency. This is due to a heterozygous deficiency for a segment of the short arm of chromosome 5. About 13 per cent of children with this syndrome carry a half-translocation of chromosome 5, received from a parent heterozygous for a reciprocal translocation between 5 and some other chromosome.

## SUMMARY AND CONCLUSIONS

Most cytological work with human chromosomes is done at metaphase.

Gross changes in nuclear chromosomes may or may not require breakage. Nonbreakage changes include single-species and multiple-species polyploidy, as well as the gain or loss of single whole

FIGURE 12-13. Metaphase containing a 15.21 half-translocation (arrow) among 46 chromosomes. This male has Down's syndrome as a result of being nearly trisomic for chromosome 21. (Courtesy of K. Hirschhorn.)

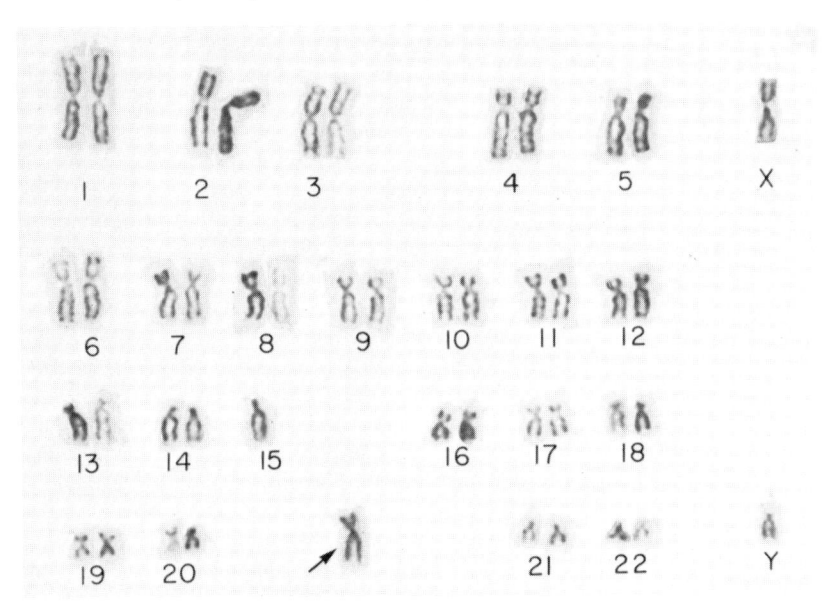

chromosomes. Whereas polyploidy does not shift gene balance appreciably, chromosome gain and chromosome loss do so—this also being the order in which they produce increasing phenotypic detriment. Monosomy and trisomy in human beings were discussed.

Breakage changes require chromosome or chromatid breakages. Most such breakages undergo restitutional union. When the unions of one or two breaks are nonrestitutional, however, a variety of gross changes are possible—including the formation of acentric or dicentric chromosomes, chromosome bridges, deficiencies, inversions, duplications, and reciprocal or half-translocations. Some examples of reciprocal and half-translocations in human beings were described.

## QUESTIONS AND PROBLEMS

**12.1** *Anagrams*: Unscramble and define each of the following: **a.** ollodippy. **b.** microsit. **c.** ticcinerd. **d.** sovirinne. **e.** bedrig.

**12.2** How does treating a cell culture with colchicine help cytologists make chromosome counts?

**12.3** Which of the following is false about polyploidy? **a.** It may occur normally or as a mutation. **b.** It increases the number of chromosome sets present. **c.** It may produce a hexaploid from a triploid. **d.** It never occurs in hybrids between different species. **e.** It upsets gene and chromosomal balance.

**12.4** Why is an otherwise healthy interspecific hybrid usually sterile?

**12.5** Assuming that the normal chromosome number is diploid, which member of each of the following pairs is expected to produce more phenotypic deteriment? **a.** Haploid vs. triploid. **b.** Monosomic diploid vs. trisomic tetraploid. **c.** Trisomic diploid vs. tetrasomic triploid.

**12.6** Which statement is correct? The most common consequence of a single chromosome break is: **a.** a breakage-fusion-bridge-breakage cycle. **b.** formation of acentrics and dicentrics. **c.** duplication of the breakage point during the S stage. **d.** restitution. **e.** healing of the broken ends.

**12.7** Which statements are correct? Down's syndrome children: **a.** always have older mothers. **b.** sometimes abort before birth. **c.** sometimes have a half-translocation. **d.** are phenotypically all alike. **e.** are never the fault of the father.

**12.8** "There are more ways to lose than to gain a chromosome." Explain.

**12.9** Given two nonhomologous chromosomes RST.UV and LMN.OPQ, where the periods indicate the position of the centromeres, draw an example of each of the following: **a.** a chromosome that will make a bridge. **b.** a chromosome that cannot move. **c.** a chromosome with a two-arm inversion. **d.** a regular reciprocal translocation. **e.** a chromosome without end.

**12.10** Given a chromosome ABCDE.FGHI, where the period represents the centromere, draw the appearance of this chromosome when synapsed with a homologous chromosome that has a: **a.** deficiency. **b.** duplication. **c.** one-arm inversion. **d.** two-arm inversion.

**12.11** What chromatids are produced after a single crossing over between two synapsed ring chromosomes?

**12.12** What chromatids are produced when a normal rod chromosome and a homologous chromosome with a one-arm inversion undergo a single

# GROSS CHANGES IN NUCLEAR CHROMOSOMES 189

crossing over: **a.** outside the inverted region? **b.** inside the inverted region?

**12.13** What happens when only one half of a regular reciprocal translocation is formed in the G1 period and is followed by: **a.** mitosis? **b.** meiosis I?

**12.14** Two haploid species have the same total chromosome DNA content. Species A has this DNA distributed among 5 chromosomes; species B, among 50. What kinds of gross chromosomal changes are equally likely in both species? More likely in species A? More likely in species B?

**12.15** Write an essay of 200 words or less to describe the events depicted in the accompanying figure.

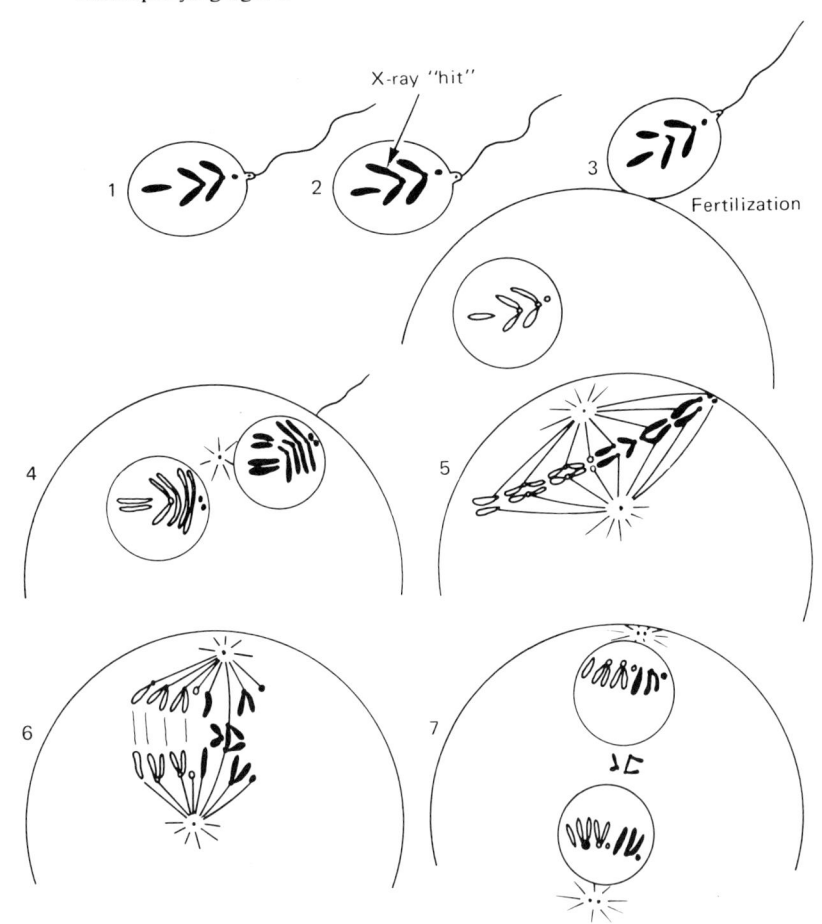

# Chapter 13

## Nonmendelian Genes in Eukaryotes

WE HAVE SEEN that eukaryotic organisms use a spindle mechanism in mitosis and meiosis to distribute nuclear chromosomes to daughter nuclei. Because these chromosomes follow the principles of segregation and independent segregation first discovered by Gregor Mendel, the genes they contain are called *mendelian genes*. Because prokaryotes such as *E. coli* do not have a spindle, their chromosomes are said to carry *nonmendelian genes*; such chromosomes use attachment to the cell membrane as the mechanism for separating daughters. This chapter deals primarily with nonmendelian genes that occur in eukaryotic cells.

**13.1** The cytoplasm or nucleus of a eukaryotic cell may contain nonmendelian or mendelian genes of other organisms.

A eukaryotic cell can be infected by (1) the naked DNA of a phage or bacterium, (2) DNA or RNA viruses, (3) prokaryotes, or (4) other eukaryotes (Figure 13-1). Such nucleic acid or organisms may occur in the nucleus, in the cytoplasm, or in various organelles located in the cytoplasm. Figure 13-2 shows diagrammatically the known, probable, and possible locations of infective and normal DNA genes in a eukaryotic cell.

When an infected eukaryotic cell divides, the free infecting nucleic acid or cell is distributed to daughter cells in a nonmendelian manner. When the infecting agent is also a eukaryotic cell, its nuclear genes are mendelian, of course. It is possible for one eukaryotic cell to be infected by another eukaryotic cell which is infected by a prokaryotic cell which is infected by a phage. This succession of infective organisms recalls the couplet:

> Big bugs have little bugs on their backs to bite 'em
> And little bugs have lesser bugs, and so *ad infinitum*.

| Infective Agent | Examples of Demonstrated Host or Host Location | Example or Phenotypic Effect |
|---|---|---|
| DNA, bacterial | Cultured mammalian cells, Ephestia, Drosophila | Transgenosis by synapsed or integrated genes |
| DNA, phage | Animal cells, tomato, Arabidopsis, cultured sycamore cell | Transgenosis by synapsed or integrated genes |
| DNA virus | | |
| 1. Pox | Cytoplasm | |
| 2. Most other | Nucleus | |
| SV40 | Nucleus | Integrated |
| Polyoma | Nucleus | Integrated |
| Epstein-Barr | Nucleus | Integrated |
| 3. ? | Chloroplast | |
| 4. ? | Mitochondrion | |
| RNA virus | | |
| 1. Poliovirus | Cytoplasm, human | |
| Killer | Cytoplasm, yeast | Killer trait |
| 2. RSV | Nucleus, chicken | Integrated as DNA |
| TMV | Nucleus, tobacco | |
| 3. TMV | Chloroplast | |
| 4. ? | Mitochondrion | |
| Prokaryote | | |
| 1. Blue-green algae | Protozoa (Glaucocystis) | |
| 2. Bacteria | | |
| Rickettsiae | Homo | Rocky Mountain spotted fever |
| Kappa | Paramecium | Killer trait |
| (?virus) | Kappa | |
| Spiroplasma | Drosophila | Male-lethal, sex-ratio |
| (?virus) | Spiroplasma | |
| Eukaryote | | |
| 1. Algae | Protozoa | Chlorella in Paramecium bursaria |
| 2. Yeast | Insects (Drosophila) | |
| 3. Protozoa | Homo | Malaria |

FIGURE 13-1. Infective DNA and RNA in eukaryotic cells.

## 13.2 Nonmendelian genetic material of viruses and bacteria may become mendelian by integrating into nuclear chromosomes.

By integrating into a nuclear chromosome, nonmendelian genetic material can become mendelian. We have already seen this in the case of Rous sarcoma virus whose nonmendelian RNA chromosome has a DNA

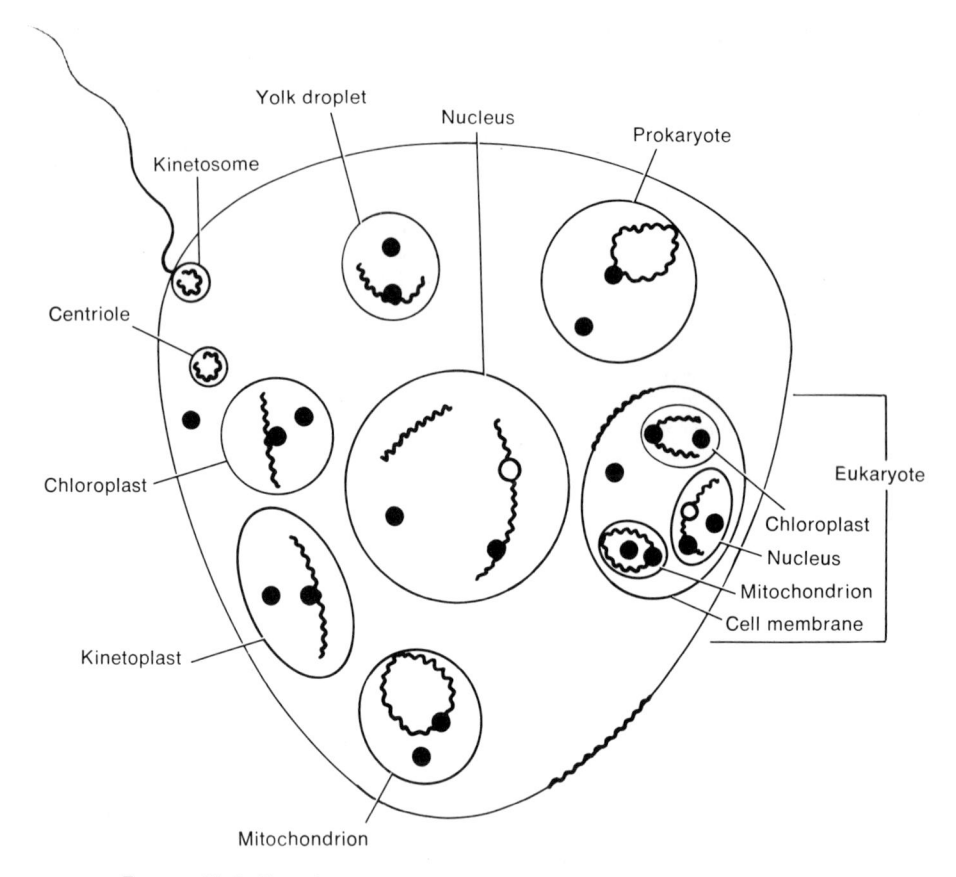

FIGURE 13-2. Some known, probable, and possible locations of DNA genes in a eukaryotic cell. Wavy lines represent DNA normally present in cell or organelle. DNA infecting a cell or organelle, indicated by solid circles, is shown both integrated and free.

reverse transcript that is integrated in a nuclear chromosome. DNA viruses such as SV40, polyoma, and Epstein-Barr apparently also integrate into nuclear chromosomes. It is believed that many DNA and RNA cancer-causing viruses have a stage in which their DNA or DNA reverse transcript becomes mendelian by integrating into a nuclear chromosome.

Genes of one organism that are transferred and expressed in another, widely separated or strange, organism may or may not become mendelian. Such *transgenosis* experiments usually involve either the uptake of naked bacterial DNA or the injection of phage DNA into eukaryotic cells. The donor DNA is marked in some genetically distinctive way, say by being prototrophic while the eukaryotic cell is auxotrophic. For example, a culture of sycamore cells that are unable to use lactose for energy can do so after they are infected with a $lac^+$-transducing phage. No transgenosis is observed if phage lacking $lac^+$ is used. In some cases, the infecting DNA may have integrated in the chromosomes of the recipient cell. In other cases, the donated DNA does not seem to have integrated because it is later subject to loss (for example, during meiosis).

## 13.3 Some of an organism's nuclear genes may become nonmendelian by mutation or by genetic programming.

Whenever mutation produces an acentric chromosome from a normal nuclear chromosome, the acentric automatically becomes nonmendelian genetic material.

Some organisms have a genetic program for synthesizing copies of certain nuclear genes and releasing these as small acentric chromosomes. This is true of nucleolus organizer DNA in the oocytes of many Amphibia (see Section 14.4). Such DNA is transcribed to yield 18S and 28S rRNA's. This is apparently true also of nucleolus organizer DNA in somatic cells of *Drosophila*. In such cells, copies of mendelian genes are nonmendelian genes. It is possible that the DNA associated with the cell membranes of lymphocytes (Section 8.7), which is derived from nuclear chromosomes, is also nonmendelian genetic material.

## 13.4 Chloroplast DNA is genetic material that codes for components of chloroplast ribosomes.

Many eukaryotic plant cells contain membrane-bound cytoplasmic bodies called *plastids*. Some of these organelles, the *chloroplasts* (Figure 13-3), are green because they contain chlorophyll; others, the *leucoplasts*, are white. The number of chloroplasts per cell varies from 1 (a giant one in the alga *Spirogyra*) to 30 to 50 (in a leaf cell). Chloroplasts lose their pigment in the dark to become leucoplasts but revert to chloroplasts upon exposure to sunlight.

The chloroplast is the site of photosynthesis, a complex anabolic process that requires the structural or enzymatic use of many proteins. The chloroplast contains RNA; it also contains 70S ribosomes that are presumably used to translate mRNA's into proteins needed for chloroplast structure or functioning. Some chloroplast proteins, including a DNA polymerase, are imported into the chloroplast, being translations of mRNA's, coded in the nucleus, on cytoplasmic ribosomes.

Most chloroplasts contain DNA. This DNA is double-stranded, not complexed with histone, usually rod-shaped, and is 15,000 or more base pairs long. Chloroplast DNA, *chl DNA*, replicates semiconservatively and is transcribed—hence it is genetic material. Each chloroplast may have two or more copies of the chloroplast chromosome. A large fraction of the chl DNA consists of redundant sequences, some of which code for the rRNA's of both subunits of chloroplast ribosomes. Other chl DNA sequences code for some chloroplast tRNA's and probably for some of the proteins in both subunits of chloroplast ribosomes.

Because of redundancy, there is probably not enough DNA information in the chloroplast to code for all ribosomal proteins; and probably few of the proteins needed for chloroplast structure or photosynthesis are coded for in chl DNA. Accordingly, most chloroplast proteins are either imported after being synthesized in the cytoplasm, or are the result of the translation of imported nuclear mRNA's on chloroplast ribosomes. That

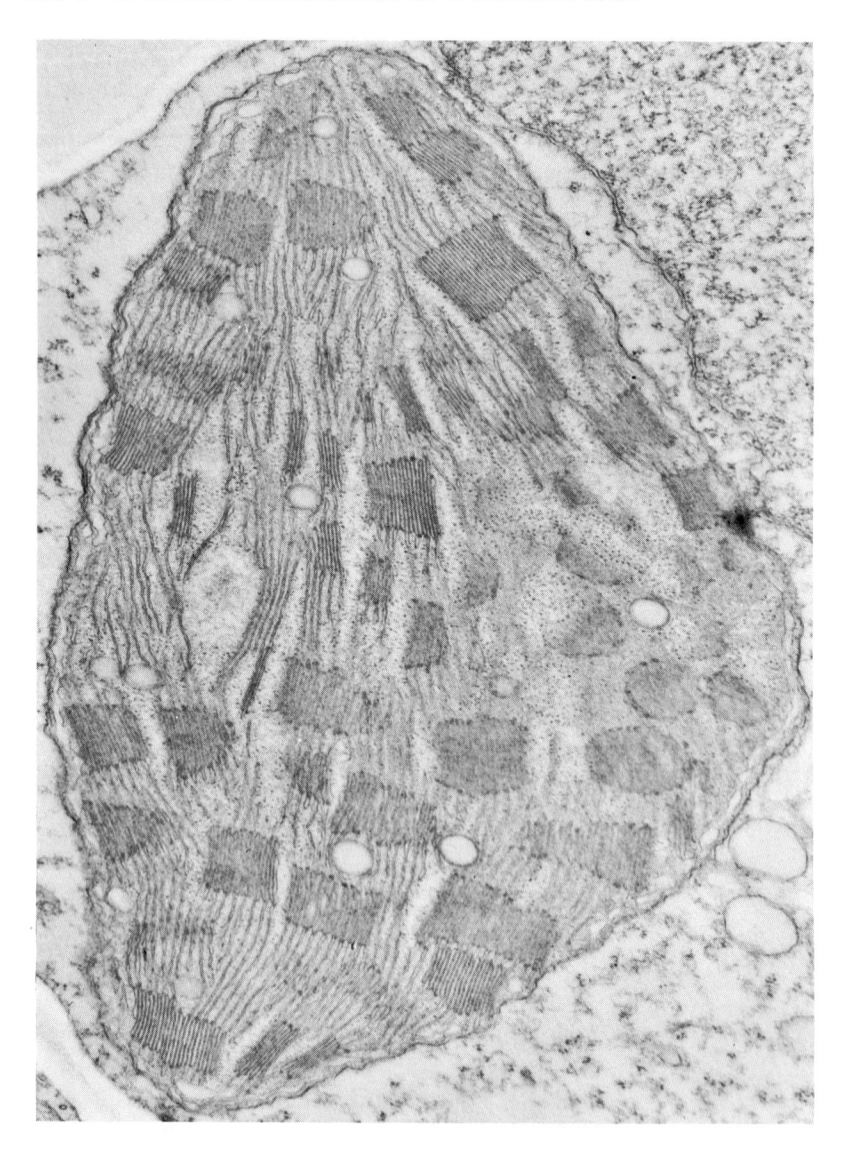

FIGURE 13-3. Electron micrograph of a cross section of a maize chloroplast. (Courtesy of A. E. Vatter.)

almost all of the proteins used in the synthesis and functioning of chloroplasts are coded, not in chl DNA but in other, probably nuclear DNA, is indicated by the absence of any detectable amount of DNA in 80 per cent of the chloroplasts of the single-celled alga *Acetabularia*, and the apparent loss of all detectable amounts of DNA in the chloroplasts of certain strains of the green single-celled organism *Euglena*.

The large subunit of the enzyme ribulose-1,5-bisphosphate carboxylase, however, is coded in chl DNA. This enzyme, composed of 8 large and 8 small subunits, is the major protein in a chloroplast and is responsible for the fixation of $CO_2$ during photosynthesis. The chloroplast contains many copies (sometimes about 75) of the gene coding the large subunit. The small subunit is thought to be coded in the nucleus. The mRNA for the small subunit is translated on cytoplasmic ribosomes. This mRNA seems to code for a signal polypeptide that enables the small subunit to cross the chloroplast membrane (Section 5.12).

## 13.5 Chloroplast DNA is nonmendelian genetic material whose loci recombine and can be mapped.

Since chl DNA is never associated with a spindle, it is nonmendelian genetic material. When a chloroplast divides, daughter chloroplasts receive copies of chl DNA perhaps by the same mechanism used in prokaryotic cells to distribute daughter chromosomes, namely, attachment to the surrounding membrane. The distribution of chloroplasts to daughter cells is also, of course, nonmendelian and more or less a matter of chance, which decreases as the amount of cytoplasm included in a daughter cell decreases. A pollen grain contains a haploid nucleus full of mendelian genes; since it contains very little cytoplasm, however, it rarely, if ever, carries a chloroplast or other plastid, and hence their nonmendelian genes.

When two green cells fuse, the fusion cell contains the chloroplasts of both parent cells. Such is the case in the single-celled alga *Chlamydomonas*, which has a single chloroplast. When *Chlamydomonas* of opposite mating types (+ and −, determined by a single pair of nuclear, mendelian genes) are mixed together, they can pair and fuse. Their single chloroplasts also fuse. Under ordinary conditions, however, only the chloroplast genes of the + parent persist and replicate in the fused cell, so only these are found in succeeding cell generations. (Perhaps the + parent has a restriction nuclease which degrades the chl DNA of the − parent.)

In rare cases, the fused *Chlamydomonas* cell spontaneously retains the nonmendelian chloroplast genes from both parent cells rather than just the + parent. (When the + parent is treated with ultraviolet light, the fused cell routinely retains and replicates the chl DNA from both parents.) When the parent cells differ by two chl DNA markers, however, occasional progeny are recombinants. For example, $ac2^+\ ac1$ and $ac2\ ac1^+$ parents produce a fusion cell bearing $ac2^+\ ac1/ac2\ ac1^+$. Recombinants with a reduced number of these loci are found, after only a few successive cell divisions, that contain only $ac2^+\ ac1^+$, or $ac2\ ac1$, and that apparently resulted from breakages between the two marker loci in each of the parental types of chromosome. The frequency of such recombination can be obtained by scoring the number of breakage recombinants per total progeny scored.

Recombination frequencies have been obtained this way for nine chloroplast loci. These frequencies show that all nine loci are linked (as expected) and, moreover, can be arranged in a circular map based on relative frequency of recombination (Figure 13-4). Recall from Section 7.5 that a circular recombination map can be obtained from a rod chromosome (such as a chloroplast chromosome) containing repeated

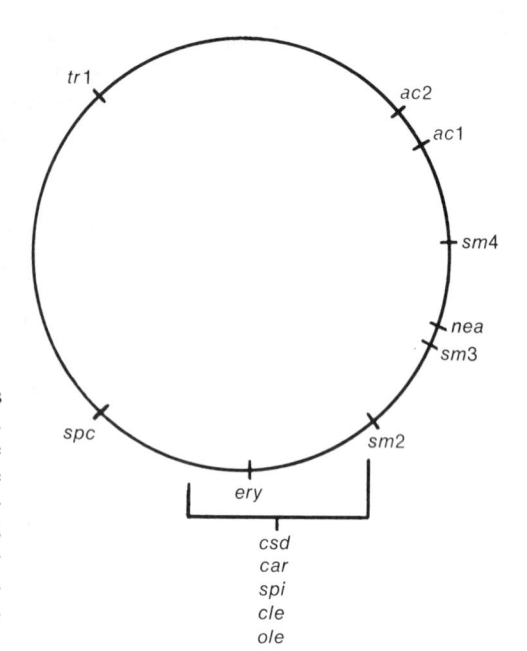

FIGURE 13-4. Circular genetic map of nonmendelian genes in *Chlamydomonas* based on recombination frequencies. (Modified from R. Sager.) Key to symbols: *ac*2 and *ac*1, acetate requirement; *sm*4, streptomycin dependence; *nea*, neamine resistance; *sm*3, low-level streptomycin resistance; *sm*2, high-level streptomycin resistance; *ery*, erythromycin resistance; *csd*, conditional streptomycin dependence; *car*, carbomycin resistance; *spi*, spiramycin resistance; *cle*, cleasine resistance; *ole*, oleandomycin resistance; *spc*, spectinomycin resistance; *tr*1, temperature sensitivity.

sequences of loci. Note that most of the chloroplast loci mapped code for resistance vs. sensitivity to different antibiotics that affect chloroplast but not cytoplasmic protein synthesis. Since antibiotic resistance is often due to a change in ribosomal proteins that protect rRNA, it is likely that different antibiotic loci in chl DNA code for different proteins in chloroplast ribosomes.

**13.6** Mitochondrial DNA is genetic material that codes for components needed for mitochondrial translation.

Mitochondria (Figure 13-5) are organelles found in all eukaryotic cells, consisting of a smooth outer membrane that is probably continuous with the endoplasmic reticulum and an inner membrane that forms double-layered folds. The outer membrane is, in general, permeable to small and moderately large molecules (with a molecular weight up to 10,000). The folds and the particles attached to them contain the enzymes that catalyze the reactions that synthesize adenosine triphosphate, the major source of chemical energy in oxygen-using cells. The number of mitochondria per cell varies from one (in a unicellular alga) to hundreds (in a kidney cell) to many thousands (in large egg cells).

Mitochondria contain mitochondrial DNA, *mit DNA*, which is double-stranded, not complexed with histone, usually circular in higher organisms, and contains about 14,000 base pairs (including 10 to 30 RNA nucleotides). This DNA seems to be located inside the inner membrane, attached to it at one point. A mitochondrion can average four to five circular mit DNA molecules. The total amount of mit DNA in the

NONMENDELIAN GENES IN EUKARYOTES 197

A

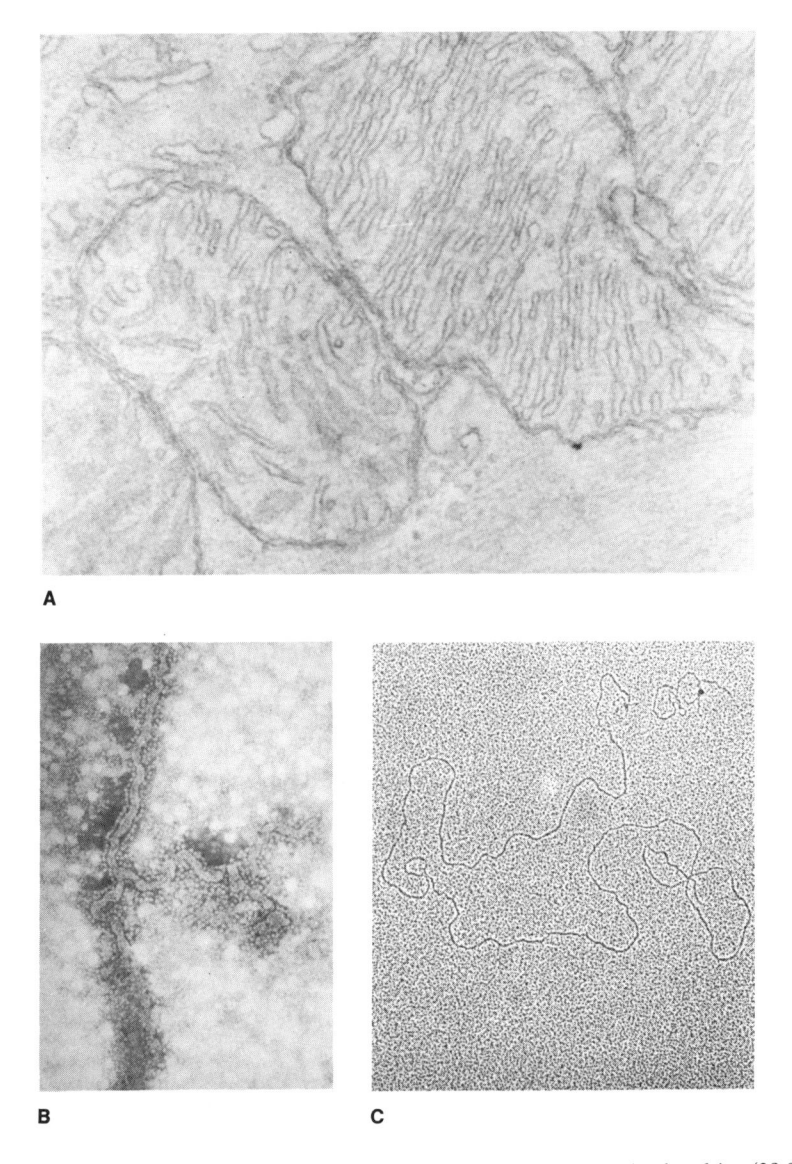

B                    C

FIGURE 13-5. Electron micrographs. **A**: Mouse heart mitochondria (38,000×). **B**: *Neurospora* mitochondria prepared so as to show folds with particles attached (35,000×). **C**: Outline of *Neurospora* mitochondrial DNA (16,000×). (Courtesy of Walther Stoeckenius.)

approximately 250 mitochondria in a mouse fibroblast cell is only about 0.15 per cent of the amount in the nucleus of that cell.

Mit DNA replicates semiconservatively, from one start point. After the circular mit DNA is isolated and nicked, its strands can be separated as heavy (H) and light (L) complements. *In vivo*, all the H and L strands are transcribed, although most of the L-strand transcript is rapidly degraded. Accordingly, mit DNA is genetic material.

Like chloroplasts, mitochondria contain ribosomes and the rest of the machinery for translation. Like chl DNA, mit DNA codes for 1) many different mitochondrion-specific tRNA's, but probably not a complete set, 2) the two rRNA's, and 3) probably some of the proteins of its 60S mitochondrial miniribosomes. In yeast, at least, mit DNA also codes for a small protein component of the inner mitochondrial membrane. Since the mitochondrion contains some 70 enzymes and numerous structural proteins, we conclude, as we did in the analogous situation with chloroplasts, that most mitochondrial structure and function is coded in the nucleus; and that the mitochondrion imports some smaller proteins (and some tRNA's) and many mRNA's for translation on mitochondrial ribosomes. In support of this conclusion is the occurrence in yeast of a defective type of mitochondrion that seems to contain no DNA.

## 13.7 Loci for rDNA's and tDNA's can be mapped in mit DNA by hybridization.

Besides being able to isolate the L and H complements of circular mit DNA, one can also isolate from mitochondria its tRNA's and two kinds of rRNA's. By mixing the rRNA's with each single-stranded DNA complement, one can detect rDNA loci from the sites in DNA that hybridize and become double-stranded. In the mit DNA of HeLa human tissue culture cells, for example (Figure 13-6), the H strand has one locus for 12S and one for 16S rRNA. The mit tRNA's are themselves too short to produce reliably detectable double-stranded regions when hybridized to the DNA. This difficulty is circumvented by first attaching to each

FIGURE 13-6. Circular map of the positions of the complementary sequences for tRNAs on the H and L strands of HeLa mit DNA. For the H strand, the positions relative to the 12S and 16S rDNA genes are shown. For the L strand, there is no reference point on the circular molecule; only the relative positions are significant. The thick lines indicating the positions of the duplex regions corresponding to the RNA–DNA hybrids are not drawn to scale. The total circumference of the mit DNA circle is 5.0 $\mu$m. (After M. Wu, N. Davidson, G. Attardi, and Y. Aloni, 1972. *J. Mol. Biol.*, **71**: 88.)

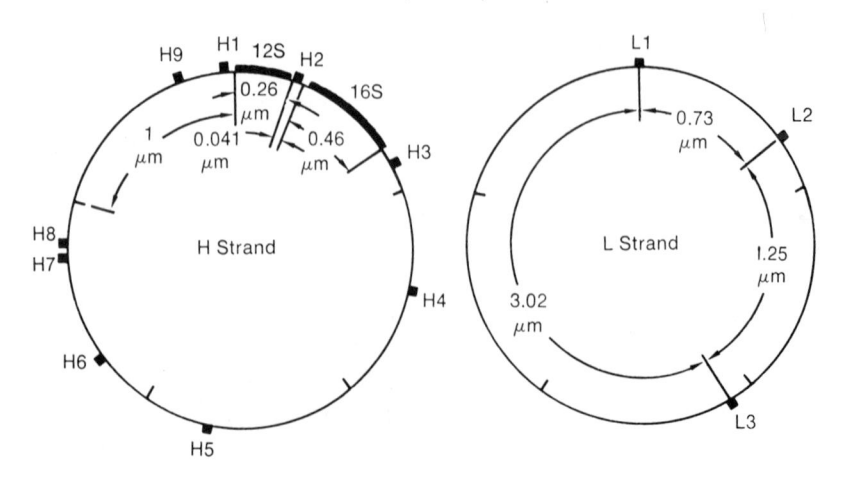

tRNA a marker molecule such as ferritin—an iron-containing molecule that is electron-opaque and hence readily seen as a granule in electron micrographs. In HeLa DNA (Figure 13-6), the H strand has nine sites complementary to tRNA; the L strand has three such sites. In the toad *Xenopus*, 15 sites of tDNA have been identified. The DNA whose function is identified by hybridization totals only about 5 per cent of the chromosome, emphasizing that the function of most of the mitochondrial chromosome (like that of most of the chloroplast chromosome) is unknown.

## 13.8 Mitochondrial DNA is nonmendelian and undergoes breakage and nonbreakage recombination.

Since mit DNA is never moved by attaching to a spindle, it is nonmendelian genetic material. Since mit DNA is attached to the inner mitochondrial membrane, daughter chromosomes probably separate to daughter mitochondria in the same way that daughter chromosomes separate to daughter cells in prokaryotes. Nonbreakage recombination of the DNA's in different mitochondria is indicated by cytological observations of mitochondria dividing or fusing together—both processes regrouping mitochondrial chromosomes.

While most mit DNA duplexes occur in the form of single circles, an appreciable amount also occurs normally in doubled condition (Figure 13-7), that is, as double circles—either as double-length rings (2 to 4 per cent) or as interlocked pairs of the single-length circle (5 to 10 per cent). Larger ring concatemers are also found. Under certain conditions, for

FIGURE 13-7. Some types of mit DNA found within an individual.

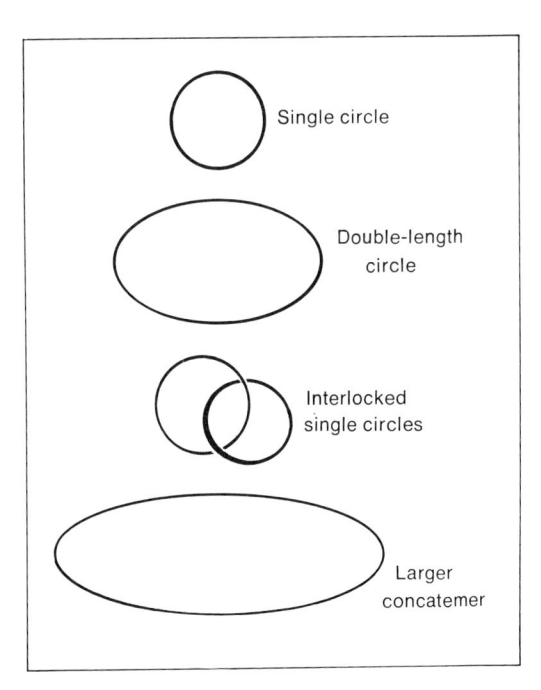

example protein starvation, as much as 70 per cent of mit DNA may occur in double circles. When normal conditions are resumed, the chromosomes return to normal double-circle frequency, indicating that the generation of circular dimers and higher multiple concatemers from single circles is reversible. This is another cytological evidence of recombination, some of which must require breakage of mit DNA.

Genetic recombination that requires breakage can be studied between mit DNA's in the same way this is studied between chl DNA's; that is, such studies usually employ markers for resistance vs. sensitivity to various antibiotics that affect mitochondrial protein synthesis but not cytoplasmic protein synthesis. The loci studied are, therefore, mit DNA loci that probably code for proteins in mitochondrial ribosomes. When two strains of yeast cells of opposite mating type differ by two such mit DNA markers, fusion cells are obtained containing both sets of markers. (Since yeast has no chloroplast or other plastids, there is no confusion as to which organelle carries the marked loci.) After a sufficiently large number of successive cell divisions has occurred, yeast cells are produced each of which carries the same genotype in all its mitochondria. Such progeny cells are then scored for recombinants that carry only one locus for each marker. The frequency of such recombinants, which required breakage to be produced, has been used to make a linkage map for these loci in mit DNA.

**13.9** **The cytoplasm may contain a variety of other normally occurring DNA's, some of which are genetic material.**

Different eukaryotic cells may normally contain in their cytoplasm one or more of a variety of other DNA's. These are listed and discussed separately.

1. KINETOPLAST DNA. Certain parasite protozoans, including *Leishmania* and *Trypanosoma*, contain a cytoplasmic organelle called a *kinetoplast* (Figure 13-8), that is involved with cell motility and is located in a modified region of a mitochondrion. The kinetoplast contains several to many copies of readily detectable double-stranded DNA that is bound to a histone-like protein. Kinetoplast DNA replicates semiconservatively and is apparently also transcribed. It is, therefore, genetic material whose function should be revealed by further investigation.

2. CELL MEMBRANE DNA. Although the cell membranes of many eukaryotic cells do not contain detectable amounts of DNA, 7S to 18S DNA's have been found bound to cytoplasmic membranes, including the cell membrane (Figure 13-9), in chick embryo cells, human kidney cells, and cultured liver cells and lymphocytes (see Section 8.7). These DNA's seem to be synthesized in the nucleus, appearing in the cytoplasm after the start of the premitotic S stage. They are linear, duplex molecules, each containing some 10,000 nucleotides in various sequences. Since we do

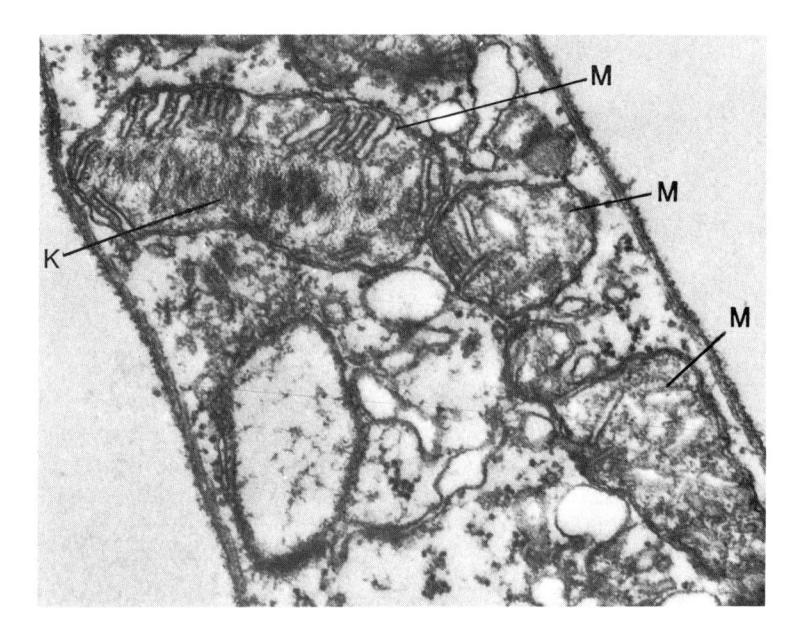

FIGURE 13-8. An electron micrograph of a longitudinal section through *Trypanosoma lewisi*, showing the kinetoplast consisting of a mass of DNA (K) within a specific region of a mitochondrion (M). (Courtesy of H. C. Renger and D. R. Wolstenholme, 1970. *J. Cell. Biol.*, **47**: 689–702.)

FIGURE 13-9. Electron micrograph of cell membranes from diploid human lymphocytes, showing associated DNA molecules. DNA was not seen when membranes were treated with DNA nuclease. Although the function of this plasma-membrane-associated DNA is still unknown, it may, perhaps, be related to antibody production. (Courtesy of R. A. Lerner, W. Meinke, and D. A. Goldstein, 1971. *Proc. Nat. Acad. Sci., U.S.*, **68**: 1212–1216.)

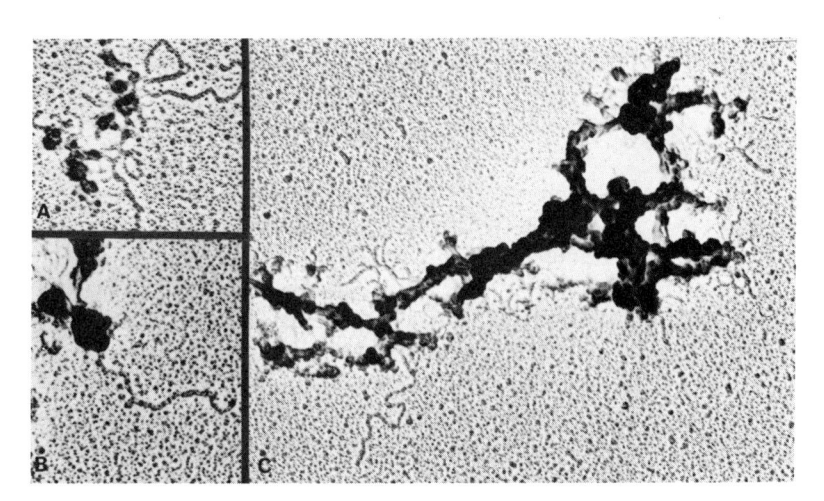

not yet know whether such cell membrane DNA is either replicated or transcribed, we do not know whether or not it is genetic material.

3. YOLK DNA. The yolk droplets found in the oocytes of amphibians such as *Xenopus* contain duplex, linear DNA about the same size as mit DNA. Since we do not know whether such yolk DNA is replicated or if it was synthesized using a nucleic acid template and is used as a template to synthesize nucleic acid, we do not know whether or not it is genetic material.

4. CENTROSOMAL AND KINETOSOMAL DNA'S. In most animal cells, including those of human beings, a clear, gel-like region called the *centrosome* is found at each pole of the spindle. (The centrosome is sometimes seen to contain a granular structure, the *centriole*.) Since centrosomes and centromeres are both attached to the spindle, and since freed centromeres mimic centrosomal behavior, these structures seem to be homologous. Moreover, like centromeres, centrosomes seem to contain double-stranded DNA. We do not know how centrosomal DNA is related to centrosomal (or centriolar) structure or functioning. The *kinetosome*, a structure found at the base of each cilium or flagellum that is responsible for ciliary and flagellar motion, seems to be derived from a centrosome (or centriole). DNA is also sometimes detected in kinetosomes. Since the amount of DNA per centrosome or kinetosome is relatively small, it will be difficult to determine whether or not it is genetic material.

## SUMMARY AND CONCLUSIONS

Most of the genetic information in a eukaryotic cell is contained in nuclear DNA that is mendelian genetic material. Other, nonmendelian genetic material is usually, probably always, present in a eukaryotic cell.

Eukaryotic cells may contain nonmendelian genetic material, introduced by infection, consisting of the DNA or RNA chromosomes of viruses or other cellular microorganisms that may replicate in the nucleus, the cytoplasm, or in cytoplasmic organelles. Although these chromosomes can replicate and undergo distribution independently of the host's chromosomes or spindle, some or all of their genes may be able to integrate with host nuclear chromosomes and thus become part of the host's mendelian genetic material.

In certain cells, some of the organism's nuclear genes become nonmendelian as a result of breakage mutations or by genetic programming.

The DNA in such normally present organelles as chloroplasts and mitochondria is nonmendelian genetic material. While the function of most of this DNA is unknown, a small fraction of it codes for components needed for intraorganelle translation. Most of the information needed to construct these organelles is coded in the nucleus and is translated in the cytoplasm or in the organelle. Recombination frequencies for nonmendelian genes have produced linkage maps for chloroplast loci and for mitochondrial loci. Mit tDNA and rDNA loci have been mapped in the mitochondrial chromosome by the hybridization technique.

Readily detected amounts of duplex DNA are also present in the kinetoplasts, cell membranes, and yolk droplets of particular cells. Kinetoplast DNA is genetic material. Relatively small amounts of DNA are detected in some centrosomes and kinetosomes. Further research is needed to determine the function any of these DNA's may have, and whether any, in addition to kinetoplast DNA, is genetic material.

## QUESTIONS AND PROBLEMS

**13.1** *Anagrams*: Unscramble and define each of the following: **a.** alminneed. **b.** spitlad. **c.** spitaltoken. **d.** cosnormeet.

**13.2** *Riddle*: What eukaryotic DNA is mendelian but not yet known to be genetic?

**13.3** Do DNA or RNA viruses integrate into nuclear chromosomes? Explain.

**13.4** How would you prove that the drug rifampsin inhibits the functions of the transcriptase in chloroplasts but not in nuclei?

**13.5** Give two pieces of evidence that at least part of a chloroplast is coded outside the chloroplast.

**13.6** How would you prove that the drug chloramphenicol inhibits translation by mitochondrial ribosomes but not cytoplasmic ribosomes, and that cycloheximide has the opposite effect?

**13.7** In *Chlamydomonas*, mating type is determined by a nuclear locus and neomycin resistance by a chloroplast locus. A cross between two haploids, $mt^+ nea^+ \times mt^- nea^-$ produces a zygote that immediately undergoes meiosis. What genotypes are expected among the meiotic products—the first four cells produced from the zygote?

**13.8** How can you show that mit DNA contains some RNA?

**13.9** When transcription and translation are inhibited in mitochondria but not elsewhere in yeast, mitochondrial transcriptase accumulates in the cytoplasm. What does this result indicate about the origin of this enzyme?

**13.10** In yeast, chloramphenicol ($c$) and erythromycin ($e$) affect protein synthesis in the mitochondrion but not in the cytoplasm. Resistance to either of these drugs is inherited in a nonmendelian manner. Where are the drug-resistant $c^r$ and $e^r$ loci located?

A cross is made between $e^s c^r \times e^r c^s$ yeast. When the progeny have divided enough times so that each contains only one type of mit DNA, the following progeny are obtained:

$e^s c^r$  46.4 per cent
$e^r c^s$  44.0 per cent
$e^s c^s$  3.3 per cent
$e^r c^r$  6.1 per cent

What conclusions can you draw from these results?

**13.11** In what respects do chloroplasts and mitochondria resemble prokaryotes?

**13.12** How do chl DNA and mit DNA differ?

**13.13** State two possible reasons why chl and mit DNA's have not been replaced by additional nuclear DNA.

**13.14** Which statement is correct? Chloroplast and mitochondrial DNA's: **a.** have not been proven to be genetic material. **b.** code primarily for some of the translational machinery these organelles contain. **c.** code for many of the proteins needed to make these organelles. **d.** are single-stranded. **e.** are translated as well as being replicated and transcribed.

*PART IV*

# HOW THE GENETIC MATERIAL CHOOSES WHICH PARTS ARE PRESENT AND FUNCTIONAL

# Chapter 14

# Programmed Gene Synthesis, Destruction, and Mutation

IN PRECEDING PARTS we have examined the structural, functional, mutational, and recombinational properties of genetic material. We have not considered, however, all the kinds of information that must be contained in the genetic material of an organism. The maintenance, growth, and reproduction of an organism requires coordination of many different cellular and often multicellular activities. This coordination requires that specific genes act in specific patterns and sequences. In this part we will first present evidence that the genetic material programs which of its parts are present, in how many replicas, and how stable these are—that is, we will discuss programmed gene synthesis, destruction, and mutation. In the next two chapters we will consider how the genetic material chooses which of the genes present are functional at different times.

# GENE SYNTHESIS

**14.1** DNA replication occurs in replication units called replicons.

Replication of DNA occurs in replication units, called *replicons*, each of which has a single gene where replication starts and then continues to completion. The DNA duplex chromosome of a prokaryote is a single complete replicon, whereas that of a eukaryote contains several to many replicons.

The DNA's of phages, plasmids, and episomes of prokaryotes also contain replicons, since they contain the information for independent replication. Not all infective DNA's behave like replicons, however. For example, the donor DNA that is not integrated in transformation or transduction is apparently not replicated (although it may be functional);

208    WHICH PARTS ARE PRESENT AND FUNCTIONAL

and the DNA of certain mutants of F cannot replicate in the free state. In each of these examples the DNA is a defective replicon.

## 14.2 Replicon action in prokaryotes is regulated by regulator genes.

Bacteria maintain a fixed chromosomal content generation after generation. In order to do so, DNA replication is coordinated with the other metabolic activities leading to cell growth. The signal to initiate DNA synthesis may be the result of cycles of synthesis and degradation of protein *initiator substances*. In *E. coli*, protein initiator substances need to interact with the replication start gene before DNA replication can begin. The initiator proteins that stimulate replicon action are said to be *regulatory proteins* coded by *regulator genes*. Replication initiation is said, therefore, to be under *positive control* by initiator proteins.

When a bacterium also contains a dispensable autonomous chromosome, the replication of this second chromosome is often controlled by a regulatory protein coded in its own DNA. For example, replication of free F (or derived F) is regulated by a protein *repressor* coded in the sex factor. This repressor permits an average of one sex particle replication per bacterial cell division. An F⁻ cell contains no repressor; so, when an F⁻ cell receives a free F, F multiplication proceeds at an accelerated pace for a time. As F functions in its new host, regulation is established by means of the repressor, so the sex factor chromosome, like the bacterial chromosome, replicates only once a generation. (Some plasmids code for a repressor that maintains 10, 20, or some other relatively constant number of its chromosomes per host cell.) Replication initiation is said to be under the *negative control* of such repressor proteins.

FIGURE 14-1. Physical and genetic map of free λ. Black arrows indicate the direction of transcription but do not necessarily define individual mRNA species. Dots indicate sites of action of λ DNA. Gray arrows indicate where the gene products act. Genes A through J are involved in formation of phage heads and tails. Key to symbols: $att^\phi_\lambda$, site of insertion of phage DNA into host DNA; *int*, *xis*, integration and excision of λ DNA; *red α* (a nuclease) and *red β*, λ recombination system; *cIII*, *cII*, required for expression of *cI* gene; *N*, required for efficient expression of early genes; *rex*, restriction of growth of T4rII; *cI*, λ repressor which binds to $O_L$ and $O_R$; *tof*, negative regulator of *cI* transcription; *O, P*, required for λ replication initiated at *ori* (origin of replication); *Q*, stimulates transcription of late genes at late gene promoter, $p^l$, between genes Q and S; *S, R*, required for lysis of host. (Courtesy of Ira Herskowitz.)

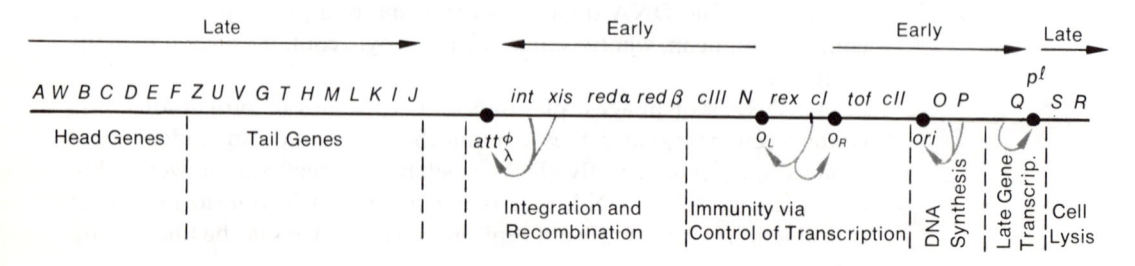

How is replication programmed when a bacterial chromosome contains two replicons—one bacterial, the other episomal? The answer is that only one of the replicons is functional in any cell. In the case of Hfr *E. coli*, the replicon of integrated F is repressed completely during cell multiplication. Replication starts at the replication start gene of the bacterial chromosome and the genes of integrated F are replicated just as any other loci in the *E. coli* chromosome. Integrated F apparently synthesizes a protein repressor that completely prevents not only the functioning of its own replicon but also that of any free F or derived F that may be present in the cell. For this reason, Hfr cells ordinarily contain but a single F. When Hfr is undergoing conjugation, however, replication starts at the locus of integrated F. In this case the repressor of integrated F is not functional because it is inactivated and/or ceases being produced.

A similar situation occurs when a bacterium is infected by a phage episome. For example, when *E. coli* is infected by $\phi\lambda$ and the $\lambda$ DNA is autonomous, its replicon is functional. The free $\phi\lambda$ chromosome has a preferred replication start gene (*ori*, in Figure 14-1), and evidence indicates that the phage codes for an initiator substance which, when joined to the DNA replication complex of *E. coli*, recognizes *ori* in free $\lambda$ DNA.

When $\lambda$ DNA is integrated, however, the $\lambda$ replicon is repressed and is replicated only when the *E. coli* chromosome replicates. In this case, repression is due to a $\lambda$-coded protein, $\lambda$ *repressor*, which is the translation product of $\lambda$'s regulator gene *cI*. Similarly, $\lambda$ repressor also represses the replication of free $\lambda$ DNA. Because of this repression, *E. coli* that are lysogenic for $\lambda$ are rendered immune to subsequent infections by $\lambda$ phages. Whether $\lambda$ acts as an active replicon (in a lytic cycle) or as an inactive replicon (in a lysogenic cycle) depends upon the balanced interaction of several stimulating and inhibiting regulator genes of the phage.

We see, therefore, that replicon action is regulated by regulatory genes whose protein products are either stimulatory or repressive.

## 14.3 Replicon action is also regulated in eukaryotes.

We know relatively little about the genetic and molecular mechanisms that regulate DNA replication in eukaryotes. Such regulation must be more complex in eukaryotes than in prokaryotes since in eukaryotes (1) the chromosomes are much longer and contain multiple replicons; (2) there are usually two or more chromosomes per nucleus, and (3) the organism is often the product of many coordinating cells. We shall describe these complexities in some detail to appreciate the kinds of control that DNA replication must be subject to in eukaryotes. Whatever is known about the mechanisms for regulating DNA replication will also be presented.

INTRACHROMOSOMAL PROGRAMS. All replicons within a chromosome must function within the S period preceding nuclear division, with the replicons of heterochromatin replicating after those of euchromatin. Completion of chromosome replication therefore requires linking together pieces of DNA synthesized at different times. Since constitutive heterochromatin remains late-replicating after being moved to other locations by inversion or translocation, it is concluded that each replicon contains information that regulates its time of replication. Nothing is known at present about the genetic mechanism for the regulation of intrachromosomal DNA replication. Histones may be part of such regulation, since they interfere with DNA strand separation and the action of certain nucleases (see also Section 15.8).

INTRACELLULAR PROGRAMS. The following observations indicate that replication of DNA is regulated at the nuclear, organellar, and cellular levels:

1. Nuclear DNA replication occurs only in the S stage of interphase. A study of one of the enzymes needed in DNA replication shows that the enzyme is synthesized when needed and is subsequently destroyed or inactivated.
2. DNA synthesis stops in most nuclei after one round of replication is completed by all chromosomes.
3. Nuclear DNA replication depends upon cytoplasmic factors, as demonstrated by the following observations:
    a. Nuclei that have stopped DNA synthesis will start to do so within an hour or two after being transplanted into the cytoplasm of cells that are starting DNA synthesis.
    b. When nuclei that are synthesizing DNA are transplanted into the cytoplasm of cells that are not, the transplanted nuclei stop synthesizing DNA.
    c. When nuclei and cytoplasm from various stages of the cell cycle are mixed, DNA synthesis is started only when G1-period nuclei are placed in S-period cytoplasm. In all the preceding cases the cytoplasmic factors have an unknown, perhaps nuclear, origin.
    d. When green cells are treated with drugs that specifically prevent transcription of chloroplast DNA or translation on chloroplast ribosomes, nuclear DNA replication ceases before chl DNA replication. This strongly suggests that nuclear DNA synthesis is regulated in part by organellar genes.
    e. The limited range in the number of chromosomes and the specific timing of DNA replication in chloroplasts, mitochondria, and kinetoplasts are evidence that organellar DNA replication is regulated at the organelle and cell levels.

INTERCELLULAR PROGRAMS. The following evidence indicates that nuclear DNA replication is regulated at a multicellular level:

1. Studies of cells in tissue culture show that the initiation of DNA synthesis is regulated at the tissue level by the arrangement of cells relative to each other. Release of a normal cell from contact with neighboring cells promotes the initiation of DNA synthesis.
2. The following results indicate that chromosome replication in different organs is coordinately regulated, even though some organs exercise a certain independence from such control. In many Diptera, including *Drosophila*, the tissue cells in certain organs of the maggot or larval stage (Figure 14-2) undergo a series of successive DNA replications, which are not followed by

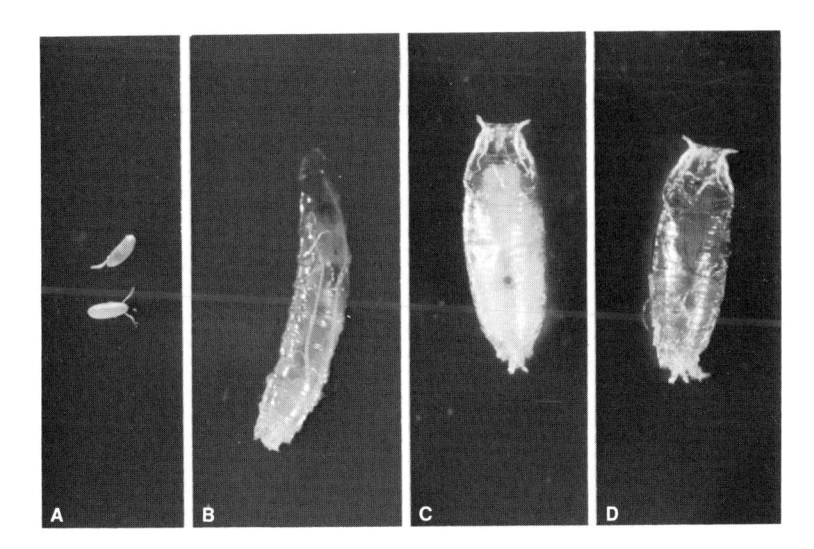

FIGURE 14-2. Egg (**A**), mature larva (**B**), early pupa (**C**), and late pupa (**D**) of *D. melanogaster*, all at the same magnification as the adults photographed in Figure 19-1. (Courtesy of L. Ehrman.)

nuclear division, and which produce interphase chromosomes containing increasing numbers of chromatids. (This process is described further in the next section.) It is found that (a) all the cells of such tissues (but not those of other tissues) undergo such DNA replications, (b) the number of such replications within a tissue is relatively narrow, (c) the replications repeatedly occur synchronously, although (d) some tissues continue DNA replication after others have ceased to do so. How this selective coordination is programmed is unknown.

We conclude from all the preceding observations that DNA replication in eukaryotes must be coordinated within a nuclear chromosome, among different nuclear chromosomes, between the nucleus and organelles, as well as within a tissue, and among different organs. A better understanding of the specific genetic programs that produce such cooperation will require extensive additional research.

## 14.4 More copies are made of some nuclear genes than of others.

Although euchromatin and heterochromatin replicate at different times, nuclear chromosomes and their parts ordinarily complete the same number of replications—one in preparation for mitosis and meiosis; or more than one in cases of polyploidy. Many exceptions are known, however, in which some nuclear chromosomes or parts replicate more often than others. Such differential or disproportionate replication of DNA in the somatic or germ line results in *amplification* of the DNA made in extra copies or *underreplication* of DNA made in too few copies. In no case do we yet know the genetic program or molecular mechanism for disproportionate DNA replication. Amplification is not known to occur generally in human beings, although one abnormal family shows a chromosome with an extra duplication of one of its arms, so that at metaphase it appears as a three-armed chromosome. We will consider three specific types of amplification that routinely occur in three different types of organism:

1. AMPLIFIED CHROMOSOME SETS. Males of the coccid *Planococcus* ordinarily have five pairs of chromosomes in the nucleus of a somatic cell. In certain cells, however, the maternally derived chromosomes undergo polyploidy while the paternally derived set, which is facultatively heterochromatic, does not. As a result, these cells come to contain up to 80 maternal-type chromosomes but only 5 paternal-type chromosomes. In this case, therefore, amplification results in differential polyploidy.

2. AMPLIFIED, INTEGRATED CHROMOSOME PARTS. As mentioned in the last section, the interphase chromosomes in cells of certain larval tissues of *Drosophila* undergo repeated DNA replications without being followed by nuclear or cell divisions. In such cells all the daughter

FIGURE 14-3. Salivary gland chromosomes of a female larva of *D. melanogaster*. (By permission of The American Genetic Association, from B. P. Kaufmann, 1939. *J. Hered.*, **30** (5): frontispiece.)

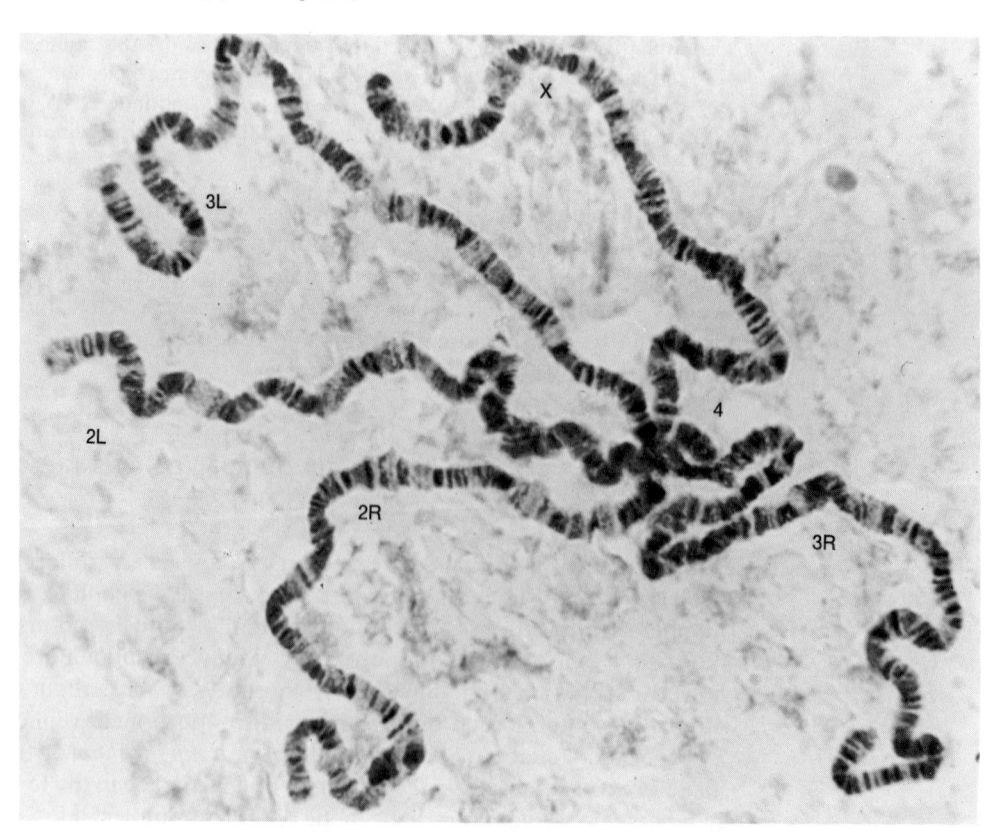

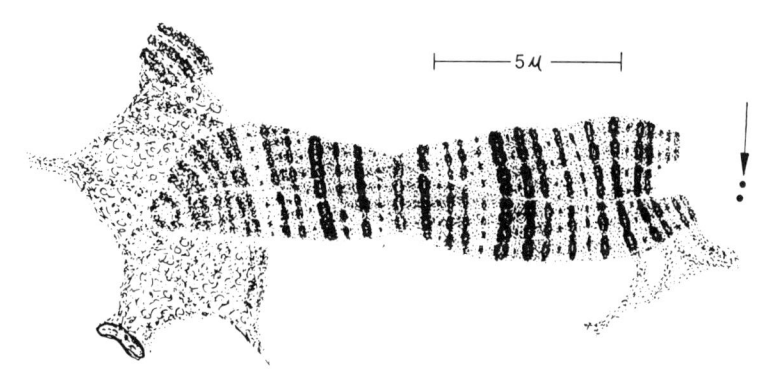

FIGURE 14-4. Pair of fourth chromosomes as seen in salivary gland nuclei (each chromosome has many chromatids) and at mitotic metaphase (arrow), drawn to the same scale. (By permission of The American Genetic Association, from C. B. Bridges, 1935. *J. Hered.*, **26**: 62.)

chromatids in a chromosome stay together. Since each such chromosome can contain up to 512 chromatids, and is therefore relatively thick (although as unwound as ordinary interphase chromosomes are), it is readily seen in the light microscope under relatively low magnification. These giant chromosomes contain a series of crossbands of varying density and thickness (Figure 14-3). A crossband results from the apposition of the same relatively coiled segment of each of the chromatids in the chromosome (Figure 14-4); whereas an interband results from the apposition of the same relatively uncoiled region of each of the chromatids in the chromosome. Since the homologous chromosomes in such cells pair with each other, double cables are produced which can contain up to 1024 chromatids. The pattern of bands is so constant and characteristic that it is possible to identify not only each chromosome but particular regions within a chromosome on the basis of the banding pattern (Figure 14-3).

Studies of such giant chromosomes in the salivary gland of the *Drosophila* larva reveal that whereas many euchromatic regions undergo the same number of successive DNA replications, certain other regions, particularly the one containing nucleolus organizer DNA, undergo fewer; while the constitutive heterochromatic regions adjacent to the centromere and at the tips of chromosomes undergo few or no replications. We see, therefore, that such chromosomes have different regions amplified to different degrees.

A model showing such intrachromosomal differences in amplification, which keeps all sister chromatids together by base pairing between complementary strands, is given in Figure 14-5. This shows the individual strands of DNA duplexes in a chromosome region that has undergone an eightfold amplification. The branch points on the drawing represent places where semiconservative replication has occurred; the extra replicas are retained as an integral part of the chromosome by means of base pairing. Figure 14-6 shows how this model can be specifically applied to the X chromosome of the larval salivary gland nucleus in *Drosophila*. In

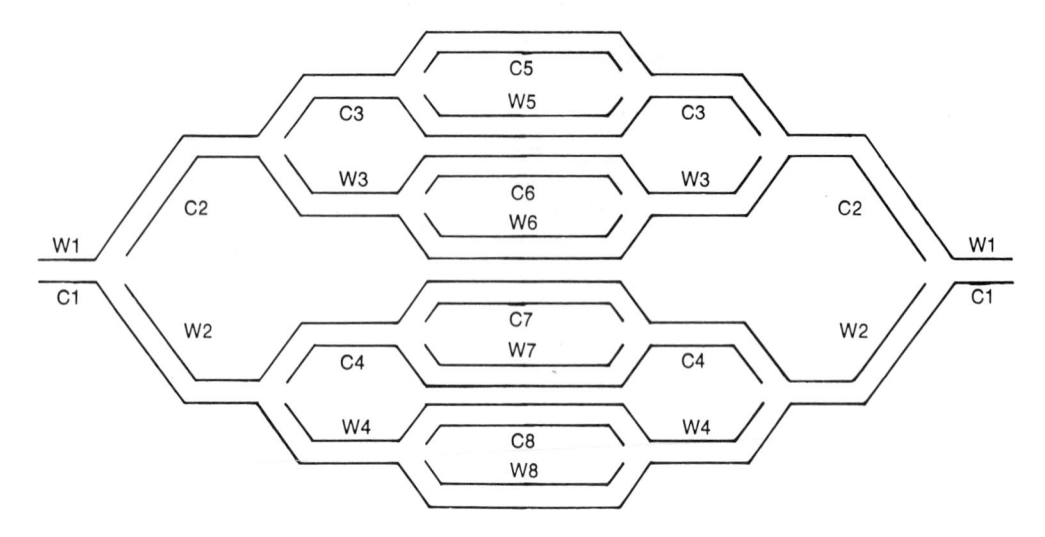

FIGURE 14-5. A region of a multichromatid chromosome showing eightfold amplification. Each line represents single-stranded DNA.

other species, individual bands in an amplified euchromatic region may undergo additional amplification, so that they contain up to 16 times as many chromatids as do other nearby euchromatic bands.

3. AMPLIFIED, FREE CHROMOSOME PARTS. The amplification of nucleolus organizer DNA occurs in oocytes of many dipterans and amphibians. This amplified DNA is either retained (at least for a time) as an integrated portion of the chromosome in dipterans, as it is in their multichromatid chromosomes, or released from the chromosome in amphibians, where it remains in the nucleus as the genetic cores of free nucleoli (Section 13.3).

In amphibians, the ordinary nucleolus organizer region contains about 450 transcription-active regions alternating with an equal number of transcription-silent spacer regions (Figures 4-5 and 4-6). By the time the oocyte has reached the middle of prophase I, it contains a dozen or so free nucleoli, each of which contains an acentric duplex ring of DNA composed of various numbers of alternating transcription-active and -silent regions. Such ring DNA's may contain as few as eight transcription-active regions or as many as a thousand or so.

We do not know how or when this rDNA is first amplified. The original amplification may have occurred in either of two general ways: replication, using DNA polymerase; or transcription of an active *and* an inactive region, followed by reverse transcription (Figure 14-7A). The free acentric rings are replicons, and subsequent amplification is accomplished at least in part by their replication, so that the mature oocyte comes to contain several hundred free extra nucleoli (Figure 14-7B). It is estimated that the total amount of amplified nucleolus organizer DNA is 2000 times as much as that contained in the normal nucleolus organizer. Since the normal nucleolus organizer DNA is about 0.1 per cent of the

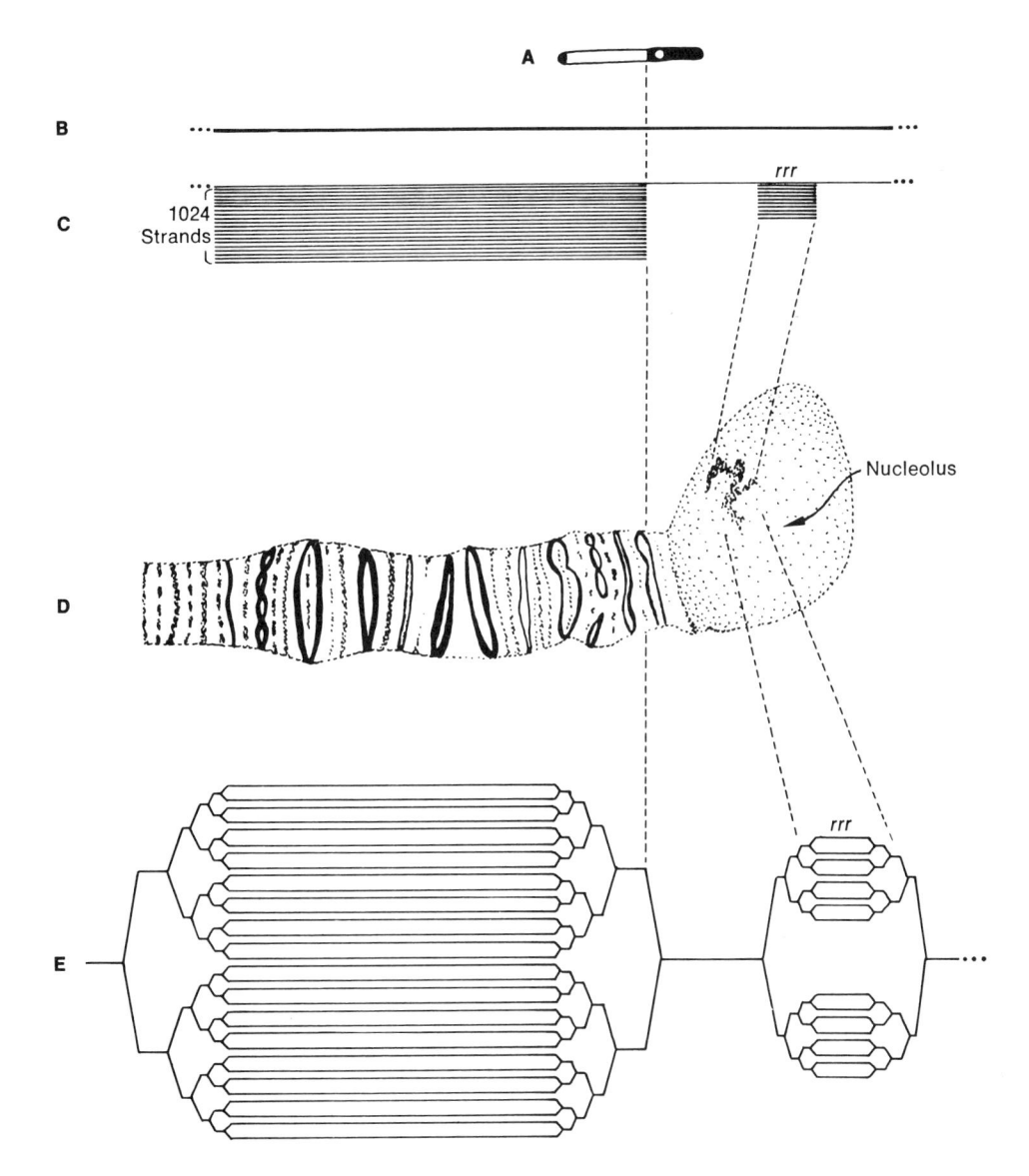

FIGURE 14-6. Model of the amplification pattern that produces a multichromatid chromosome in *Drosophila*. Each line in **B**, **C**, and **E** represents double-stranded DNA. **A**: Metaphase X chromosome showing euchromatin (white bar), centromere (white spot), and constitutive heterochromatin (black). **B**: Segment of unwound single DNA duplex in **A**. Euchromatin is to the left, and heterochromatin to the right, of the dotted line. **C**: A mature multichromatid chromosome with the euchromatin amplified up to 512 duplexes, the nucleolus organizer DNA (*rrr*) amplified to a lesser amount, and the rest of the heterochromatin not at all. **D**: The corresponding larval salivary gland banding pattern for the segment shown in **C**. **E**: The duplexes present in **D** make a continuous DNA molecule, the branch points representing replication forks. The left end, which is shown unamplified, represents the left heterochromatic end of the X. (After C. D. Laird, 1973.)

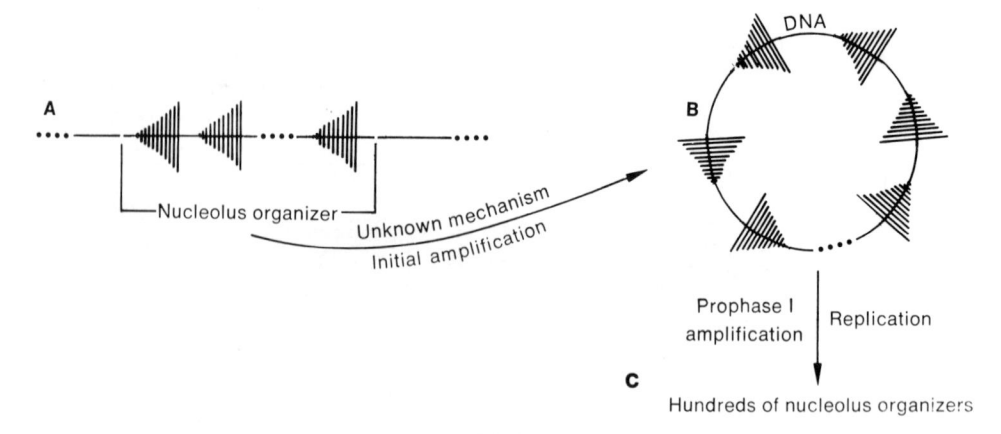

FIGURE 14-7. Amplification of nucleolus organizer DNA in amphibians. **A**: Portion of a chromosome in a somatic or germ nucleus showing 3 of about 450 transcription-active regions in the nucleolus organizer. **B**: One of the first ring chromosomes produced by amplification. **C**: Additional replication of freed DNA greatly amplifies nucleolus organizer DNA.

total amount in a diploid nucleus, there is twice as much free nucleolus organizer DNA present in the oocyte at full amplification as there is DNA in a normal somatic nucleus. (It should be noted that amplification of rDNA does not occur in somatic tissues of amphibians, at least in the embryo and red blood cells. On the other hand, such rDNA amplification is reported to occur in the regular somatic nuclei of *Drosophila*.) Since the amplified DNA is acentric, any present will be lost at the next nuclear division.

The preceding observations suggest that when particular cells require larger amounts of certain proteins or rRNA than can be produced by a diploid genotype, the genes coding for such substances are sometimes programmed to undergo amplification.

# GENE DESTRUCTION

**14.5** Gene destruction in eukaryotes is sometimes genetically programmed.

Many eukaryotic organisms genetically program the destruction of their own genetic material. We have already noted two examples of this, one in the last section—the loss of acentric ring chromosomes in oocytes—and the other in Section 10.6—the destruction of up to three of the polar nuclei produced during meiosis in females. We give below some other examples in eukaryotes of normally programmed destruction of whole nuclei, whole chromosomes, or chromosome parts in single cells, or in the numerous cells of tissues, organs, and systems.

1. ORGAN-SYSTEM LEVEL

   a. In insect development the change from a larva to an adult requires the digestion, during the pupal stage, of many larval organs—including those, such as the salivary gland, that have multichromatid chromosomes.
   b. In frog development, the change from a tadpole to an adult requires the digestion of the tadpole tail.
   c. In human development, the change from the embryonic to the adult circulatory system requires the degradation of certain arteries.

2. TISSUE OR CELL-GROUP LEVEL

   a. In insects, the oocyte is connected to *nurse cells* by cytoplasmic canals (Figure 14-8). As development proceeds, the nurse cells are destroyed and their nuclear contents are digested and transferred through the canals to the growing oocytes.
   b. In human beings and most mammals, the maturation of a red blood cell is accompanied by the loss of its nucleus.
   c. In seed plants, sieve cells function in the transport of liquids after they have lost their nuclei.

3. INTRANUCLEAR LEVEL

   a. In the fungus gnat *Sciara*, whole chromosomes are eliminated in the somatic line but are retained in the germ line.
   b. In the roundworm *Ascaris*, the chromosomes in somatic cells are programmed to fragment and the acentric pieces are lost; fragmentation and hence such loss does not occur in the germ line.

We expect that any acentric chromosomes or chromosome fragments that are produced will be lost in the next nuclear division. We do not yet

FIGURE 14-8. Photomicrograph of a section through a *Drosophila* egg chamber. At this stage, a single layer of cuboidal follicle cells surrounds the oocyte (which occupies a position at the lower left corner of the chamber) and the 15 nurse cells, 9 of which are evident in this section. One canal connects the oocyte with a nurse cell while two others interconnect three nurse cells. Note the particulate material which appears to have been fixed during its passage through the canals. (Courtesy of E. H. Brown and R. C. King, 1964. *Growth*, **28**: 41–81, Fig. 3.)

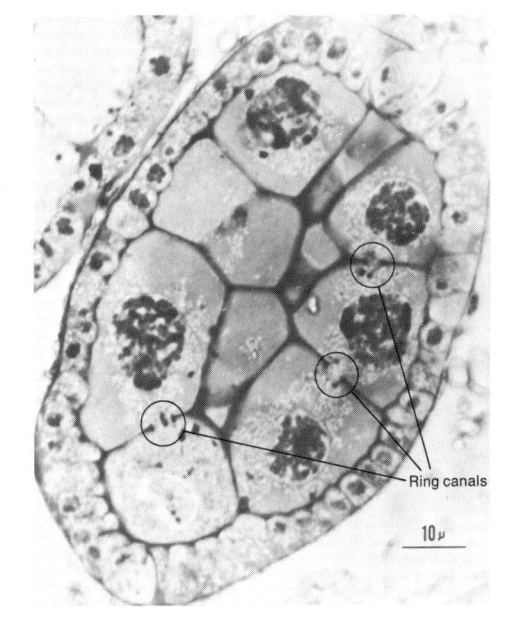

know, however, how whole nuclei or whole chromosomes are genetically programmed for selective destruction.

# MUTATION

### 14.6 The mutation frequency in prokaryotes is controlled genetically.

We have already described in Section 6.8 how the mutation frequency in prokaryotes depends upon mutation-producing and mutation-preventing products of the cell's metabolism. One evidence that the mutation rate in prokaryotes is genetically regulated is that, although the amount of genetic DNA in prokaryotes and the simpler fungi varies by a factor of 1000, the mutation rate per chromosome set is constant (0.5) from one duplication to the next. This must mean that the mutation rate in the different species is genetically regulated (probably suppressed) so that cells will have the same number of mutations (most of which are detrimental) despite differences in DNA content. If the mutation rate were *not* genetically regulated, one would expect the number of mutants occurring per chromosome set to increase with the amount of DNA per set.

It should be noted in this connection that the DNA of certain temperate phages, such as $\phi$Mu-1, can integrate within almost any locus in *E. coli*. Such integration interrupts the sequence of the locus and produces a mutation. To the extent that bacteria can control attachment and infection by such phages, they are regulating their mutation rate.

### 14.7 The mutation frequency in eukaryotes is controlled genetically.

The occurrence of mutations is genetically programmed in eukaryotes also. For example, different strains of the same species of *Drosophila* have different spontaneous mutation frequencies as a result of different alleles, which cause as much as a tenfold difference in mutation frequency. Evidence has also been obtained that cold-blooded species regulate their mutational response to temperature. Thus, in one species of *Drosophila*, the difference in spontaneous mutation frequency between a tropical and a temperate strain is less than one would expect on the basis of the temperature difference. Apparently, the tropical strain has genetically suppressed (or the temperate strain has genetically enhanced) its mutational response to temperature.

Other factors that are determined genetically affect mutation frequency in eukaryotes. These factors include mitotic frequency, the amount and distribution of constitutive heterochromatin, the number and arrangement of chromosomes in a nucleus, as well as the metabolic activity of the cell. To the extent these factors are genetically programmed, so is mutation frequency. Finally, many viruses, but not all, cause host chromosomes to break, leading to the production of dicentrics, rings, and chromosome loss. To the extent that eukaryotes, including human beings, control such viral infection, they are regulating their own mutation frequency.

## SUMMARY AND CONCLUSIONS

DNA synthesis occurs in replication units called replicons. The prokaryotic chromosome is a single replicon, as is F and the chromosomes of many temperate phages. The initiation of replication is regulated by regulatory proteins coded in regulator genes. Some regulatory initiator proteins are needed to stimulate the initiation of replication in the chromosome of *E. coli* or of $\phi\lambda$. The initiation of replication is inhibited by other regulatory proteins such as the repressors of F and $\lambda$. When an episome is integrated (1) replication starts in the replicon of the host, except in a conjugating Hfr, where it starts in the replicon of integrated F, and (2) an episome-coded repressor inhibits the replication of free episomes of the same type. Eukaryotic chromosomes contain multiple replicons whose actions are regulated intrachromosomally, intracellularly, and intercellularly.

Cells that need large quantities of certain gene products sometimes make extra copies of the genes concerned. Such amplification may result in differential polyploidy; or in amplified chromosome parts that remain integrated, as occurs in the giant multichromatid chromosomes in the larvae of Diptera; or in amplified chromosome parts that are free in the nucleus, as occurs for nucleolus organizer DNA in the oocytes of amphibians.

Destruction of nuclear DNA is sometimes genetically programmed at the organ-system, tissue or cell-group, and intranuclear levels to degrade whole nuclei, whole chromosomes, or parts of chromosomes.

The mutation frequency is controlled genetically both in prokaryotes and eukaryotes.

## QUESTIONS AND PROBLEMS

**14.1** *Anagrams*: Unscramble and define each of the following: **a.** incloper. **b.** resporers. **c.** flipamy. **d.** circetan.

**14.2** *Riddle*: When can a phage determine bacterial sex?

**14.3** *Riddle*: Where can one find more rings of the same kind and of different sizes than one finds at the usual jeweler?

**14.4** What is the advantage of having a locus where replication initiates?

**14.5** What do you conclude from the observation that when $\phi$T4 chromosomes lacking one third of their length infect bacteria, they initiate replication but do not initiate a second replication?

**14.6** How can you explain the finding that a mutant strain produces 10 to 20 times as much of a modified $\phi$T4 DNA polymerase than the wild-type strain produces normal polymerase?

**14.7** What is the advantage of $\lambda$ repressor to $\phi\lambda$? Of F repressor to F?

**14.8** When is an F replicon functional? Repressed?

**14.9** Is the mechanism for regulating DNA replication expected to be more complex in eukaryotes than in prokaryotes? Explain.

**14.10** State one piece of evidence that events at the cell membrane affect DNA replication in the nucleus.

**14.11** When is the loss of a nucleus during red blood cell formation a mutation? When is it not?

**14.12** What is not amplified in: **a.** *Planococcus*? **b.** *Drosophila*? **c.** amphibians?

**14.13** What is the advantage of some cells not undergoing amplification although others in the same organism do so?

**14.14** What advantages do the larval salivary gland chromosomes of *Drosophila* have in cytological studies of genetics?

**14.15** Compare qualitatively and quantitatively the nuclear DNA content of a prophase I oocyte and a cell at mitotic prophase in the same amphibian.

# Chapter 15

# Regulation of Gene Action

IN CHAPTER 14, we discussed how organisms regulate the amount and quality of their genetic material. Regulation of protein synthesis is also required for various metabolic activities that occur at specific times and places in a cellular organism. In this chapter we consider how an organism chooses which genes are to be functional in the transcription–translation sequence. Such regulation of gene action is discussed first in prokaryotes, then in eukaryotes.

## IN PROKARYOTES

**15.1** The functioning of the promoter region is under positive control; that is, it is enhanced by regulatory protein.

*E. coli* transcriptase contains a $\sigma$ polypeptide chain (Section 4.2) that is released from the transcriptase once transcription starts. (Hence, $\sigma$ is not needed to continue transcription.) Since transcriptase without a $\sigma$ chain will not bind to the promoter, $\sigma$ is apparently needed only for transcriptase binding. Thus, the functioning of the promoter is under positive control, since the binding of $\sigma$ to transcriptase enhances transcriptase binding to the promoter. Thus, $\sigma$ is a regulatory protein coded by a regulator gene.

Although all *E. coli* promoters are under the positive control of $\sigma$, some of them are also under another type of positive control. To better understand how this other control system operates, let us examine the actual base sequence of the promoter region of the *lactose* (*lac*) *operon* in *E. coli* (Figure 15-1). Since the *lac* promoter, $p^+$, contains two regions that serve different functions, it is actually composed of two different genes. One region (or gene) contains a palindrome that seems to be recognized and bound by a regulatory protein called *CAP* (which itself

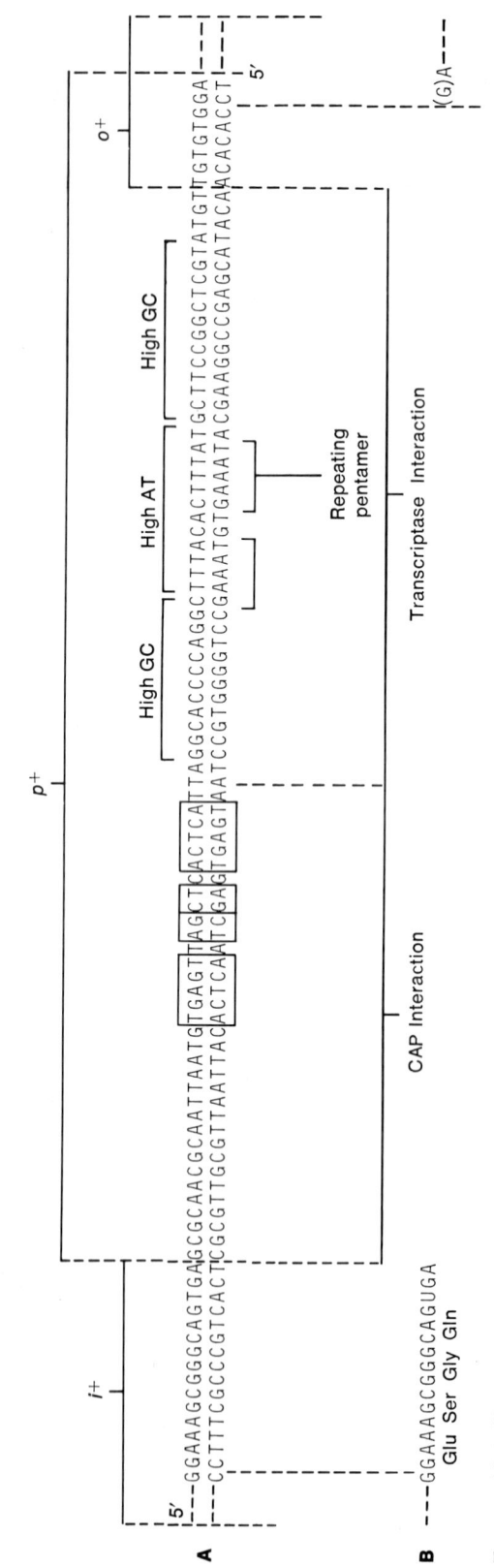

FIGURE 15-1. The promoter region of the *lac* operon of *E. coli*. **A**: The base sequences of the promoter and of portions of the $i^+$ gene that precedes it and of the $o^+$ gene that follows it. Boxed base pairs indicate a palindrome. **B**: The mRNAs and translational products of the portions of the $i^+$ and $o^+$ base sequences indicated.

has been activated by combining with a particular nucleotide called *cAMP*). This region is called the *CAP interaction region*. By binding to the DNA the activated CAP is thought to enhance the functioning of the adjacent region, the *transcriptase interaction region*, the site where transcriptase binds.

The actual binding site for transcriptase is an AT-rich region, about 12 base pairs long which, in the *lac* transcriptase interaction region, is bordered by two GC-rich regions. Since transciptase uses single-stranded DNA as a template, the strands in the duplex AT-rich region apparently need to separate. It is believed that strand separation in the AT-rich sequence is hindered by the greater H-bond stability of the bordering GC-rich sequences; but that the binding of activated CAP to the CAP interaction region destabilizes the nearby GC-rich region, permitting the strand separation of the AT-rich sequence and hence the binding of transcriptase. We see, therefore, that the *lac* promoter is also under positive control because the binding of transcriptase is enhanced by the binding of activated CAP.

FIGURE 15-2. Some simple types of positive (**A**) and negative (**B, C**) control of transcription in operons. Gene symbols: *R*, regulator gene; *P*, promoter; *O*, operator; *A, B, C*, protein-coding genes.

**A** Positive control

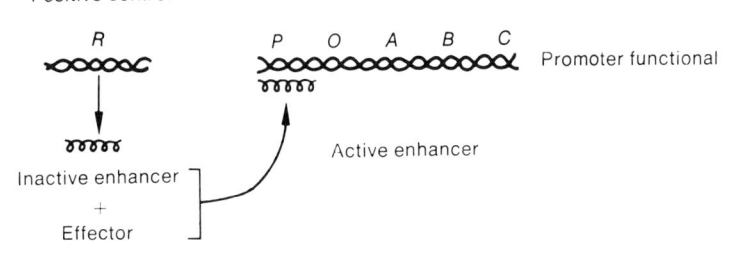

**B** Negative control; inducible operon usually off

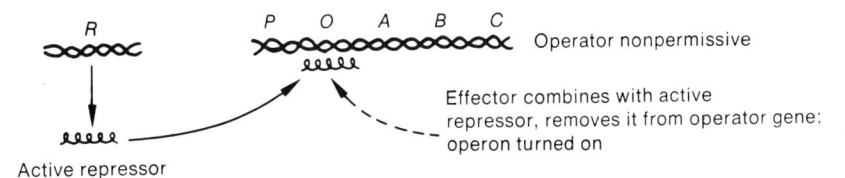

**C** Negative control; repressible operon usually on

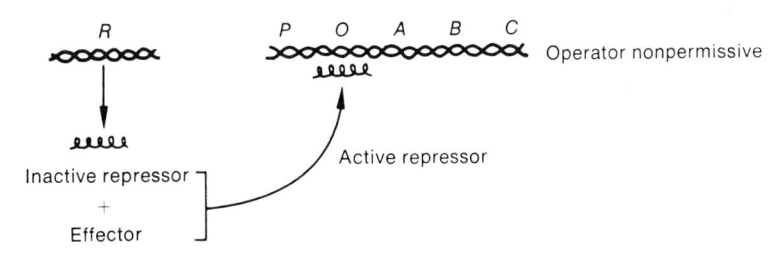

In the present case, a CAP-coding regulator gene produces an *inactive enhancer* (Figure 15-2A) which, after being activated by combining with an *effector* (a small, usually nonprotein molecule; in this case cAMP), helps the promoter to be functional. It should be noted again that some promoters, such as those preceding tDNA's, do not have a CAP interaction site. In addition, although all transcriptase-binding sequences are AT-rich and about the same length, they can differ in base sequence. Different base sequences in such regions will have different affinities for transcriptase, and hence help program the production of different numbers of transcripts from different regions of the chromosome.

## 15.2 The functioning of the operator gene is under negative control.

As indicated in Section 4.2, the operator gene permits transcription of some or all of its base sequence only when it is unbound by a protein. In other words, the functioning of the operator gene (and hence the rest of the operon) is under the negative control of a repressor protein. The repressor protein is coded by a regulator gene located outside the operon it regulates.

Operons whose protein products are used in catabolic processes have operator genes that are negatively controlled by regulator genes that code for *active repressors* (Figure 15-2B). Such is the case for the *lac* operon whose products are enzymes needed for the entry into the cell and the degradation of lactose. In the *lac* operon (Figures 15-1 and 15-3), the regulator gene $i^+$ is directly adjacent to the promoter, $p^+$. The product of $i^+$, lac *repressor*, recognizes a palindrome in the operator gene, $o^+$ (Figure 15-3), and binds to at least 12 base pairs in $o^+$. Since *lac* repressor is a large protein, when bound to $o^+$, it prevents transcriptase from binding to $p^+$, thus blocking the initiation of transcription. When, however, lactose is present as a nutrient, this sugar undergoes a slight molecular rearrangement (to allolactose) and by binding to *lac* repressor causes the repressor to be released from the operator (Figure 15-2B). In this manner lactose acts as an *inducer* for the *lac* operon to synthesize the enzymes needed for lactose catabolism. When all the available lactose is digested, the active repressor will occur once again and bind to the operator, stopping the production of transcripts for enzymes no longer needed. The *lac* operon provides a model for other instances where large amounts of the enzymes needed for the catabolism of a nutrient are induced by the appearance of the nutrient in the food medium, that is, of *induced enzyme formation*.

On the other hand, operons whose protein products are used in anabolic processes have operator genes that are negatively controlled by regulator genes that code for *inactive repressors* (Figure 15-2C). Such is the case in *E. coli* for the *tryptophan (trp) operon*, whose products are enzymes needed for the synthesis of the amino acid tryptophan. In this case, the *trp* operon is transcribed until there is an oversupply of tryptophan (either synthesized by the cell or present in the nutrient medium).

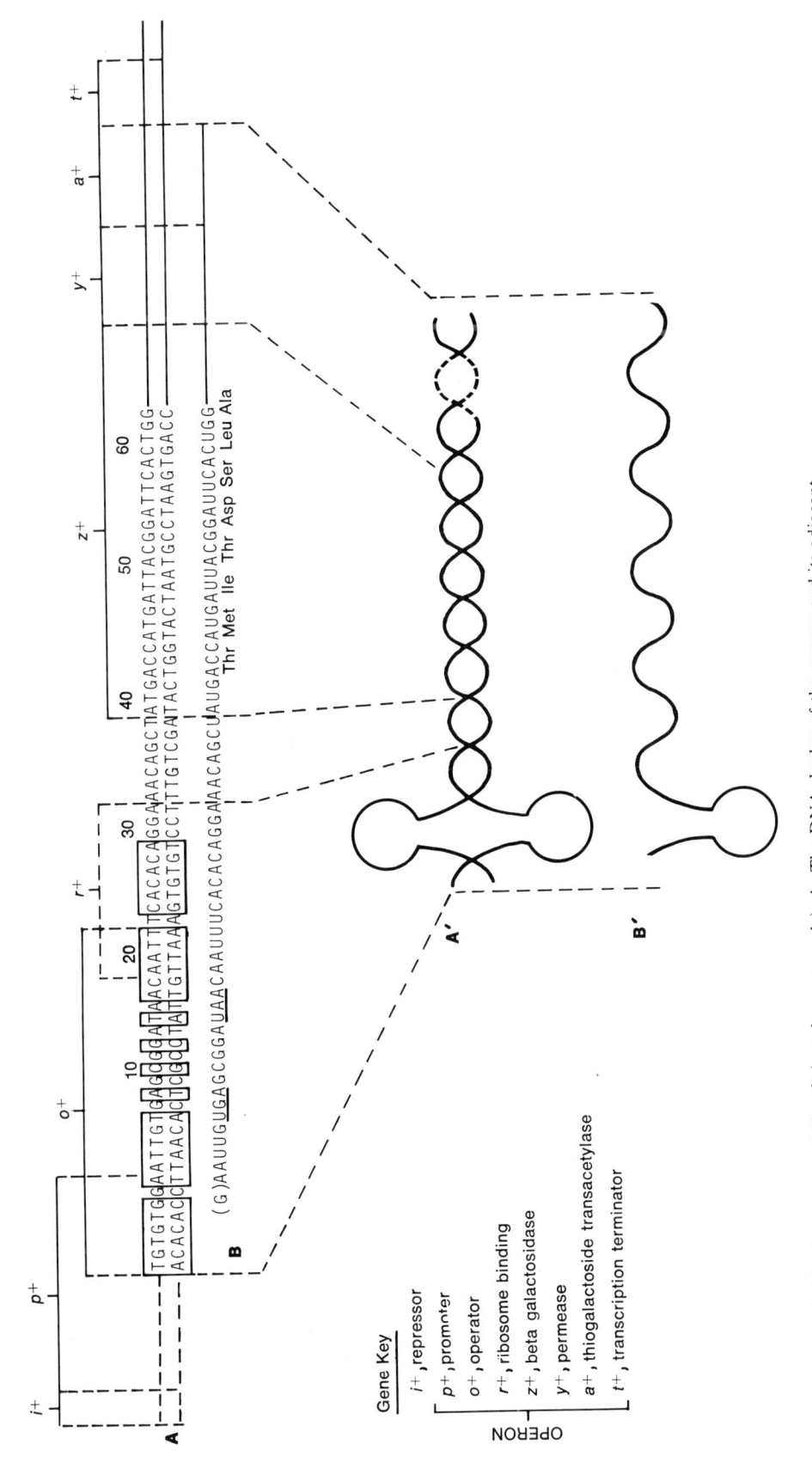

FIGURE 15-3. The *lac* operon of *E. coli* (not drawn to scale). **A**: The DNA duplex of the operon and its adjacent regulator gene, giving the base sequence in the $o^+$ region, whose palindrome is indicated by boxed base pairs. **B**: mRNA base sequence that shows part of the $o^+$ gene is transcribed; 15 per cent of the time transcription starts with G; two terminator codons are underlined; the amino-terminal amino acid sequence of $z^+$ is indicated. **A′**: Sector of **A** showing possible double-hairpin loops formed by base pairing of internally symmetrical regions. **B′**: Sector of **B** showing possible hairpin loop formed by base pairing.

At that time, tryptophan combines with the inactive repressor coded by the regulator gene, *trp R*, converting it to the active repressor form that can bind to the *trp* operator, and thus shut off transcription. Operons such as the *trp* operon are, therefore, said to be *repressible* by their end products acting as effectors. When, for example, tryptophan is once again in short supply, it will be removed from the active repressor, leaving inactive repressor. Since inactive repressor cannot bind to the operator, the *trp* operator will be unblocked and transcription of the operon will thereby be derepressed.

The preceding demonstrates the advantage of the operon, namely that the action of a sequence of genes, all serving a related metabolic need, can be coordinately turned on or off at the level of transcription as metabolism dictates.

## 15.3 The functioning of the transcription terminator gene is under positive control.

As mentioned in Section 4.2, transcription in *E. coli* is terminated at the transcription terminator gene. The functioning of the terminator gene is under positive control by a protein factor $\rho$ (*rho*). $\rho$ seems to be composed of six identical subunits that surround the DNA of the terminator gene, thereby blocking transcriptase and stopping transcription. The role of $\rho$ can be demonstrated *in vitro*. When $\phi\lambda$ DNA is used for transcription, in the presence of $\rho$ the mRNA's made *in vitro* are the same short types as are made *in vivo*; in the absence of $\rho$, the mRNA's made *in vitro* are too long!

We conclude our consideration of the regulation of transcription in prokaryotes by noting that we have only considered the simplest arrangement of promoter and operator genes. In a more complicated case, two promoters (operons) share a common operator; and in another still more complicated case, the transcription of a single operon is regulated by a series of promoters having different affinities for transcriptase and a series of operators having different affinities for active repressor.

## 15.4 Differences in mRNA leader sequences may be used to regulate translation.

Just as all transcribed genes are potentially subject to regulation at the transcriptional level, so are they all potentially subject to regulation at the translational level. One source of translational regulation is the leader sequence of mRNA. In the case of the *lac* operon in *E. coli*, the mRNA starts with an untranslated leader sequence of 38 nucleotides (Figure 15-3B). This sequence includes the transcript of the ribosome-binding gene, $r^+$, covering nucleotides 18 to 31. Let us compare these sequences

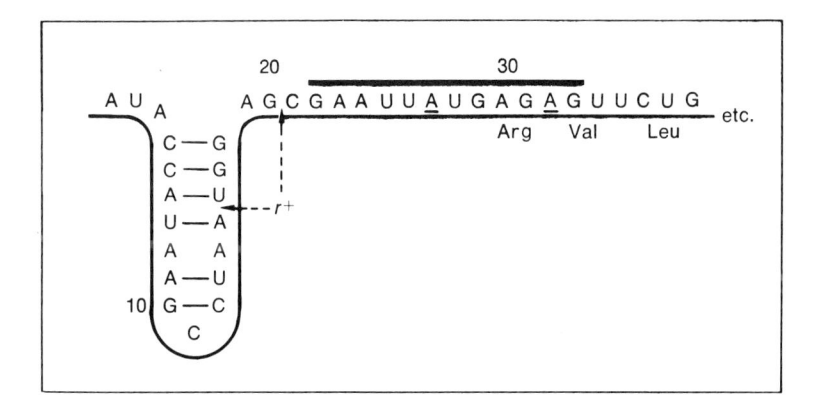

FIGURE 15-4. The mRNA sequence of the *E. coli gal* operon. Translation probably starts at position 27, so bases 1 to 26 comprise the leader sequence. The leader sequence contains a loop of six base pairs and a ribosome-binding sequence, $r^+$, as does the leader in *lac* mRNA (Figure 15-3). In addition, the 12-base sequence overlined is identical to the start of the *lac* operator sequence, except for the two bases that are underlined.

with those for the *gal* operon in *E. coli* (Figure 15-4). In this case the leader sequence is probably only 26 nucleotides long, and the ribosome-binding gene transcript contains a sequence of five nucleotides which is not matched in the *lac* leader. We see, therefore, that the *gal* and *lac* operons differ both in the length and sequences of the bases in their leader sequences and in the sequences of the bases in their ribosome-attachment genes. The differences between leader sequences may make the mRNA's differ in susceptibility to degradation by nuclease. As a result, the two mRNA's may have different longevities and, therefore, may be translated for different lengths of time. The difference in sequence in their $r^+$ genes may result in different ribosome-binding affinities and, therefore, differences in the number of translational starts in a given period of time. Additional research is needed to determine whether these possibilities for regulating translation are actualities.

It should be noted that in *E. coli*, as was observed in viruses such as $\phi$X174 (see Section 4.6), a given nucleotide may be part of more than one gene. Such *nucleotide sharing* occurs between promoter and operator genes (Figure 15-3), between operator and ribosome-binding genes (Figure 15-3), and between the operator gene and the first polypeptide-coding gene in the *gal* operon (Figure 15-4). One nucleotide has been found to be shared *between* two polypeptide-coding genes in the *trp* operon. In this case, the A in the mRNA base sequence ... UGAUG ... is used both as part of the UGA terminator codon for *trp B* protein and as part of the AUG initiator codon for *trp A* protein.

Nucleotide sharing increases the efficient use of the genetic material, particularly in viruses and prokaryotes, which have few or no redundant base sequences.

**15.5 Translation may be regulated in a variety of other ways.**

Regulation of translation in prokaryotes may be actually or potentially achieved in several other ways.

1. An mRNA often contains one or more extra places where ribosomes can drop off or attach. Extra dropoff sites mean that more copies will be made of the first-coded proteins in the mRNA than of those coded later. Extra attachment sites produce the opposite effect. Additional ribosome attachment and dropoff sites are used, therefore, to produce different amounts of the different proteins coded in a single mRNA.
2. It has been found that the types and amounts of tRNA in a prokaryote change with its culture conditions. It is a real possibility, therefore, that different proteins may be synthesized in different amounts because their mRNA's use different codons for the same amino acid and the charged tRNA's used to translate these codons are present in different amounts.
3. Theoretically, the total rate of translation may be regulated by controlling the availability of any of the nine proteins needed for protein initiation, elongation, or termination.
4. DNA phages use a variety of mechanisms to control translation in the host so that phage mRNA is translated preferentially or exclusively. These mechanisms include (a) modifying host tRNA's or aminoacyl-tRNA synthetases, (b) coding for phage-specific tRNA's, (c) coding for phage-specific nucleases that degrade host RNA's, and (d) modifying host ribosomes so that these preferentially translate phage mRNA.

# IN EUKARYOTES

**15.6 Transcription is suppressed in heterochromatin.**

It seems reasonable to suppose that one might correlate the cytological appearance of a chromosomal region in the ordinary light microscope with its potential for transcription. One would expect that a DNA chromosome or chromosome region that is highly coiled, clumped, condensed, or packed does not have its complementary strands separated and therefore is not a suitable template for transcriptase. Conversely, one that is uncoiled, unclumped, or diffuse is or can be strand-separated and can be transcribed, provided that its sense strand is suitably coiled (having neither too much nor too little space between nucleotides) and the surrounding space is not occupied by some other substance.

This idea can be tested *in vivo* by studying the relative ability of clumped chromatin (heterochromatin) and diffuse chromatin (euchromatin) to synthesize RNA in an interphase nucleus. RNA synthesis is determined by the incorporation into the nucleus of radioactive RNA raw material. Radioautographs show silver grains (sites of RNA synthesis) chiefly over the euchromatic regions (Figure 15-5). As expected, metaphase chromosomes (which are completely heterochromatic) do not synthesize RNA *in vivo* and do so only poorly *in vitro*.

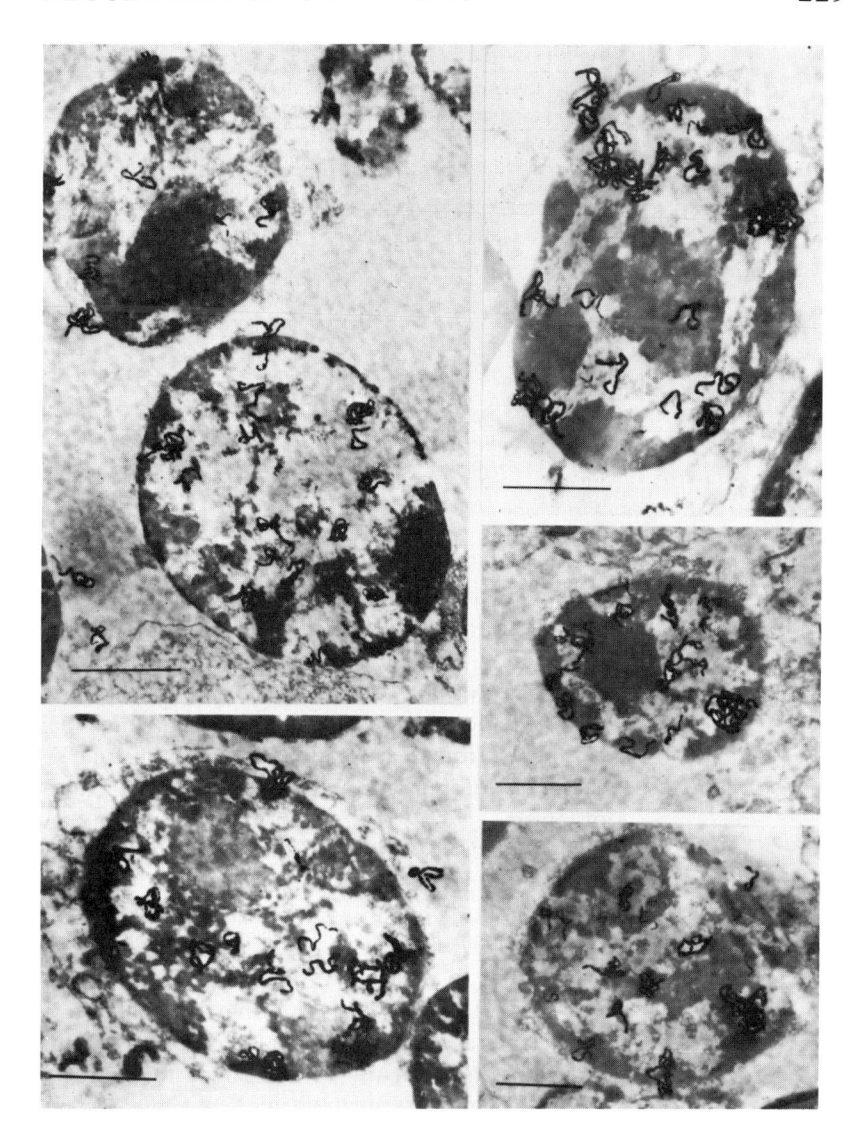

FIGURE 15-5. Transcription in isolated calf thymus nuclei. Radioautographs show silver grains chiefly over diffuse regions of the nuclear chromatin after incubation in a radioactive RNA precursor. The line in the lower left corner of each photograph is 1 μm. (Courtesy of V. C. Littau, V. G. Allfrey, J. H. Frenster, and A. E. Mirsky, 1964. *Proc. Nat. Acad. Sci., U.S.*, **52**: 97.)

We conclude that transcription is suppressed by the heterochromatic state.

## 15.7 Binding to histone causes DNA to supercoil and fold.

Histones are basic proteins that contain large amounts of the hydrophilic basic amino acids arginine and lysine (Figure 5-5). There are three main

| Class | Fraction | Lys/Arg Ratio | Number of Amino Acids | Molecular Weight | Modifications |||
|---|---|---|---|---|---|---|---|
| | | | | | Phosphorylation | Acetylation | Methylation |
| Lysine-rich | I | 20 | ~215 | 20,000-22,000 | + + + + | | |
| Moderately lysine-rich | IIb₁ | 2.5 | 129 | 13,000-15,000 | + | + | |
| | IIb₂ | 2.5 | 125 | 13,774 | + | + + + + | |
| Arginine-rich | III | 0.8 | 135 | 15,324 | + | + + + + | + |
| | IV | 0.7 | 102 | 11,282 | + | + + + + | + |

FIGURE 15-6. Classes of histones in eukaryotic nuclear chromosomes.

classes of histone (Figure 15-6), two of which have two fractions. Therefore, all in all, histones are of five major types. Although there are some minor variants of histone, there being some variability in the types found in different individuals, the diversity of histones has been restricted during evolution. This limitation in diversity of the five major types of histone is indicated by the amino acid sequences being nearly identical for histone III from the cow and carp and for histone IV from the cow and pea seedling, despite the great evolutionary distances between these organisms.

Histone mRNA is transcribed only during the S period of interphase, from 400 or so redundant histone DNA sequences, some of which are clustered but separated by transcription-silent regions. After synthesis the mRNA, which is 7 to 9S and does not carry poly A, is quickly transferred to the cytoplasm, where it is translated only during the S period. Histone synthesis is intimately related to DNA synthesis. When DNA synthesis is arrested, histone synthesis declines rapidly because the polyribosomes that are synthesizing histones are preferentially disrupted.

Newly made histone molecules enter the nucleus and bind to the DNA very close to the sites where DNA has just replicated. The histone is bound to double-stranded DNA in a special, regular manner. Two molecules of histones IIb₁, IIb₂, III, and IV combine to form a *histone cluster* (Figure 15-7A); and the DNA duplex is wrapped around evenly spaced histone clusters to produce DNA–histone cluster beads (Figure 15-7B). Each bead thus produced has a diameter of about 125 Å and contains about 200 base pairs. The supercoiling shortens the length of the DNA five- to sixfold. Histone I, which is not part of a histone cluster, probably binds to regions of DNA that are not bound in beads, thereby causing the supercoiled DNA to fold (Figure 15-7C).

The interaction between DNA and histone is primarily an interaction between DNA's acidic phosphates and histone's basic amino acids. These amino acids are mostly clustered near the beginning of the histone molecule. DNA is released from bound histones by either degradation or chemical modification of the histones. Most modifications are the addi-

# REGULATION OF GENE ACTION

tion of either phosphate or acetate groups (Figure 15-6), which decrease the basicity of the histones and, therefore, their capacity to bind to DNA. Upon losing these modifications, histones can once more bind to DNA.

It is clear from the preceding that one of the functions of histone must be to package nuclear DNA so that it will be short enough for mitotic and meiotic divisions.

## 15.8 Histones act as nonspecific repressors of transcription.

The chemically bound DNA in chromatin is a much poorer template for transcription than pure DNA. This is shown by the number of sites that bind calf thymus transcriptase being 12 times greater for pure calf thymus DNA than it is for calf thymus chromatin. That the repression of transcription is due largely to histones is indicated by histones preventing

FIGURE 15-7. Hypothesized relationship between histones and chromosome form. **A**: Unfolded and uncoiled chromosome. Diameter 30 to 40 Å, typical of functioning euchromatin. **B**: Unfolded and coiled chromosome. DNA is periodically wrapped around histone clusters, forming beads of DNA. Diameter about 125 Å, typical of nonfunctioning euchromatin. **C**: Folded and coiled (supercoiled) chromosome. Coiled DNA has folded, apparently because histone 1 (not shown and not part of a histone cluster) has bound to bead-free DNA. Diameter 250 to 300 Å, typical of nonfunctioning heterochromatin.

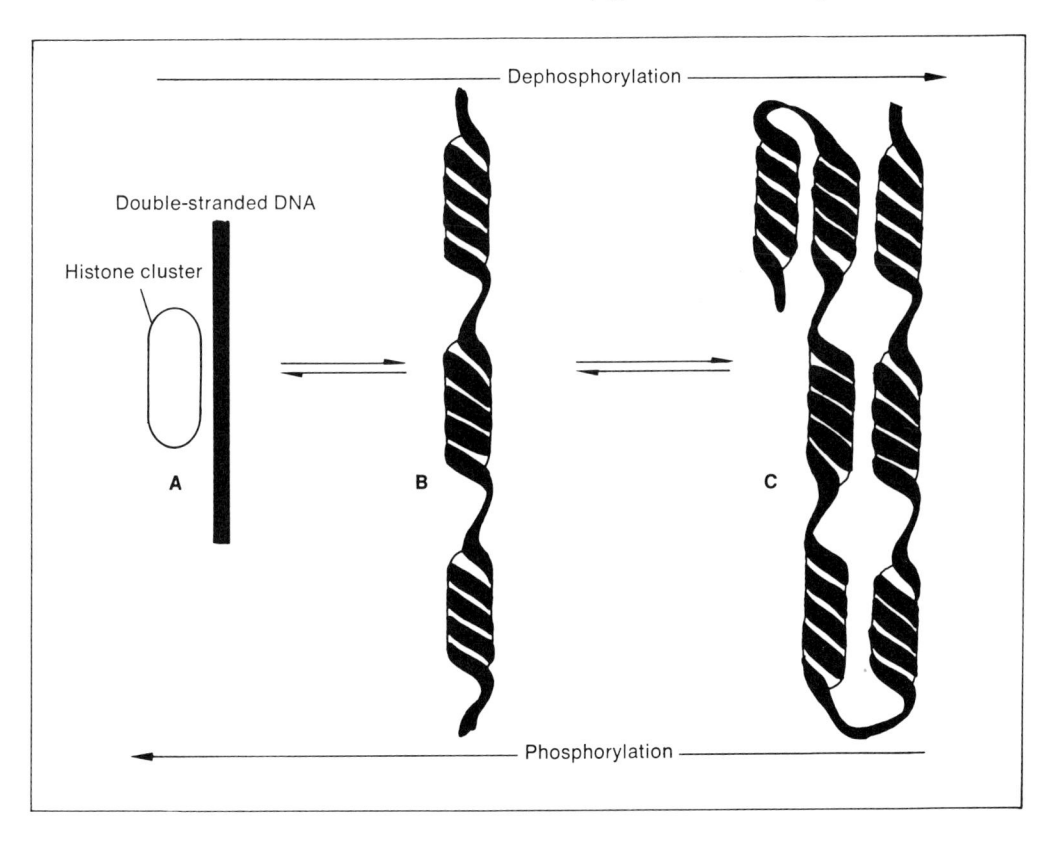

DNA strand separation (as mentioned in Section 14.3), probably by binding in the larger space surrounding the duplex.

Studies of isolated nuclei and of isolated chromatin show that histones in general strongly suppress transcription. In one such study, the transcriptive ability of nuclei isolated from calf thymus is observed. *Addition of histones to the incubation mixture inhibits RNA synthesis.* (Histones from any source will bind to duplex DNA from any source.) *Removal of almost all the histone (but little nonhistone material) permits a three- to fourfold increase in RNA synthesis.* Moreover, when histone is added to such histone-deficient nuclei, RNA synthesis is immediately repressed. Similar results have been obtained with pea chromatin. We conclude, therefore, that histones also act as nonspecific repressors of transcription.

## 15.9 Nonhistone proteins activate specific loci for transcription.

Nonhistone nuclear proteins have a high content of acidic amino acids. There are about 200 nonhistone proteins, most of which have yet to be identified. Nonhistone proteins include a fraction that will bind to pure DNA sequences of the same eukaryote, but not to those of a prokaryote. That nonhistone proteins can activate specific loci for transcription is shown by the results of *chromatin reconstitution* experiments, in which chromatin is separated into various components, then put together in different combinations that are tested for their efficiency as templates for transcription.

In the chicken, globin DNA, which codes for globin, the protein of hemoglobin, is normally transcribed in the *reticulocyte* (R) (the immature red blood cell) but not in either the nucleated *erythrocyte* (E) (the mature red blood cell) or the liver cell (L). When R chromatin is incubated *in vitro*

FIGURE 15-8. The *in vitro* synthesis of globin mRNA in chromatin reconstitution experiments. (After T. Barrett, D. Maryanka, P. H. Hamlyn, and H. J. Gould, 1974.)

| Chromatin | Reconstitution |||  Globin mRNA |
| | DNA | Histone | Nonhistone | |
| Reticulocyte | | | | + |
| | Reticulocyte | Reticulocyte | Reticulocyte | + |
| | Erythrocyte | | | − |
| | Erythrocyte | Reticulocyte | | − |
| | Erythrocyte | Reticulocyte | Liver | − |
| | Erythrocyte | Erythrocyte | Reticulocyte | + |

with RNA raw materials and transcriptase, globin mRNA is synthesized (Figure 15-8). Globin mRNA is also made when R chromatin is reconstituted from its separated DNA, histone, and nonhistone fractions. E DNA will not serve as a template for globin mRNA synthesis, however, unless R nonhistone is added; addition of L nonhistone does not work. Apparently, therefore, R and L nonhistones differ and only the former can specifically activate the transcription of globin DNA. In other reconstitution experiments it is found that S-period nonhistones are better than G1 or metaphase nonhistones in stimulating transcription of histone mRNA. We conclude, therefore, that nonhistone proteins activate specific loci for transcription *in vitro* and, probably, *in vivo*.

We can only speculate at present on how nonhistones activate transcription. Perhaps nonhistones free a DNA template by combining with histones or RNA transcripts, thereby removing them from the DNA; perhaps nonhistones act as positive regulators of transcription by binding to promoters alone or in combination with hormones that act as effectors (see Section 20.5).

**15.10** Translation in eukaryotes may be regulated in a variety of ways.

In contrast to prokaryotes, in eukaryotes (nuclear) transcription is spatially separated from (cytoplasmic) translation. Since the machinery for translation is very similar in both kinds of organism, however, we expect the same or a greater range of mechanisms for the genetic regulation of translation in eukaryotes as in prokaryotes. We will list several examples of translation regulation, including some of viruses.

1. RIBOSOMAL MEMBRANE BINDING, REPRESSION, OR ACTIVATION. We have already noted in Section 5.11 that translation is regulated in that membrane-bound and membrane-free ribosomes translate different sets of mRNA's. It is also found that at metaphase and prior to fertilization, proteins combine with ribosomes and thereby repress translation, the inhibition being due to the inability of the ribosomes either to attach to or to stay attached to mRNA. Upon the enzymatic removal of such protein, the ribosome functions normally. Translational control is also reported to be due to the activation of ribosomes by tissue-specific factors.

2. AVAILABILITY OF CHARGED tRNA's. The pattern of tRNA's and of aminoacyl-tRNA synthetases differs in different tissues of a eukaryote. It is possible, therefore, that the differential availability of charged tRNA's may be used to regulate translation if mRNA's differ in the codons they use for the same amino acid (as already mentioned for prokaryotes in Section 15.5).

3. SUPPRESSION OF mRNA. Translation is sometimes regulated by suppressing the functioning of mRNA. In the chick, for example, globin

mRNA appears at a very early stage of development (the midprimitive streak) and is not translated for about 6 hours (until the seven- to eight-somite stage). How this suppression of mRNA is achieved is unknown.

4. LONGEVITY OF mRNA. As might be expected, different mRNA's in a eukaryotic cell have different longevities before they are degraded. Translation may thus be controlled in this manner.

5. SPEED OF POLYPEPTIDE SYNTHESIS. Toadfish that are accustomed to living in warm water adapt when they are transferred to cold water. Acclimation to the lower temperature is apparently made possible, at least in part, by an appropriate increase in the speed with which amino acids are added to growing polypeptides, so that proteins are synthesized rapidly despite the low temperature.

6. VIRUS CONTROL OF TRANSLATIONAL MACHINERY. Some DNA viruses, such as herpes simplex virus, code for virus-specific tRNA's that are apparently used to preferentially translate viral mRNA's.

## SUMMARY AND CONCLUSIONS

Gene-action regulation in prokaryotes occurs at the transcriptional level by (1) the positive control of the promoter by regulatory proteins such as $\sigma$ and CAP, (2) the negative control of the operator gene by active repressors, and (3) the positive control of the transcription terminator gene by $\rho$.

Catabolic operons are inducible by the substrate catabolized; induction is accomplished by the inducer inactivating the active repressor coded by a regulator gene. Anabolic operons are repressible by the substance synthesized; repression is accomplished by the substance synthesized activating the inactive repressor coded by a regulator gene. In both types of operons, inactivation of the active repressor frees the operator and permits transcription.

Gene-action regulation in prokaryotes may occur at the translational level due to (1) differences in mRNA leader sequence, which may affect mRNA ribosome-binding capacity and mRNA longevity; (2) extra ribosome-binding or ribosome-dropoff sites, which cause different proteins coded in the same mRNA to be translated unequally; (3) different concentrations of different charged tRNA's; and (4) variations in other parts of the translation machinery.

In eukaryotes gene action is generally suppressed at the transcriptional level by supercoiling and folding the DNA, as occurs in heterochromatin. The molecular basis for such suppression is the binding of histone clusters to DNA, which produces beads, and the binding of histone I to DNA, which probably folds the DNA. Transcription is activated at specific loci by nonhistone proteins.

Translation in eukaryotes may be regulated in a variety of ways (similar to those in prokaryotes), including those that affect ribosomes, tRNA's, mRNA's, or other parts of the translation machinery.

## QUESTIONS AND PROBLEMS

**15.1** *Riddle*: What string of beads has no holes in the beads?

**15.2** What do you suppose is the consequence of some Thr in *E. coli* transcriptase having phosphate added by an enzyme coded in $\phi$T7?

**15.3** Knowing that glucose freely enters an *E. coli* cell and that together with galactose it makes up lactose, what advantage to *E. coli* is the fact that glucose inhibits the synthesis of cAMP?

**15.4** Will exposure to a sugar not normally found on earth induce *E. coli* to synthesize the enzymes needed for its catabolism? Explain.

**15.5** List the genes whose mutation will affect the transcription of the $z^+$ gene in the *lac* operon.

**15.6** Differentiate between induction and derepression.

**15.7** What proteins are used for what purpose in the positive or negative regulation of the functioning of an operon?

**15.8** Give two examples of nucleotide sharing within an operon.

**15.9** How is mRNA structure known or expected to be related to the regulation of translation in prokaryotes?

**15.10** Name two different genes in a DNA phage that help in the preferential translation of phage mRNA.

**15.11** What cytological feature do you know or expect to be associated with the absence of transcription in the interphase nucleus of: **a.** a human neuron? **b.** a *Drosophila* larval salivary gland cell?

**15.12** What are the usual chemical components of chromatin?

**15.13** What unsolved problems exist regarding histones?

**15.14** What evidence can you give that chromosome coiling and folding is accomplished in much the same way in nuclear chromosomes of most eukaryotes?

**15.15** What evidence is there that histone synthesis and DNA synthesis are coordinated?

**15.16** Cite two evidences that histones are general suppressors of transcription.

**15.17** Which fraction do you suppose best stimulates pure human DNA to synthesize histone mRNA *in vitro*? Why? **a.** Histone from S period. **b.** Histone from G1 period. **c.** Nonhistone protein from metaphase. **d.** Nonhistone protein from S period. **e.** RNA from S period.

**15.18** State one way each of the following may regulate translation in eukaryotes. **a.** ribosome. **b.** tRNA. **c.** mRNA. **d.** virus.

# Chapter 16

## Heterochromatization in Eukaryotes

THIS CHAPTER DEALS with the regulation of the action of specific genes or groups of genes in eukaryotes. Most of the regulation discussed is the prevention of transcription; this is manifested cytologically by changing chromatin from a euchromatic to a heterochromatic condition. We have already noted in Section 10.5 that such facultative heterochromatin occurs; it is said to be the result of *heterochromatization*—the clumping (supercoiling and folding) of euchromatin. The discussion includes examples from human beings and other mammals, *Drosophila*, and maize.

## HUMAN BEINGS AND OTHER MAMMALS

**16.1** Many X-limited genes produce the same phenotypic effect whether in single or double dose.

Not all the genes on sex chromosomes have a phenotypic effect on sex. In diploid species with X and Y sex chromosomes, X-limited genes not concerned with sex occur in single dose in one sex and in double dose in the other. The organism often controls these genes so that their level of activity is equal in both sexes; this is known as *dosage compensation*. For example, the enzyme glucose 6-phosphate dehydrogenase (G6PD) is coded in human beings by an X-limited gene; erythrocytes from males and from females have the same amount of G6PD activity, even though the female has twice the dose of the gene that the male has. Many other X-limited loci in human beings also show dosage compensation, although the locus for red-green colorblindness does not.

**16.2** Dosage compensation in human beings is effected by permitting the expression of only one gene in the two-dose condition.

In human females who are heterozygous for an allele that codes for defective G6PD, some erythrocytes have complete G6PD activity and others no activity at all; no cells of intermediate activity are found. It appears that some of these erythrocytes are derived from cells in which the normal gene is functional and the defective allele is nonfunctional, while the others come from cells in which the defective allele is functional and the normal allele is nonfunctional. Apparently, only one of any two G6PD alleles present in a given cell is expressed in a female. Since the female heterozygote has some cells with G6PD activity and some without, she is a *functional mosaic* for the G6PD locus. At least six other X-limited loci give rise to functional mosaicism when heterozygous, so it appears that dosage compensation in human females is routinely accomplished by completely suppressing the action of one of two alleles present.

**16.3** Dosage compensation is due to the heterochromatization of one gene in the two-dose condition.

Many of the interphase nuclei in somatic cells of human females have a large clump of chromatin that touches the nuclear membrane (Figure 16-1a). Although the fraction of cells showing such clumps differs in different somatic tissues, such heterochromatic clumps are not found in normal males, so the material is called *sex chromatin* or the *Barr body* (after its discoverer). Sex chromatin is absent in the germ line and first appears, in females, at about the twelfth day of development; so sex chromatin is facultative heterochromatin produced by heterochromatization. For each two sets of autosomes all X's but one are heterochromatized. Thus, a diploid cell containing XX, XXX, or XXXXY may have one, two, or three Barr bodies as in Figure 16-1a, b, and c, respectively; and a tetraploid cell containing XXXX may have two Barr bodies (Figure 16-1d, and lower section).

Heterochromatization is the mechanism of dosage compensation in human beings and other mammals. In organisms that have a Barr body, all the inactive loci are probably in the same heterochromatized X chromosome. That heterochromatization prevents transcription is indicated by RNA synthesis in the clumped Barr body, being only 18 per cent of the amount in a similar volume of unclumped X chromosome. Since dosage compensation does not seem to occur for loci present in the short arm of the human X chromosome (see Section 19.7), a Barr body may not contain the entire X chromosome.

Although mice (unlike most other mammals) have no Barr bodies, one of the female's X chromosomes condenses early during mitosis, showing that it is facultatively heterochromatic at least at that stage. Moreover, when X-limited loci in the mouse are heterozygous, a functional mosaic is

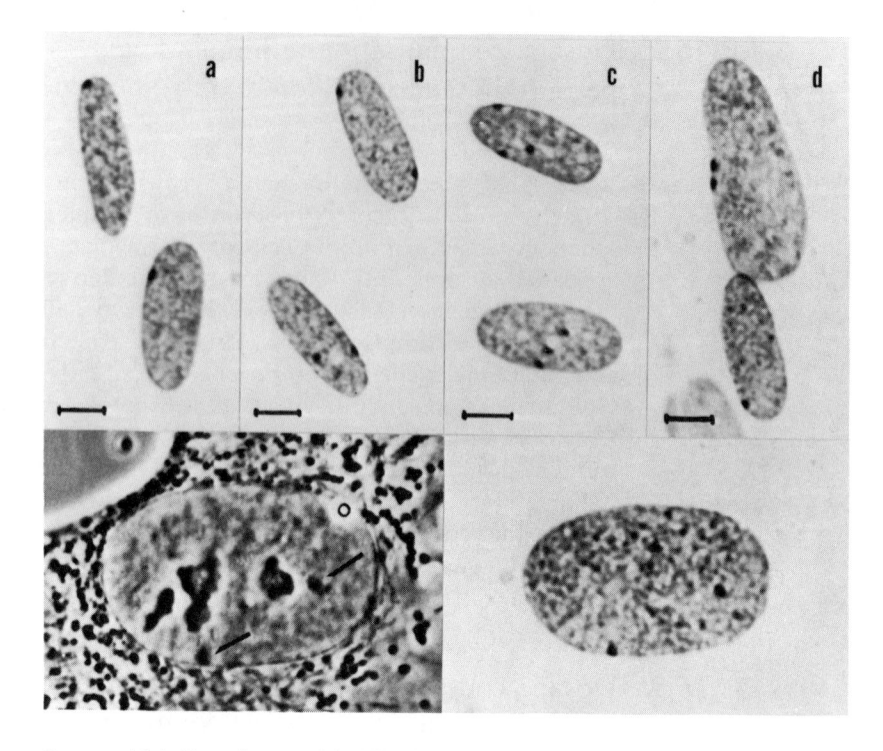

FIGURE 16-1. Interphase nuclei of fibroblast cells cultured from the skin of humans with different numbers of X chromosomes and then stained with the Feulgen reaction, which is specific for DNA. **a**: XX female, with one Barr body. **b**: XXX female with two Barr bodies. **c**: XXXXY male, with three Barr bodies (**a**, **b**, and **c** depict diploid nuclei). **d**: Diploid XX nucleus (lower) and first interphase of a tetraploid nucleus (upper) derived from an XX diploid cell. The bottom row shows that the Barr bodies are visible in living cells and that condensation of the X is not a technical artifact. *Left*: Nucleus of a living tetraploid cell viewed with the phase-contrast microscope. The cell's cytoplasm and nucleoli are visible. Pointers indicate condensed (X) chromosomes. *Right*: The same nucleus stained with the Feulgen reaction and showing two Barr bodies at positions corresponding to the condensed regions visible in the living nucleus. (Courtesy of R. DeMars.)

produced (for all loci tested but one). It thus appears that one X chromosome of the female mouse is largely inactive transcriptionally in interphase, although its clumping is not extensive enough to be detected cytologically.

## 16.4 Autosomal loci translocated to an X chromosome are sometimes inactivated when the X is.

Does an X chromosome that is inactivated as the result of dosage compensation have any effect on the phenotypic expression of autosomal loci that have become linked to it by means of translocation? The answer is obtained by studying female mice that are heterozygotes for a reciprocal translocation between the X chromosome and an autosome (Figure

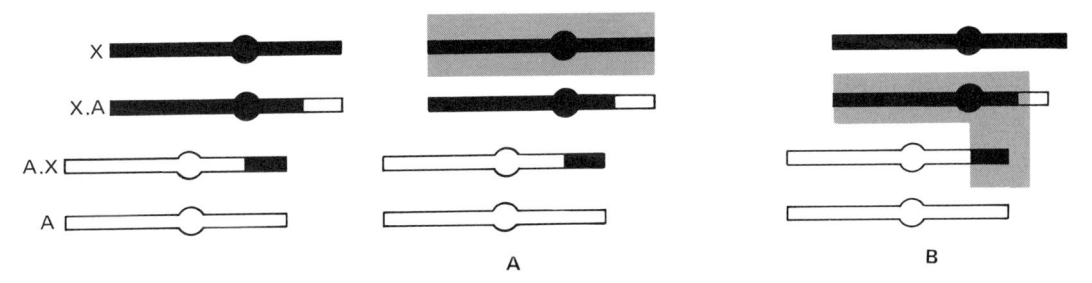

FIGURE 16-2. The spread of heterochromatization is hypothesized to explain the position effect observed in the female mouse for some autosomal loci translocated to the X chromosome. The grey box indicates the X chromosome that is heterochromatized.

16-2). In those cells of the female where the nontranslocated X is the one inactivated for dosage compensation, all the autosomal loci are expressed normally (Figure 16-2A). In those cells where the translocated X is the one inactivated, however, autosomal loci near the point of attachment to the X are often inactivated (Figure 16-2B). Thus, the dosage compensation inactivation spreads to autosomal loci that have become attached to an inactivated X. The inactivation of an autosomal locus is less frequent, however, the farther away it is from its connection to the X. In cells with inactivated autosomal loci, therefore, only the nontranslocated alleles of the inactivated genes are functional. When the autosomes under discussion are suitably marked genetically, one observes functional mosaicism for autosomal loci. It is most likely that the inactivation of translocated-to-X autosomal genes is due to the prevention of transcription.

We see from the preceding that the expression of certain autosomal loci depends on their positions in the chromosome set. Any gene whose functioning is altered when its position in the genetic material is altered is said to exhibit a *position effect*. In prokaryotes, all the polypeptide-coding genes in an operon are subject to a position effect, since if they were removed from their operon and relocated elsewhere in the chromosome, they might function as part of another operon or not at all. In eukaryotes that have Barr bodies, we expect position effects to occur for autosomal loci translocated to the X due to their heterochromatization (provided they are close to the breakpoint in the X) and hence inactivation when that X segment becomes part of a Barr body.

## 16.5 The X inactivated for dosage compensation is the same in all descendant nuclei.

In the early stages of development of female mammals, neither X is inactivated. When inactivation starts, for example in human beings and mice, there seems to be an equal probability of inactivating either of two normal X's in different somatic cells. We do not know what factor determines which X shall be inactivated in somatic line cells or that neither shall be inactivated in germ line cells. It is known, however, that X's that are structurally abnormal or carry certain X-linked alleles are preferentially inactivated. In female kangaroos and probably in other marsupials, it is the paternal X that is inactive in somatic cells.

Heterochromatization in interphase is ordinarily permanent, although it may be modified by age or experimental means. The inactivation of an X in a Barr body prevents its transcription but not its replication, which is delayed, indicating that the clumped condition interferes with DNA synthesis as well as RNA synthesis. In the next and each succeeding mitosis, it is the same X that is inactivated. We know this because in a functional mosaic each patch of tissue is phenotypically uniform—there are no patches within patches. When autosomal genes are translocated to the X, each patch of tissue that shows a position effect also is phenotypically uniform. Such a uniform effect is called a *constant* or *C-type position effect*.

# DROSOPHILA

### 16.6 Dosage compensation in *Drosophila* permits the equal expression of both genes in the two-dose condition.

Dosage compensation for X-limited loci that are not related to sex also occurs in *Drosophila*. Contrary to the situation in mammals, however, the alleles in both X chromosomes of the *Drosophila* female are apparently equally functional. This conclusion is based on two observations:

1. In multichromatid chromosomes such as occur in the larval salivary gland, a region undergoing transcription unwinds, strand-separates, and together with its enzymes and nascent transcripts produces a localized puffed region in the chromosome. When the two multichromatid X's in a female are examined cytologically, they seem identical in their bands, interbands, and puffs.
2. In a female that is heterozygous for a particular X-limited gene, each allele can contribute one polypeptide to form a hybrid enzyme.

That dosage compensation in *Drosophila* is accomplished at the transcriptional level is indicated by the finding in salivary gland cells that although the single multichromatid X in the male and the pair of such X's in the female have DNA in the expected ratio of 1:2, cells with the single and double dose of X's contain just about the same amount of RNA. It appears, therefore, that dosage compensation is accomplished at the transcriptional level without heterochromatizing one of the two X's present. Two mechanisms for such dosage compensation have been proposed. According to one hypothesis, the level of transcription of both doses in the female is repressed to the level of transcription of the single dose in the male; according to the other hypothesis, the level of transcription of the single dose in the male is enhanced to match the level of two doses in the female. Since both hypotheses have received strong support, perhaps they are both correct in different cells or at different times in the same cell.

## 16.7 Shifting the relative positions of euchromatin and constitutive heterochromatin often produces position effects.

Euchromatin and constitutive heterochromatin differ in cytological appearance, base content, distribution in the chromosome, time of replication, crossing-over frequency, and amplification in multichromatid chromosomes. They also differ in their coding function, since euchromatin contains protein-coding genes, whereas constitutive heterochromatin does not. Since the wild-type arrangement of euchromatin and constitutive heterochromatin is doubtless advantageous to the cell, and since the rearrangement of euchromatin and facultative heterochromatin produces position effects in mammals, it would not be surprising if the rearrangement of euchromatin and constitutive heterochromatin gave rise to position effects in *Drosophila* and many other eukaryotes. In fact, gross chromosomal rearrangements in *Drosophila* often produce position effects; these are expressed in the form of mosaic or variegated traits. For example, when the allele for dull-red eye color, $w^+$, which lies in the euchromatin of the X chromosome, is placed in or adjacent to constitutive heterochromatin by means of a breakage rearrangement, the result is that the eyes are mosaic for color, being composed of a mixture of dull-red and white patches. No change in base sequence of the $w^+$ gene is involved, since no break occurred within the $w^+$ gene and the gene resumes producing the nonvariegated dull-red eye color when returned to euchromatic neighbors by means of another rearrangement or of crossing over. The variegated phenotype is therefore a position effect due to a change in the functioning of the $w^+$ gene by neighboring constitutive heterochromatin. Since other genes in euchromatin that are placed in or near constitutive heterochromatin are also expressed in a variegated manner, this type of position effect is called a *variegated* or *V-type position effect*.

## 16.8 The phenotypic variegation in a V-type position effect is due to a variegation in heterochromatization.

Genes located some distance from a point of breakage in a rearrangement sometimes show V-type position effects. This spreading effect is similar to the spreading of the C-type position effect to autosomal euchromatic genes translocated near facultative heterochromatin, as seen in the mouse. Both types of position effect are also similar in that the presence of heterochromatin represses the phenotypic expression of genes. The C-type position effect is attributed to gene inactivation at the transcriptional level, even though no direct cytological evidence of heterochromatization is available. In *Drosophila*, however, the giant multichromatid chromosomes of the larva offer a unique cytological opportunity to detect the heterochromatization of very short segments of euchromatin. In such chromosomes constitutive heterochromatin is less amplified than euchromatin and so is neighboring chromatin that is producing a V-type position effect. One can compare the degree of phenotypic mosaicism for a gene with its degree of heterochromatization in giant chromosomes. Different rearrangements of the same gene, for example $w^+$, produces different degrees of variegation, for example different amounts of whiteness in the mosaic. It is found that as the fraction of a mosaic expressing the repressed condition of the gene increases, so does the frequency of salivary gland nuclei that show the band containing that gene to be underreplicated, that is, to be heterochromatized.

Let us consider further the implications of mosaicism in phenotype and heterochromatization. Phenotypic mosaicism means that although each somatic cell has the same genotype, in some cells the gene is functioning normally (say $w^+$ produces dull-red color), whereas in some others it is completely repressed (producing white). Heterochromatization mosaicism means that in some cells the gene is amplified as usual, whereas in others it is amplified less than usual. In a V-type position effect for $w^+$, the parallel is between gene-product mosaicism in the pupal eye cells and the heterochromatization mosaicism in larval salivary cells. This result indicates that the normal arrangement of chromatin is adaptive and has been selected on an organismal basis, so that when a chromosome rearrangement disturbs this adaptive arrangement, heterochromatization occurs in many tissues. It is presumed that the heterochromatization that is expressed by underreplication in larval salivary chromosomes is expressed by clumping in the ordinary chromosomes of other tissues.

## 16.9 In *Drosophila*, the frequency of heterochromatization depends upon parental derivation, sex, and other factors.

The frequency with which a locus in a particular rearrangement is heterochromatized depends upon several factors.

1. PARENTAL DERIVATION. In individuals of the same sex, a locus is heterochromatized much more frequently when received from the father than when received from the mother. Somehow the germ line must be *setting* the level of heterochromatization in the next generation. This setting must be subject to *erasure* and resetting, since a locus set for a low frequency of heterochromatization (having been received from the mother) must be reset higher when it passes through the germ line of a male.

2. SEX. When derived from the same parent, a locus is less often heterochromatized in males than in females.

3. AMOUNT OF CONSTITUTIVE HETEROCHROMATIN. When additional constitutive heterochromatin is added to the nucleus, heterochromatization (and phenotypic variegation) is reduced. We do not know the genetic or molecular basis for such effects.

As implied earlier, the same locus is more often heterochromatized the closer it is placed near constitutive heterochromatin and the larger this piece of heterochromatin is. The smaller the piece of euchromatin inserted into constitutive heterochromatin, the more often it is heterochromatized. Finally, the insertion of constitutive heterochromatin into a euchromatic region causes it to amplify more than it normally would; in other words, its undergoes *euchromatization*.

# MAIZE

## 16.10 Controlling genes in maize are transposable and may function by heterochromatizing neighboring genes.

We have already mentioned in Section 14.6 that $\phi$Mu-1 can integrate at almost any locus in the *E. coli* chromosome. Prokaryotes contain other types of genetic elements, called *Insertion Sequences*, which are not phages, that like $\phi$Mu-1 can jump or *transpose* from one locus in the chromosome to another. Transposable genetic elements also occur in eukaryotes, including *Drosophila* and maize.

In maize the transposable genes, which are normal components of the nuclear chromosomes, are called *controlling genes* because they control the functional state or the level of activity of neighboring genes. Controlling genes are constitutively heterochromatic and produce a position effect on neighboring euchromatic genes by suppressing their expression, probably at the level of transcription, possibly by means of heterochromatization. That this suppression is a spreading position effect like that of the C- and V-type position effects already described is supported by the finding that controlling genes can suppress the functioning of genes that are as much as five crossover map units away.

The behavior and phenotypic effects of controlling genes can be illustrated by their effects in corn kernels on the expression of neighboring genes whose functioning is needed to produce pigment. For example, the functioning of the protein-coding gene $A_1$, needed for the production of anthocyanin pigment, has been found to come under the regulation of two types of controlling gene. The controlling gene that is inserted near the $A_1$ locus can have any one of a number of settings, each of which results in a different amount of suppression of the action of $A_1$. When no other $A_1$ allele is present in the genotype, kernels have a uniform pigmentation that ranges from full or almost-full pigmentation to white (top row of Figure 16-3). In the absence of the other type of controlling gene, the controlling gene neighboring $A_1$ retains its setting and is not transposable. In the presence of the other type of controlling gene, whose locus may be on any one of the chromosomes, the setting of the first type is changed, after which it may transpose. The controlling gene neighboring $A_1$ is, therefore, a *responding gene* that reacts to signals of the other controlling gene serving as a *signaling gene*. Apparently, the heterochromatic signaling gene is transcribed (and perhaps the transcript is translated) to produce the signal to which the responding gene reacts. In fact, in the presence of a particular signaling gene, the responding gene is first reset so that $A_1$ produces even less pigmentation. Then the signaling gene may signal the responding gene to transpose, that is, to leave its site near $A_1$ and relocate elsewhere in the chromosome set. Once the responding gene transposes, the $A_1$ gene is no longer suppressed and produces its full anthocyanin phenotypic effect. As a result of the presence of the responding gene and the signaling gene, therefore, a kernel has overall less pigmentation than it has in the absence of the signaling gene, and contains speckles of full pigmentation whose size depends upon the number of cell divisions occurring after the responding gene has transposed (bottom row of Figure 16-3).

Signaling genes also suppress the functioning of adjacent euchromatic genes; moreover, they are able to transpose themselves. Once transposed, a controlling gene suppresses the functioning of its new euchromatic neighboring genes. It should be noted, finally, that different controlling systems in maize employ from one to four controlling genes, and that extra doses of some signaling genes increase, whereas extra doses of others decrease, the frequency of transposition. Accordingly, controlling genes can produce a large variety of phenotypic responses during development.

FIGURE 16-3. Different phenotypic effects of $A_1$ due to the presence of a nearby responding gene. *Top row*: Kernels produced in the absence of the signaling gene. *Bottom row*: Corresponding kernels produced in the presence of one signaling gene. See text for explanation. (Courtesy of B. McClintock, 1965. *Brookhaven Sympos. Biol.*, **18**: 172.)

## SUMMARY AND CONCLUSIONS

Dosage compensation occurs in eukaryotes with X and Y sex chromosomes for loci not concerned with sex that are paired in one sex but not the other. Such gene-action regulation is apparently accomplished at the level of transcription. In human beings and most other mammals, transcription is inhibited during interphase by heterochromatization, which produces facultative heterochromatin from euchromatin. This explanation probably also holds for the mouse, where facultative heterochromatin is observed at prophase, but is insufficiently clumped to be cytologically detectable at interphase. In *Drosophila*, dosage compensation is accomplished not by heterochromatization but by having both genes in the two-dose condition functional to the same extent as one gene is in the one-dose condition.

In mammals, a suppressive C-type position effect spreads to autosomal loci linked by means of breakage rearrangement to an X inactivated by dosage compensation. Such position effects probably occur at the level of transcription by means of cytologically undetected heterochromatization. In *Drosophila*, a suppressive V-type position effect occurs on euchromatic genes brought by breakage rearrangement near constitutive heterochromatin. In this case, because of the advantageous presence of multichromatid chromosomes, it is possible to show that the amount of the mosaic exhibiting the suppressed phenotypic effect parallels the frequency of heterochromatization of the genes showing the position effect. In maize, transposable controlling genes exert a spreading suppressive position effect on neighboring genes. Since controlling genes are constitutively heterochromatic, it is likely that they suppress the transcription of adjacent euchromatic genes by cytologically undetected heterochromatization.

## QUESTIONS AND PROBLEMS

**16.1** What kinds of gene are dosage-compensated?

**16.2** Give the genotype of a human male with a Barr body.

**16.3** How much G6PD activity do you expect in: **a.** a diploid cell that is X0, XXX, or XXXY?  **b.** a tetraploid cell that is XX, XXX, or XXXY?

**16.4** How can you tell whether a phenotypic change is due to a position effect rather than to an intragenic mutation?

**16.5** In human beings, in what cells of a normal diploid XX individual is there no dosage compensation?

**16.6** How does dosage compensation differ in human beings and *Drosophila*?

**16.7** Distinguish between a C-type and a V-type position effect.

**16.8** How is the heterochromatization of $w^+$ expressed in a *Drosophila*: **a.** larval salivary gland cell?   **b.** compound eye cell?

**16.9** Offer a hypothesis to explain why the addition of extra heterochromatin reduces a V-type position effect in *Drosophila*.

**16.10** Give an example of setting and erasure in: **a.** kangaroos.  **b.** *Drosophila*.  **c.** maize.

**16.11** Under what conditions is heterochromatin euchromatized in *Drosophila*?

**16.12** Do all genes have a stable locus in the chromosome set? Explain.

**16.13** In what respects are signaling and responding genes in maize similar?

**16.14** Give two explanations for color striping found in maize leaves.

**16.15** Which one of the following statements about dosage compensation is false? **a.** It occurs for X-limited loci not concerned with sexual traits. **b.** It occurs in human females by completely suppressing transcription of one of two alleles present. **c.** It produces up to three Barr bodies in tetraploid nuclei. **d.** It is accomplished in human beings by heterochromatization. **e.** It affects both alleles equally in a *Drosophila* female.

# PART V

# HOW GENE PRODUCTS INTERACT AND THE PHENOTYPIC CONSEQUENCES OF GENE ACTION

# Chapter 17

## Phenotypic Effects of Environment, Genotype, and Single Loci

THE SUCCESS OF AN ORGANISM does not depend only on the physical–chemical quality and quantity of its genetic material, nor on how this material is replicated, transcribed for translation, varied, recombined, transmitted, and regulated. Although these subcellular features are important in determining biological success, the phenotype expressed at the level of the cell is a major factor determining whether an organism is successful. (No matter how good a muscle cell is in all genotypic respects, it is a failure if, phenotypically, it cannot contract.) This means that, in order to understand the contributions of genes to organisms, we need to know the principal ways that the products of genes interact with each other and with the environment to produce the phenotypes of cells, tissues, organs, organ systems, and the integrated whole individual.

This chapter deals with the phenotypic effects of the interaction between the environment and the products of the genotype, with emphasis on the effects of single loci.

**17.1** Twin studies tell us the relative importance, but not the nature, of genotype and environment in the production of a trait.

The extent to which a given human trait is the result of the genotype and to what extent it is a result of the environment has always been a question of considerable interest. The *relative* contributions of genotype (nature) and environment (nurture) in the production of a human trait can be determined from the phenotypes of the two kinds of twins. Twins of one kind are derived from a single fertilized egg that subsequently produced two separate masses of cells, each of which developed into an individual. Such one-egg twins are *monozygotic twins*, and since, barring mutation, they are genetically identical, they are also called *identical twins* (Figure 17-1). Twins of the other kind are derived from two fertilized eggs, and

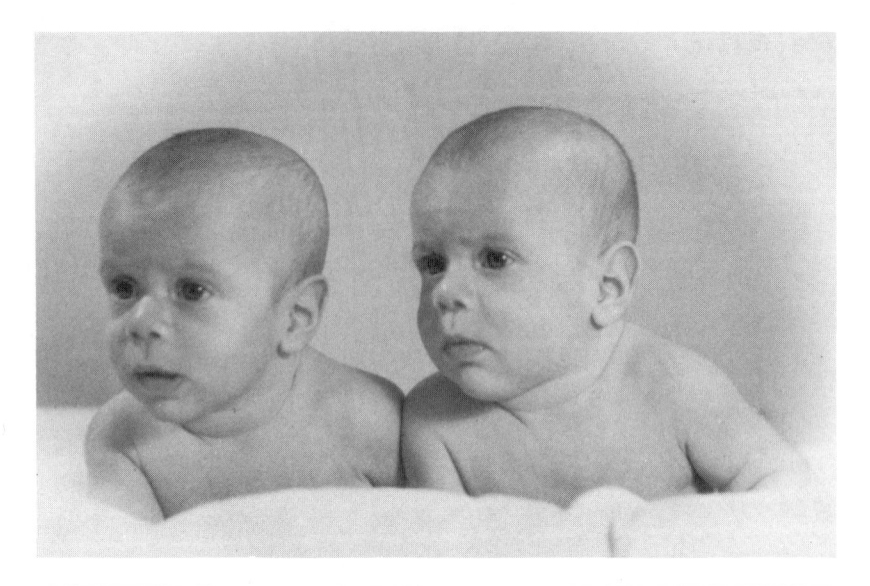

FIGURE 17-1. Monozygotic twins, Ira and Joel, at $3\frac{1}{2}$ months and at 19 years of age. (Courtesy of Mrs. Reida Postrel Herskowitz.)

are no more similar genetically than any two siblings (children of the same parents). Such two-egg twins are called *dizygotic* or *nonidentical twins* (Figure 17-2).

In twin studies mutation is usually ignored as being too rare to affect the conclusion. Since monozygotic twins are genetically identical, phenotypic differences between them are attributed to environmental differences. To obtain an indication of the effect of the environment, therefore, one can compare the average phenotypic differences between monozygotic twins reared together and those reared apart. To obtain an indication of the effect of genotype, one can compare the average phenotypic differences

# PHENOTYPIC EFFECTS

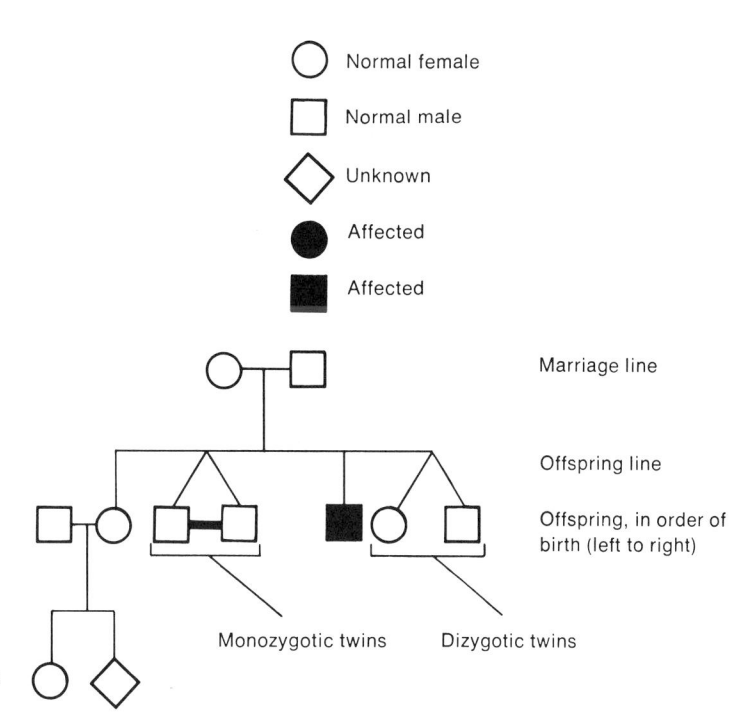

FIGURE 17-2. Symbols used in human pedigrees.

between monozygotic twins reared together and dizygotic twins reared together.

Twin studies of differences in physical and mental traits show, in general, that they have both a genetic and an environmental basis, and that the relative importance of each varies for different traits. For example, having or not having (1) the same ABO blood type is determined 100 per cent by genotype and 0 per cent by environment; (2) clubfoot is determined 29 to 32 per cent by genotype and 68 to 71 per cent by environment; and (3) tuberculosis is determined 46 to 74 per cent by genotype and 26 to 54 per cent by environment. It should be noted, however, that twin studies tell us nothing about the specific genes or the nature of the environmental factors responsible, or their time or mode of action and interaction.

## 17.2 The phenotype is affected by the genetic and nongenetic environment.

The phenotype of a human being depends upon the genotypes of other organisms in its immediate environment, that is, upon the *genetic environment*. Some infecting organisms may be slightly, moderately, or greatly harmful to the human phenotype because of the diseases they produce, whereas others, such as the bacterial flora in the intestines, are distinctly beneficial. The baby developing in the mother's uterus and the kernels growing in an ear of maize receive products of gene action of the parent nurturing them.

FIGURE 17-3. Male Siamese cat, grown under temperate conditions, showing the pigmentation pattern also found in the Himalayan rabbit. (Courtesy of Joan Wolk.)

The phenotypic expression of a genotype requires a suitable *nongenetic environment*. For example, the expression of the coat color in the Siamese cat (Figure 17-3) depends upon the outside temperature. In a cool environment, the temperature of the skin of the main part of the body is high enough to inactivate an enzyme needed for pigment formation; at the extremities (paws, ears, and tail), however, the skin temperature is lower, the enzyme is functional, and pigment is produced. In a very cold environment, the skin temperature is lower, and the coat is completely black. In many human beings, for another example, the amount of sunlight determines the phenotype with regard o tan or freckles.

When the genotype is uniform we can often attribute differences in phenotypes to differences in specific features of the genetic or nongenetic environment.

**17.3** Some genes have a sometimes undetectable or a variable phenotypic effect.

Provided that conditions are compatible with existence, some genes express themselves phenotypically in the same way despite differences that occur in the genetic or nongenetic environments or in the remainder of the genotype, that is, in the *genetic background*. Such is the case for the single gene pair that determines ABO blood type. Genes that always express themselves phenotypically are said to show 100 per cent *penetrance*, and penetrant genes that produce invariant phenotypic effects are said to show uniform *expressivity*.

## PHENOTYPIC EFFECTS

Other genes have less than 100 per cent penetrance and/or variable expressivity. In the case of the Siamese cat, for example, the homozygous gene pair responsible for coat color mosaicism is not penetrant in very hot climates, where the coat is completely white; and its expressivity when penetrant is variable, the coat color getting darker as the environmental temperature drops. When the same genes in different individuals show incomplete penetrance and/or variable expressivity, this may be due to the environment or the genetic background varying in some uncontrolled (sometimes unknown) manner in different individuals. This is the case for *polydactyly*, a rare condition in which human beings have more than five digits on a limb due to the presence of allele *P* (Figure 17-4). The left side of the pedigree shows a polydactylous daughter whose father carried a *P* that was not penetrant. The different numbers of extra fingers and toes (and their different sizes) indicate the variable expressivity of penetrant *P*. In this case we do not know any of the factors responsible for the differences in penetrance and expressivity in different individuals. Differences in penetrance or expressivity are sometimes shown by essentially duplicate parts of the same individual; for example, one hand has five and the other has six digits, or one hand has a small extra digit and the other a large one. In these cases one can be reasonably certain that such differences are based on environmental differences, and not on differences in the organism's genotype in different parts of the body.

### 17.4 The same trait may be produced by different alleles acting in different environments.

Genotypically different individuals that are phenotypically different in the same environment may become phenotypically similar when their environments differ. For example, the Siamese cat and a genetically black cat are phenotypically different when both are grown at a moderate temperature. When the Siamese cat is grown in the cold, however, its coat is all black. In this case, the black Siamese cat is a phenotypic imitation or *phenocopy* of the genetically black cat.

FIGURE 17-4. Pedigree of polydactyly in human beings.

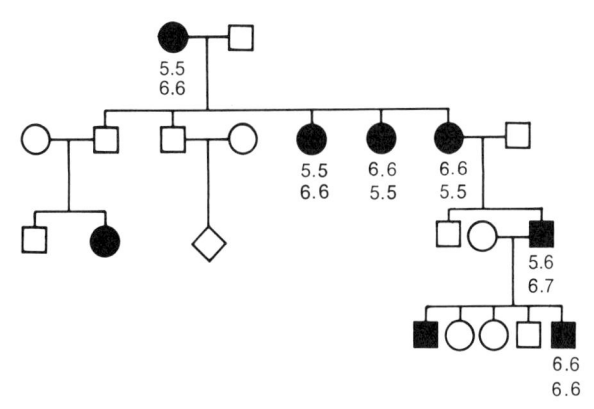

Both normal and abnormal phenotypes can be phenocopied: (1) *phenocopying normal*: persons who are genetically diabetic and take insulin are phenocopies of genetically normal persons who do not take insulin; (2) *phenocopying abnormal*: genetically normal human embryos whose mothers are exposed to the drug thalidomide may develop into babies with two or four limbs missing or reduced to stumps. Such individuals are phenocopies of abnormal persons (grown without thalidomide) having the genetic disease *phocomelia*, which is caused by a single gene in homozygous condition.

It is obvious that the most dependable genetic markers are those that do not phenocopy each other within the range of genetic or environmental conditions to which the individuals under study are subject.

## 17.5 A single gene usually has phenotypic effects on many traits.

A gene that codes for a polypeptide has only one immediate phenotypic effect, the production of the polypeptide. Even when the polypeptide has only a single primary structural or enzymatic function, however, it will require the many chemical reactions that lead up to its synthesis as well as affecting many chemical reactions after its synthesis. Since any protein influences metabolism at many biochemical steps, many of which ultimately influence somewhat different aspects of the phenotype, we expect that an allele that alters a translational product will affect many phenotypic traits; that is, it will have *multiple phenotypic effects*.

In human beings, the abnormal allele that, when homozygous, causes *sickle-cell anemia* shows multiple phenotypic effects. The normal allele $\beta^A$ codes for a globin called the $\beta^A$ chain that has Glu as the sixth amino acid from the beginning. The abnormal allele $\beta^S$ probably resulted from a transversion from AT to TA. This mutation caused U to substitute for A as the middle base in the codon for the sixth amino acid, which, therefore, codes for Val instead of Glu. The globin of the abnormal allele is designated $\beta^S$.

In genetically normal $\beta^A \beta^A$ individuals, two $\beta^A$ chains and two $\alpha^A$ chains (coded at a different locus, $\alpha^A$) join with four iron-containing *heme* groups to produce *hemoglobin A* (Hb-A), $\alpha_2^A \beta_2^A$. Homozygotes for the abnormal allele, $\beta^S \beta^S$, produce abnormal *hemoglobin S* (Hb-S) composed of $\alpha_2^A \beta_2^S$.

The direct chemical effect of the abnormal allele also has direct physical effects in that the abnormal hemoglobin is modified in electrical charge and solubility and has a slightly lower oxygen-carrying capacity. There are secondary effects of the abnormal allele as well. In $\beta^S \beta^S$ individuals the shape of many erythrocytes is changed from disc-like to sickle-like (Figure 17-5). Sickling seems, at least sometimes, to be due to the stacking of deoxygenated hemoglobin S molecules which make long stiff polymers that push against the cell membrane and distort the shape of the cell. The stacking of Hb-S may be associated with a bonding of the

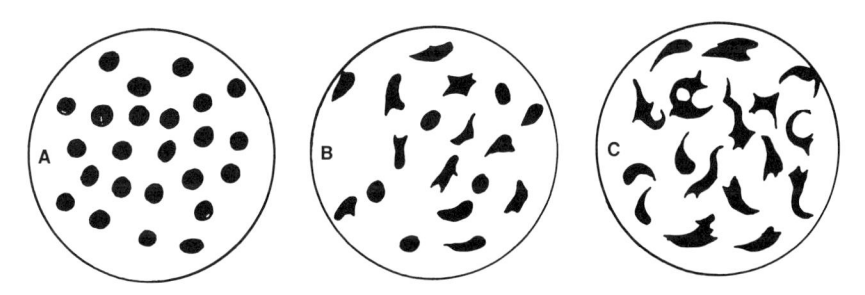

FIGURE 17-5. Silhouettes showing various types of human erythrocytes. **A**: Normal, in normal homozygote. **B**: Sickle-cell trait, in $\beta^S$ heterozygote. **C**: Sickle-cell anemia, in $\beta^S$ homozygote.

Val at position 6 with the Val normally present at position 1. The sickle cells have three main effects:

1. The most common effect is that sickle cells catch or hook onto each other as they try to pass through capillaries, thereby clogging the passageways and preventing blood flow. This clogging, which can occur in any capillary of the body, causes pain, swelling, and tissue death. As a consequence of the local failures in blood supply, various organs and organ systems are adversely affected (Figure 17-6).
2. The sickle shape makes it more likely that the erythrocyte will break or be destroyed, resulting in anemia. The anemia has various detrimental effects on the physical and mental abilities of the affected person (Figure 17-6).
3. The volume of blood retained in the spleen, which is enlarged and becomes fibrous, may increase so much that the decrease in circulating blood can be fatal. In fact, unless some of the detrimental effects are alleviated medically, the $\beta^S$ homozygote usually dies before sexual maturity.

We see, therefore, that the multiple phenotypic effects of the $\beta^S$ allele occur at various levels of organization—organismal, organ system, organ, tissue, cell, and molecular—all of them traceable to the primary effect, the production of an abnormal $\beta$ chain.

## 17.6 Some of the multiple phenotypic effects of one allele can mask those of a different allele.

Let us consider the relationship between the phenotypic expressions of two alleles in the heterozygous individual. The $\beta^A$ and $\beta^S$ alleles are instructive in this respect. In the $\beta^A\beta^S$ heterozygote, both alleles are transcribed, and translation produces both $\beta^A$ and $\beta^S$ chains, neither allele influencing the transcription or translation of the other allele or its mRNA. When the heterozygote is examined for globins—the primary translation products—as expected, the two alleles are found to have separate, detectable phenotypic effects. When the heterozygote is examined for some of the multiple, posttranslational phenotypic effects of the $\beta^S$ allele, however, the result is different. The heterozygote ordinarily shows no anemia, or other tissue, organ, or organ-system

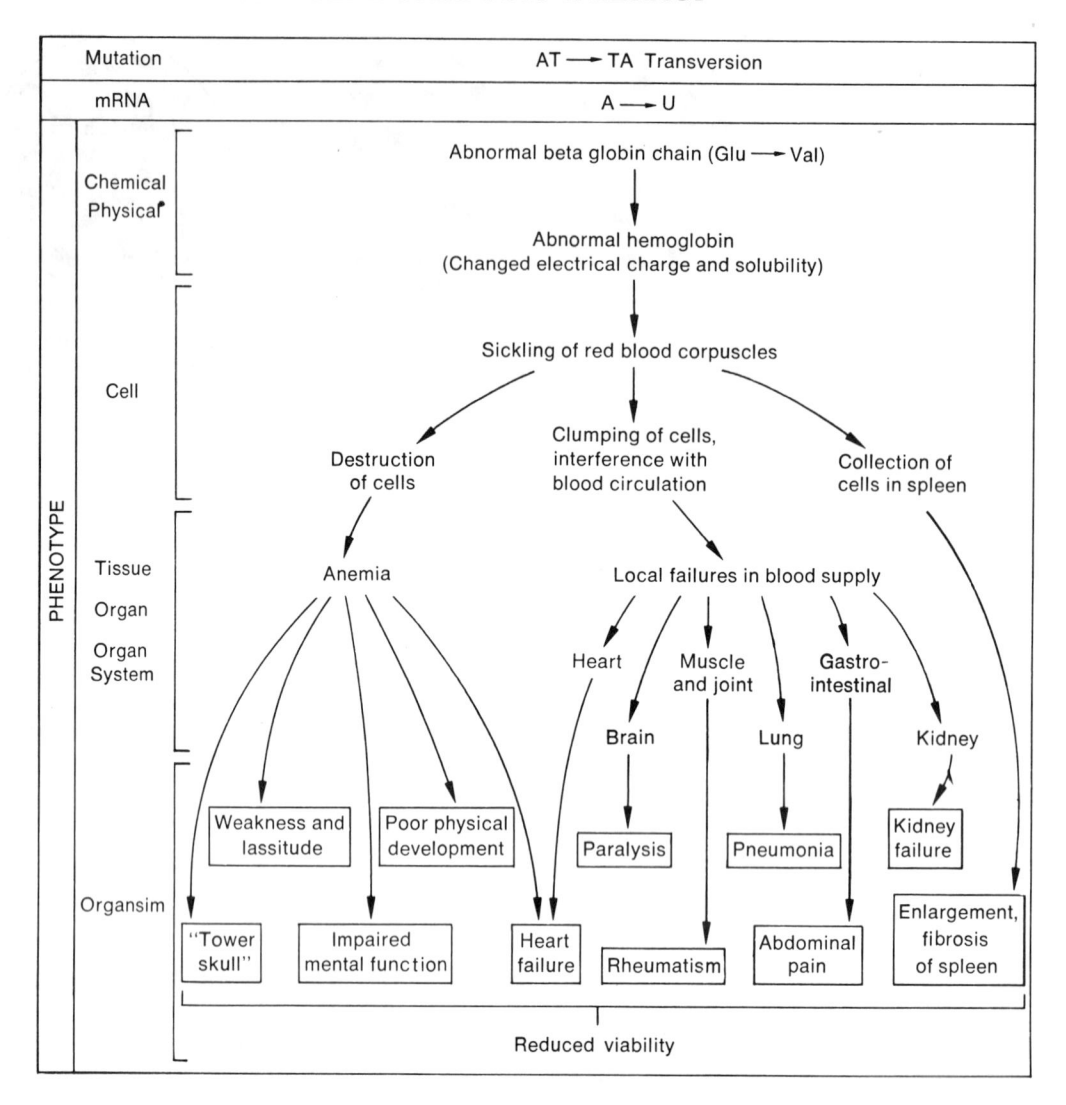

FIGURE 17-6. A "pedigree of causes" of the multiple effects of the abnormal beta chain of hemoglobin S. (After J. V. Neel and W. J. Schull, 1954. *Human Heredity*. Chicago: University of Chicago Press.)

disability that occurs in the $\beta^S$ homozygote (Figure 17-7). For these traits, therefore, the phenotypic effect of $\beta^A$ is *dominant* to $\beta^S$, which has a *recessive* phenotypic effect ($\beta^A$ shows dominance and $\beta^S$, recessiveness in the heterozygote). When, as in the present case, the heterozygote has exactly or nearly the same phenotypic effect as one homozygote, dominance is complete or almost complete.

Consider the sickling trait. The heterozygote shows some sickled cells under reduced oxygen tension, but fewer than are found in the $\beta^S$ homozygote. For this trait, $\beta^S$ is partially dominant to $\beta^A$.

Finally, the heterozygote is more resistant to a certain kind of malaria, falciparum malaria, than the $\beta^A$ homozygote. With respect to this trait,

FIGURE 17-7. Dominance relations in the $\beta^A\beta^S$ heterozygote.

|  | $\beta^A\beta^A$ | $\beta^A\beta^S$ | $\beta^S\beta^S$ | Dominance |
|---|---|---|---|---|
| Globin chains | $\beta_2^A$ | $\beta_2^A + \beta_2^S$ | $\beta_2^S$ | None |
| Anemia; Tissue, organ, organ system disability | − | − | + | $\beta^A > \beta^S$ |
| Sickling | − | + | ++ | Partial |
| Malaria resistance | − | + | + | $\beta^S > \beta^A$ |

therefore, $\beta^S$ is dominant to $\beta^A$. Because $\beta^S\beta^S$ is so often lethal, it is hard to determine its malarial resistance and hence the degree of dominance of $\beta^S$ in the heterozygote. Genes such as $\beta^S$, which are lethal when homozygous or are so rare that homozygotes are never found, are simply called dominant if they produce phenotypic effects in heterozygotes.

We see, therefore, that at the globin chain level there is no masking of the phenotypic effect of $\beta^S$ by $\beta^A$ or the reverse—there is no dominance; that at the level of tissue-organ disability, $\beta^A$ is completely or almost completely dominant to $\beta^S$; that for sickling trait $\beta^S$ is partially dominant to $\beta^A$; and that for malarial resistance $\beta^S$ is dominant to $\beta^A$.

In cases of complete or almost complete dominance, where the primary gene products are not known (the usual situation), one generally cannot tell from the phenotype whether an individual is homozygous or heterozygous for the dominant allele. This can be decided, however, from the phenotypes of a sufficiently large number of offspring produced by mating such an individual with one (or more) that is homozygous recessive for the genes involved (Figure 17-8). It should be noted that when a

FIGURE 17-8. Results of sample test crosses ($P_1$) involving one pair (column **A**) or two (independently segregating) pairs (column **B**) of mendelian genes showing complete dominance. $T$ (tall) is completely dominant to $t$ (short); $R$ (round) is completely dominant to $r$ (wrinkled).

| A | B |
|---|---|
| $P_1$   T?   X   tt<br>     Tall      Short | $P_1$   T? R?    tt rr<br>    Tall, round   Short, wrinkled |
| $F_1$   Some short *(t t)*<br>     Some tall   *(T t)*<br><br>Unknown $P_1 = Tt$ | $F_1$   Some short, wrinkled *(t t rr)*<br>     Some short, round    *(t t Rr)*<br>     Some tall, wrinkled    *(Tt rr)*<br>     Some tall, round      *(Tt Rr)*<br>Unknown $P_1 = Tt\ Rr$ |
| $P_1$   T?   X   tt<br>     Tall      Short | $P_1$   T? R?   X   tt rr<br>    Tall, round   Short, wrinkled |
| $F_1$   Many tall *(T t)*<br>Unknown $P_1 = TT$ | $F_1$   Many tall, wrinkled    *(Tt rr)*<br>     Many tall, round      *(Tt Rr)*<br>Unknown     = $TT\ Rr$ |

heterozygote has a phenotype (such as gray) that is exactly intermediate between the two homozygotes (black and white), there is no dominance. Since this seems to happen rarely for posttranslational traits, some degree of dominance is expected and usually found.

**17.7** Most mutants are less adaptive and recessive.

Most present-day wild-type alleles occur at loci that have been in existence a long time—long enough for natural selection to have retained as the wild type those alleles which arose by mutation and were most adaptive for the organism's success. What phenotypic effect do we expect in a series of mutant alleles of a protein-coding, wild-type gene that produces substitutions or deletions of one, two, or many amino acids? Since mutations occur without regard to the translational effects they produce, the larger the number of amino acids changed, the less efficient the protein is expected to be in performing the structural or enzymatic functions of the wild-type allele. As long as the mutant allele produces some detectable protein product, both normal and mutant gene products can be detected in the heterozygote, and no dominance is involved. When, however, the heterozygote is scored with respect to one of the other phenotypic traits that characterizes the wild type, the phenotypic effect of the mutant is usually relatively small and, therefore, recessive. The reason for this small, recessive effect is that the phenotypic effect of a single wild-type allele is already near the optimum effect (for reasons discussed in the next section).

**17.8** In the diploid, the phenotypic effect of a single wild-type gene tends to be near, although somewhat short of, the optimum effect.

There must be an optimum amount of gene product that can be made and used phenotypically in a given range of environments. In haploid organisms or haploid stages of life cycles, natural selection retains as the wild type the allele, of those made available by mutation, that most closely produces the optimum phenotypic effect. In diploid organisms or diploid stages of life cycles, however, natural selection retains as the wild-type allele the allele that *best falls short* of producing the optimum phenotypic effect when present in *single* dose. The advantage of such selection in diploids is that if the second allele is also a wild-type allele, the optimum phenotypic effect is attained or more closely approached; if it is different and less adaptive, it will probably do little harm but can be selected against; whereas if it is different and more adaptive than the wild-type allele in bringing the gene product closer to the optimum amount, selection can favor and establish it as the wild-type allele. If natural selection favored alleles whose single dose produced the optimum phenotypic effect, the diploid would be riddled with completely recessive alleles that could only be eliminated in the homozygous condition. It is more efficient, however, to get rid of a harmful mutant soon after it arises, when it is in heterozygous condition, than it is to wait until it has spread, when it is occurring in homozygous condition (see also Section 23.6). We see, therefore, that by favoring alleles whose phenotypic effect in single dose approaches but falls somewhat short of the optimum effect, natural selection gains the advantage of being able to select against heterozygotes.

## 17.9 Different alleles can affect viability to different degrees.

Different alleles can produce large, small, or undetectable effects on viability. Genes that kill an individual before sexual maturity are called *lethal genes* or *lethals*. Lethals that kill when homozygous are *recessive lethals*; without medical intervention the allele for sickling, $\beta^S$, is a recessive lethal. Lethals that kill when heterozygous are *dominant lethals*; *retinoblastoma*, a type of cancer of the eye, is a dominant lethal in the absence of surgical treatment. Lethals can act at any stage of development. Sometimes a lethal is produced not by one gene or a pair of genes, but by the combined effect of several nonallelic genes. In such a *synthetic lethal*, some of the nonalleles are contributed by each parent, and the offspring dies because the nonalleles, viable when separate, are lethal when together.

Different alleles, recessive or dominant, have viability effects that range from those that are lethal, through those that are greatly or slightly detrimental, to those that are apparently neutral or even beneficial (Figure 17-9). Mutants with small viability effects occur much more frequently than those with large effects. For example, mutants that lower viability but do not kill when homozygous are at least three to five times more frequent than mutants that are recessive lethals. As expected, alleles that are detrimental or lethal in homozygous condition usually also have a detrimental effect on viability when heterozygous, the heterozygous effect being only of the order of a few per cent of the homozygous effect.

## SUMMARY AND CONCLUSIONS

The phenotype of an organism is the result of the expression of its genotype in its genetic and nongenetic environment. Even in the absence

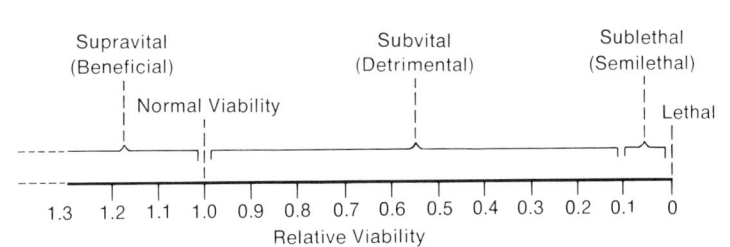

FIGURE 17-9. Classification of effects that alleles have on viability.

of knowledge of specific determining factors, the relative importance of genotype and environment in the production of a trait can be estimated from studies of twins. Such twin studies show that differences in mental and physical traits have both a genetic and an environmental basis, whose relative importance varies for different traits.

The penetrance and expressivity of a specific gene or group of genes depend upon the genetic and nongenetic environment as well as the genetic background. Different environments can cause phenocopying by different alleles.

Individual genes usually have multiple phenotypic effects. Whereas there is no dominance among protein-coding alleles at the protein level, there is usually some dominance for some of their posttranslational phenotypic effects. Most mutant alleles are less adaptive than, and recessive to, their wild-type alleles. In the diploid, natural selection favors as the wild type the allele whose phenotypic effect in single dose falls somewhat short of the optimum effect.

Different alleles have a spectrum of effects on viability. Most mutants are detrimental when homozygous, and are less so when heterozygous.

## QUESTIONS AND PROBLEMS

**17.1** *Anagrams*: Unscramble and define each of the following: **a.** zigdocity. **b.** ponychope. **c.** mandition. **d.** servesice. **e.** tellha.

**17.2** *Play*: "Your Arms Are Too Short to Box with God." What genetic application comes to mind?

**17.3** *Song*: "Bend Me, Shape Me." What genetic application has this song?

**17.4** "Nature and nurture are important in the production of every phenotypic trait." Do you agree? Explain.

**17.5** Can human identical twins ever be of opposite sex? Explain.

**17.6** How could you get a Siamese cat to grow the word LOVE on its back: **a.** in black on a white background? **b.** in white on a black background?

**17.7** Differentiate between penetrance and expressivity.

**17.8** Describe the multiple phenotypic effects of the gene for sickling erythrocytes.

**17.9** Give an example of the same gene being beneficial or detrimental depending upon the: **a.** genotype. **b.** environment.

**17.10** Is a recessive allele affected by the presence of a dominant allele? Explain.

**17.11** Why are most mutants less adaptive and recessive to their wild-type alleles?

**17.12** Why does one wild-type allele usually not produce exactly the same phenotypic effect as two wild-type alleles?

**17.13** Compare the relative detriment caused by a recessive allele in two species—one is diploid, and the other is its single-species tetraploid.

**17.14** In mice, matings between yellow-haired individuals produce $F_1$ in the ratio 2 yellow : 1 nonyellow. Explain. How could you test your explanation?

**17.15** In the snapdragon a cross of two pale green plants produces seedlings in the ratio 1 green : 2 pale green : 1 white. Is a recessive lethal involved? Explain.

**17.16** Two normal people marry and have a child polydactylous on one hand only. How can you explain this?

**17.17** A certain type of baldness is due to a gene that is dominant in men and recessive in women. A nonbald man marries a bald woman and they have a bald son. Give the genotypes of all three. What are the genotypic and phenotypic expectations for the daughters of these parents?

**17.18** Determine for each pedigree the method of inheritance of the trait in question; and, as far as possible, determine for that trait the genotype of each individual in the pedigree.

   **a.** Feeble-mindedness

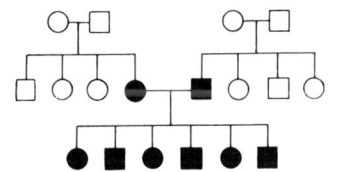

   **b.** Short-sightedness

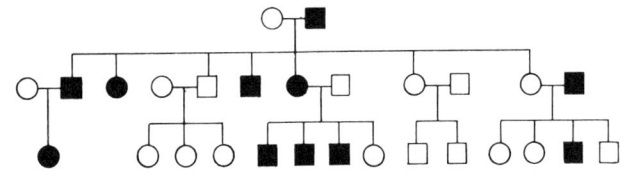

**17.19** When were identical twins not twins?

# Chapter 18

## Phenotypic Interactions of Two or More Loci

THE PHENOTYPIC INTERACTIONS between the gene products of alleles were discussed in Chapter 16. This chapter deals primarily with the phenotypic consequences of interactions between the gene products of two or more nonalleles.

**18.1** Some proteins act individually on the phenotype; others act in combination.

Some functional gene products are single polypeptides coded by a single gene. In such cases the phenotypic effects of different alleles and non-alleles result from the independent action of different proteins. Other functional gene products, however, are proteins composed of two or more polypeptides coded by one, two, or more loci. For example, in human beings, some abnormal hemoglobin molecules are composed of four beta chains, $\beta_4$, coded by a single locus, although the usual hemoglobin A, $\alpha_2^A \beta_2^A$, is coded by two nonallelic loci. The promoter-binding *E. coli* transcriptase $\alpha_2\beta\beta'\omega\sigma$ is coded by five genes, $\alpha$, $\beta$, $\beta'$, $\omega$, and $\sigma$. In this transcriptase the two beta chains are always different; but there is no general rule that the members of a dimer, trimer, and so on, shall be the same or different. In other cases, the two members of a dimer that make up certain enzymes, coded for at a single locus, may be the same or different; as a result, the heterozygote for such a locus produces the "hybrid" as well as both "pure" types of enzymes (Section 16.6).

A similar situation occurs in human beings and other animals, where the enzyme *lactic dehydrogenase* is a tetramer. In somatic cells, each of the four polypeptide chains in a molecule of this enzyme can be derived from either of two unlinked loci, $A$ and $B$, that code for polypeptides of the same length but different amino acid sequence. As a consequence, five lactic dehydrogenase molecules are normally produced—$A_4$, $A_3B_1$, $A_2B_2$, $A_1B_3$, and $B_4$—which differ in their reaction to different substrates, and therefore may have slightly different phenotypic effects. In this case two loci normally produce five functional products of gene action (rather than the two that would result if tetramers could contain only one type of polypeptide). Moreover, since rare alleles of $A$ (and $B$) are known, hybrid individuals may form as many as 12 different kinds of lactic dehydrogenase. In the germ line, a third locus, $C$, is functional that codes for another polypeptide that can be one or more of the chains in the enzyme tetramer. This locus further increases the potential number of different lactic dehydrogenases that can be formed in the germ line.

A functional gene product composed of more than one polypeptide can thus be formed from the polypeptides coded by one or more loci. In such cases, different polypeptides interact soon or immediately after translation, before they are put to metabolic use, and subsequent phenotypic effects result from their action in fixed combinations.

## 18.2 Dominance can cause the number of phenotypic classes to be less than the number of genotypic classes.

When one or both parents are heterozygotes for mendelian genes, we expect the progeny to occur in certain genotypic ratios. Often, however, the observed phenotypic ratios do not approximate the expected genotypic ratios for any one of several reasons, other than sample size being too small. In this section we discuss complete (or almost complete) dominance as a cause of such an effect. Other kinds of gene-product interaction which also modify phenotypic ratios will be discussed in the next section.

In the cross between hybrids, $AA' \times AA'$, the expected genotypic ratio in the $F_1$ is 1 $AA$:2 $AA'$:1 $A'A'$ (see Figure 11-1). If $A$ is completely dominant to $A'$, however, the phenotypic ratio becomes 3 $A$ (composed of 1 $AA$ and 2 $AA'$):1 $A'$ individuals, and the three genotypic classes become only two phenotypic classes.

Consider next the genotypic expectation from the cross between hybrids for two independently segregating loci, $AA' BB' \times AA' BB'$. Each parent produces four equally frequent types of haploid gamete (Figure 18-1). Since male gametes fertilize female gametes at random, we can find all the possible zygotes and their frequencies by using the

FIGURE 18-1. Genotypes of gametes formed by $AA' BB'$, a hybrid for two gene pairs segregating independently.

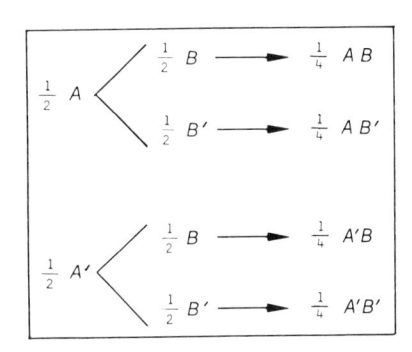

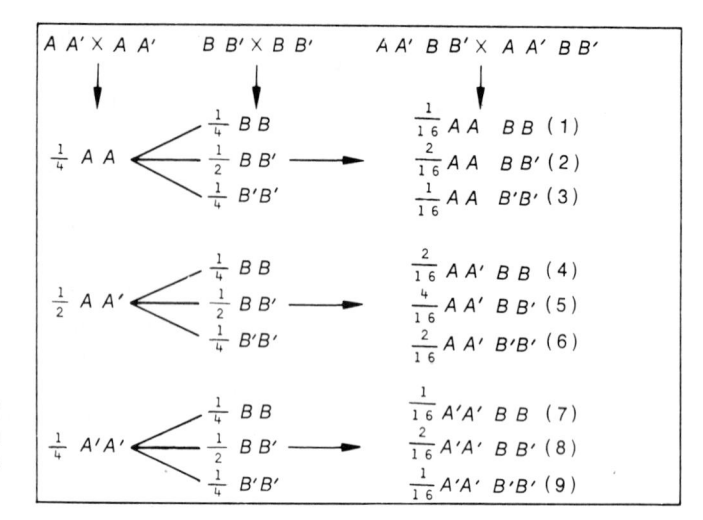

FIGURE 18-2. Genotypic results of segregation and random fertilization which occur independently for two gene pairs in the cross $AA'\,BB' \times AA'\,BB'$.

checkerboard method of combining gametes at random (as was used in Figure 11-2). Figure 18-2 shows how to obtain the same zygotic products a different way: the segregation and random fertilization of $AA' \times AA'$ and of $BB' \times BB'$ are shown occurring independently and the results are then combined at random. In this way we randomly combine the 1:2:1 ratio for one gene pair with the 1:2:1 ratio for the other gene pair.

Among every 16 offspring, on the average, there will be nine different genotypes in the ratio 1:2:1:2:4:2:1:2:1. If neither gene pair shows dominance, and if each pair acts both independently and on different traits, every genotype will have a different phenotype, and the phenotypic ratio will match the genotypic ratio (Figure 18-3A, column N). This is the phenotypic result expected in the children of parents both of whom have thalassemia minor, a mild type of anemia due to the gene pair $Tt$, and MN blood type, due to $L^M L^N$. In this case, the $TT$ individual is phenotypically normal, whereas $tt$ has thalassemia major (a severe anemia, also called Cooley's anemia); the $L^M L^M$ individual is phenotypically blood type M, whereas $L^N L^N$ is phenotypically blood type N. The genotypic and phenotypic ratios in the progeny are, therefore:

| 1 | $TT$ | $L^M L^M$ | normal, M |
| 2 | $TT$ | $L^M L^N$ | normal, MN |
| 1 | $TT$ | $L^N L^N$ | normal, N |
| 2 | $Tt$ | $L^M L^M$ | t. minor, M |
| 4 | $Tt$ | $L^M L^N$ | t. minor, MN |
| 2 | $Tt$ | $L^N L^N$ | t. minor, N |
| 1 | $tt$ | $L^M L^M$ | t. major, M |
| 2 | $tt$ | $L^M L^N$ | t. major, MN |
| 1 | $tt$ | $L^N L^N$ | t. major, N |

When one of the two pairs of genes in our model cross in Figure 18-2 shows complete dominance, however, two different genotypes will produce the same phenotype, and fewer than nine phenotypes will be found. Thus, referring to Figure 18-2, if $B$ is completely dominant to $B'$,

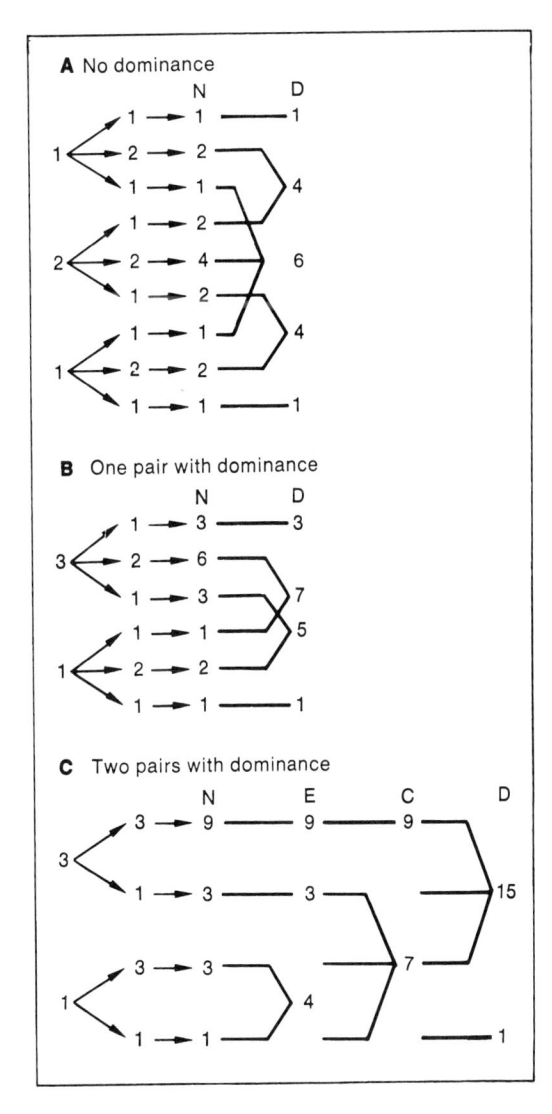

FIGURE 18-3. Phenotypic ratios (columns N, D, E, and C) produced by crossing identical hybrids for two independently segregating gene pairs under different conditions of dominance. N, expected; D, duplicate genes; E, epistasis; C, complementary genes. See the text for explanation.

genotypes 1 and 2 are expressed as a single phenotype, genotypes 4 and 5 as another, and 7 and 8 as a third, so the phenotypic ratio becomes 3:6:3:1:2:1. This ratio results from the random combination of a 3:1 ratio for one gene pair and a 1:2:1 ratio for the other gene pair (Figure 18-3B, column N). This is the phenotypic ratio expected in the children of parents both of whom have MN blood type and are heterozygous for albinism $(A\,a)$—$A\,A$ and $A\,a$ persons are phenotypically normal, whereas $a\,a$ is albino.

When both gene pairs show complete dominance, one phenotype has both dominants expressed (genotypes 1, 2, 4, 5 in Figure 18-2), another phenotype expresses one dominant (genotypes 3 and 6), a third expresses the other dominant (genotypes 7 and 8), and a fourth expresses neither dominant (genotype 9), these phenotypes occurring in a 9:3:3:1 ratio. This ratio results from the random combination of two 3:1 ratios (Figure

18-3C, column N). This is the phenotypic ratio Mendel observed in the progeny of self-fertilizing garden peas heterozygous for two gene pairs.

We see, therefore, that when hybrids for one or more pairs of genes are crossed, dominance will cause the number of phenotypic classes to be less than the number of genotypic classes.

## 18.3 When more than one pair of genes affect the same trait, the number of phenotypes may be less than the number of genotypes.

The same trait is sometimes affected by the protein products of two (or more) pairs of mendelian genes. In these cases, the number of phenotypes observed may be less than the number of genotypes. This reduction in phenotypic classes can be due to gene pairs that are (1) epistatic–hypostatic, (2) complementary, or (3) duplicate. Each of these relationships is discussed with respect to its effect on the phenotypic ratios expected from the model cross, $A A' B B' \times A A' B B'$, described in the last section and Figure 18-2.

1. EPISTATIC–HYPOSTATIC GENES. Sometimes different gene pairs act independently on the same trait, but the expression of one pair prevents the expression of the other pair. The gene pair whose phenotypic expression is hampered by a nonallelic gene pair is said to be *hypostatic*, whereas the gene pair whose phenotype is expressed is said to be *epistatic*. In human beings, for example, the expression of phocomelia (where digits are absent) would be epistatic to the expression of thumb shape or of polydactyly, which would be hypostatic; and complete baldness would be epistatic to hair color or texture, which would be hypostatic. There need be no relationship between the dominance of a gene over its allele and the ability of the gene to be epistatic or hypostatic.

The effect of epistasis–hypostasis on phenotypic ratios is illustrated with an example from *Drosophila* (Figure 18-4). Here $A'$ is a recessive allele that reduces the wing to a stump, the dominant allele $A$ producing the normal large wing. The recessive allele $B'$ causes the wing to be curled, whereas the dominant allele $B$ produces the normal straight wing. Our model cross does not produce the expected 9:3:3:1 ratio, but rather 9 flies with normal wings : 3 with curled wings : 4 whose wings are stumps (of which one fourth would have had curled wings if the full wing had formed) (Figure 18-3C, column E). Here the expression of the no-wing phenotype by one gene pair is epistatic to any expression of the other gene pair.

2. COMPLEMENTARY GENES. In another case of interaction between nonalleles, either of two pairs of genes may prevent a given phenotype from occurring. Suppose that the dominant alleles $A$ and $B$ are *complementary genes*; that is, they each independently contribute something different but essential for the production of a phenotypic effect, whereas their corresponding alleles $A'$ and $B'$ fail to make the respective

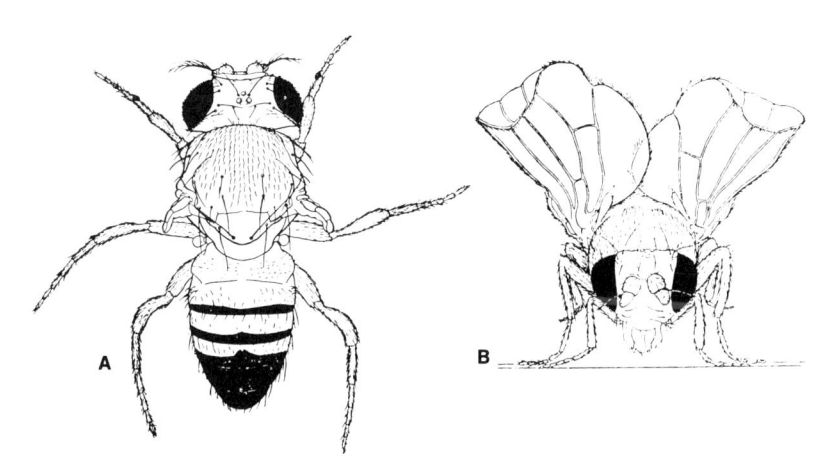

FIGURE 18-4. *D. melanogaster* mutants showing the no-wing (left) and the curled wing (right) phenotypes. (Drawn by E. M. Wallace.)

independent contributions to the phenotypic effect. Our model cross will produce 9 with the phenotype (say red flowers): 7 without the phenotype (say colorless flowers) composed of three homozygotes for $A'$ only, three homozygotes for $B'$ only, and one homozygote for both $A'$ and $B'$ (see Figure 18-3C, column C).

Complementary genes are involved when each of a series of chemical reactions is catalyzed by a different enzyme. For example, in the chemical reaction sequence $A \xrightarrow{1} B \xrightarrow{2} C$, the occurrence of phenotype C requires the functioning of enzymes 1 and 2, which are coded by complementary genes. If either enzyme is not produced because the locus coding it is homozygotes for $B'$ only, and one homozygote for both $A'$ and $B'$ (see in which products A and B combine to produce C; any gene that is required to produce A is a complementary gene to one that is required to produce B, if C is the phenotype under observation.

3. DUPLICATE GENES. Diploids sometimes contain two pairs of genes that contribute equally toward the expression of a phenotype. We will consider two cases of such *duplicate genes*, one in which neither pair shows dominance, and the other in which both pairs show dominance.

a. *No dominance.* The difference in human skin color between whites and blacks depends largely on two pairs of genes. In this case, $A$ and $B$ seem to produce equal amounts of melanin pigment, and $A'$ and $B'$ produce none, dominance being absent. Our model cross between two mulattoes yields the phenotypic ratio 1 black (type 1, Figure 18-2, with four unprimed letters): 4 dark (types 2 and 4, with three unprimed letters): 6 mulatto (types 3, 5, and 7, with two unprimed letters): 4 light (types 6 and 8, with one primed letter): 1 white (type 9, with no unprimed letter). Figure 18-3A shows how the nine possible phenotypic classes (column N) are combined into five actual phenotypic classes (column D) in this case. It should be noted that more than five color classes are actually observed, owing in part to the presence of two or three additional pairs of genes that produce melanin and of other pairs that modify the expression of melanin-producing genes.

b. *Dominance.* Other cases are known in which *A* and *B* show complete dominance, either allele producing the full phenotypic effect. If in our model cross the trait is flower color, two parents with colored flowers will produce offspring in the ratio of 15 colored types (types 1–8, Figure 18-2) : 1 colorless (type 9; see also Figure 18-3C, column D).

Still other combinations of parts of the genotypic ratio from our model cross are possible by other kinds of interactions between nonalleles, producing, for example, phenotypic ratios such as 9:6:1; 10:6; 12:3:1; and 13:3. When we observe abbreviated ratios of the kinds mentioned, therefore, we attribute them to (1) dominance and/or (2) epistatic–hypostatic, complementary, or duplicate genes, or to similar interactions between the products of nonalleles.

**18.4** Whereas the alternative phenotypes of discontinuous (qualitative) traits can be separated into discrete classes, those of continuous (quantitative) traits cannot.

Up to this point, most of the traits in our examples of gene interaction in eukaryotes have had clear-cut, qualitatively different alternatives, as do wing length in *Drosophila* and colorblindness in human beings. These are called *discontinuous* or *qualitative traits* because in each case an individual belongs clearly to one phenotypic class or another. Although the interaction of many genes may ultimately be required for the appearance of a given phenotype, the alternative phenotypes considered so far have been determined primarily by only one or a few pairs of genes. Moreover, in most of the cases considered, the usual environment had little or no effect upon the phenotypic differences observed.

Other traits, such as height or intelligence of human beings, are called *continuous traits* because there are so many grades that individuals are not separable into discrete types or classes. Such traits are also called *quantitative traits* because the continuous range of phenotypes observed requires that an individual be measured in some way in order to be classified. The remainder of this chapter deals with quantitative traits, their genetic and environmental basis, and expression in individuals and populations.

**18.5** Quantitative traits are due to many gene pairs interacting with each other and with an environment that has a greater phenotypic effect than any one gene pair.

Whereas qualitative traits are determined largely by the action of one or a few gene pairs, quantitative traits are due largely to the action of many gene pairs. Each gene pair contributes only slightly toward the expression of a quantitative trait, and the effect of the environment is relatively larger than that of any single gene or gene pair. The significant effects of

# PHENOTYPIC INTERACTIONS

diet upon the height of human beings illustrates the importance of the environment in the expression of a quantitative trait. As a result of the interaction of many gene pairs and the environment, the number of phenotypic classes is so great as to form a virtual continuum. In quantitative traits, therefore, it is usually impossible to detect the phenotypic effect of any single gene or gene pair.

It should be noted that a given trait may be determined qualitatively in certain respects and quantitatively in other respects. In human beings, for example, one pair of genes may determine whether an individual is normal or dwarf, or is normal or has a serious mental deficiency, although normal individuals have a height or mental ability that varies in a continuous way as a result of interactions of the environment and a large number of genes.

**18.6** The larger the number of heterozygous gene pairs determining a quantitative trait, the smaller the chance of obtaining a phenotype that differs greatly from the mean phenotype.

Consider the effect on the types and frequencies of phenotypes produced in the progeny when a particular trait is determined one, two, three, or many heterozygous gene pairs in the parents. Let the trait be size, and the

FIGURE 18-5. Dependence of number of phenotypic classes upon number of gene pairs in the absence of dominance and linkage. Horizontal axis shows classes, vertical axis indicates relative frequencies of occurrence.

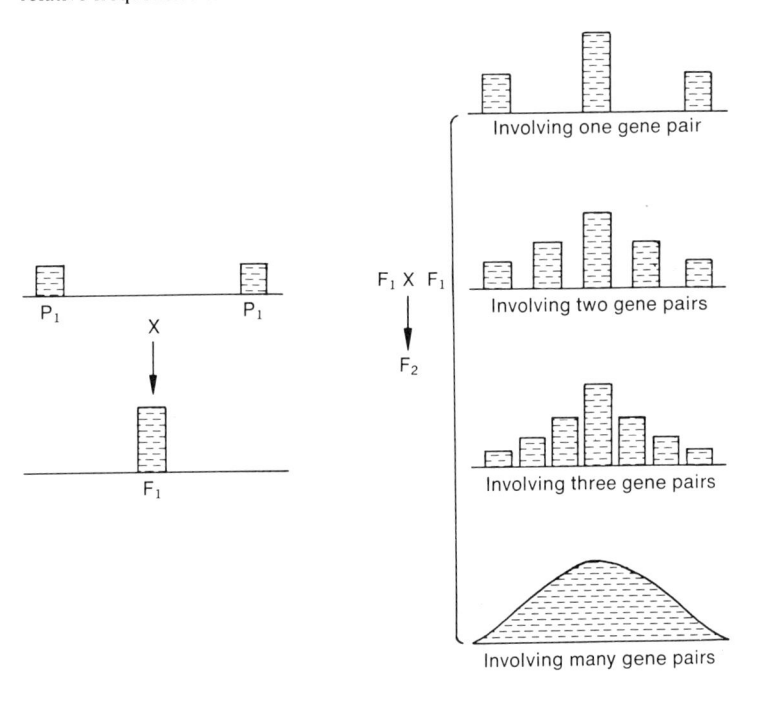

P₁ mating large × small. Assume for now that there is no dominance or linkage, so that the $F_1$, which are phenotypically intermediate (medium) between the $P_1$ parents, are uniformly heterozygous for nonalleles that will segregate independently. Figure 18-5 shows the phenotypic results of matings between $F_1$ (by cross- or self-fertilization) that are heterozygous for one, two, three, or many gene pairs. As the number of determining gene pairs increases, so does the number of $F_2$ classes; when there are many gene pairs, environmental action displaces many individuals from the phenotypic class, so they fall into the space between classes or into adjacent classes, forming a continuum of phenotypes. Note that as the number of gene pairs increases, the fraction of all $F_2$ resembling either $P_1$ becomes smaller. Thus, with one pair of genes, $\frac{1}{2}$ of the $F_2$ are large or small; with two pairs, $\frac{1}{8}$; with three pairs, $\frac{1}{32}$; and so on (left side of Figure 18-6). Consequently, as the number of gene pairs increases from 3 to 10, and from 10 to 20 or more, the continuous distribution of phenotypes gives rise to an $F_2$ curve that becomes narrower and narrower. In other words, the chance of obtaining in $F_2$ any given phenotype that differs greatly from the mean phenotype decreases as the number of heterozygous gene pairs increases. It is possible through the use of statistical methods to estimate the number of heterozygous gene pairs determining

FIGURE 18-6. Dependence of the number of phenotypic classes and their relative frequencies upon the number of gene pairs (each potentially contributing equally toward the phenotype) and the absence or presence of complete dominance. All crosses are between identical heterozygotes. Numbers in parentheses represent the number of genes (in cases of no dominance) or of gene pairs (in cases of dominance) contributing toward the left extreme phenotype. Numbers in brackets are the sums of all but the two extreme classes. *Note*: As the number of gene pairs increases, (1) so does the number of phenotypic classes, (2) a smaller fraction of all progeny is at the extremes, and (3) dominance tends to reduce both these effects.

| Number of Heterozygous Gene Pairs | Dominance Absent | Dominance Complete |
|---|---|---|
| 1 | $\frac{1}{4}$ (2)    $\frac{1}{2}$ (1)    $\frac{1}{4}$ (0) | $\frac{3}{4}$ (1)    $\frac{1}{4}$ (0) |
| 2 | $\frac{1}{16}$ (4)  $\frac{4}{16}$ (3)  $\frac{6}{16}$ (2)  $\frac{4}{16}$ (1)  $\frac{1}{16}$ (0) | $\frac{9}{16}$ (2)  $\frac{6}{16}$ (1)  $\frac{1}{16}$ (0) |
| 3 | $\frac{1}{64}$ (6)  $\frac{6}{64}$ (5)  $\frac{15}{64}$ (4)  $\frac{20}{64}$ (3)  $\frac{15}{64}$ (2)  $\frac{6}{64}$ (1)  $\frac{1}{64}$ (0) | $\frac{27}{64}$ (3)  $\frac{27}{64}$ (2)  $\frac{9}{64}$ (1)  $\frac{1}{64}$ (0) |
| 4 | $\frac{1}{256}$ [ $254/256$ ] $\frac{1}{256}$ (8)(7)(6)(5)(4)(3)(2)(1)(0) | $\frac{81}{256}$ [ $174/256$ ] $\frac{1}{256}$ (4)(3)(2)(1)(0) |

a quantitative trait from the shape of the curve describing the frequencies of different phenotypes in a population.

**18.7** **Dominance reduces the number of phenotypic classes and produces more extreme phenotypes for a quantitative trait.**

Since most genes for qualitative, and presumably for quantitative, traits show complete or partial dominance, consider the effect of dominance on the distribution of phenotypes for a quantitative trait in a population. Dominance reduces the number of phenotypic classes (Section 18.2) and causes a greater proportion of the population to express an extreme phenotype. This can be seen in Figure 18-6, where the frequencies of different phenotypes are given for our model $F_2$ populations produced by parents having 1, 2, 3, or 4 heterozygous gene pairs determining the trait (say, size); when there is no dominance (left side), and when each gene pair has an allele that shows complete dominance (right side) for the same extreme phenotype (say, large size). As the result of dominance, all the distributions in the figure are skewed in favor of the left extreme phenotype (large size).

When many gene pairs determine a trait, however, it is possible that in some of them the dominant allele will be for one extreme phenotype, and in others it will be for the other extreme phenotype. This will reduce the skewness of the distribution, so that the general effect of dominance will be to cause more phenotypes to occur at both extremes than would occur were there no dominance. In other words, dominance causes the distribution curve of phenotypes to be wider than it would be in the absence of dominance. Since the statistical estimate of the number of gene pairs responsible for a quantitative trait is inversely related to the width of the distribution curve, dominance may cause one to underestimate the number of gene pairs involved.

**18.8** **Dominance causes parents that are phenotypically extreme for a quantitative trait to have progeny that are, on the average, less extreme.**

When no dominance occurs among the genes determining a quantitative trait, the offspring of parents all having the same extreme phenotype will be, on the average, just as extreme as their parents. (The environment will cause the offspring to fluctuate somewhat around the extreme phenotype.) If one selects parents having the same extreme phenotype, therefore, it only takes one or a few generations to establish a population whose mean is the extreme phenotype. In the case of dominance, however, some of the extreme parents will be heterozygotes, and some of their progeny will become homozygous for recessive alleles. As a result, the offspring will be, on the average, somewhat less extreme in phenotype than their parents, although somewhat more extreme than the original

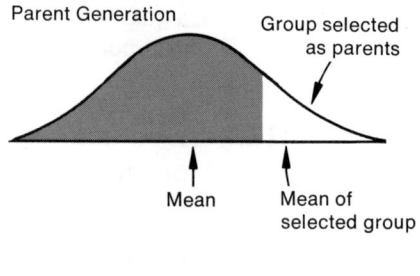

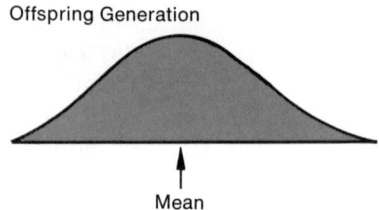

FIGURE 18-7. Selection for a quantitative character determined by genes showing dominance.

mean (Figure 18-7). As one continues to select appropriately extreme individuals as parents, the offspring in successive generations will, on the average, approach more and more closely the extreme phenotype desired. In cases of dominance, therefore, to obtain a population whose mean is the extreme phenotype, requires selecting extreme parents for several to many successive generations.

## SUMMARY AND CONCLUSIONS

Some alleles and nonalleles interact phenotypically by each contributing a polypeptide chain to a complex protein molecule that has an enzymatic or structural function in metabolism. The polypeptides thus contributed by nonalleles may be different (as in hemoglobin synthesis) or the same or very similar (as in lactic dehydrogenase synthesis).

When little or no interaction occurs between the products of alleles and nonalleles, observed phenotypic ratios directly represent expected mendelian genotypic ratios. The occurrence of such interaction, however, between alleles (dominance) or nonalleles (including epistatic–hypostatic, complementary, or duplicate genes) can reduce the observed number of phenotypic classes relative to the expected number of genotypic classes.

Quantitative traits result from the phenotypic interaction of many gene pairs, each of which has a phenotypic effect that is small and often matched or exceeded by the action of the environment. The variability of a quantitative trait is such that the larger the number of heterozygous genes determining it, the narrower is the distribution curve and therefore the smaller the chance of obtaining either of the extreme phenotypes in the offspring. When genes are heterozygous, dominance has the effect of reducing the number of phenotypic classes and of placing proportionally more offspring in extreme classes. Consequently, dominance may cause one to underestimate the number of gene pairs determining a quantitative trait. Since dominance produces progeny that are, on the average,

# PHENOTYPIC INTERACTIONS

less extreme than phenotypically extreme parents, selection must often be continued for a large number of generations to obtain a line that approaches the desired extreme phenotype.

## QUESTIONS AND PROBLEMS

**18.1** *Movies*: "Show Boat." What are the genetic implications of this movie?

**18.2** Name three proteins coded by two or more loci.

**18.3** Is dominance a help or a hindrance in determining genotypes from phenotypes? Explain.

**18.4** What genotypic and phenotypic conclusions can you draw from from the following phenotypic ratios? **a.** 1:2:1. **b.** 3:1. **c.** 1:2:1:2:4:2:1:2:1. **d.** 9:3:3:1. **e.** 9:7. **f.** 15:1. **g.** 63:1.

**18.5** Two chickens with walnut-shaped combs mate and produce offspring whose combs occur in the following ratio: 9 walnut:3 rose:3 pea:1 single. How can you explain these results genetically?

**18.6** When pure White Leghorn poultry are crossed with pure White Silkies, all the $F_1$ are white. Matings between $F_1$ whites produce offspring that occur in the ratio 13 white:3 colored. Explain these results genetically.

**18.7** If all gene pairs are on different chromosome pairs, what fraction of the progeny of the crosses shown below are expected to have a genotype containing *only* small letters? **a.** $Rr\ Ss \times rr\ Ss$. **b.** $Mm\ Nn\ Pp \times Mm\ nn\ Pp$. **c.** $Aa\ Bb\ cc\ Dd\ ee \times aa\ Bb\ cc\ Dd\ Ee$.

**18.8** In each of two different strains of maize, plants occur which, when self-fertilized, produce about $\frac{3}{4}$ normal green:$\frac{1}{4}$ white seedlings. If two such white-producing plants, one from each strain, are crossed, the $F_1$ are all green, but certain $F_1$ when self-fertilized produce seedlings that are approximately $\frac{9}{16}$ green:$\frac{7}{16}$ white. Explain, giving genotypes.

**Note:** In summer squashes the allele for white fruit color, $W$, is epistatic to that for yellow, $Y$; $W$- $Y$- and $W$- $yy$ fruits are white, $ww$ $Y$- are yellow, and $ww$ $yy$ are green.

**18.9** What are the phenotypic expectations with regard to fruit color from the following crosses: **a.** $Ww\ Yy \times Ww\ yy$. **b.** $ww\ YY \times Ww\ yy$. tall. Assuming independent segregation occurs for all gene pairs in the

**18.10** Give the genotypes of two white-fruited squash plants whose offspring occur in a ratio approximating $\frac{3}{4}$ white:$\frac{3}{16}$ yellow:$\frac{1}{16}$ green.

**18.11** Give the genotypes of a white-fruited squash crossed to a green-fruited one whose offspring are approximately $\frac{1}{2}$ white:$\frac{1}{2}$ yellow.

**18.12** Suppose that each gene represented by a capital letter causes a plant to grow an additional inch in height, $aa\ bb\ cc\ dd\ ee$ plants being 12 inches tall. Assuming independent segregation occurs for all gene pairs in the mating $Aa\ BB\ cc\ Dd\ EE \times aa\ bb\ CC\ Dd\ Ee$: **a.** how tall are the parents? **b.** how tall will the tallest $F_1$ be? **c.** how tall will the shortest $F_1$ be? **d.** what proportion of the $F_1$ will be the shortest?

**18.13** Suppose that the phenotypes for a certain trait are distributed with frequencies like that of a normal (bell-shaped) curve. How will the shape of the curve change if: **a.** the genes showing dominance stopped doing so? **b.** the genes not showing dominance started doing so? **c.** selection of one extreme phenotype was carried out for five generations?

**18.14** A normal woman of M blood type has a baby with thalassemia minor and MN blood type. What genotypes are possible for the father?

**18.15** What is the darkest child produced from the marriage between a white and a: **a:** light? **b.** mulatto? **c.** dark? **d.** black?

**18.16** Two parents are heterozygous for four pairs of alleles on different chromosomes. Each parent is $A\,a\;B\,b\;C\,c\;D\,d$. What proportion of their offspring are expected to be $A\,A\;b\,b\;C\,C\;D\,d$?

**18.17** Deaf mutism in human beings is due to the presence of either or both of the completely recessive nonallelic genes $a$ and $b$ in homozygous condition. These genes are located in different autosomes. **a.** Give the genotypes of a deaf mute man and a deaf mute woman who can produce only normal children. **b.** A deaf mute of genotype $a\,a\;b\,b$ has normal parents. Give the genotypes possible for his parents.

**18.18** A single pair of parents had the offspring shown in the accompanying figure. Give a genetic explanation for these results.

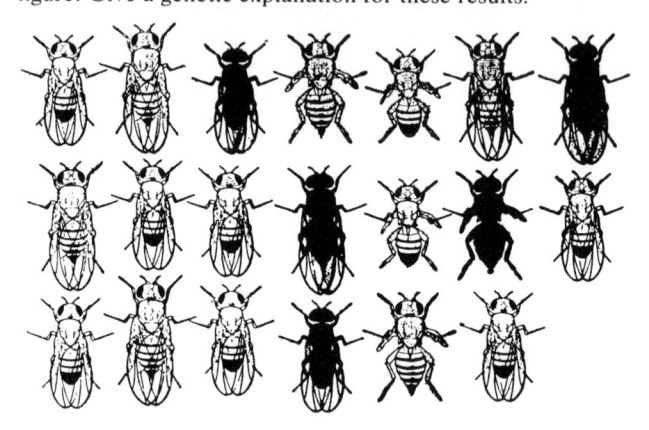

*Chapter 19*

# Determination of Sex in Eukaryotes

MEIOSIS AND THE SUBSEQUENT FUSION of gametic nuclei during fertilization are the two most important features of the sexual mechanism for genetic recombination in eukaryotes. This mechanism requires the production of phenotypically different gametes and/or sexes—the cellular and organismal vehicles for accomplishing meiosis and nuclear fusion.

In this chapter we examine the relative importance of genotype and environment in the production of sexual phenotypes in various eukaryotic plants and animals, including human beings. The examples show in general that although sex-type potentiality is specified genetically at the start of an organism's existence, the sex phenotype actually developed depends to various degrees upon the environment.

# YEAST

**19.1 Mating type in yeast seems to be determined by transposable genes.**

We have already noted in Section 13.5 that the mating type of *Chlamydomonas* is determined by a single pair of nuclear genes having + and − alleles. Haploid individuals contain either + or − alleles and mating occurs between individuals of opposite mating type. Mating type in haploids of yeast, *Saccharomyces cerevisiae*, is also determined at a single nuclear locus (on chromosome 3) that normally carries either of two alleles, *a* and *α*. In the presence of a nonallelic gene, *ho*, mating type is stable, and the conversion of mating type *a* to *α* or the reverse occurs about once per million individuals. In the presence of the allele *HO*, however, mating type interconversion, $a \rightleftharpoons \alpha$, occurs as often as each generation. *HO*, the gene responsible for the high frequency of mating type interconversion, promotes a change at the mating type locus, the new mating type being stable after *HO* is removed by recombination. A clue to the mechanism of *HO* action comes from the finding that in addition to the locus for mating type on chromosome 3 that is phenotypically expressed, there are other loci that contain unexpressed or silent copies of *a* and *α*. The latter loci are silent presumably because they are not preceded by a promoter or a ribosome-binding gene. The currently accepted explanation is that *HO* directs a copy of one of the silent mating type loci to substitute for the resident information at the expressed mating type locus; or, in other words, *HO* changes the cassette playing mating type information. Using the terminology of controlling genes, *HO* is a signalling gene that directs the transposition of a mating type gene, the responding gene, from a silent locus to an active locus where it replaces the resident information.

# DROSOPHILA

**19.2** In *Drosophila*, sex type is determined by the balance between genes located on the X chromosome and those located on autosomes.

In *Drosophila*, the usual female (Figure 19-1) has two X chromosomes and the usual male has an X and a Y chromosome in addition to two sets of autosomes (Figure 19-2). Viable *Drosophila* that have abnormal chromosome numbers occur as the result of segregation in triploid females, or of germ-line chromosome nondisjunction or loss in diploids. Since otherwise diploid flies that have an X as their only sex chromosome, being X0, are sterile males, it is clear that, in *Drosophila*, the Y chromosome is needed not for sex type but for male fertility. Since diploids that are X0, XY, or XYY are male, whereas those that are XX,

FIGURE 19-1. Normal (wild-type) *Drosophila melanogaster* male (**A**) and female (**B**). (Drawn by E. M. Wallace; photographs courtesy of L. Ehrman.)

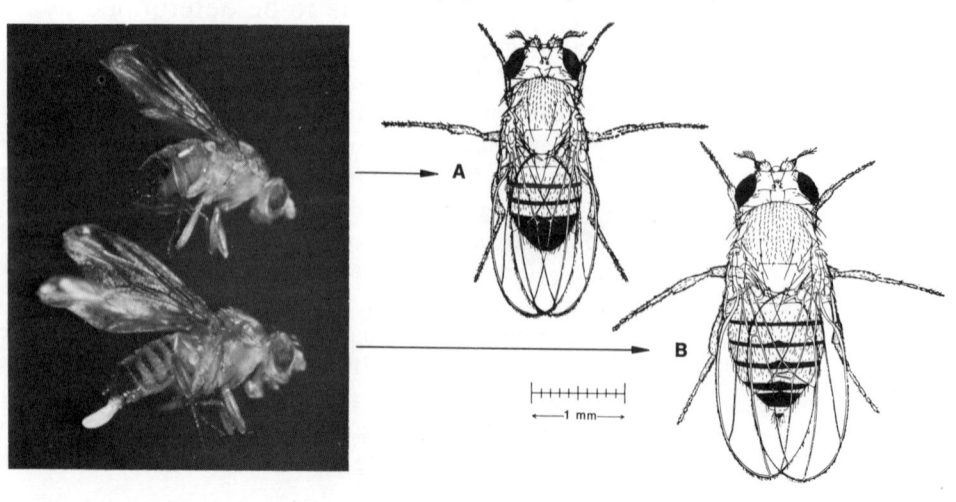

# DETERMINATION OF SEX IN EUKARYOTES

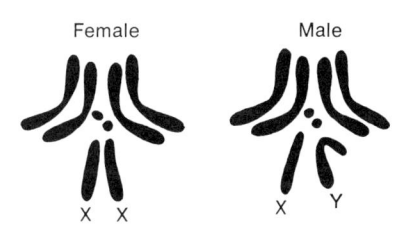

FIGURE 19-2. Silhouettes of condensed mitotic chromosomes of *D. melanogaster*.

| Phenotype | | Number of X Chromosomes | Number of Sets of Autosomes (A's) | Sex Index (X/A Ratio) |
|---|---|---|---|---|
| Superfemale | | 3 | 2 | 1.5 |
| Normal female | Tetraploid | 4 | 4 | 1.0 |
| | Triploid | 3 | 3 | 1.0 |
| | Diploid | 2 | 2 | 1.0 |
| | Haploid | 1 | 1 | 1.0 |
| Intersex | | 2 | 3 | 0.67 |
| Normal male | | 1 | 2 | 0.50 |
| Supermale | | 1 | 3 | 0.33 |

FIGURE 19-3. Sex index and sexual type in *D. melanogaster*.

XXY, XXYY are female, it appears that sex type is determined by the ratio of X's to sets of autosomes (Figure 19-3). When the ratio of the number of X's to the number of sets of autosomes is $\frac{1}{2}$, the result is a male. When the ratio is 1, the result is female, whether the ratio comes from a diploid individual or from a haploid, triploid, or tetraploid one. Apparently, therefore, it is the balance between X's and autosomes and not their numbers that determines sex type.

The chromosome balance explanation of sex determination leads to two predictions: (1) *Drosophila* having ratios other than $\frac{1}{2}$ or 1 should be of abnormal sex type, and (2) loci on autosomes should affect sex type. Both expectations are confirmed by the occurrence of individuals of three abnormal sex types (Figure 19-3). One of these, the *intersex*, has a ratio of $\frac{2}{3}$ and a phenotype that is sterile and overall intermediate between a normal male ($\frac{1}{2}$) and a normal female (1). The other two abnormal sex types are *supersexes* (really *infrasexes*, since they are sterile): one is the sterile *supermale* with a ratio of $\frac{1}{3}$ whose phenotype expresses certain male characteristics even more strongly than the normal male does; the other is the sterile *superfemale* with a ratio of $\frac{3}{2}$ whose phenotype expresses certain female characteristics even more strongly than the normal female does. As expected, the three abnormal sex types have abnormal, but appropriate ratios. That decreasing the ratio from $\frac{1}{2}$ to $\frac{1}{3}$ by the addition of a set of autosomes changes sex type proves that autosomes affect sex type.

Sex type in *Drosophila* is a quantitative trait that is the result of the interaction of many genes that are completely penetrant and have a uniform expressivity. In other words, the environment normally encountered does not affect sex type. These genes occur in two groups. One group of genes is autosomal and overall tends to make for maleness; the other group of genes is X-limited and overall tends to make for femaleness. When the ratio is $\frac{1}{2}$, the single X cannot overpower the maleness of two sets of autosomes, and the sex type develops in the direction of femaleness only as far as the normal male phenotype. When the ratio is $\frac{2}{2}$, the two X's overpower the two sets of autosomes and drive the phenotype further toward femaleness as far as the normal female phenotype. When the ratio is $\frac{3}{2}$, the three X's drive the femaleness tendency beyond normality to produce the superfemale phenotype. When, however, the ratio is $\frac{1}{3}$, the extra set of autosomes drives the phenotype further toward maleness to produce the supermale phenotype.

Since sex type is a multigenic trait, individual genes in the wild-type *Drosophila* are expected to make only small contributions to sexuality. The dominant autosomal gene, $tra^+$, is apparently one such gene ordinarily found in the wild-type individual. When its rare allele *tra* is homozygous, however, it produces a maleness tendency that is so strong that it transforms XX diploids into sterile males. This sex-transforming allele, *tra*, has no effect when homozygous on XY or X0 diploids. Although *tra* is not important in nature, it has a spectacular result and illustrates that autosomal loci favor maleness.

**19.3** In *Drosophila* and other insects, mosaicism in sex chromosome constitution is paralleled by a mosaicism in sexual phenotype.

In insects hormones play a relatively small role in determining sexual or other phenotypes. Accordingly, each body part develops largely or completely according to the genotype it contains. In *Drosophila*, for example, one of the first two nuclei produced by mitotic division of an XX zygote may lose an X, producing one XX and one X0 nucleus. Subsequent normal mitotic divisions will produce a genotypically mosiac individual, about half of which is XX and phenotypically female, and the remainder X0 and phenotypically male. Individuals that are phenotypic mosaics with regard to sex—some parts typically male are clearly demarcated from other parts typically female—are called *gynandromorphs* (Figure 19-4). Thus, genotypic mosaicism for X chromosomes is expressed in the phenotypic mosaicism of gynandromorphs. If the loss that produces the X0 cell occurs at the second or third mitosis in development, a correspondingly smaller portion of the gynandromorph will be male.

Gynandromorphs also occur in insects such as moths, where XX is male and XY is female. In the moth, for example, the gynandromorph

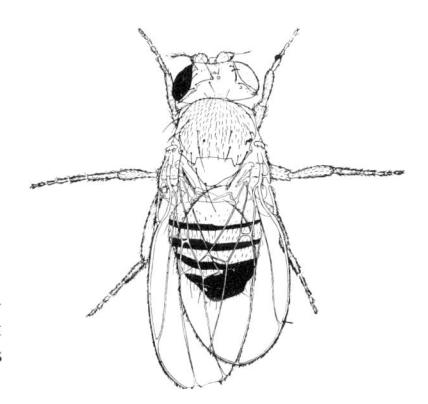

FIGURE 19-4. *D. melanogaster* gynandromorph whose left side is female and right side is male. The zygotic genotype was $X^{\omega+}X^{\omega}$. (Drawn by E. M. Wallace.)

starts out as an XX male zygote and loss of an X gives rise to X0 female parts. A left–right moth gynandromorph has the large, beautifully patterned wing of the male on one side and the small stump of the female wing on the other side.

# HYMENOPTERA

**19.4** In Hymenoptera, the sexes usually differ in number of chromosome sets.

In Hymenoptera, which include bees, ants, wasps, and sawflies, unfertilized eggs develop as haploid males and fertilized eggs usually develop as diploid females. Haploid males produce haploid sperm by suitable modifications of the meiotic process, and all gametes of males and females have chromosomal compositions that appear identical (but which may differ in allelic content). In bees, the type of nutrition determines whether the diploid develops into a fertile queen or a sterile worker.

In one parasitic wasp, *Microbracon*, when the parents are closely related, some sons are haploid whereas others are diploid and have 10 pairs of chromosomes like their sisters. These diploid males, which are relatively inviable and semisterile, seem to be homozygous for a sex-determining locus that has many alleles. Females are, therefore, heterozygous at this locus, and normal males are haploid. In this case, as in many other Hymenoptera, sex determination occurs before the start of development, is uninfluenced by the environment, and is based not only upon the haploid vs. diploid chromosome number but upon the heterozygosity vs. homozygosity of a particular locus in the diploid.

## HUMAN BEINGS

**19.5** In human beings, genotypic sex type is decided at fertilization; phenotypic sex type depends upon the action of hormones and the environment.

Genotypic sex type is determined at the time of fertilization in human beings, as it is in insects: XY zygotes are genetic males; XX zygotes, genetic females. The sex type actually expressed phenotypically in human beings differs from that in insects in being dependent upon the action of hormones and influenced by the environment.

In early human development the sex organs or *gonads* of all individuals look alike. They are composed of two regions: an outer *cortex* and an inner *medulla*. As development proceeds (Figure 19-5), in XY individuals the cortex degenerates and the medulla forms a testis; whereas in XX individuals the medulla degenerates and the cortex forms an ovary. Once the testis or ovary is formed, it takes over the regulation of further sexual development by means of the hormones it produces. The hormones direct the development or degeneration of various sexual ducts, the formation of genitalia, and secondary sexual characteristics.

Since sexual differentiation is largely controlled by the sex hormones, it is not surprising that genotypically normal persons are phenotypically variable with regard to sex. Any change in the environment that can upset the production of sex hormones or tissue responses to them can produce effects that modify the sex phenotype. So, the phenotypes considered normal male or normal female show some variability as a result of both allelic and environmental differences—each adding to the spice of life.

FIGURE 19-5. The relation in human beings between sex phenotype and the differentiation of the early gonad.

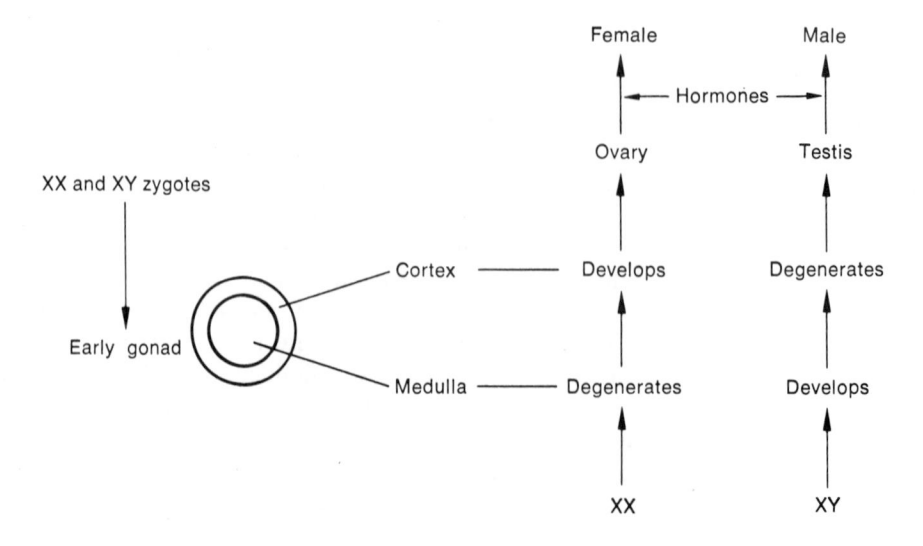

Genetically normal persons exposed to abnormal environmental conditions, however, can produce phenotypes that lie between the two normal ranges of sex type and, therefore, are intersexual in appearance. Intersexual phenotypes due to environmental factors usually begin as genotypic males but develop partially along female lines. Although it is sometimes easy to classify a person as an intersex who is phenotypically clearly between the two norms, other persons at the extremes of normality cannot be labeled normal, or intersex, or supersex.

We should also realize that phenotypic sex is a combination of physical and behavioral traits, both of which are subject to environmental change. For example, physical phenotype may not match genotypic sex type after accidents, purposeful mutilation, or surgery; and behavioral sex may be the opposite of genotypic sex in cases where boys are brought up as girls, and vice versa.

## 19.6 Human beings with an abnormal number of sex chromosomes often have sexual abnormalities.

Zygotes that are otherwise diploid may have abnormal numbers of sex chromosomes.

MONOSOMICS. Of the two types with a single sex chromosome, the Y0 is always lethal, proving that an X chromosome is needed to be viable. The X0 type is fairly common among abortuses, perhaps 1 in 50 surviving and being born. Among live births, the X0 female occurs about once per 3000 females. (In a population of 100 million females, there would be about 33,000 X0 females.) About 70 per cent of X0 females contain a maternally derived X, so the missing sex chromosome is in most cases the paternal one. This origin for the X0 condition can be recognized genetically, for example, by a red-green colorblind father having an X0 daughter of normal vision. The paternal sex chromosome might be lost during meiosis or in the fertilized egg.

X0 females typically have a set of phenotypic abnormalities called *Turner's syndrome* (after its discoverer) that prevents them from maturing physically and sexually. Turner-type females usually do not develop breasts, ovulate, or menstruate. They are, therefore, almost always sterile. They usually have a webbed-neck and low-set ears; they are always short in stature. Although X0 females have IQ's in the normal range, they are defective in the ability to orient themselves in space and to perceive forms. Considerable phenotypic variation occurs in X0 females as a result of variations in the specific alleles present in the X and autosomes as well as in the environment, including medical treatment. X0 females given hormone therapy often phenocopy normal females, and some are happily married.

TRISOMICS. Otherwise-diploid persons having three sex chromosomes are of three types: (1) XXX is female and occurs about once per 1000 female live births. Such females are fertile and relatively tall. Although some are mentally retarded, many are normal. (2) XYY is male and occurs about once per 700 male live births. Such males are fertile and relatively tall. (Any excess of sex chromosome, especially the Y, makes a person taller than expected from the family pedigree.) XYY males are about 20 times as likely to be found in mental or penal institutions as XY males. However, since only about 4 per cent of XYY's are so institutionalized, 96 per cent of them are as normal as uninstitutionalized XY's. (3) XXY is male and occurs about once per 800 male live births. Such males typically have a set of phenotypic abnormalities called *Klinefelter's syndrome* (after its discoverer). Klinefelter-type males usually have undersized sex organs and may develop various secondary sexual characteristics of women, such as sparse body hair, enlarged breasts, and wide hips. All are sterile and relatively tall. As expected, they show some phenotypic variability. Although sterile, some show normal sex drive and behavior; and although many are mentally retarded, some are not.

It should be noted that the live-birth frequency of any one of the three types of person with an extra sex chromosome is larger than that for the X0 condition, in accordance with the view that the gain of a chromosome is less detrimental than its loss. The frequency of XXX and especially XXY individuals increases with mother's age, as in the case of the trisomy producing Down's syndrome (Figure 12-4). Most of these sex chromosome trisomics are, therefore, due to maternal nondisjunction. Trisomy due to paternal nondisjunction does occur, however, and can be recognized genetically. For example, owing to a meiotic nondisjunction in the father, red-green colorblind women can have XXY Klinefelter sons of normal vision.

POLYSOMICS. Otherwise-diploid persons containing two or more extra sex chromosomes are also known. They include the following types: XXXX (♀); XXXY (♂); XXYY (♂); XXXXX (♀); XXXXY (♂); XXXYY (♂). All these individuals are abnormal; those with one or more Y's are sterile and may have some other features of Klinefelter males; and the greater the number of extra X's, the greater the mental retardation.

MOSAICS. Otherwise-diploid persons may contain different sex chromosome constitutions in different somatic tissues. Such sex-chromosome mosaics have been found with the following mixed constitutions: XX/XY, XXX/X0, XX/X0, XY/X0, XXY/XX, XXXY/XY. Such mosaic genotypes are usually due to one or more errors in chromosome distribution among the daughter nuclei produced after fertilization. Most, if not all, XX/XY mosaics, however, come from double fertilization of eggs containing two haploid nuclei, or from embryo fusion. Sex-chromosome mosaics often have sexual abnormalities, and some may have one ovary-like and one testis-like gonad. Nevertheless, because sex hormones circulate throughout the body, their overall phenotype is uniform and may be intersexual or infrasexual, but is not mosaic in external sexual characteristics; that is, the phenotype is not gynandromorphic. Thus, the half man/half woman in a circus is a fake.

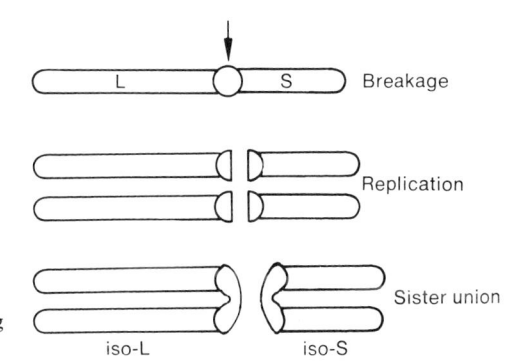

FIGURE 19-6. Formation of isochromosomes following centromeric breakage and chromosome replication.

## 19.7 Human beings with rearranged sex chromosomes often have sexual abnormalities.

Human beings that are otherwise diploid may have sex chromosomes that have undergone any one of several types of breakage rearrangement. A single, nonrestituting break may occur near or within the centromere of a sex chromosome (Figure 19-6). If sister ends fuse after chromosome replication, *isochromosomes*, chromosomes composed of identical lengthwise halves, are produced. The transmission of such centromere-containing chromosomes is expected to be normal, or almost normal, in human beings as it is in other species having isochromosomes, or chromosomes that contain partial deletions of the centromere, or two centromeres close together.

Persons are known (Figure 19-7) who carry a normal X and either the long-arm isochromosome of X ($X^L.X^L$, the period representing the centromere) or the short-arm isochromosome of X ($X^S.X^S$). Both types are sterile females, showing that each arm has sexual effects. The $X^L.X^L$ carrier has all the features of Turner's syndrome, whereas the $X^S.X^S$ carrier does not have certain nonsexual features of Turner's syndrome (short stature, neck webbing, and so on). Therefore, the nonsexual features of Turner's syndrome are attributed to the loss of the genetic material in $X^S$ (Figure 19-8A).

Isochromosomes $Y^L.Y^L$ and $Y^S.Y^S$ also occur for the Y chromosome. Absence of $Y^S$ (in X $Y^L.Y^L$ individuals) produces sterile phenotypic females, with no nonsexual features of Turner's syndrome, demonstrating that the genes for male sex determination are mainly in $Y^S$. Absence of $Y^L$ (in X $Y^S.Y^S$) produce phenotypic males with many of the nonsexual features of Turner's syndrome (Figure 19-8B). This result indicates that some of the genes in $Y^L$ are also located in $X^S$—that is, $Y^L$ and $X^S$ seem to have homologous regions.

If two breakages occur in constitutive heterochromatin, one in $Y^S$ near the centromere and the other near the tip of the X chromosome, a reciprocal translocation of the small pieces might make such a small difference in chromosome length as to be cytologically undetectable (Figure 19-9). Persons who receive an X from their mother and the half-translocation bearing $Y^S$ from their father (Figure 19-7) would be

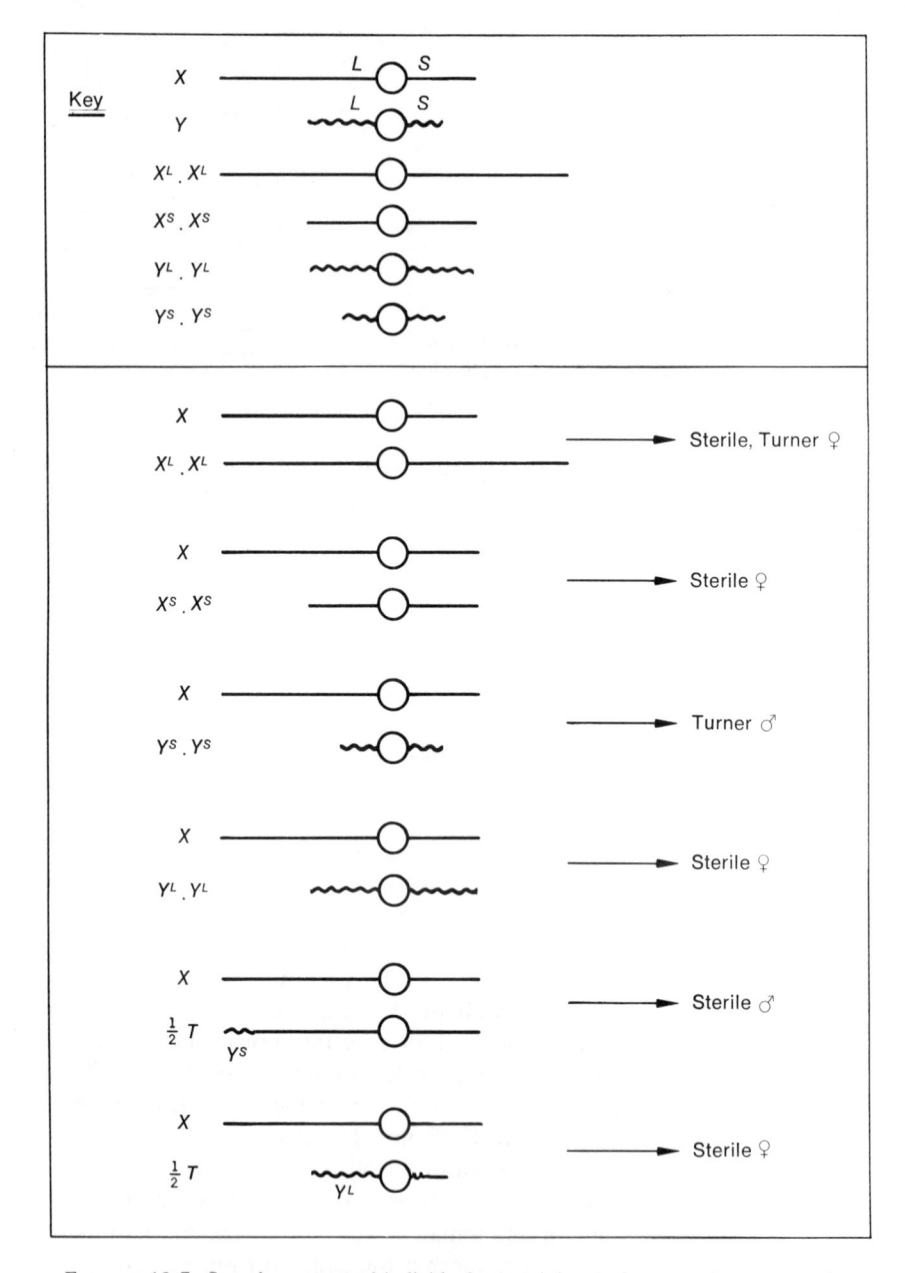

FIGURE 19-7. Sex phenotypes of individuals containing various grossly rearranged sex chromosomes. $\frac{1}{2}T$ = half-translocation.

phenotypically Klinefelter males who *seem* to have an XX constitution. Persons who receive an X from their mother and the half-translocation containing $Y^L$ would be phenotypically sterile females who *seem* to be XY. Both types of person have been observed.

Still other rearranged X chromosomes occur that are rings or have deletions in the short arm or long arm. Any cytologically detectable loss from the X seems to cause sterility in a female. The size of the Y

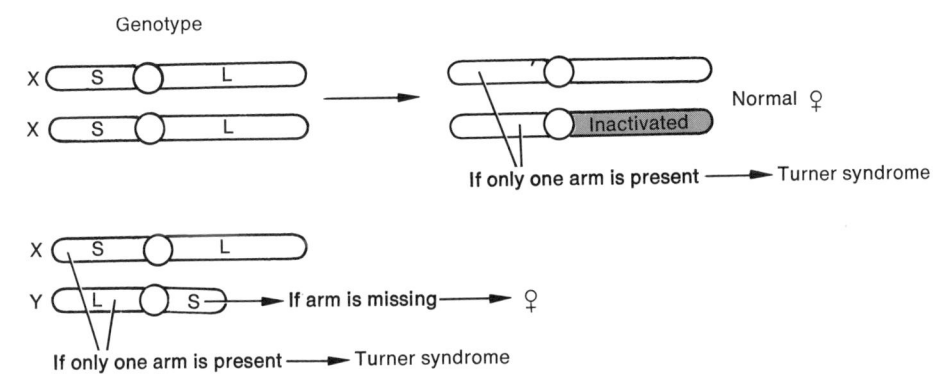

# DETERMINATION OF SEX IN EUKARYOTES

FIGURE 19-8. Mainly the genes that are located in one $X^L$ are hypothesized to be inactivated in normal human females. L, long arm; S, short arm of the sex chromosome.

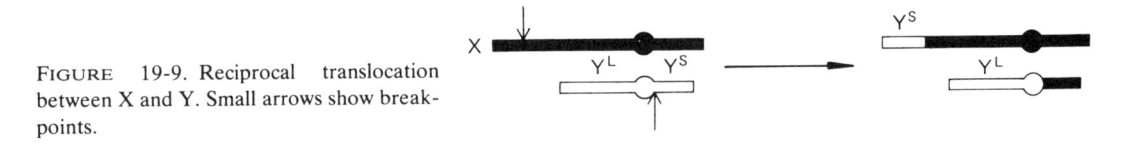

FIGURE 19-9. Reciprocal translocation between X and Y. Small arrows show breakpoints.

chromosome varies in phenotypically normal men—there being no correlation between size and virility.

The $Y^S$ versus no-$Y^S$ sex-determining mechanism in human beings means that male sexuality is primarily determined by $Y^S$-limited loci. Autosomal loci also affect male sexuality, however, since trisomic Down's syndrome males are sterile. Since the absence of an X (XX vs. X0) or an extra X (XY vs. XXY) causes sterility, one X must function in balance only with one or more X's or one or more Y's plus two sets of autosomes, and Y's must function in balance with one X and two sets of autosomes. In other words, the idea that sex is normally the result of a balance of action among genes located in sex chromosomes and autosomes applies also to human beings (and other mammals).

# ENVIRONMENTALY DETERMINED SEX

**19.8** When one genotype codes for both kinds of gametes or sexes, the environment determines the sexual phenotype.

In certain animals and plants, male and female gametes are produced in the same individual. For example, in the snail *Helix*, eggs and sperm are produced in different positions in the same gonad; in the earthworm, eggs and sperm are produced in separate gonads; and in certain haploid mosses, egg- and sperm-like gametes are produced in separate gonads. In these cases, a single genotype codes for the production of two kinds of gamete, and the type of gamete produced depends upon the different positions that cells have within a single organism; that is, gametic phenotype is determined by the nature of the internal environment.

In other cases, the internal environment changes with the growth of an organism and, thereby, changes sexual phenotype. For example, in the marine worm *Ophryotrocha*, where the two sexes are in separate individuals, sex type is determined by the size of the organism. When the animal is small, because of youth or because it was obtained by amputation from a larger organism, it manufactures sperm; when larger, the same individual shifts to the manufacture of eggs.

Finally, sexual differences may depend upon the presence of a chemical messenger in the external environment. This is true in another marine worm, *Bonellia*, that has separate sexes radically different in appearance and activity: females are walnut-sized with a long proboscis, while males are microscopic ciliated forms that live as parasites in the body of the female. Fertilized eggs grown in the absence of adult females (and hence in the absence of a chemical messenger they make) develop as females; they develop as males in the presence of either adult females or simply an extract of the female's proboscis.

Nothing has been stated about the specific genetic basis for sexuality when different sexes or gametes are determined by internal and external environmental differences acting upon a uniform genotype. Nevertheless, the genotype must play an important role in all such cases by making possible different sexual responses to variations in the environment.

## SUMMARY AND CONCLUSIONS

The formation of different types of gametes or sexes always has a genetic basis. In some species, a given genotype produces different gametes or sexes in response to differences in internal or external environment. In other species, different genotypes produce different gametes or sexes independently of normal variations in the internal or external environment. In single-celled eukaryotes, for example, individuals of different mating type carry different, sometimes transposable, alleles at a single functional locus. In many multicellular eukaryotes, for example, genetic differences between sexes are correlated with cytologically detectable differences in chromosome content. In these cases genes responsible for sexual phenotypes are located in autosomes as well as sex chromosomes; and sex type depends upon the balance in the dosages of such genes.

The importance of gene balance in sex determination is illustrated in *Drosophila* and human beings by the occurrence of sexual abnormalities that accompany the gain or loss of whole chromosomes or parts of chromosomes. Such shifts in chromosomal balance produce intersexes or infrasexes. Sex chromosome mosaicism produces gynandromorphs in insects but not in organisms, such as mammals, that synthesize sex hormones. In human beings, both sex hormones and the environment influence the phenotypic sex type.

## QUESTIONS AND PROBLEMS

**19.1** *Anagrams*: Unscramble and define each of the following: **a.** permaluse. **b.** texiners. **c.** damlule. **d.** donga. **e.** camios.

**19.2** *Songs*: **a.** "Boy Named Sue."   **b.** "Show Me a Rose." What are the genetic implications of one or both songs?

**19.3** What evidence have we that sex in yeast is under the influence of a controlling gene?

**19.4** Differentiate between a supersex and an intersex.

**19.5** Give one evidence that chromosomal balance is important in sex determination in: **a.** *Drosophila*.   **b.** human beings.

**19.6** Give one evidence that autosomes influence sex determination: **a.** *Drosophila*.   **b.** human beings.

**19.7** Are all gynandromorphs sex chromosome mosaics? Is the reverse true? Explain.

**19.8** How does sexual differentiation differ in insects and human beings?

**19.9** Discuss the basis for the determination and differentiation of sex in fish such as the grouper, which are female when young and male when old.

**19.10** What determines the phenotypic difference between: **a.** a worker and a queen bee?   **b.** a male or female wasp?   **c.** a male or female *Ophryotrocha*?   **d.** a male or female *Bonellia*?

**19.11** How many Barr bodies are present in otherwise diploid: **a.** X0 females?   **b.** XXY males?   **c.** XXX females?

**19.12** What is your opinion of sweatshirts that are labeled X0 or XXY?

**19.13** Distinguish between genotypic sex and phenotypic sex in human beings. Name a factor other than those mentioned in the text that can result in phenotypic sex not matching genotypic sex.

**19.14** How may one identify the presence of a $Y^s$ in a human male who appears to have two X's and no separate Y?

**19.15** Should chromosome counts be taken of all newborns, and parents informed when their children have an abnormal sex chromosome constitution? Explain.

**19.16** What is the basis for sex determination in the worm *Dinophilus*, where small eggs produce males and large eggs, females?

**19.17** Explain how a homozygous normal vision mother married to a red-green colorblind father could have a red-green colorblind daughter.

**19.18** Explain how a red-green colorblind mother married to a normal vision father could have a normal vision son.

**19.19** When genetic male and female calves develop together as twins, the female is often sexually imperfect and sterile. How can you explain such females, called *freemartins*?

**19.20 a.** In what respects would you say that fly A and fly B are abnormal?
**b.** Offer a genetic explanation for these phenotypes.

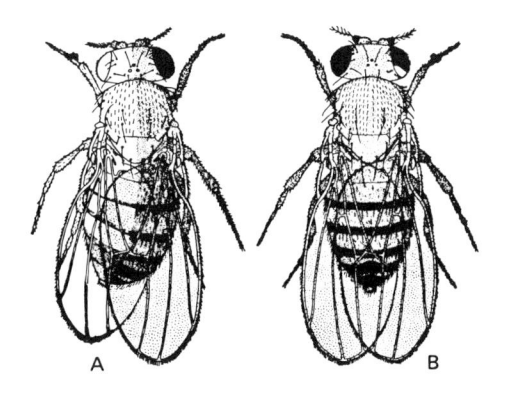

288  HOW GENE PRODUCTS INTERACT

**19.21** Suppose that you found among the *Drosophila* F$_1$ one additional fly (A or B or C or D). What explanation would you offer in each case?

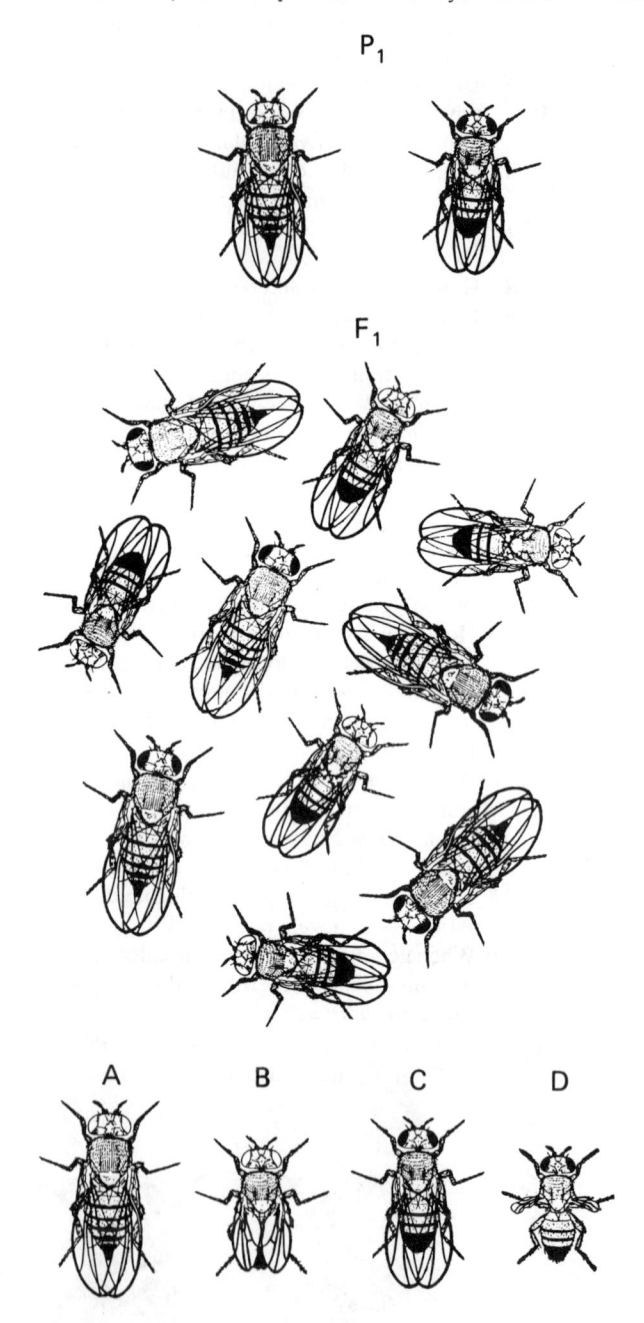

P$_1$

F$_1$

A   B   C   D

# Chapter 20

# Differentiation and Development

THIS CHAPTER STARTS with a general consideration of the role that genes have in development and differentiation. In the broadest sense, *development* encompasses the entire life history of an organism, and *differentiation* includes all the directional (noncyclical) and cyclical changes that distinguish different parts of an organism. The chapter continues by discussing specific ways that the genetic material is used to direct early development and the differentiation of specific tissues in multicellular animals.

**20.1** Most differentiation is due to differential gene transcription.

Some differentiation is associated with changes in the quality or quantity of the organism's genetic material. For example, the loss of chromosomes or chromosome parts is associated with cell lines of certain organisms (Section 14.5). In other cases, certain gene sequences, chromosomes, or whole chromosome sets are amplified during differentiation (Section 14.4). Differentiation may also be affected by foreign nucleic acids that come from (1) foreign mRNA's and tRNA's, (2) viruses, and (3) cellular symbionts and parasites. Most differentiation, however, occurs in cells whose genes are present in full sets and in unamplified numbers. This is supported by the following kinds of evidence:

1. As revealed by DNA–DNA hybridization studies, the same families of DNA sequences are present in the mouse in all undifferentiated and differentiated tissues at all stages of development.
2. Barring mutation, the gross amount of DNA and the gross and fine structure of chromosomes ordinarily remain unchanged throughout the life of different cells despite mitotic divisions or differentiation.
3. Nuclei of differentiated cells retain the total scope of information that was present in the zygote. For example, when transplanted into an egg whose

nucleus has been removed, the nucleus of an intestine cell can direct the complete, normal development of a toad, and a single root cell of a carrot can develop into a complete, normal plant.
4. Cells can change differentiated states. For example, macronuclear leucocytes can change to macrophages that can change to fibroblasts.
5. Cells specialized to produce large quantities of particular proteins, such as the globins of hemoglobin, ovalbumin, or silk fibroin, do not show any evidence of amplification of the genes coding them.

From the preceding we conclude that differentiation is usually accomplished by the differential activity of complete sets of unamplified genes. Although differential gene activity is regulated to some extent by the regulation of translation in prokaryotes (Sections 15.4 and 15.5) and in eukaryotes (Section 15.10), most differentiation seems to be regulated by differential transcription. This conclusion is supported by the following kinds of evidence:

1. RNA–DNA hybridization studies of prokaryotes and eukaryotes show that the mRNA's present during different stages of growth and differentiation are derived from distinctly different groups of loci (Figure 20-1).
2. RNA–DNA hybridization studies and studies of specific sites undergoing transcription in multichromatid chromosomes show that in multicellular organisms different kinds of completely differentiated cells have less than 10 per cent of their single-copy DNA sequences transcribed, this fraction coming from different loci in different kinds of cells.
3. In general, when specific eukaryotic mRNA's are being translated in the cytoplasm, the genes coding them are being transcribed in the nucleus; and when the same or another cell is not translating these mRNA's, they are not transcribing them (see the related discussion with regard to globin synthesis in Section 15.9).

We conclude, therefore, that most differentiation is due to the differential transcription of complete sets of unamplified genes.

FIGURE 20-1. Typical variations in the level of transcription during differentiation.

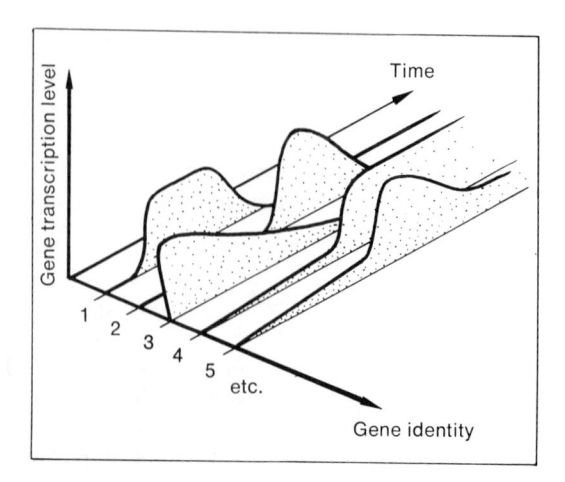

# SURVIVAL DURING EARLY DEVELOPMENT

Multicellular animals that start life as fertilized eggs go through an early developmental period during which their cells must divide and begin to differentiate before such organisms can obtain food from their environment. Survival during early development therefore requires the stockpiling of energy and materials prior to fertilization, specifically in the oocyte. How is genetic material involved in this stockpiling?

**20.2** Oocytes stockpile the products of gene action of their own and other cells.

Almost all oocytes are surrounded by a layer of *follicle cells* (Figure 14-8). These cells are active in RNA and protein synthesis; and proteins, especially those of the yolk and chorion, made in them or elsewhere are transported through or from them into the oocyte. In some cases there is no cytoplasmic continuity between them and the oocyte, in other cases there is, and in still others the oocyte engulfs entire follicle cells.

In species that do not have a prolonged prophase I, an additional class of cells often aids in stockpiling. These are *nurse cells*, which are characteristically joined to the oocyte by means of cytoplasmic bridges that form canals through which whole mitochondria, polyribosomes, DNA, and other preformed materials pass into the oocyte (Figure 14-8). Some oocytes even stockpile rRNA and other RNA's transcribed by the sperm nucleus while it is waiting for the oocyte to finish meiosis.

It is common for oocytes to stockpile yolk and mitochondria. As a result, oocytes commonly contain hundreds of times more DNA in their cytoplasm than in their nucleus. Although some of this DNA is yolk droplet DNA, of unknown function (Section 13.9), most of it is mit DNA. Whereas mit DNA in an ordinary somatic cell is only about 0.15 per cent of the diploid nuclear amount, 40,000 or more times this amount can be found in the oocyte.

In oocytes, such as occur in amphibians, that have a prolonged prophase I, transcription produces all classes of RNA. Large amounts of the two larger types of rRNA are synthesized, often aided by the amplification of nucleolus organizer DNA. In amphibians, 5S rRNA synthesis is enhanced by transcription of oocyte 5S rDNA. These rRNA's are combined with ribosomal proteins to produce a store of ribosomes. Some 5S rRNA is stored together with newly synthesized tRNA. The oocyte also transcribes a complex set of mRNA's from about 5 per cent of the chromosomal sequences. Some of these mRNA's are stored—for example, those for histones, spindle tubule protein, and certain enzymes. Some mRNA's are translated to produce, for example, histones, spindle tubule protein, DNA and RNA polymerases, $o^+$ protein (to be discussed in the next section), and specific enzymes, all of which are stored and later used in the development of the embryo.

**20.3** During early development, stockpiled materials are used and new transcripts and their products are made in anticipation of their need.

By the time they reach maturity, most oocytes are transcriptionally and translationally inactive. Upon fertilization, translation resumes (Section 15.10) using maternal ribosomes and maternal mRNA's. Most development up to gastulation uses maternally transmitted proteins, tRNA's, mRNA's, and ribosomes. Although new transcripts are produced during the cleavage stage and affect cleavage in mammals, they and their translational products have no effect in the sea urchin until gastulation. It would seem generally true that the developmental needs of one stage are usually anticipated by transcription, translation, and storage at a previous stage.

Two specific consequences of stockpiling in oocytes upon subsequent development are cited. The stockpiling of mitochondria in the sea urchin oocyte is adaptive since transcription of mit RNA is important in the cleavage stage, although it is relatively unimportant in subsequent embryonic stages. In amphibians, the oocyte synthesizes a large, slightly acidic protein, $o^+$ *protein*, and stores it in the cytoplasm. At the midblastula stage, $o^+$ enters the nucleus, where it activates the transcription of genes whose products are needed for the developing individual to enter the gastrula stage. After midblastula, $o^+$ is not present in detectable amounts. $o/o$ females, who are homozygous for a defective allele of $o^+$, produce no $o^+$ and their eggs cannot gastrulate; such eggs probably make little or no RNA after midblastula. In this case, oocyte gene activity determines blastula gene activity, which in turn determines gastrulation. Since $o^+$ is widespread in amphibians and seems to control the activation of a group of genes, it seems to be a eukaryotic regulator gene that acts as a master switch for a key stage in development.

It should be noted that materials are not stored or distributed at random in the cytoplasm of the oocyte or of cells of later stages of development. Since the nature of the cytoplasm affects DNA replication (Section 14.3) and gene activity, differentiation and development depend to some extent upon differences in the localization of gene products.

# THE COORDINATION OF BODY PARTS

Different body parts are able to develop, differentiate, and function coordinately due to interactions brought about by chemical messengers and nerve impulses. We have already mentioned in Section 14.3 the occurrence of genetic programs that regulate DNA replication through intercellular action. We consider here the genetic basis for the coordination in differentiation that results from the interaction of adjacent tissues and from the interaction of nonadjacent tissues through the mediation of hormones.

## 20.4 Differentiation and development depend upon the intercellular effects of gene action.

Many genes produce intercellular effects that are important for development and differentiation. Various experiments with embryos have shown that specific cells in an early embryo will become a specific kind of tissue in the mature organism only after interaction with adjacent cells. In this process of *induction* (Figure 20-2), the inducing tissue affects the differentiation of an adjacent responding tissue. Induction between adjacent tissues and the effects of hormones on nonadjacent tissues can be adversely affected by abnormal alleles. As we shall see in the examples that follow, such abnormal alleles produce a profound detrimental effect on development. In each case we can assume that the normally occurring process is at least in part the phenotypic result of the action of the normal allele.

1. *Abnormal alleles can cause a loss of inductive capacity.* In the chicken the normal wing develops from two layers of cells in the embryo, ectoderm covering mesoderm. In the genetically *wingless* chicken, however, the mesoderm has lost the inductive capacity to maintain the ectoderm, the ectoderm degenerates, and wing development stops.
2. *Abnormal alleles can prevent contact between inducing and responding tissues.* In the normal mouse, the ectoderm is induced to form a lens by the optic vesicle. In genetically *lensless* mice, the optic vesicle sometimes does not contact the ectoderm, in which case no lens is formed.
3. *Abnormal alleles can cause the loss of the ability to respond to induction.* In the normal mouse, future notochord tissue induces surrounding mesoderm to differentiate into cartilage and vertebrae. In homozygous *Brachy* mice, however, the mesoderm is unable to respond to this inductive stimulus, no cartilage and vertebrae form, and the individual dies.
4. *Abnormal alleles can cause a general slowdown of growth.* A dominant abnormal allele in human beings induces a slowdown in growth at a certain time in development. The structures most affected seem to be those that normally are growing most rapidly at the time, namely the long bones of the arms and legs. The result is a disproportionate dwarfism, that is, normal head and trunk size but shortened arms and legs, called *achondroplastic (chrondrodystrophic) dwarfism*. Homozygotes for this allele have phocomelia (Section 17.4). Similar alleles produce similar phenotypes in other organisms, for example, dachshund dogs, Creeper chickens, and Ancon sheep.

FIGURE 20-2. Interaction between adjacent tissues and its prevention by mutants.

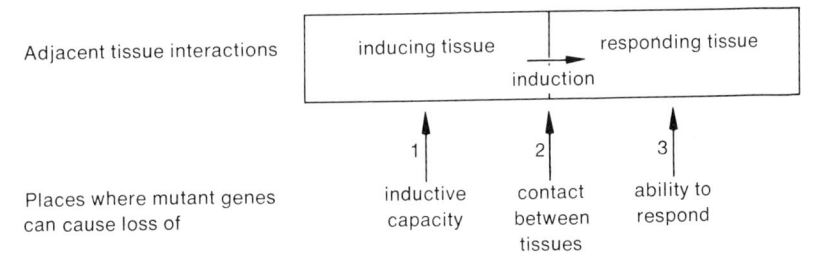

Adjacent tissue interactions: inducing tissue → responding tissue (induction)

Places where mutant genes can cause loss of:
1. inductive capacity
2. contact between tissues
3. ability to respond

When homozygous, an abnormal allele in the mouse prevents the synthesis of a growth hormone by the anterior pituitary gland. As a result, after starting at a normal rate, all growth suddenly stops, producing proportionally dwarf mice.

## 20.5 Some hormones coordinate differentiation by affecting transcription or translation.

The characteristic that all hormones have in common is that very small amounts of them produce great metabolic effects in the target cells, the cells that respond to them. Hormones differ from one another, however, in chemical composition; some are steroids, whereas others are polypeptides or derivatives of a single amino acid (Figure 20-3). Hormones also differ in the mechanisms by which they affect gene action.

1. TRANSCRIPTIONAL EFFECT. Many hormones, especially those of steroid type, actually enter their target cells. In this case they act as an effector by binding to a regulatory protein. The complex then binds to specific DNA sequences which are activated to undergo transcription (Section 15.9). This pathway of hormone action is followed, for example, by estrogen, thyroid hormone, and the plant hormone auxin.

Many polypeptide hormones do not seem to enter their target cells. Such hormones exert their effect at the cell membrane by stimulating the production of the nucleotide cAMP. cAMP, which stimulates transcription in prokaryotes (Section 15.1), also seems to do so in eukaryotes.

Other hormones, such as hydrocortisone, seem to inhibit RNA synthesis. The molecular basis for such inhibition is unknown.

2. TRANSLATIONAL EFFECT. Some animal hormones of steroid and amino acid-derivative type affect translation *in vitro*. Perhaps they also do so *in vivo*. Plant hormones such as gibberellic acid, abscisic acid, and cytokinin also seem to affect translation.

FIGURE 20-3. Structural formulas of a steroid hormone (**A**) and an amino acid-derivative hormone (**B**).

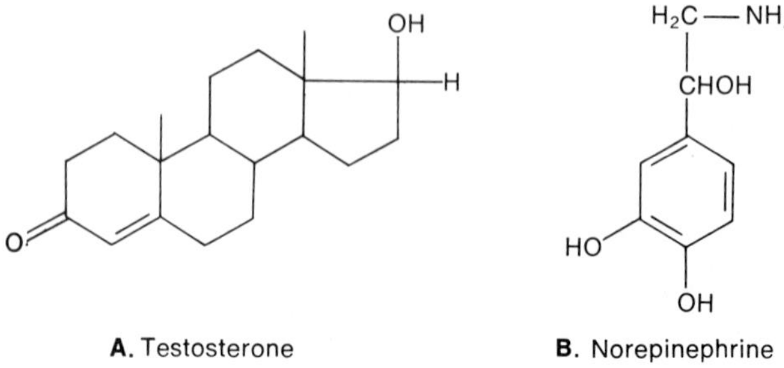

**A.** Testosterone

**B.** Norepinephrine

# THE SYNTHESIS OF SPECIFIC TISSUE PROTEINS

Differentiation of different tissues often requires the synthesis of special proteins. These proteins are often required in large quantities (as in erythrocytes and muscle tissue) or in large variety (as in the white blood cells that make antibodies, and perhaps in nerve cells). The former type of protein is made largely independently of changes in the environment; the latter type is made in response to changes in the environment. We shall describe the genetic basis in human beings for the synthesis of two kinds of special proteins: the hemoglobins in erythrocytes and the antibodies in white blood cells.

# HEMOGLOBIN SYNTHESIS

**20.6** Hemoglobins of different kinds and amounts are synthesized at different times in development.

Our attention is confined to the synthesis of the protein portion of hemoglobin, globin (Section 17.5). Six different kinds of globin are normally produced, each about the same length, and each coded at a separate locus. The genes at these loci are called by the first six letters of the Greek alphabet—alpha ($\alpha$), beta ($\beta$), gamma ($\gamma$), delta ($\delta$), epsilon ($\varepsilon$), and zeta ($\zeta$). The same symbol is used for the gene and the polypeptide it codes for. Each person, however, carries two separate $\gamma$ loci, $\gamma^G$ and $\gamma^A$, which code for slightly different polypeptides; many persons also have two separate, apparently identical, $\alpha$ loci. Figure 20-4 shows the most likely arrangement of these loci and the promoter genes, $p$, that precede them. Notice that the $\alpha$, $\zeta$, and $\varepsilon$ loci are recombinationally unlinked to each other and to the remaining loci, which are linked to each other.

Different globin genes are normally functional at different times in development to produce different kinds of hemoglobin (Figure 20-5). (All hemoglobins contain two pairs of polypeptides, of which one pair is always of $\alpha$ type except in early embryonic hemoglobin.) The $\zeta$, $\varepsilon$, and $\alpha$ genes are activated very early in embryonic or fetal development and make early embryonic and embryonic hemoglobins. Although the $\alpha$ loci

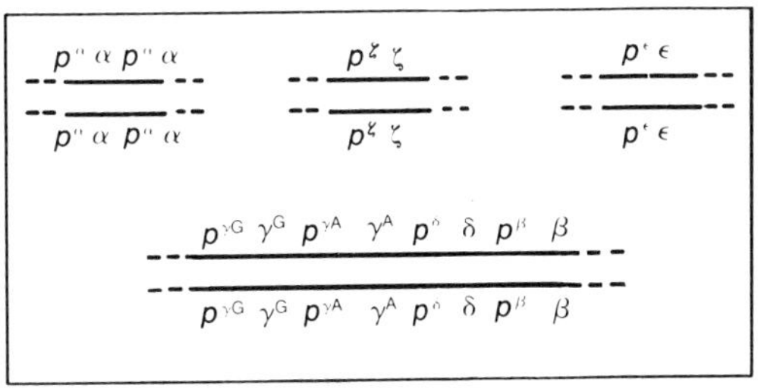

FIGURE 20-4. Most likely arrangement of promoter and structural genes for hemoglobin (in diploid condition) in many human beings.

| Genes | Hemoglobins (Hb's) | | |
|---|---|---|---|
| | Chain Composition | Name or Symbol | Principal Period of Synthesis |
| $\zeta \longrightarrow \gamma?$ | ? $\zeta_2 \gamma_2?$ | Gower 2 Portland 1 | Early embryonic |
| $\alpha \Big\langle \begin{array}{l}\epsilon \\ \gamma^G \\ \gamma^A\end{array}$ $\alpha$ $\begin{array}{l}\beta \\ \delta\end{array}$ | $\alpha_2 \epsilon_2$ | Embryonic | Embryonic |
| | $\alpha_2 \gamma_2^G$ $\alpha_2 \gamma_2^A$ | F | Fetal |
| | $\alpha_2 \beta_2$ | A or A$_1$ | Postnatal |
| | $\alpha_2 \delta_2$ | A$_2$ | |

FIGURE 20-5. The genes for different types of human hemoglobin.

remain active for the rest of life, the $\zeta$ and $\epsilon$ genes become inactive before the third month of development. At about this time (Figure 20-6), the $\gamma$ loci become active and contribute to the formation of *fetal hemoglobin* (Hb-F), $\alpha_2\gamma_2$. Before birth, the $\gamma^G$ locus is about three times more active

FIGURE 20-6. Percentages of $\alpha$, $\beta$, $\gamma$, $\delta$, and $\epsilon$ hemoglobin polypeptide chains present during prenatal and early postnatal life. (After E. R. Huehns, N. Dance, G. H. Beaven, F. Hecht, and A. G. Motulsky, 1964. *Cold Spring Harbor Sympos. Quant. Biol.*, **29**: 327–331.)

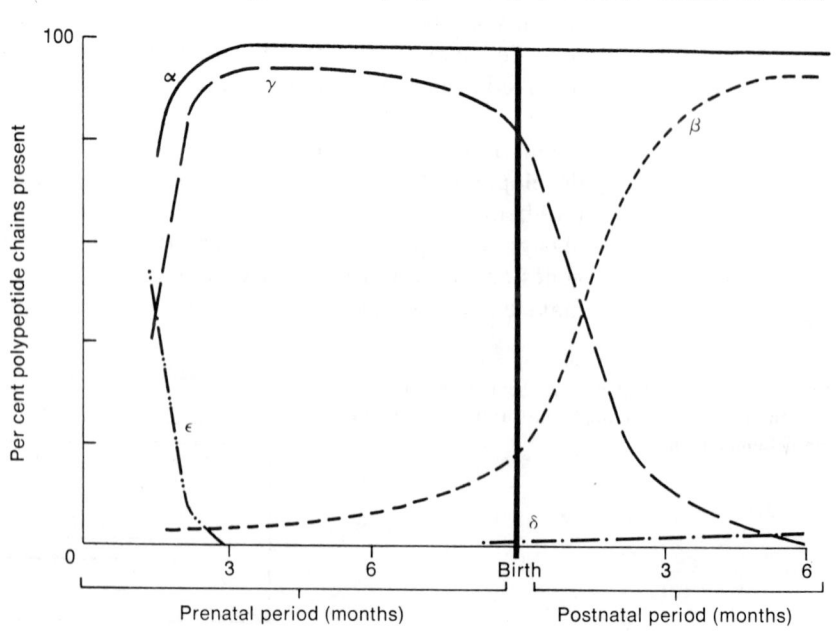

than $\gamma^A$. Hb-F, which has a higher affinity for oxygen than adult hemoglobin (Hb-A), decreases in relative amount in the last months before birth. At birth or shortly before, the synthesis of $\gamma$ chains is drastically reduced, so that by 6 months of age less than 1 per cent of hemoglobin is Hb-F (Figure 20-6). This low level of Hb-F persists throughout adulthood, during which time the $\gamma^G$ locus is somewhat less active than $\gamma^A$. Although the $\beta$ globin has been detected as early as the eighth week of development, large quantities are not synthesized until birth, producing Hb-A, $\alpha_2\beta_2$. The $\beta$ locus remains active throughout the rest of life, so that more than 90 per cent of hemoglobin in the adult is Hb-A. The $\delta$ gene seems to start to function shortly before birth and contributes to the formation of Hb-A$_2$, $\alpha_2\delta_2$, which comprises about 2.5 per cent of the total hemoglobin present in the adult. The time and level of globin gene action changes, therefore, during development.

## 20.7 Globin gene action is regulated at the transcriptional and translational levels.

Various lines of evidence indicate that globin gene action is regulated largely at the level of transcription. Although other mechanisms have been proposed for such transcriptional regulation, we shall describe only the hypothesis of *sliding activation* as it applies to the globin loci that are known to be linked to each other (Figure 20-4). It is supposed that, in the embryo, which makes no $\gamma$, $\beta$, or $\delta$ amino acid chains, the entire euchromatic region coding for these chains is repressed by heterochromatization (Figure 20-7A). In the fetus, a stretch of DNA covering approximately $1\frac{1}{2}$ globin genes at the left of the gene sequence is derepressed or activated by assuming the euchromatic state (Figure 20-7B). As a result, $\gamma^G$ transcription is uninhibited and $\gamma^A$ transcription (whose promoter is available to transcriptase) is moderately inhibited, while $\delta$ and $\beta$ transcription remains completely or greatly inhibited. After birth, however, the same length of activation is assumed to slide, like a narrow wave, to cover approximately the $1\frac{1}{2}$ globin genes at the right of the sequence (Figure 20-7C). As a result, $\beta$ transcription is uninhibited and $\delta$ transcription (whose promoter is not readily available to transcriptase) is largely inhibited, while $\gamma^G$ and $\gamma^A$ are almost completely inhibited, although they are now transcribed in more equal amounts.

Globin gene activity is also regulated at the level of translation. We have already mentioned the temporary storage of globin mRNA in untranslated condition during early chick development (Section 15.10). Other evidence of translation regulation includes (1) the stimulation of $\alpha$ mRNA translation by free heme, and (2) the stimulation of $\beta$ mRNA translation by free $\alpha$ chains. Such regulation may serve to assure the synthesis of correct quantities of the heme, $\alpha$ chains, and $\beta$ chains that make up Hb-A.

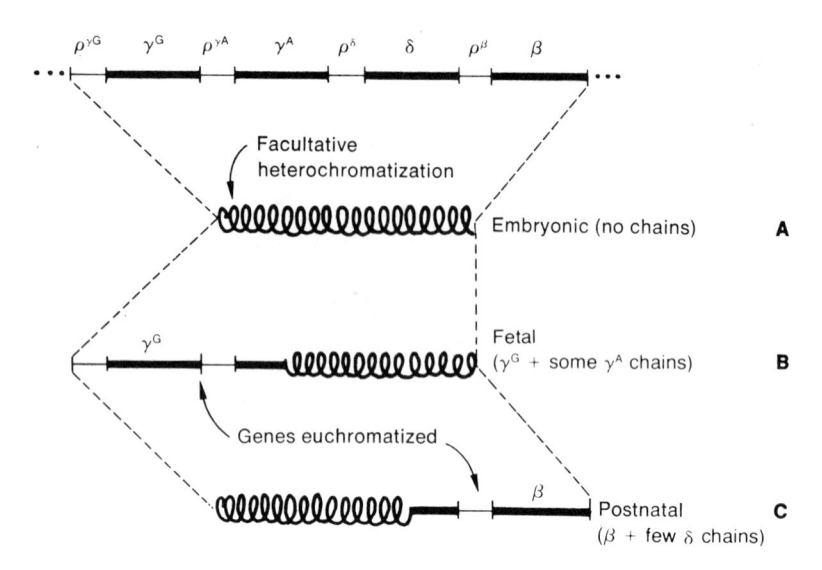

FIGURE 20-7. The sliding activation hypothesis for the regulation of transcription of four linked globin-coding genes. The extent of each globin-coding gene is indicated by a thickened line. **A**: In the embryo, the entire chromosome region is inactive due to heterochromatization. **B, C**: In the fetus and after birth, a length of $1\frac{1}{2}$ globin-coding genes is euchromatized. In **B**, the euchromatized region is the leftmost portion of the gene sequence. In **C**, the euchromatized region shifts and is the rightmost portion of the gene sequence.

# ANTIBODY SYNTHESIS

### 20.8 Large numbers of different antibodies are synthesized by plasma cells.

Human beings have two systems of immunity based on two classes of white blood cells. Our attention is confined to the system in which lymphocytes differentiate into *plasma cells* that secrete proteins called *antibodies* or *immunoglobulins* (*Ig*). Antibodies combine with *antigens*, foreign molecules that are often proteins but can also be polymers of sugars, nucleic acids, or other giant molecules. The antigen–antibody complex is more readily destroyed by the body than is the antigen alone, thereby conferring immunity.

There are millions of different plasma cells each of which is capable of producing a different kind of antibody. Each kind of plasma cell has some feature at its cell surface, perhaps related to the cell membrane DNA it contains (Section 13.9), that recognizes a particular antigen and is stimulated to synthesize antibodies against it. A mature plasma cell synthesizes only one kind of antibody and secretes about 2000 molecules of it per second for a few days.

An antibody is composed of two pairs of polypeptides bonded together to form a Y-shaped molecule (Figure 20-8). Each member of the shorter pair is called a *light* (*L*) chain; each member of the longer pair is called a *heavy* (*H*) chain. At the end of each arm of the Y the interacting L and H

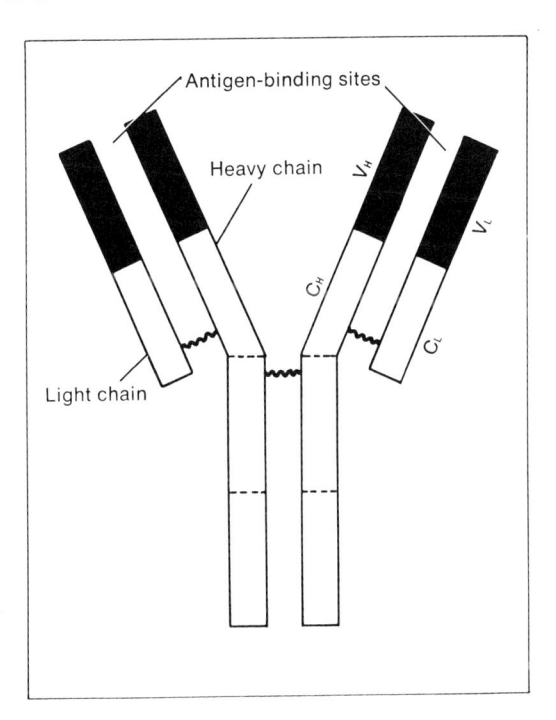

FIGURE 20-8. The immunoglobulin (Ig) molecule is composed of two identical light (L) and two identical heavy (H) chains held together by disulfide bonds (zigzag lines). Each type of chain has many different amino acid sequences in its N-terminal region (shaded), called the variable region ($V_H$ and $V_L$), and relatively few different sequences in the remaining region, called the constant region ($C_H$ and $C_L$). The molecule has two antigen-binding sites, each formed by joining $V_H$ with $V_L$. $V_L$ and $C_L$ seem to be homologous; so do $V_H$ and the three regions in $C_H$ separated by dashed lines.

chains form an *antigen-binding site*. Each antibody, therefore, contains two identical antigen-binding sites. An antigen-binding site is composed of the *variable regions* of the L and H chains, called $V_L$ and $V_H$. The remaining parts of the antibody are composed of the *constant regions* of the L and H chains, called $C_L$ and $C_H$. If $V_H$ and $V_L$ chains each occur in 1000 different amino acid sequences, there could be 1000 × 1000, or a million, different antigen-binding sites.

Each L or H chain is produced by translation of one mRNA. Each such mRNA is, however, the combined transcript of two different genes, one *V* and one *H*. Three loci of the same length code for the constant regions of all L chains and ten loci of the same length code for the constant regions of all H chains. Based on certain data, one possible arrangement of these loci and of single genes that code for the variable regions of these chains is shown in Figure 20-9. The figure shows one (of possibly thousands) of the $V_H$ genes preceding the ten different linked loci for $C_H$ chains. It also shows one (of possibly thousands) of the $V_L$ genes that precede each of two unlinked groups of $C_L$ loci. Evidence indicates that the *V* and *C* loci that cooperate to produce a single L or H chain are close but not adjacent to each other in the antibody-producing plasma cell. The same loci, however, are further apart in an embryo cell. Some kind of somatic rearrangement must occur, therefore, during plasma cell differentiation to bring particular *V* and *C* loci closer together. This rearrangement may result from a crossing over-like or a transposition-like event.

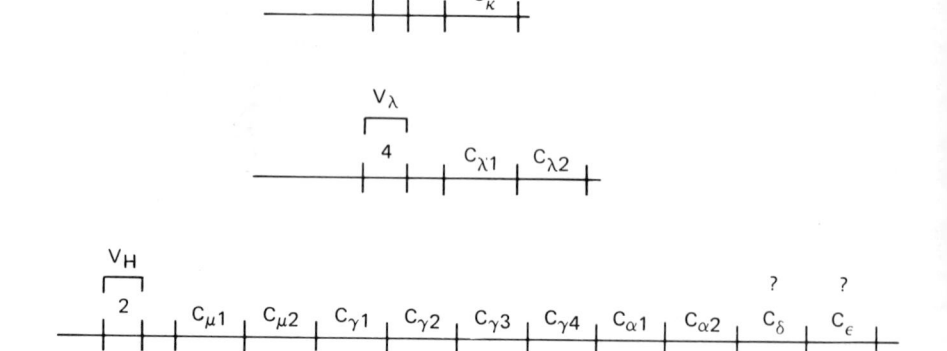

FIGURE 20-9. One possible arrangement of the structural genes for Ig synthesis in human beings.

## 20.9 Antibody synthesis is regulated at the transcriptional and translational levels.

When each lymphocyte differentiates, all the heavy chains it or its descendant cells produce carry the same $V_H$ region. The first heavy chains synthesized contain constant regions coded by $C_{\mu 1}$ or $C_{\mu 2}$ (Figure 20-9); the constant regions of later chains are all coded by one of four $C_\gamma$ loci; and still later ones are all coded by one of two $C_\alpha$ loci. Since the same $V_H$ transcript is joined to different $C_H$ transcripts at different times in differentiation, it is clear that antibody synthesis is regulated at the level of transcription.

Although several mechanisms have been proposed for such transcriptional regulation, we shall describe only two. The sliding activation hypothesis, already presented in Section 20.7, is applied to explain the synthesis of H chains. Before differentiation, all the euchromatic loci for H chain components are assumed to be heterochromatized (Figure 20-10A). When a cell is stimulated to synthesize antibodies, a $V_H$ gene that has relocated near the start of the sequence of $C_H$ loci is derepressed or activated to assume the euchromatic state. A second, narrow band of activation occurs in the same chromosome and slides (to the right in the Figure) as a wave until it includes a single $C_\mu$ gene (Figure 20-10B). With the wave stopped at this point, H chains are synthesized containing a $C_\mu$ region. Subsequently the wave of activation continues to the right and stops at a $C_\gamma$ locus (Figure 20-10C), so the H chains made now contain the same $V_H$ region but a $C_\gamma$ region. Still later the wave continues to the right, stopping next at a $C_\alpha$ gene, so the H chains produced will contain a $C_\alpha$ region. This hypothesis proposes (Figure 20-10B, C) that a heterochromatized region of a chromosome can supercoil or fold in such a way that two euchromatic regions are juxtaposed. This juxtaposition would then permit transcriptase to make a single mRNA from a template DNA

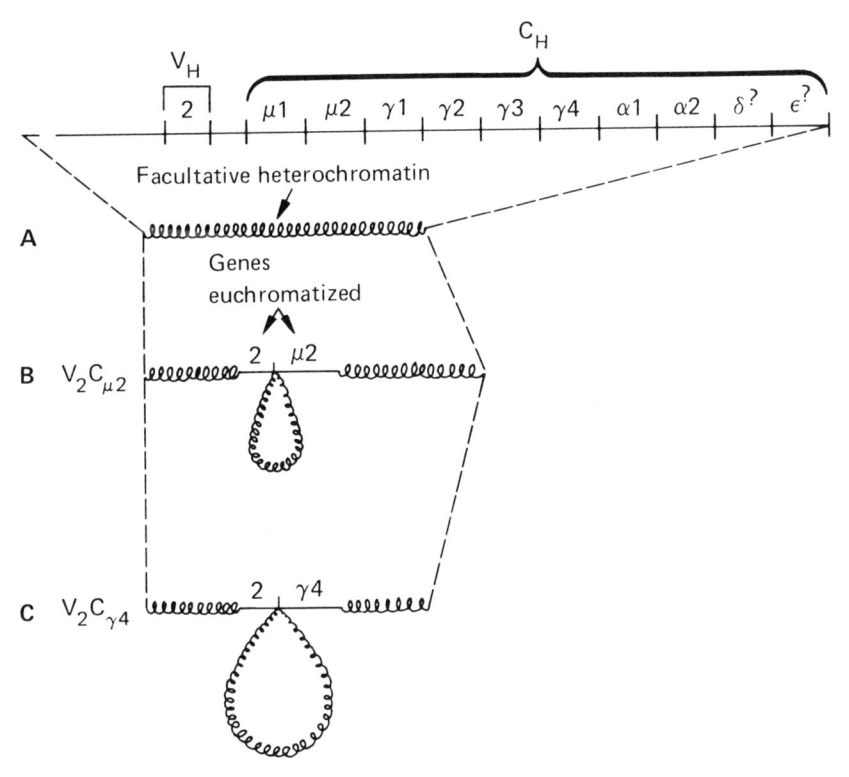

FIGURE 20-10. Sliding activation for the formation of Ig H (and L) chains. **A**: Chromosome region that codes for an H chain is inactive due to heterochromatization. **B**: A $V_H$ gene (2) and a $C_H$ gene ($\mu2$) become adjacent after being euchromatized. Transcription yields $V_2 C_{\mu2}$ chains. (C) The same $V_H$ gene becomes adjacent to a different $C_H$ gene ($\gamma4$) when the euchromatization in this region slides to the right. Transcription yields $V_2 C_{\gamma4}$ chains.

that contains an interruption. According to this sliding activation hypothesis the region between $V_H$ and $C_H$ (known in one case to be 1250 nucleotides long) is not transcribed.

The *excision-splicing hypothesis* assumes that the DNA between the $V_H$ and the $C_H$ loci being expressed is transcribed, but is subsequently excised when its bordering segments are spliced to produce mature H chain mRNA. This hypothesis applies the excision-splicing process normally used to produce eukaryotic mRNA's (Section 4.4) to two juxtaposed $V$ and $C$ loci. As differentiation continues and the $C_H$ gene shifts from $\mu$ to $\gamma$ to $\alpha$, the transcript produced would have to lengthen and the segment to be excised would have to increase correspondingly.

Evidence is available that antibody synthesis is also regulated at the level of translation. For example, antibodies inhibit proliferation of the clone of cells that produce them; and they inhibit H-chain synthesis by binding to H-chain mRNA.

## SUMMARY AND CONCLUSIONS

Some differentiation is due to changes in the quality or quantity of normally present or infective genetic material and to the regulation of translation. Nevertheless, most differentiation seems to be due to the differential transcription of complete sets of unamplified genes.

Multicellular animals that start life as fertilized eggs are able to survive early development because as oocytes they stockpiled the products of gene action of their own and of other cells, such as follicle cells and nurse cells. Besides storing ribosomes, yolk, and mitochondria, oocytes transcribe and store all classes of RNA; they store a variety of proteins, such as spindle tubule protein, histones, and DNA and RNA polymerases, needed in early developmental stages. Amphibian oocytes also store o$^+$ protein that must stimulate transcription midblastula before gastrulation can occur. As the fertilized egg develops, it uses the stockpiled materials and, in turn, transcribes, translates, and stores products of gene action in anticipation of their use in a subsequent stage of development.

Adjacent body parts differentiate coordinately due to gene-specified induction-response systems. Nonadjacent body parts differentiate coordinately, owing to the mediation of hormones that affect transcription or translation in their target cells.

Differentiation of erythrocytes during development requires switching from making large quantities of a few types of hemoglobin to making large quantities of a few different types of hemoglobin. The required switching of gene action from certain globin-coding loci to other linked globin-coding loci seems to occur at the level of transcription. This switch is hypothesized to occur by a wave-like shift of a transcription-active region in facultative heterochromatin. Hemoglobin synthesis is also regulated at the level of translation.

Lymphocytes synthesize large quantities of up to a million different antibodies in response to antigens. Any single lymphocyte, however, synthesizes only one type of antibody. The two L chains and the two H chains in each antibody are composed of V and C regions; the V and C regions of an H or L chain are coded by separate genes. As differentiation continues, a lymphocyte or its descendants make successive changes in the type of $C_H$ portion of the H chain it produces, while keeping the same $V_H$ portion. This switch in gene action occurs at the level of transcription or of RNA tailoring. Antibody synthesis is also under translational control.

## QUESTIONS AND PROBLEMS

**20.1** *Anagrams*: Unscramble and define each of the following: **a.** roomhen. **b.** duncoinit. **c.** tindobay. **d.** mehe. **e.** gianten

**20.2** *Riddle*: Why do Ancon sheep cause insomnia?

**20.3** *Books*: **a.** *Dracula*. **b.** *Gulliver's Travels*. What are the genetic implications of one or both books?

**20.4** List three reasons to believe that most differentiation is not accompanied by differential gene loss.

**20.5** What does an amphibian oocyte stockpile?

**20.6** Is stockpiling more important for earlier than for later developmental stages? Explain.

**20.7** Offer a possible explanation for the observation that the fraction of *Acetabularia* (a one-celled alga) mRNA that is translated *in vitro* by the *E. coli* translation machinery decreases as the alga differentiates.

# DIFFERENTIATION AND DEVELOPMENT

**20.8** How can the differentiation of one tissue depend upon an adjacent tissue?

**20.9** How can such small quantities of hormones have such a large metabolic effect?

**20.10** Name three general, if not universal, chemical characteristics of all types of human hemoglobin.

**20.11** What kinds of hemoglobin are made two months: **a.** before birth? **b.** after birth?

**20.12** Describe how the sliding activation of transcription hypothesis applies to the regulation of hemoglobin synthesis.

**20.13** Give one bit of evidence that hemoglobin synthesis is regulated at the level of translation.

**20.14** Do you expect that some RNA's are antigenic? Explain.

**20.15** Name three chemical characteristics of all human antibodies.

**20.16** Describe how the sliding activation of transcription hypothesis applies to the regulation of antibody synthesis.

**20.17** Why do we believe that some mRNA's are single transcripts of two separate genes?

**20.18** Give one piece of evidence that antibody synthesis is regulated at the level of translation.

**20.19** What question about antibodies would you most like to have answered by future research?

*PART VI*

# HOW THE PRECEDING CAME ABOUT IN INDIVIDUALS AND POPULATIONS

# Chapter 21

# The Origin and Evolution of Genetic Material

IN THE PRECEDING CHAPTERS, we learned the characteristics of the genetic material of organisms that exist on the earth today. To fully understand the genetic material of today and the future, however, it is necessary to understand how today's genetic material came into being. In other words, we need to understand the history—the evolution—of genetic material. The first chapter of this part of the book deals with the origin of genetic material and its evolution in individuals. The remaining chapters deal with the evolution of genetic material in groups of individuals, especially those that reproduce sexually.

**21.1** The origin and early evolution of proteins and nucleic acids occurred in the absence of genetic material.

The synthesis of the first genetic material on earth is believed to have been preceded by a chemical evolution that first produced amino acids, organic bases, simple sugars, and phosphoric acid. From these components chemical evolution subsequently produced polypeptides, nucleotides, and polynucleotides. These two pregenetic stages in chemical evolution are considered in more detail.

About 4 billion years ago the earth's atmosphere was apparently rich in methane, carbon monoxide, ammonia, water, and hydrogen, but was poor in oxygen and carbon dioxide. Chemical reactions among these materials were spurred by a large variety of energy sources: sunlight, cosmic rays and other penetrating radiations, lightning, volcanic heat, and meteorites. Some of the first molecules formed were probably formaldehyde, hydrogen cyanide, and cyanoacetylene. These, in the presence of water and ammonia, produced amino acids, purines, pyrimidines, and sugars. Since phosphoric acid appears to have been readily available, all the components of proteins and nucleic acids could have been synthesized spontaneously on primitive earth. The accumulation of these components, especially in the oceans, produced an "organic soup."

The union of components to synthesize a polypeptide, a nucleotide, or a polynucleotide requires the removal of water molecules (Section 1.2). Evaporation or freezing concentrated the organic soup. Components were also concentrated by being adsorbed to the surfaces of certain clays or droplets. Once concentrated, a source of energy such as heat from the sun or volcanic activity or ATP can readily combine components by this water-removal process. The product of polymerizing amino acids this way is called a *proteinoid*. Proteinoids are very similar to natural proteins. In water they often form microscopic spheres whose surface provides a favorable site where various chemical reactions can take place; in other words, proteinoids act as rudimentary catalysts.

This nonenzymatic synthesis of proteinoids on primitive earth was very likely paralleled by a similar nonenzymatic synthesis of nucleotides and polynucleotides. Once polynucleotides were formed, however, they apparently functioned as templates for the nonenzymatic synthesis of complementary polynucleotides. In this way, through the formation of complementary intermediates, nucleic acids became able to replicate themselves.

## 21.2 The evolution of proteinoids and nucleic acids became interdependent, leading to the formation of the first organism.

The original process of nucleic acid replication was probably very slow and inefficient. At first replication meant that a pool of nucleotides would more likely be synthesized into polynucleotides resembling a given polynucleotide and its complement than into polynucleotides whose bases are incorporated at random. Nucleic acids would then undergo an evolution in which the most abundant polynucleotide was the one that combined the greatest stability with the quickest replication. In other words, there would be a chemical selection among nucleic acids that is analogous to the natural selection that would later occur among organisms.

The original nucleic acids may have had arabinose as the sugar. Evolution of stability would favor replacement of arabinose by the more stable ribose, and then the replacement of ribose by the more stable deoxyribose. Evolution of nucleic acid stability must also have been aided by proteinoids also present on the early earth. By combining with nucleic acids proteinoids protect nucleic acids from degradation. Since proteinoids have catalytic abilities, one can imagine the spontaneous occurrence of a proteinoid that catalyzed some aspect of nucleic acid replication. Obviously, the replication rate of nucleic acid would be increased as the number of such molecules increased. There would be, therefore, chemical selection in favor of any nucleic acid that aided the synthesis of proteinoids that stabilized or catalyzed the replication of nucleic acids. We expect, therefore, that chemical evolution progressed from (1) nucleic acids whose stability and self-replication were independent of proteins through (2) those associated with proteins that originated independently of nucleic acid to (3) those dependent upon proteins whose existence depended upon nucleic acid. The result of this last evolutionary step was the production of a mutually interdependent nucleic acid–protein unit that we recognize as an *organism* containing nucleic acid genetic material.

## 21.3 The genetic code and the translation machinery have undergone evolution.

The interdependency between proteins and nucleic acids seems to have originated through chemical reactions that occurred between various-sized pieces of one class of molecule with various-sized pieces of the other class. The occurrence of some sort of preferential interaction between the two classes of molecule may have been the original basis for establishing a genetic code, that is, a means whereby information in nucleic acid corresponds to amino acid information. (Even today the specific binding of nucleotide sequences to amino acid sequences is illustrated in the affinity of the promoter for activator protein and transcriptase, of the operator for repressor protein, and so forth.) The machinery first used for translation was probably relatively simple and inefficient, making many errors.

It seems likely that the genetic code started as, or soon became, a triplet code for a small number of amino acids. It was selectively advantageous, however, to reduce the chance of making wrong proteins due to mutation or to errors in transcription or translation. One way to avoid making the wrong protein is to retain very few of the triplet codons as nonsense codons and to have the remaining sense codons highly degenerate. If most codons make amino acid sense, some changes that occur prior to translation will not modify amino acid composition; other changes will make amino acid substitutions that permit the full, or almost full, functioning of the protein; as a result, such changes have little or no detrimental phenotypic effect. Moreover, as the code and the translation machinery evolved, when changes did produce amino acid substitutions, the chance was maximal that an amino acid was replaced by a functionally similar one, thereby minimizing detrimental phenotypic effects. Once the genetic code and translation machinery evolved to a state that produced the fewest and most tolerable errors, both became essentially stable and universal.

## 21.4 Directed by natural selection, organisms have increased in biochemical complexity during evolution.

The organic soup in which the first organisms arose was rich in the raw materials needed for their reproduction. After a time, however, one or more required raw materials became limited in supply, slowing down the rate of reproduction. If, for example, raw material Z came to be in short supply, natural selection would favor any genetic material that coded for a protein that catalyzed the chemical conversion of a similar raw material, Y, into the required raw material. When, in turn, the similar Y material became limited in supply, natural selection would favor genetic material that could convert another raw material, X, into Y. In this manner, sequences of biochemical reactions such as $X \rightarrow Y \rightarrow Z$ could have become established, where each step was catalyzed enzymatically by proteins coded in genetic material. Thus, whereas the original organisms obtained from their environment all the raw materials they needed in directly usable form, organisms (such as prokaryotes) were eventually required to depend more and more upon their own genotypes to convert simple and abundant substances in their environment into needed components. When energy-containing compounds in the environment were depleted, it became selectively advantageous, perhaps about 3 billion years ago, to establish a genetic basis for photosynthesis of energy-containing substances.

We see, therefore, that the biochemical phenotypes of early organisms became more and more complex as time progressed. Natural selection must also have favored other gene-based phenotypic changes that protected an organism from unfavorable environments and/or permitted it to search for and travel to favorable environments. Attainment of such advantages increased the complexity of the structural as well as the biochemical phenotype.

## 21.5 Genetic material has undergone a quantitative evolution, primarily by small duplications.

Since the structural and functional complexity of organisms has increased during the course of evolution, the amount of information, and hence the amount of genetic material, needed to specify an organism has also increased. One way to increase the amount of genetic material is by polyploidy. Single-species polyploidy may have been successful among early organisms. It does not seem to have been a major mechanism for increasing gene number in sexually reproducing organisms, however, because it often produces gametes with unbalanced chromosome numbers. Multiple-species polyploidy is also not a major means of increasing gene number, although as much as one fourth of presently existing species of flowering plants originated this way. The gain of single whole chromosomes likewise does not seem to be a major mechanism for increasing the amount of genetic material because of the genetic imbalance it produces. The main means of increasing gene number seems to be the duplication of small chromosome regions that produce phenotypic changes small enough to be tolerated by the organism. Duplications can result either from breakage and exchange union or from crossing over between repeated base sequences that have undergone mismatched synapsis (Figure 21-1).

Several lines of evidence indicate that duplication of short chromosomal regions has played an important role in increasing gene

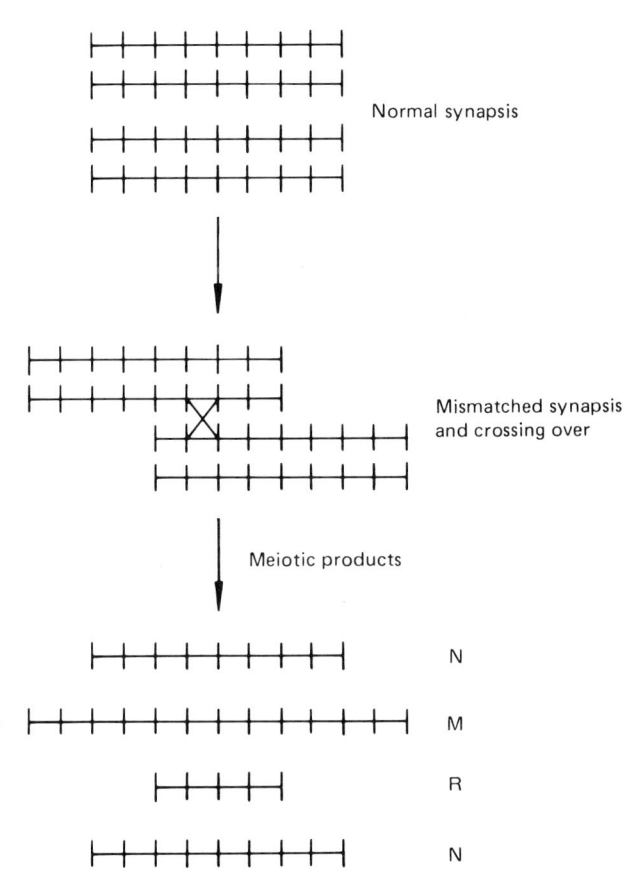

FIGURE 21-1. Increasing the number of repeated regions in a chromosome by crossing over after mismatched synapsis. N, normal sequence with eight repeats; M, magnified sequence with twelve repeats; R, reduced sequence with four repeats.

number, after which other mutations caused repeated regions to become somewhat different.

1. *Similarities in cytology.* Multichromatid chromosomes have many adjacent paired bands.
2. *Similarities in phenotype.* Adjacent loci often have similar phenotypic effects.
3. *Similarities in nucleotide sequence.* A sequence of 10 nucleotides at each end of 5S rRNA seems to correspond to a similar sequence within tRNA.
4. *Similarities in amino acid sequence.* Since the degeneracy of the genetic code is limited, similar amino acid sequences must be due to similar nucleotide sequences. Two or more similar or identical amino acid sequences have been detected within many proteins, including myoglobin (a relative of hemoglobin found in muscle cells), cytochrome, and the antibody H chain whose C region contains three repeated regions (Figure 20-8). This indicates that internal duplications have occurred in the evolution of many protein-coding genes. Similar amino acid sequences have also been observed in different proteins coded at different loci, for example, between myoglobin and the $\alpha$ chain of hemoglobin, between the $\alpha$, $\beta$, $\gamma$, and $\delta$ chains of hemoglobin, and between the L and H chains of an antibody. This indicates that separate loci that code for similar proteins today arose as duplications of an ancestral gene.
5. *Similarities in protein configuration.* The configurations of three different regions in a particular protein are nearly identical, suggesting that they result from the triplication of an ancestral nucleotide sequence.

## 21.6 Evolutionary trees can be constructed for single genes.

The complete nucleotide sequence can be determined for the same rRNA or tRNA in a wide range of organisms. One can then estimate how different their genes have become during evolution from the *mutational distances* between them. The term *mutational distance* is the number of nucleotides that would need to be altered to convert the RNA in one organism to the corresponding RNA of another organism. One can then construct an *evolutionary tree* in which the genes for a given RNA in two or more different organisms are located at the ends of branches whose lengths are equal to the mutational distance between the genes.

The complete amino acid sequence can be determined for the same protein in a wide range of organisms. Since the genetic code is universal, we can use such sequences to determine the mutational distance between the genes coding for this protein. In this case the mutational distance is the minimal number of nucleotides that would need to be altered to convert the protein in one organism to the corresponding protein of another organism. Figure 21-2 shows an evolutionary tree (inverted) for cytochrome c from a large number of species using such mutational distances. Evolutionary trees have also been constructed for myoglobin and hemoglobin chains as well as other proteins.

Since sequencing the bases in DNA is now relatively easy, we expect that evolutionary trees made directly from DNA differences will eventually be made for various nontranscribed as well as transcribed sequences.

## 21.7 Genetic material has undergone a functional evolution.

The genetic material of early organisms may have been used more or less directly as a template to synthesize protein required for the maintenance and/or replication of the genetic material. As evolution progressed and the amount of genetic material per organism increased, however, mutations modified many genes, enabling them to code for (1) proteins having different structural or enzymatic functions or (2) RNA's, such as rRNA, tRNA, and ribosome-binding RNA, that assist in the translation process. In addition, there arose genes whose function was not to be transcribed but was to recognize proteins, such as DNA polymerase, transcriptase, and $\rho$, in order to start and stop replication or transcription. Genes also arose that recognized protein enhancers or repressors of replication or transcription, thereby serving to regulate these processes. It is likely that most of the highly and moderately redundant DNA in eukaryotes, including most of the DNA transcribed into giant RNA and the nontranscribed spacer DNA's, also serve in as-yet-unknown ways to regulate the functioning of genetic DNA. That repetitive sequences seem to be important in evolution is indicated by the observation that differences between species are sometimes correlated more closely with differences in repetitive non-protein-coding sequences than with differences in nonrepetitive, protein-coding sequences.

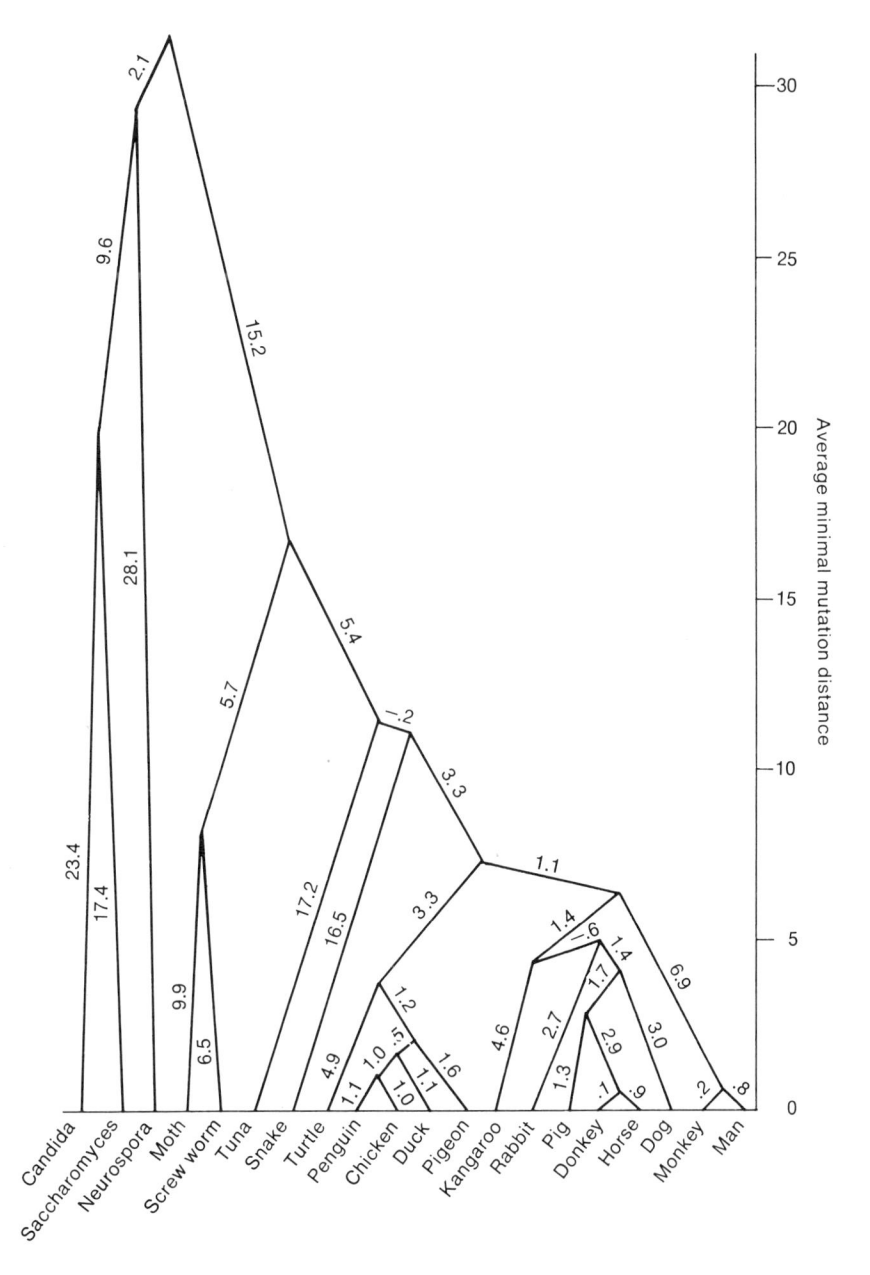

FIGURE 21-2. Inverted evolutionary tree for the cytochrome c gene. Each number is the best-fitting mutation distance, each apex being placed at an ordinate value representing the average of the sums of all mutations in the lines of descent from that apex. (After W. M. Fitch and E. Margoliash, 1967.)

## SUMMARY AND CONCLUSIONS

The earth has undergone a preorganismal chemical evolution that resulted in the formation of polypeptides and polynucleotides. Once these two kinds of polymers became so interdependent that the maintenance and replication of nucleic acid became dependent upon protein, whose existence in turn depended upon the nucleic acid that it helped maintain and replicate, the first organism containing nucleic acid genetic material came into being.

Organismal or biological evolution then proceeded in favor of organisms that most efficiently used the environment to perpetuate themselves. This efficiency was attained by the evolution of a genetic code and transcription–translation machinery that became essentially universal, by increasing the quantity of genetic material, usually by small duplications, and by subsequent qualitative genetic changes to produce different kinds of genes that have different functions. The path of evolution that the same gene takes in different organisms can be seen in evolutionary trees based upon mutational distances.

## QUESTIONS AND PROBLEMS

**21.1** Why do you suppose all organisms on earth had a single rather than a multiple origin?

**21.2** If the basis of natural selection of organisms is the survival of the fittest, what would be the basis of natural selection among chemical substances prior to the occurrence of organisms?

**21.3** List five different sources of the energy needed for chemical reactions on the prebiotic earth.

**21.4** What characteristic does nucleic acid have that protein does not, that explains why proteins are not genetic material?

**21.5** What distinguishes proteinoids from proteins?

**21.6** What evidence can you offer from today's organisms that proteins and nucleic acids may have once "coded" for each other directly?

**21.7** Assuming that life exists on other planets in the universe, what chemical basis for genetic material would you expect there?

**21.8** Is it natural to expect that the biochemical complexity of organisms will increase during evolution? Explain.

**21.9** What two specific types of change in quantity of genetic material are most adaptive in today's organisms? Why?

**21.10** How can we trace the path of past evolution using nucleotide sequences of present-day organisms?

**21.11** Do organisms regulate their own rates of evolution? Explain.

**21.12** What can you conclude is the basis of the main evolutionary difference between two genera of amphibians that have a sevenfold difference in nuclear DNA content?

**21.13** Why are interplanetary missiles sterilized?

**21.14** Ozone, produced from oxygen, absorbs ultraviolet light. Why do we believe that UV was the most important mutagen of early organisms?

*Chapter 22*

# Population Genotypes and Mating Systems

ALTHOUGH GENETIC MATERIAL OCCURS in individual organisms, evolution proceeds in groups of organisms. It becomes necessary, therefore, to understand the fate of genetic material in groups of organisms. Because of the prevalence of sexual reproduction, we shall give most of our attention to the genetics of populations, a *population* being all the interbreeding members of the same kind of organism. Because a chromosome set usually contains a large number of genes that can occur in many alleles, it is not possible to trace the population history of complete genotypes. For simplicity, we shall consider first the population behavior of a single mendelian locus as affected by random and nonrandom mating.

## RANDOM MATING

**22.1** If matings are random, allele frequencies can be used to determine genotype frequencies in populations.

Consider the genotypes produced in a diploid population by a single locus with only two alleles. Suppose that the locus in question determines blood type in human beings. The two alleles are $L^M$ and $L^N$ and, since they show no dominance, three blood type phenotypes are possible—M, MN, and N. There is, therefore, no ambiguity in determining the genotypes from the phenotypes present in a population, and the frequency of the two alleles is readily calculated. For example, in a population of 1000 persons of which 358 have M blood type, 484 MN, and 158 N (Figure 22-1), the allele frequency, $p$, of $L^M$ in the population is found to be 0.6 and that of $L^N$, $q$, is 0.4.

Let us suppose that matings in this population occur at random, that is, without regard to blood-type genotype. If so, it would seem reasonable

| Phenotype | M | M N | N |
|---|---|---|---|
| Genotype | $L^M L^M$ | $L^M L^N$ | $L^N L^N$ |
| Number (Total = 1000) | 358 | 484 | 158 |
| Frequency (Total = 1.0) | 0.358 | 0.484 | 0.158 |

Frequency of $L^M = p = 0.358 + \frac{1}{2}(0.484) = 0.6$
Frequency of $L^N = q = 0.158 + \frac{1}{2}(0.484) = 0.4$
$p + q = 0.6 + 0.4 = 1.0$

FIGURE 22-1. Phenotype, genotype, and allele frequencies in a specific population.

that random mating could be represented by a random combination of gametes. By having gametes combine randomly, the genotype frequencies expected with regard to MN blood type can be obtained (Figure 22-2), using for both eggs and sperm the allele frequencies determined above. The expected genotype frequencies closely approximate those actually found in the sample population.

The random union of gametes whose $p = 0.6$ and $q = 0.4$ produces diploid genotypes that are also expected to contain these alleles in these frequencies (Figure 22-2). Thus, the allele frequencies of the $F_1$ are identical to those of the $P_1$. Furthermore, the allele frequencies of the next ($F_2$) generation and all subsequent generations will be the same.

## 22.2 The generalization of the statements made in Section 22.1 is called the Hardy–Weinberg principle.

The preceding analysis can be expressed in more general terms by letting $p$ equal the frequency of gametes in the population that carry $A$, and $q$ equal the frequency of those that carry $A'$, where $p + q = 1$. Figure 22-3 gives the results of random union of these gametes. The frequencies of homozygotes are given by $p^2$ and $q^2$, and the frequency of heterozygotes is $2pq$. The offspring population, then, is

$$p^2 A A + 2pq A A' + q^2 A' A'.$$

The frequencies of $A$ and $A'$ among the gametes produced by the $F_1$ population are

$$\left. \begin{array}{l} A = p^2 + pq = p(p+q) = p \\ A' = q^2 + pq = q(q+p) = q \end{array} \right\} \text{ since } p + q = 1.$$

Thus, the allele frequencies in the $F_1$ are the same as in the previous generation. Likewise, all future generations will have the same allele frequencies and the same relative frequencies of diploid genotypes. This is known as the *Hardy–Weinberg principle.*

The Hardy–Weinberg principle applies only when mating is random and nothing else occurs to differentially modify the reproductive ability of

# POPULATION GENOTYPES AND MATING SYSTEMS 317

|  | Female Gametes | |
|---|---|---|
|  | 0.6 $L^M$ | 0.4 $L^N$ |
| Male Gametes 0.6 $L^M$ | 0.36 $L^M L^M$ | 0.24 $L^M L^N$ |
| 0.4 $L^N$ | 0.24 $L^M L^N$ | 0.16 $L^N L^N$ |

Population Genotype Frequencies

|  | $L^M L^M$ | $L^M L^N$ | $L^N L^N$ |
|---|---|---|---|
| Expected | 0.36 | 0.48 | 0.16 |
| Actual (Figure 22-1) | 0.358 | 0.484 | 0.158 |

The expected allele frequencies

$p = 0.36 + 0.24 = 0.6$

$q = 0.16 + 0.24 = 0.4$

FIGURE 22-2. Allele frequencies and genotype frequencies in a population where random union occurs between gametes.

|  | Eggs | |
|---|---|---|
|  | $pA$ | $qA'$ |
| Sperms $pA$ | $p^2 AA$ | $pq AA'$ |
| $qA'$ | $pq AA'$ | $q^2 A'A'$ |

FIGURE 22-3. Types and frequencies of genotypes produced by random union of gametes in a population containing $pA$ and $qA'$.

any genotype. Under these conditions, the principle predicts that allele frequencies will be at equilibrium immediately upon establishment of a population, while diploid genotype frequencies are stabilized in one generation and are at equilibrium thereafter.

**22.3** The Hardy–Weinberg principle is readily applied to cases where dominance occurs.

When a Hardy–Weinberg equilibrium exists, the allele frequencies can be determined despite the presence of dominance. If complete dominance causes two genotypes ($AA$ and $Aa$) to have the same phenotype, allele frequencies can be determined from the frequency of individuals with the recessive phenotype ($aa$). The frequency of $aa$ individuals must be equal to the square of the frequency of the recessive allele, $q$. For example, if $q = 0.7$, $q^2$ would be $(0.7)^2$, or 0.49. Working backward, if the recessive phenotype comprised 0.49 of the population ($q^2 = 0.49$), the frequency of the recessive allele would be $\sqrt{0.49}$, or 0.7. The frequency of the dominant allele, $p$, would equal $1 - q$—in the present example, $1 - 0.7$, or 0.3. The equilibrium genotypes in our example would also include $p^2 = 0.3 \times 0.3$, or 0.09, homozygous dominant individuals, and $2pq = 2(0.3)(0.7)$, or 0.42, heterozygous individuals.

The Hardy–Weinberg principle can also be applied to cases involving multiple alleles and X-limited genes, as well as cases involving two or more loci.

# NONRANDOM MATING

The alleles at most loci are probably not in an unchanging Hardy–Weinberg equilibrium. The value of the Hardy–Weinberg principle is, therefore, not that it has wide application but that it permits one to measure the imbalance produced in such an equilibrium by various disturbing factors. In the rest of this chapter we consider the effect nonrandom mating has on a population that would otherwise be in a Hardy–Weinberg equilibrium. The next chapter discusses other factors that disturb a Hardy–Weinberg equilibrium.

**22.4 Allele and genotype frequencies are little affected if homozygotes for a rare allele do not mate.**

In determining the frequencies of genotypes in a population at equilibrium, matings have so far been assumed to occur randomly with respect to the trait under consideration. What happens if the different genotypes do not mate at random? Consider the disease *phenylketonuria* that occurs in human beings homozygous for a certain recessive allele. Affected individuals cannot convert phenylalanine to tyrosine and, as one of the multiple phenotypic effects, may suffer mental retardation. The frequency of the normal allele ($A$) is 0.99; of the abnormal allele ($a$), 0.01. In a population at Hardy–Weinberg equilibrium, therefore, $AA:Aa:aa$ individuals should occur with frequencies of 9801/10,000 : 198/10,000 : 1/10,000, respectively. Notice that $Aa$ individuals are 198 times more frequent than $aa$, and contain 99 per cent of all $a$ genes in the population.

Individuals who are $AA$ and $Aa$ apparently marry at random, but affected persons who are mentally retarded do not marry at all. Even if no phenylketonurics had children, the frequency of the recessive allele in the next generation would be decreased only by 1 per cent. Genotypic frequencies would also be little affected, since most $Aa$ individuals result from matings of $Aa$ with $AA$ or $Aa$, and most $aa$ individuals result from matings between two $Aa$ individuals. We see, therefore, that the allele and genotype frequencies for rare, recessive alleles will change only slightly each generation if homozygotes for the allele reproduce less or not at all.

**22.5 Inbreeding increases the frequency of homozygous individuals and the variability of the genotypes in a population without changing allele frequencies.**

Mating can be nonrandom because of *inbreeding* and *assortative mating*. Inbreeding refers to matings in which the mates are more closely related than randomly chosen mates. Whereas inbreeding involves matings of individuals who are genotypically similar, assortative mating refers to matings of individuals who are phenotypically similar. To the degree that phenotypic similarities are based upon genotypic similarities, assortative mating has much the same consequences as inbreeding. Accordingly, we will confine our attention to inbreeding.

The genotypic consequences of inbreeding with respect to nonrare, common alleles can be illustrated by a simple example. Suppose that a trait is due to a single pair of genes and a population is made up of individuals with $AA$, $AA'$, and $A'A'$ genotypes. Assume that each individual can reproduce only by *self-fertilization*, the closest form of inbreeding. In this event, all progeny of the $AA$ and $A'A'$ types will be like their parents. The progeny of the $AA'$ type will be, however, on the average, $\frac{1}{4} AA$, $\frac{1}{2} AA'$, and $\frac{1}{4} A'A'$. As a consequence, the proportion of heterozygotes in the population in the next generation will be only one half as large as it was before, and the proportion of the two types of homozygote will increase accordingly. If only self-fertilization again occurs in the next generation, the same thing will happen—the proportion of heterozygotes will again be reduced by one half. Figure 22-4 shows what happens when a population composed of 100 per cent of $AA'$ individuals practices self-fertilization for four successive generations. Note that the allele frequencies ($p = 0.50$, $q = 0.50$) do not change—only the relative number of homozygotes and heterozygotes. If self-fertilization were continued for seven generations, more than 99 per cent of the population would be homozygous. Having started with 100 per cent $AA'$ individuals, we see that self-fertilization simultaneously increases the variability of the genotypes in a population while it increases the frequency of homozygotes.

Individuals are often heterozygotes at many loci. In a population mating at random and at equilibrium, matings that reduce heterozygosity are balanced by those that increase heterozygosity. If this population were to practice self-fertilization for one generation, half of all heterozygous genes would become homozygous because of inbreeding. The total amount of homozygosity expected would be the frequency normally occurring in the random mating population plus $\frac{1}{2}$ the frequency of heterozygosity. Whereas the fraction of all heterozygous genes caused to become homozygous is $\frac{1}{2}$ for self-fertilization, it is $\frac{1}{4}$ for *brother-sister (sibling) matings*, and $\frac{1}{16}$ for *(first) cousin marriages*.

FIGURE 22-4. The effect of successive generations of self-fertilization on the percentage of different genotypes in a population originally composed of 100 per cent of $AA'$.

| Generation | Genotype (%) | | |
|---|---|---|---|
| | $AA$ | $AA'$ | $A'A'$ |
| 1 | 25 | 50 | 25 |
| 2 | 37.5 | 25 | 37.5 |
| 3 | 43.75 | 12.5 | 43.75 |
| 4 | 46.875 | 6.25 | 46.875 |

INDIVIDUALS AND POPULATIONS

*Outbreeding* and *disassortative mating* are the opposite of inbreeding and assortative mating. As such they have the opposite population effects; they increase the heterozygosity and reduce the variability of the population genotypes.

## 22.6 The homozygosity caused by inbreeding occurs for detrimental as well as beneficial alleles.

All forms of inbreeding increase homozygosity. Consider the effect of (first) cousin marriage upon the frequency of a disease such as phenylketonuria (Figure 22-5). The frequency of $Aa$ heterozygotes per 10,000 people is 198 (Section 22.4). Cousin marriage reduces the number of heterozygotes by $\frac{1}{16}$. If all 10,000 persons married cousins, the number of heterozygotes would be reduced by 12 (198 ÷ 16), of which 6 would be expected to be normal ($AA$) and 6 affected ($aa$). Since the population would normally contain 1 affected individual per 10,000, cousin marriages would bring the total number of phenylketonurics to seven. Accordingly, there is a sevenfold greater chance for phenylketonuric children from cousin marriages than from marriages between unrelated parents. Since cousin marriage is so infrequent, however, it does not significantly change the genotype frequencies in the population.

Another example of how cousin marriages increase the risk of birth defects comes from a study of a Japanese population (Figure 22-6) in which congenital malformations, stillbirths, and infant deaths were 28 to 48 per cent higher among the children of cousins than among the children of unrelated parents. Since defects such as these are sometimes due to recessive genes in homozygous condition, these results support the view that homozygosis resulting from inbreeding can produce detrimental effects.

Although inbreeding produces homozygosis that can lead to the appearance of defects, it must not be inferred that inbreeding is disadvantageous under all circumstances. As the result of inbreeding just as

FIGURE 22-5. Pedigree showing the occurrence of phenylketonuria among the offspring of cousin marriages (denoted by thick marriage lines). Open circles and squares indicate normal females and males, respectively; filled circles and squares, affected females and males.

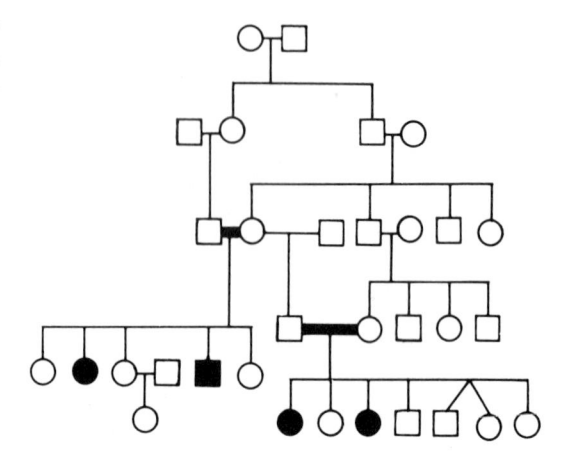

|  | Frequency from Unrelated Parents | Increase in Frequency with Cousin Marriage | Per cent Increase |
|---|---|---|---|
| Congenital malformation | 0.011 | 0.005 | 48 |
| Stillbirths | 0.025 | 0.006 | 24 |
| Infant deaths | 0.023 | 0.008 | 34 |

FIGURE 22-6. Increased risk of genetic defect with cousin marriages. (Data from Hiroshima and Nagasaki.)

many individuals become homozygous for the normal allele as become homozygous for detrimental genes. No obvious disadvantage seems to have accompanied the brother–sister matings practiced for many generations by the Pharoahs of ancient Egypt. In fact, the success of self-fertilizing species is testimony to the general advantage of homozygosity in some populations.

## SUMMARY AND CONCLUSIONS

The Hardy–Weinberg principle predicts the allele and genotype frequencies of mendelian genes in successive generations of randomly mating populations. The Hardy–Weinberg equilibrium that produces static allele and genotype frequencies is usually upset by various factors discussed in detail in Chapter 23. Dominance has no effect on a Hardy–Weinberg equilibrium and nonrandom mating of homozygotes for rare recessive alleles has a very small effect per generation. Although nonrandom mating involving common alleles has no effect on allele frequencies, it does change genotype frequencies. For example, inbreeding and assortative mating increase homozygosity and the variability of genotypes in the population. It is noted that homozygosis that results from inbreeding occurs alike for detrimental and normal (beneficial) alleles.

## QUESTIONS AND PROBLEMS

**22.1** What genetic applications and implications are found in *The Bible*?

**22.2** The frequency of blood types in a sample of 300 persons is: M, 42.7 per cent; MN, 46.7 per cent; N, 10.7 per cent. Is this sample drawn from a population at Hardy–Weinberg equilibrium?

**22.3** Suppose that trypsin occurs in only two alternative forms, A and B, in a human population at Hardy–Weinberg equilibrium. If the locus for trypsin is autosomal, and 16 per cent of the population has only type A, what percentage of the population is expected to produce both A and B types?

**22.4** What would be the Hardy–Weinberg expectation for the frequency of *b* if 99 per cent of people were phenotypically like *B*, the only other, dominant allele?

**22.5** Assuming that the Hardy–Weinberg principle applies, what is the frequency of gene *M* if its only allele *M'* is homozygous in the following percentages of the population: **a.** 81 per cent? **b.** 4 per cent? **c.** 25 per cent? **d.** 49 per cent?

**22.6** Predict the phenotype of the girl Kareem-Abdul Jabbar brings home to meet his folks.

**22.7** What frequencies of $p$ and $q$ give the greatest proportion of heterozygotes?

**22.8** Suppose that the frequencies of alleles $A$ and $a$ are 0.4 and 0.6, respectively, in a population at Hardy–Weinberg equilibrium. **a.** What per cent of the population is composed of homozygotes? **b.** What would be your answer to part a. after one generation of mating hybrids only with hybrids? **c.** How would the condition in part b. affect allele frequencies?

**22.9** Suppose that $B$ is a completely dominant and penetrant autosomal gene for brown eyes, and its only allele is $b$ for blue eyes. Suppose that a Martian colony is started with 10 per cent blue-eyed and 90 per cent homozygous brown-eyed people. If a Hardy–Weinberg equilibrium is established, what will be the frequency of **a.** heterozygotes? **b.** each type of homozygote?

**22.10** Suppose that 100 per cent of the initial population in the preceding question had blood group MN. What will be the blood types present at Hardy–Weinberg equilibrium?

**Note:** About 70 per cent of Americans get a bitter taste from the drug phenylthiocarbamide (PTC); 30 per cent do not, being taste blind. Tasting is due to a single dominant, autosomal allele with complete penetrance.

**22.11** What is the frequency of the taster gene $T$ and of its nontaster allele $t$, assuming a Hardy–Weinberg equilibrium?

**22.12** What proportion of marriages between tasters and nontasters have no chance of producing nontaster children among Americans?

**22.13** An investigator found 278 taste blind children in a total of 761 children from marriages of tasters with nontasters. Is this sample drawn from a population at Hardy–Weinberg equilibrium?

**22.14** What is the effect of outbreeding on a Hardy–Weinberg equilibrium?

**22.15** How much more likely is the occurrence of a homozygous recessive from cousin marriage than from random mating, if the random mating population has 4 per cent homozygotes for this allele?

# Chapter 23

# Factors That Affect Gene Frequencies in Populations

IN THE PRECEDING CHAPTER, we saw how inbreeding upsets the Hardy–Weinberg equilibrium by changing genotype frequencies but not allele frequencies in populations. In this chapter we examine the factors that change gene frequencies in populations. These factors are *selection*, *mutation*, *migration*, and *genetic drift*.

# SELECTION

**23.1** By favoring certain phenotypes and hence genotypes, selection changes population gene frequencies.

Since all organisms tend to overproduce offspring, selection works to preserve individuals that give the population the greatest fitness. Since selection acts on whole phenotypes, not single traits, it conserves genotypes and not single genes. Selection occurs at all stages in the life cycle, including the haploid and diploid stages of the same individual. A relatively fit genotype at one stage of the life cycle may be relatively unfit at another. An individual's overall fitness is the sum of the fitness of all its phenotypes at different stages of its life. Because selection favors genotypes that produce the maximal fitness of the population as a whole, it often occurs that some members of a population contain genotypes that are undesirable from the individual's standpoint but that are desirable from the population's standpoint. For example, in many animal populations some individuals are genetically programmed to be sterile, to work, or to fight to the death for the good of the population. When certain genotypes are favored by selection, a Hardy–Weinberg equilibrium will not occur, and the frequencies of certain alleles in the population will increase while those of others will decrease.

## 23.2 The frequency of a deleterious allele depends upon its effect on fitness when homozygous and heterozygous.

From a biological standpoint fitness refers to the ability to reproduce. The symbol, $w$, for *fitness*, refers to the ability of a genotype to reproduce; it is defined to be 1.0 for the most fit genotype, and less for relatively less fit genotypes. The symbol $s$ is a measure (from 0 to 1) of the unfitness of the genotype and, therefore, of the selection against it. That is, $s$ is the *selection coefficient* that measures the selective disadvantage of one genotype relative to that of another. In general, $w = 1 - s$.

We wish to see what happens to allele frequency in successive generations of a population when an allele being selected against is (1) completely dominant, (2) completely recessive, or (3) shows partial or no dominance. To compare the results of these different conditions we will always start with a population in which the frequency of the deleterious gene $A$ is 0.95, and its normal allele, $A'$, is 0.05 (Figure 23-1).

SELECTION AGAINST A COMPLETE DOMINANT. The rapidity with which an allele's frequency will change will depend upon how detrimental it is; that is, its $s$ value. If, for example, $A$ is a dominant lethal, $AA'$ heterozygotes have $s = 1$, and the $A$ allele is removed from the population after one generation of selection. Since $AA$ homozygotes cannot occur, all genotypes present after one generation are $A'A'$, and $A'$ allele frequency is 1.0.

If, on the other hand, $A$ is a completely dominant allele that produces a 50–50 chance of reproducing, $AA$ and $AA'$ each have $s = 0.5$. Figure 23-1, curve $a$, shows the change in frequency of such an allele in successive generations. $A$ frequency decreases until the entire population is composed of the homozygous recessive, $A'A'$. Note that when $A$ comprises about 95 per cent of the genes, few homozygotes for the recessive allele are present and selection can increase $A'$ frequency only slowly. Once the recessive allele reaches a frequency of 0.15, however, selection against the dominant allele becomes rapid in subsequent generations.

FIGURE 23-1. Changes in gene frequency (starting at 95 per cent) when allele $A_1$ is selected against in different genotypes and to different degrees. Curve $a$, complete dominance: $s = 0.5$ for $AA'$. Curve $b$, complete recessive: $s = 1.0$ for $AA'$. Curve $c$, complete recessive: $s = 0.5$ for $AA$. Curve $d$, no dominance: $s = 0.2$ for $AA$; $s = 0.1$ for $AA'$.

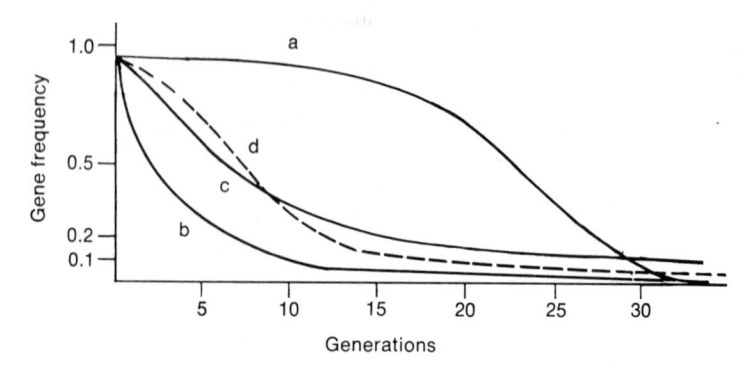

SELECTION AGAINST A COMPLETE RECESSIVE. Suppose that $A$ is a completely recessive lethal, so that $AA$ has $s = 1$. In this case, selection causes the frequency of $A$ to fall very sharply in the first several generations (Figure 23-1, curve $b$), but decreases slowly when, after 15 or so generations, the gene is less frequent. Figure 23-1, curve $c$, shows a similar, although slower, decrease in frequency when $A$ is completely recessive with $s = 0.5$. The results obtained after the fifteenth generation illustrate the inefficiency of selection against homozygotes for rare recessive genes (recall the case of phenylketonuria, Section 22.4), at least insofar as lowering the frequency of such alleles is concerned.

SELECTION WHEN DOMINANCE IS ABSENT OR INCOMPLETE. When the detrimental allele $A$ shows no dominance or incomplete dominance, it will be selected against both when it is homozygous and when it is heterozygous. In such a case (Figure 23-1, curve $d$), even when $A$ has relatively small $s$ values (say 0.20 for $AA$ and 0.10 for $AA'$), its frequency will fall essentially as fast as it did for the complete recessive with $s = 0.50$ that was selected against only when homozygous.

## 23.3 Heterosis occurs when the $s$ for a heterozygote is smaller than for either homozygote.

In the preceding section the frequency of $A$ was always driven toward zero because it always produced a detrimental effect in any genotype in which it was expressed. As implied in Section 23.1, some alleles are selected against in some combinations but are selected for in other combinations. Many examples are known where $A$ and $A'$ are recessive lethals ($s = 1.0$ for $AA$ and $A'A'$) but $AA'$ is viable ($s = 0.0$). A population with such alleles, called *balanced lethals*, is permanently heterozygous. The greater fitness of the heterozygote compared to either homozygote is an example of *hybrid vigor* or *heterosis*.

A second, somewhat less extreme, example of heterosis has already been mentioned in Section 17.6. In this case, persons who are heterozygous, $\beta^A \beta^S$, are more resistant to falciparum malaria than are $\beta^A \beta^A$ persons. Since untreated $\beta^S \beta^S$ individuals usually die before maturity, they have $s = 1.0$. In certain malarial environments, therefore, the heterozygote has $w = 1$, and the two homozygotes have smaller $w$ values. As in the case of balanced lethals, selection will maintain the $\beta^S$ recessive lethal in the population when its heterozygote is more adaptive than either homozygote. In the populations of falciparum malarial countries, the frequency of the $\beta^S$ allele may be as high as 0.10. In nonmalarial environments, however, $\beta^S$ confers no antimalarial advantage and is selected against both in heterozygous and homozygous condition. Under these circumstances, selection will rapidly decrease the frequency of $\beta^S$ in the population until it is rare.

**23.4** Heterosis also occurs in the hybrid whose parents were homozygous for different recessive detrimental genes.

In addition to being due to heterozygosity at a single locus, as described in the preceding section, heterosis can also be due to heterozygosity at two to many loci. The second type of heterosis can be illustrated by crossing two different homozygous lines having genotypes $AA\ bb\ CC\ dd$ and $aa\ BB\ cc\ DD$. Uppercase letters represent dominant alleles that raise fitness and lowercase letters recessive alleles that reduce fitness. The $F_1$ is genetically $Aa\ Bb\ Cc\ Dd$ and is more fit (with dominant alleles at all four loci) than either parent (each with dominant alleles at only two loci). The $F_1$ shows heterosis, therefore, because the dominant alleles received from each parent cover the detrimental effect of recessive alleles received from and expressed in the other parent.

# MUTATION

**23.5** Mutation changes allele frequencies and increases the genetic diversity of a population.

If a population had $A$ as the only allele at a locus, mutation at that locus would have two effects: it would decrease the frequency of $A$, thereby changing allele frequency, and produce new alleles, thereby increasing genetic diversity. Since selection acts upon phenotypic diversity that is often based on genotypic diversity, mutation provides the genetic raw material for selection.

1. NONRECURRENT MUTATIONS. When a complex mutation such as a reciprocal translocation or an inversion occurs, it is unlikely that the identical mutation will occur in the population again. Such mutations are, therefore, essentially unique and nonrecurrent. They originate in the population as heterozygotes. If the heterozygote is more fit than the other alternatives present, the old alternatives will decrease in frequency somewhat, as shown in Figure 27-1, curve $a$. If the heterozygote reduces fitness, it will eventually be lost.

2. RECURRENT MUTATIONS. Simple mutations, such as those that change but a single locus, however, usually occur repeatedly in a population at a rate that is relatively constant. If we consider the mutation from $A$ to any other allele $A^x$, it is clear that the relative increase in the frequency of $A^x$ is greater when $A$ is frequent than when it is rare. In the absence of reverse mutation from $A^x$ to $A$ (and in the absence of any differential selection), of course, all alleles will eventually become $A^x$. If the reverse mutation occurs, it will have negligible effect while the frequency of $A^x$ is small but will be significant when the frequency of $A^x$ is larger. An equilibrium for the frequencies of $A$ and $A^x$ will occur in the population when the gain of $A$ (from reverse mutation by $A^x$) equals the

GENE FREQUENCIES IN POPULATIONS 327

loss of $A$ (from mutation of $A$). The equilibrium frequency of $A(p)$ can be calculated as follows:

$$\frac{\text{rate of gain of } A}{\text{rate of gain of } A + \text{rate of loss of } A} = \frac{\mu}{\mu + v}.$$

For instance, if $\mu = 0.00002$ and $v = 0.00006$, $p = 0.25$.

**23.6** The frequency of a rare detrimental allele will be at equilibrium when its rate of increase by mutation equals its rate of loss by selection.

As already shown, allele frequencies can reach equilibrium in a population when a single disequilibrating factor operates in two opposite directions—selection for and against an allele, or mutation to and from an allele. When two different factors have opposite effects on allele frequency, as often occurs in natural populations, an equilibrium may occur between them. Equilibrium between selection and mutation is considered next.

A rare allele $A$ will reach equilibrium when the rate with which it is gained by mutation ($\mu$) equals its rate of loss by selection. (Since $A$ is rare its rate of loss due to mutation can be neglected.) The rate of loss by selection must equal the selection coefficient ($s$) multiplied by the frequency of the genotypic class being selected against, multiplied by the fraction of its genes that are $A$ (that is $\frac{1}{2}$ for heterozygotes and 1 for homozygotes). That is, at equilibrium,

$$\mu = (s)(2pq)(\tfrac{1}{2}) \quad \text{for } A A^x \text{ heterozygotes}$$

$$\mu = (s)(p^2)(1) \quad \text{for } A \text{ homozygotes.}$$

We consider this equilibrium for dominant and recessive alleles.

1. DOMINANT LETHAL. A dominant lethal $A$ has $s = 1$. When $A$ is at equilibrium, $\mu = (s)(2pq)(\tfrac{1}{2}) = (1)(2pq)(\tfrac{1}{2})$. That is, the mutation rate is $\frac{1}{2}$ the frequency of affected individuals.

2. DOMINANT DETRIMENTAL. In the case of achondroplastic dwarfism (Section 20.4), which is due to a fully penetrant dominant allele ($A$) in heterozygous condition, all the unknowns in our equilibrium equation were determined independently in a particular population. Since the frequency of such dwarfs in the entire population ($2pq$) was 10 dwarf this value, or 0.000042. Such dwarfs ($A A^x$) produced only 20 per cent as many children as normal persons; therefore, $w = 0.2$ and $s = 0.8$. The frequency of such dwarfs in the entire population ($2pq$) was 10 dwarf babies in 94,075 consecutive births scored without regard to parental size, equal to a frequency of 0.000106. Substituting in our equilibrium equation,

$$\mu = (s)(2pq)(\tfrac{1}{2})$$

$$0.000042 = (0.8)(0.000106)(\tfrac{1}{2}).$$

Since the values for both sides of the equation are almost identical, it can be concluded that in this population $A$ is at an equilibrium determined solely, or principally, by mutation and selection. Note, in this population at equilibrium, that the frequency of $A$ heterozygotes is more than twice the mutation frequency; the frequency of $A$ (which is roughly $\frac{1}{2}$ the frequency of $A$ heterozygotes), however, is not very much larger than the mutation frequency, demonstrating the efficiency of natural selection in eliminating such alleles from the population.

3. RECESSIVE LETHAL OR DETRIMENTAL. The allele for *juvenile amaurotic idiocy* ($A$) is a recessive lethal that has no striking phenotypic effect when heterozygous. Affected individuals have $s = 1$ and occur with a frequency of 2 per 100,000, or 0.00002. Assuming that the population is at equilibrium, this rate of removal of $A$ must equal $\mu$, its rate of entry into the population by mutation. The equilibrium equation

$$\mu = (s)(p^2)(1)$$

becomes

$$0.00002 = (1)(p^2)(1).$$

The frequency of $A$ in the population, $p$, must be equal to $\sqrt{0.00002}$, or about 0.004, and that of the $A^x$ alleles must be 0.996. Note from Figure 23-2, as expected for a rare allele, that heterozygotes (*carriers*) are 400 times as frequent as affected homozygotes and carry more than 99 per cent of all the $A$ genes in the population.

By rearranging the terms of the equation, $\mu = (s)(p^2)(1)$, the frequency of any recessive allele in the population at equilibrium can be expressed as $p = \sqrt{\mu/s}$, where $s = 1$ for a recessive lethal. If the recessive had $s = \frac{1}{4}$ instead of 1, $p$ would be twice as large.

4. PARTIALLY DOMINANT. Many alleles that are detrimental when homozygous also produce a few per cent of such a detrimental effect when heterozygous (Section 17.9). In other words, most recessive alleles have a weaker dominant effect, and so are subject to selection in both heterozygous and homozygous condition. When an allele is rare, as in Figure 23-2, most copies of it are present in heterozygotes. Suppose that the heterozygote for juvenile amaurotic idiocy suffers only a 1 per cent reduction in fitness, $s$ being 0.01 (whereas it is 1 in $AA$ individuals). Since there are 400 times as many $AA^x$ as $AA$ individuals, (400)(0.01), or 4 $A$ genes would be eliminated from $AA^x$ individuals for every 2 eliminated from $AA$ individuals. Therefore, in establishing an equilibrium between mutation and selection, most of the elimination of detrimental alleles by selection usually occurs in the heterozygous state.

FIGURE 23-2. Juvenile amaurotic idiocy. See the text for explanation.

|  | $p^2\ AA$ | $2pq\ AA^x$ | $q^2\ A^x A^x$ |
|---|---|---|---|
| Frequency at equilibrium | $(0.004)^2$ | $2(0.004)(0.996)$ | $(0.996)^2$ |
|  | 0.00002 | 0.008 | 0.992 |

# MIGRATION AND GENETIC DRIFT

**23.7** Migration and genetic drift are other factors that can change allele frequencies in populations.

Gene flow or *migration* can also change allele frequencies in populations. If the genotypes of emigrants from or immigrants into a population do not contain, on the average, the same allele frequencies as the resident population, the population allele frequencies will change. The single generation change in frequency of an allele A due to immigration, $\Delta p$, can be calculated as follows: the fraction of all the alleles involved that are contributed by the immigrants ($m$) is multiplied by the difference between the frequency of A in the immigrants ($p_m$) and in the natives ($p_x$). In symbols, therefore, $\Delta p = m(p_m - p_x)$. For example, if a population is composed of 10 per cent immigrants with 0.6 A and 90 per cent natives with 0.4 A, $\Delta p = 0.1(0.6 - 0.4)$, or 0.02. In this case, the frequency of A in the population will increase from 0.40 to 0.42 in a single generation due to immigration.

Selection, mutation, and migration ordinarily act to change allele frequencies progressively and directionally. Allele frequencies can also change nondirectionally, that is, at random, owing to the chance selection of the gametes that serve as the bridge between generations. The allele frequencies of the progeny are exactly like those of the parents only if the population size is infinite; the smaller the population, the greater is the probability and the extent to which progeny allele frequencies will shift by chance from the parental values. Nondirectional change in allele frequencies due to chance variations is called *genetic drift*.

Genetic drift an be illustrated with an extreme example. Such a diploid population contains 0.5 A and 0.5 A' at Hardy–Weinberg equilibrium. If only a single pair of individuals is selected by chance to be the parents of the entire next generation (Figure 23-3), there is only a $\frac{6}{16}$ probability the allele frequency will be unchanged. There is, however, a $\frac{1}{16}$ chance of permanently fixing allele frequency at 1.0 A and 0 A', a $\frac{1}{16}$ chance of the opposite fixation, and an $\frac{8}{16}$ chance of shifting the allele frequencies to 0.75 A and 0.25 A' or the reverse. Of course, allele frequencies will also drift if, by chance, different parents contribute unequally to the next generation. Because it can fix allele frequencies at 1.0 or 0, genetic drift tends to reduce allele and genotypic variability in a population.

Different alleles that, owing to degeneracy, code for the same polypeptide or that code for different polypeptides in which certain amino acids are substituted by others with similar properties may have the same (or almost the same) fitness. In this case they exhibit no difference in their selection coefficients and no difference in their adaptiveness; that is, their difference is adaptively neutral. Since they are not differentially subject to selection, the relative frequencies of such adaptively neutral alleles in the population is determined only by mutation, migration, and genetic drift.

| Parental Mating | Probability | Gene Frequency in Progeny Population ||
|---|---|---|---|
| | | A | A' |
| AA × AA | $\frac{1}{4} \times \frac{1}{4} = \frac{1}{16}$ | 1.0 | 0 |
| AA × AA'<br>AA' × AA | $\left.\begin{array}{l}\frac{1}{4} \times \frac{1}{2}\\ \frac{1}{2} \times \frac{1}{4}\end{array}\right\} = \frac{4}{16}$ | 0.75 | 0.25 |
| AA × A'A'<br>A'A' × AA<br>AA' × AA' | $\left.\begin{array}{l}\frac{1}{4} \times \frac{1}{4}\\ \frac{1}{4} \times \frac{1}{4}\\ \frac{1}{2} \times \frac{1}{2}\end{array}\right\} = \frac{6}{16}$ | 0.50 | 0.50 |
| AA' × A'A'<br>A'A' × AA' | $\left.\begin{array}{l}\frac{1}{2} \times \frac{1}{4}\\ \frac{1}{4} \times \frac{1}{2}\end{array}\right\} = \frac{4}{16}$ | 0.25 | 0.75 |
| A'A' × A'A' | $\frac{1}{4} \times \frac{1}{4} = \frac{1}{16}$ | 0 | 1.0 |

FIGURE 23-3. Genetic drift when only one pair of parents contributes to the next generation. The parent population is at equilibrium, containing 0.5 A and 0.5 A'.

## SUMMARY AND CONCLUSIONS

Allele frequencies, and therefore genotype frequencies, can be shifted from their Hardy–Weinberg equilibrium values by selection, mutation, migration, and genetic drift.

Selection favors those phenotypes and hence genotypes that give the population the greatest fitness. Selection may occur in opposite directions in the same individual at different times; it may occur in the same direction with different strengths for homozygotes and heterozygotes; and it may favor heterozygotes for one or many loci more than their homozygotes to produce heterosis.

Mutation not only changes allele frequency but provides the genotypic variability upon which selection acts. Allelic diversity is maintained by mutation and reverse mutation. Rare detrimental alleles reach an equilibrium frequency in the population when their rate of increase by mutation equals their rate of loss by selection.

Just as selection and mutation, migration can change allele frequencies directionally. Genetic drift, however, shifts allele frequencies at random, nondirectionally. In very large populations genetic drift is not important. As the size of the population decreases, however, genetic drift is increasingly important and the greater is its tendency to reduce allele and genotypic diversity.

## QUESTIONS AND PROBLEMS

23.1 *Anagrams*: Unscramble and define each of the following: **a.** cliontees. **b.** sniftes. **c.** tirehoses. **d.** aimorting.

23.2 *Song*: "Only the Strong Survive." Do you agree? Explain.

# GENE FREQUENCIES IN POPULATIONS

**23.3** What are the four main factors responsible for changes in allele frequency in populations? Which of these factors make such changes in a directional or nonrandom way?

**23.4** If a population is at Hardy–Weinberg equilibrium at one locus, is it automatically true at another locus? Explain.

**23.5** What effect do you think the Napoleonic wars had on the average stature of the French people? To what would you attribute this?

**23.6** What is the equilibrium frequency of $R$ when it is lost by mutation at a frequency of 0.00004 and is gained by mutation at a frequency of 0.00008?

**23.7** Under what specific circumstances has allele $\beta^S$ $w = 1$ and $w = $ less than 1?

**23.8** Multiple-gene heterosis has been likened to wearing two moth-eaten bathing suits at the same time. In what respects is this a good or a bad simile?

**23.9** What is the $s$ value of heterozygotes for rare dominant gene $X$ whose increase in frequency due to mutation is 0.000018 when the heterozygotes occur with a frequency of 0.000054 at equilibrium?

**23.10** What is the expected frequency at equilibrium of a completely recessive allele whose selection coefficient is 0.09 and whose increase by mutation occurs at a frequency of 0.00036?

**23.11** What would $w$ be for a completely recessive gene whose increase in frequency due to mutation was 1 per million and whose equilibrium frequency in the population was **a.** 0.0001?  **b.** 0.01?

**23.12** How frequent would be the homozygous recessives in each case in the previous question?

**23.13** What would be the mutation rate at equilibrium that increases the frequency of a completely recessive allele expressed in 0.02 per cent of persons whose fitness is: **a.** 0.1?  **b.** 0.5?  **c.** 0?

**23.14** A lunar population of 80 persons with frequencies of 0.7 $A$ and 0.3 $a$ receives 20 immigrants with frequencies of 0.3 $A$ and 0.7 $a$. What is the immediate change in the frequency of the two alleles in the expanded population? What are the new allele frequencies?

**23.15** A tropical island paradise has a human population in which $A$ has a frequency of 0.5 before receiving immigrants, and 0.56 after. If immigrants make up 20 per cent of the total population, what was the frequency of $A$ among them?

# Chapter 24

# Genetic Variability of Populations and Speciation

WE HAVE DESCRIBED the four factors that are the principal causes of changes in allele frequencies in a population: mutation, selection, migration, and genetic drift. Since the evolution of a population is based upon the evolution of its genotypes, these four factors are the principal causes of biological evolution.

The present chapter deals primarily with the extent of genetic variability in a population, its disadvantages and advantages, and its role in the evolution of new species.

**24.1** Natural populations contain a great deal of genetic variability.

At first glance all the wild-type individuals of a single population—be it of flies, oak trees, or giraffes—seem phenotypically alike. Such individuals are, however, not genotypically alike. On the contrary, a population usually contains a great deal of concealed genetic variability. This genetic variability is revealed three ways:

1. *Cytologically.* Microscopic examination of chromosomes shows that many populations contain a variety of gross chromosomal rearrangements, including reciprocal translocations, inversions, deletions, and duplications.
2. *Effect on fitness.* Wild-type individuals can be tested to determine how often they are heterozygotes for recessive alleles that reduce fitness when homozygous. The results of one such study of *Drosophila* are typical (Figure 24-1). When autosomes were tested, about 25 per cent carried a recessive lethal or semilethal allele, about 40 to 90 per cent carried subvital alleles, and 4 to 14 per cent carried alleles causing sterility. Obviously, natural populations carry a large number of recessive detrimental alleles.
3. *Effect on proteins.* A population may contain two or more amino acid sequences for the same protein, coded by an equal number of different alleles. The different proteins can be separated from each other, and hence

FIGURE 24-1. Recessive detrimental mutants in natural populations of *D. pseudoobscura*. (After Th. Dobzhansky.)

| Mutant Type | Percentage of Chromosomes |||
|---|---|---|---|
|  | 2 | 3 | 4 |
| Lethal or semilethal | 25 | 25 | 26 |
| Subvital | 93 | 41 | 95 |
| Female sterile | 11 | 14 | 4 |
| Male sterile | 8 | 11 | 12 |

counted, by electrophoresis—by being placed on a gel and subjected to an electric field—provided that they differ slightly in molecular weight and/or electric charge. A study of xanthine dehydrogenase, for example, revealed that 33 different forms, identified by electrophoresis and temperature sensitivity, occur in certain natural populations. In different investigations, 43 and 58 per cent of all the protein-coding loci tested electrophoretically had two or more alleles. The individuals in such studies were heterozygous for about 12 per cent of their protein-coding genes.

The preceding discussion demonstrates that natural populations exhibit a great deal of variability in the chromosomal arrangements and alleles that they carry.

## 24.2 Factors that decrease the fitness of a population produce a genetic load that causes genetic death.

Various factors operate to reduce the adaptive genes and genotype frequencies of a population. Any factor that reduces the fitness of a population is said to produce a genetic burden or *genetic load*. The ultimate origin of all genetic loads is mutation. Mutations are constantly occurring in a population and increasing genetic variability. Since most mutations lower fitness, mutation usually increases the genetic load.

The reduced fitness caused by genetic load can be expressed as an increased chance of *genetic death*, the failure to reproduce. Natural selection operates to remove detrimental alleles from the population by genetic death. The number of generations required to remove a detrimental allele from the population is inversely related to its selection coefficient. For example, a person who carries a dominant lethal allele has $s = 1$ and this gene is eliminated from the population in the generation it arises by mutation. It persists, therefore, only one generation. On the other hand, a dominant detrimental allele with $s = 0.2$ will persist for five generations, on the average, before suffering genetic death. That is, if the population size is constant in successive generations, in each generation the heterozygotes have a 20 per cent chance of not transmitting the detrimental allele. When such an allele arises by mutation, it sometimes fails to be transmitted in the very first generation. It may suffer genetic death in the fifth generation or in the tenth, but, on the average, the allele persists for five generations. This principle of *persistence* is valid even though genetic drift or migration may cause fluctuations.

Each rare, detrimental allele is equally harmful to a constant-sized population, in the respect that each eventually causes a genetic death. Thus, a gross chromosomal abnormality with $s = 1$ causes a genetic death after one generation, and a base substitution mutant with $s = 0.001$ causes a genetic death after persisting, on the average, 1000 generations.

**24.3** **The genetic variability of a mutational load is adaptive for the future evolution of a population.**

Under a given set of environmental conditions, the great majority of mutations in a population are detrimental. Perhaps only 1 mutant in 1000 increases the fitness of its carrier. Yet, provided the mutation rate is not too large and sufficient genetic recombination occurs, these rare, beneficial mutants offer the population the opportunity to become better adapted to their environment.

Mutations also provide the raw materials needed to extend the population's range to different environments, whether such environments are a question of space or of time. A population already well adapted to its immediate environment is appreciably harmed by the occurrence of mutation. But environments differ, and any given environment will eventually change. A nonmutating population, although successful at one time, will, in the normal course of events, eventually face extinction. A mutational load, therefore, is the price paid by a population for future fitness to the same or different environments.

**24.4** **A population whose genetic constitution differs significantly from that of other populations of the same species is called a race.**

Any wild population is genetically variable for many loci, and does not become genetically pure or homozygous with the passage of time. Thus, different populations of the same species that are located in different parts of the world may differ both in the types and frequencies of the genetic alternatives they carry. Two populations can be called *races* when two more or less isolated populations of the same species have significantly different, characteristic genetic constitutions.

Since a race does not have a uniform genotype, it is defined by the relative frequencies of the genetic alternatives it contains. (Without a uniform genotype, a race cannot have a uniform phenotype. Accordingly, it is futile to try to picture a typical member of any race.) In practice, the number of races recognized is a matter of convenience.

## 24.5 The genetic variability that differentiates races is often adaptive.

The environment is not uniform in different parts of the territory occupied by a species. Clearly, then, no single genotype will be equally fit in all the different environments encountered within the range of a species. One way in which a species can achieve and maintain maximum fitness is to remain genetically variable and to separate into geographical populations or races that differ genetically and that are adapted to different environments. An example of the adaptiveness of genetic differences between races follows.

The British peppered moth, *Biston betularia*, can be divided into races, a light-colored one and a dark-colored one. The dark pigmentation is due to a single allele whose dominance is modified by the action of nonalleles. The dark-colored race is prevalent in central England, where high industrialization pollutes the air with soot, as well as in eastern England, which is downwind. In the less industrialized north and south of England, however, the light race predominates. The color of the light moth is very much like that of lichens growing on tree trunks in unpolluted areas. The color of the dark form is very much like that of soot-covered trees whose lichens have been killed by pollution. When body color matches the tree background, the moth is protected from bird predators. These deductions have been substantiated by experiments in which equal numbers of light and dark moths were released in unpolluted and in soot-polluted woods. Predation patterns were visually observed, and counts were taken of moths that were later recaptured. In unpolluted woods, many more light moths survived; in polluted woods, the situation was reversed.

As one might expect, the frequency of the dark race in a given region has varied with the amount of air pollution. In the Manchester area, for example, the dark race was less than 1 per cent of the population about the year 1850. As the result of increased pollution, however, it accounted for more than 99 per cent in 1900. The recent increase in the frequency of the light race in Manchester is correlated with efforts at smoke control.

## 24.6 Different races of a species may undergo genetic changes that cause partial reproductive isolation between them.

A species usually consists of races that are adapted to the different environments of the territories that they occupy. All the races are kept in genetic continuity by interracial breeding. Sometimes so much interracial breeding occurs that two races can become a single race that occupies a larger territory. Othertimes, most of the matings are intraracial. In this event, in the course of time, the differences in the genetic constitutions of different races resulting from mutation, selection, and genetic drift may increase faster than interracial hybridization erases them.

As such changes accumulate, the alleles that adapt each race to its territory may, by their multiple phenotypic effects, make matings between two races less likely or may cause the hybrids of such matings to be less fit than the members of either parental race. In this way, partial reproductive isolation between races may be initially an accidental or incidental by-product of the adaption of genotypes to a given environment.

### 24.7 Different races become different species when reproductive isolation is sufficient to keep their genetic constitutions separate.

The accumulation of alleles that tend to reproductively isolate races is usually aided first by space and time barriers between races. Such barriers to interracial mating include:

1. *Geography.* Water, ice, mountains, deserts, wind, earthquakes, and volcanic activity may separate races.
2. *Ecology.* The habitats of different races may differ in temperature, humidity, sunlight, food, predators, and parasites.
3. *Seasons.* Different races may become fertile at different seasons.

Races that become genetically adapted to these conditions then tend to become reproductively isolated in the following ways:

4. *Form.* Different races may not interbreed because their sex organs are incompatible or because they use different body marks as sexual clues.
5. *Behavior.* One race may be active only during the day and another only at night, so that they never encounter each other.
6. *Physiology.* Failure of one race's cells to fertilize those of another, so that the hybrid zygote is formed infrequently.
7. *Hybrid inviability.* Even when formed, the development of hydrbid zygotes may be so abnormal that it can rarely be completed.
8. *Hybrid sterility.* Even when hybrids complete development and are hardy, they may be sterile.

As these reproductive barriers help different races to diverge genetically, so that they are more and more reproductively isolated from each other, two races may eventually attain such separate and different genetic constitutions that instead of being two races of the same species, they are two different species. A cross-fertilizing *species* can be defined as all populations that can interbreed with each other but that maintain a genetic constitution different from all other groups by means of gene-based reproductive isolation. Note that such species formation, or *speciation*, is almost always an irreversible process. Once a species is established, it rarely loses its identity, even though it may be able to cross-breed with another species. The formation of species by reproductive isolation of races seems to be the most common method of speciation by cross-fertilizing organisms.

A study of various closely related species shows that any two are separated by several to many reproductive barriers, each of which is usually due to different alleles at many loci. Since difference in form is only one isolating mechanism, it is not surprising that closely related species may still look very much alike. Finally, when the territory of two closely related species is shared in whole or in part, natural selection favors genotypes that prevent mating between them, thereby reducing the waste of their gametes.

### 24.8 Some species are founded by one or a few individuals.

Not all species of cross-fertilizing organisms arise from the differentiation of populations into races, then species. Under special circumstances some species are founded by one or a few individuals. This *founder principle* seems to apply to terrestrial species that have formed on volcanic islands. For example, many species of *Drosophila* in the Hawaiian Islands seem to have originated from single fertilized females that came from other islands or from the mainland. Such founder individuals contain, of course, only a portion of the genetic variability of their parent populations, and may not be particularly well adapted to their new habitat. Nevertheless, since the new habitat is permissive, the population quickly swells (and then shrinks because of overpopulation). As a result of geographic isolation and genetic drift, a new species could occur soon after the founding event, after which mutation and selection could act to increase the fitness of the new species.

### 24.9 Speciation can occur by single-species polyploidy.

Single-species polyploids are highly sterile because the gametes they produce contain unbalanced sets of chromosomes. Nevertheless, if such individuals are plants that are perennials or that can be propagated asexually by budding or grafting, they can successfully form new species. The chromosome numbers of different species of the same genus indicate whether single-species polyploidy has occurred. For example, in the genus *Chrysanthemum*, species occur with chromosome numbers of 18, 36, 54, 72, and 90, indicating that a set of 9 chromosomes has undergone polyploidy. In the genus *Solanum* (the nightshades, including the potato), species occur with chromosome numbers of 24, 36, 48, 60, 72, 96, 108, and 144 chromosomes, indicating that a set of 12 chromosomes has undergone polyploidy.

### 24.10 Speciation can also occur after interspecific hybridization.

New species may originate not only from races or individuals of a single species, but after hybridization between individuals from different species. The interspecific hybrid, however, particularly in plants, may be converted into a stable, phenotypically intermediate species that is reproductively isolated from the parental species by three methods, as follows:

1. MULTIPLE-SPECIES POLYPLOIDY. The production of a multiple-species polyploid from an interspecific hybrid has been described in

Section 12.1. That section and Section 21.5 notes the success this mechanism has had in producing new species of flowering plants.

2. STABILIZING RECOMBINATIONS. If the chromosomes of the two species forming the hybrid are very similar, they may be able to synapse in pairs in the $F_1$ hybrid. In this case, the $F_1$ will be fertile. Segregation, independent segregation, and crossing over may yield recombinant progeny of the hybrid that are stable and isolated from both parental species. In the larkspur genus, for example, *Delphinium gypsophilum* arose as a hybrid between *D. recurvatum* and *D. hesperium*—all three species having 8 pairs of chromosomes.

3. INTROGRESSION. In this process the interspecific hybrid crosses with one of the parental species. As the result of such *introgression* the recombinant progeny favored by natural selection may contain genetic material from both species, may be true-breeding, and, eventually, may become a new species. The evolution of maize through artificial selection was aided by genes incorporated by introgression from teosinte, *Zea mexicana*.

## SUMMARY AND CONCLUSIONS

Natural populations contain a great deal of genetic variability. This variability originated ultimately in mutation. Since most mutations decrease fitness, they produce a mutational load. Natural selection operates to reduce this load from the population by genetic death. The persistence of a detrimental allele is inversely related to its probability of causing genetic death. Despite the detriment associated with a mutational load, genetic variability is adaptive for different races and for the future evolution of a population.

Races are populations of the same species that have characteristic genetic variabilities. Different races may differentiate genetically, aided by reproductive barriers, and will become different species when reproductive isolation is complete enough to keep their genetic constitutions separate.

Although most species have a racial origin, some are founded after one or a few individuals are isolated in a new habitat; others result from single-species polyploidy. Still other species occur when interspecific hybridization is followed by polyploidy, by selection of recombinants, or by selection of individuals produced after introgression.

## QUESTIONS AND PROBLEMS

24.1 *Movies*: "Star Wars." What are the genetic applications or implications of this movie?
24.2 *Movies*: "Rosemary's Baby." What are the genetic applications and implications of this movie?
24.3 *Movies*: "King Kong." What is the genetic implication of this movie?
24.4 Name three different methods of studying the genetic variability of populations.

# GENETIC VARIABILITY

**24.5** Can the mutation rate be too high to provide a population with the genetic variability it needs for future adaptability? Explain.

**24.6** What is the difference between the specific environments that the dark and light races of the British peppered moth are adapted to?

**24.7** Does a genetic death necessarily produce a corpse? Explain.

**24.8** Does an allele whose $s = 0.1$ cause less, equal, or more total suffering than one whose $s = 0.8$? Explain.

**24.9** Do pure races of cross-fertilizing organisms occur? Explain.

**24.10** Are races and/or species artificial or natural classifications? Explain.

**24.11** Do races have to be kept apart for a period of time before they can become different species? Explain.

**24.12** State two different ways that new species may originate without being derived by the differentiation of races.

**24.13** In the early 1880s, a species of American marsh grass (diploid number = 62) was accidentally introduced in Europe, where it grew near a species of European marsh grass (diploid number = 60). By the early 1900s a new marsh grass species (chromosome number = 120, 122, or 124) appeared that looked intermediate between the two older species. How can you explain the origin of this new species?

**24.14** Mexican Indians still plant a little teosinte with their maize, believing it to bring good luck. What do you think about this tradition?

# PART VII

# THE PRESENT AND FUTURE CONSEQUENCES OF GENETICS

*Chapter 25*

# Applications to Agriculture and Ecology

THE PRECEDING PORTIONS of this text dealt with the basic principles of the science of genetics. They have not discussed the very important contributions that genetics has made, is making, and will continue to make to our lives, our civilization, and our planet. Some of the ways that genetics applies to agriculture and ecology are described in this chapter. The next two chapters will discuss some of the ways genetics applies to other fields.

## GENETICS AND AGRICULTURE

**25.1** Domesticated crops have been, are, and probably will be our main source of nutrition.

The farmers of 10,000 years ago were women who grew all the cereals and many of the legumes we grow today (Figure 25-1). About that time the human population on earth was about 10 million. It was relatively easy to grow enough food for one's own family even though the yield of a crop was relatively small. The main concern was to make sure there was something to harvest—to be sure no calamity befell the entire crop. The farmers learned that the best way to assure long-term survival of a crop was to maintain adequate diversity so that, on the average, local populations were adapted to survive the local conditions of soil, climate, and

FIGURE 25-1. Crops grown in prehistoric times.

Cereals
    wheat, rice, sorghum, barley, potato, cassava, taro, yam, sweet potato

Legumes
    bean, soybean, peanut, pea, chickpea, lentil

pests. In today's terms, the original farmers had no need for inbreeding or other artificial selection but encouraged hybridization.

It took the Stone Age population thousands of years to double, eventually yielding our present population of 3.8 billion persons. In the present U.S. population one farmer, plus two off-farm support workers, produces food for about 48 people. The great increase in crop productivity during the present century is due largely to the application of selective breeding and artificial selection, increased acreage, use of fertilizer, and better management (harvesting, storage, and transportation).

The present population of 3.8 billion will double in about 35 years—the most overpopulated and poorest nations will double their population in 21 to 25 years, whereas a relatively rich nation like Sweden will take 140 years to do so. In a relatively rich nation, the fraction of the total income spent on food has steadily decreased during this century (Figure 25-2). A rise in the price of food is undesirable but not as critical in a rich nation as it is in poor Third World nations, where 1 billion people spend 80 per cent of their income on food. Food or fertilizer supplies need decrease only slightly to produce drastic price increases. For example, a shortage of 3.8 per cent drove up the price of wheat from $1.78 to $6.40 per bushel. Persons in the United States have recently had a drastic increase in the cost of orange juice and coffee due, at least in part, to crop failures.

Many nongenetic steps can be taken to increase crop productivity and hold down the cost of food. These include increasing acreage (but space on earth is limited), bettering management, and increasing fertilizer production. We can readily understand the importance of fertilizer since it is mostly nitrates (needed to supply the nitrogen to make each amino acid and organic base) and phosphates (needed to make each nucleotide). Unfortunately, it requires fuel to make fertilizer, and fertilizer demand

FIGURE 25-2. Per cent of income spent on food in the United States.

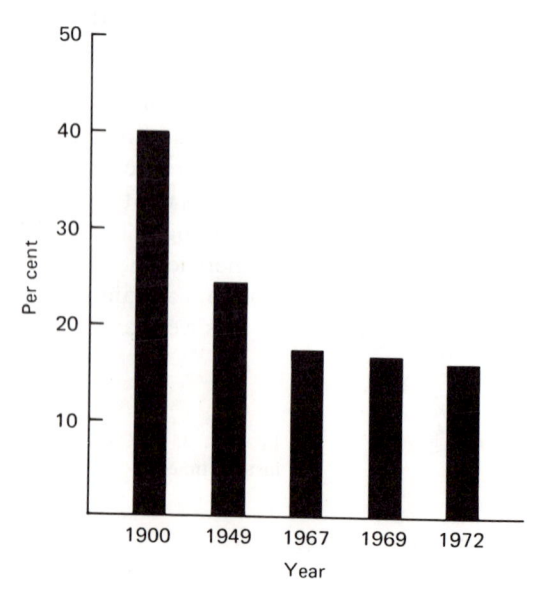

has increased faster than supply. In the next few sections we will describe how selective breeding and artificial selection have increased the productivity (or beauty) of plants in recent times and how these and other genetic principles can be applied to plants in the future. Since it takes 3 to 10 pounds of grain to produce a pound of farm animal meat, plants have been and will continue to be the cheapest source of food. Nevertheless, animals are important as food, as workers, and as hobbies and pets; and we will also describe the use of genetic principles in the breeding of domesticated animals. Of course, one way to help meet the impending food shortage is to decrease the size of the human population by birth control.

**25.2** **Based on naturally occurring genetic variability, controlled breeding and artificial selection have increased the usefulness of many plants.**

Various principles and methods of genetics have been applied in the production of plants useful to human beings. In most cases, the genetic variability of a species in the wild is so great that controlled breeding and artificial selection can be used to increase the organism's usefulness. A common system for increasing the usefulness of a species is to combine moderately close inbreeding with fairly rigorous artificial selection. This produces many distinct and genetically uniform varieties, each of which contains the desired (and perhaps some undesired) traits. In some cases plant breeders have gone one step further, crossing varieties to produce genetically uniform progeny in which undesirable traits are masked and hybrid vigor is displayed. A few examples of the success of breeding systems in plants are given next—only a minor portion of the gains cited being attributable to better nutrition or management.

1. MAIZE. Indian corn is normally a self-fertilizing species. The pollen produced by tassels located at the top of the plant fall on the silks leading to an ovary located lower down on the same plant. The ovary then develops into an ear. Simple selection of self-fertilized plants has been used to obtain lines that differ in oil content. After 50 generations of selection for low and high oil content, the percentage of oil was lowered in the low line from 5 to 2 per cent and raised from 5 to 15 per cent in the high line.

The early maize breeders, recognizing the advantages of cross-fertilization, used to remove the tassels of some plants by hand to assure some cross-fertilization. Today hand detasseling has been replaced by introducing genes for male sterility into some lines. The heterosis obtained by cross-fertilizing maize became systematized when the "double-cross" method of producing hybrid corn was introduced in 1917.

At that time, the corn produced by hybridizing local varieties differed in different regions in their good and bad qualities with respect to traits such as ear size, kernel taste and texture, development time, resistance to wind, drought, rusts, smuts, molds, and insects. Four strains from

346   CONSEQUENCES OF GENETICS

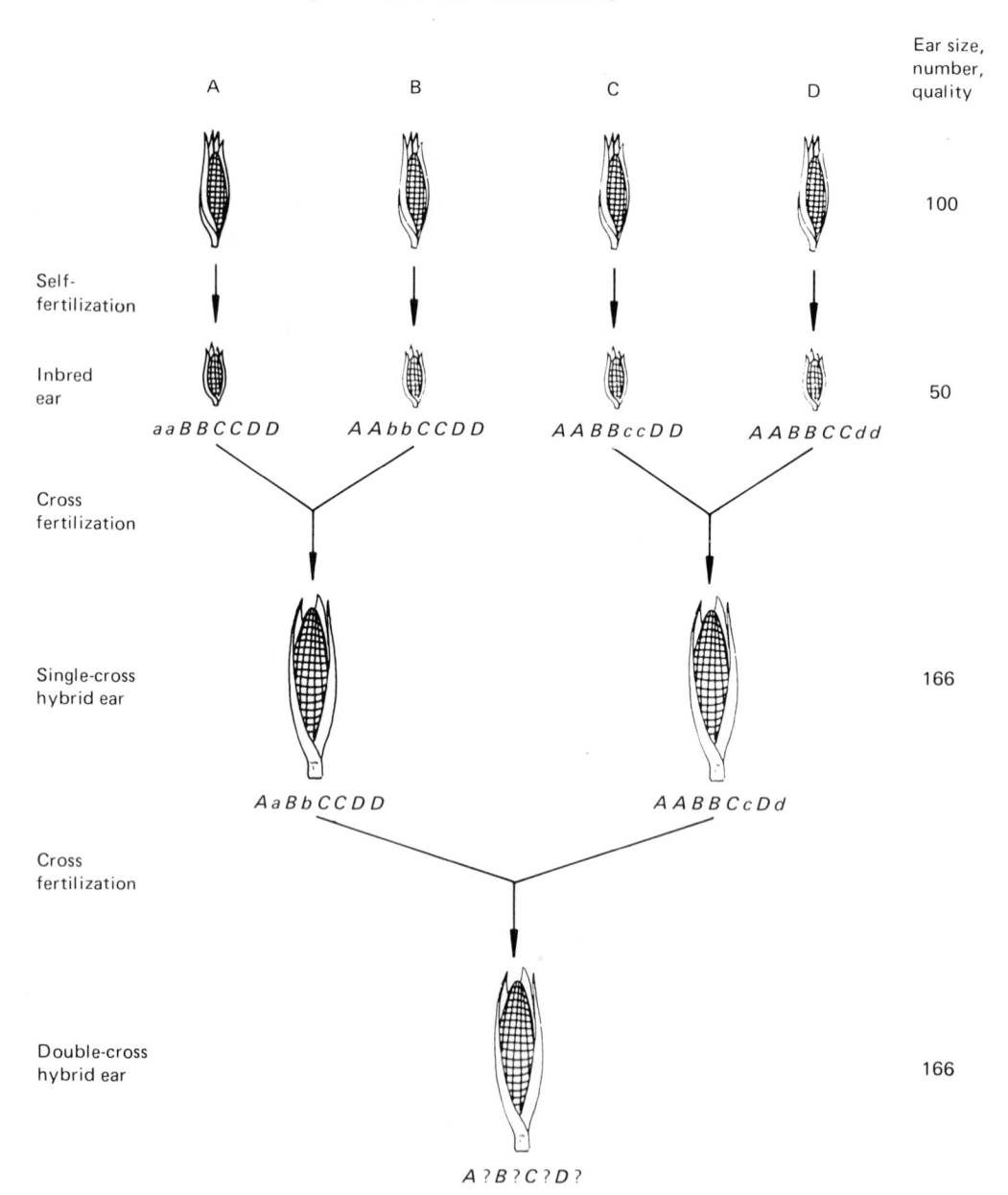

FIGURE 25-3. Production of commercial hybrid maize seed by the "double-cross" breeding procedure.

different regions were chosen (A, B, C, and D, Figure 25-3) and permitted to self-fertilize for a number of generations to become homozygous for genes for different specific desirable traits. The result was four inbred lines whose ear size, number, and quality was only 50 per cent of the original (Figure 25-3)! (This result is an advance made by going backward.) This inbreeding had produced four uniform lines homozygous for different dominant, desirable genes as well as being homozygous for different undesirable recessive genes. One way to overcome the detrimental effects of the recessives while at the same time obtaining genotypic uniformity is to cross-fertilize inbred A with inbred B and inbred C with inbred D, to produce single-cross hybrid ears (Figure 25-3). These single-cross hybrid ears were 166 per cent better than the original strains because of multiple-gene heterosis (Section 23.4). This can be readily appreciated by assuming that each of the four inbred lines was homozygous for a different detrimental recessive gene ($a, b, c, d$). (In fact, of course, each inbred line was homozygous for many different detrimental recessives.) All single-cross hybrid progeny would show heterosis because they contained dominant alleles at all four loci (being $Aa\ Bb\ CC\ DD$ or $AA\ BB\ Cc\ Dd$).

Although the kernels grown from single-cross hybrid ears are vigorous and uniform, they come from ears grown on a much less vigorous inbred line. For this reason, hybrid seeds are not sufficiently numerous and hence are too expensive to sell commercially. This difficulty is overcome by crossing the two single-cross hybrids to each other to produce double-cross hybrid ears (Figure 25-3). These ears will be just as heterotic as their parents since, as in our example, they too will express none of the detrimental recessives. Moreover, since these ears are produced on a vigorous single-cross hybrid parent, their seeds are plentiful and can be sold inexpensively. By 1949, 78 per cent of all corn grown in the United States was hybrid corn.

The production of such hybrid corn seed has had some very important consequences. It helped teach farmers in the United States (and some foreign governments, too) to change not only their breeding systems but their irrigation and fertilizer programs with respect to maize and other crops. Hybrid corn helped the United States win World War II by producing the same amount of food with fewer farmers, by helping to feed her allies, and by being used to make alcohol, synthetic rubber, and explosives. By providing food after the war, hybrid corn had the political effect of preventing some European governments from going communistic. In recent years, the heterosis of hybrid corn has increased the annual yield 1 billion bushels in the United States alone. Our excess corn is either sold to foreign countries such as the USSR, whose crops have had a shortfall, or is given away as emergency food to poor countries or in cases of natural disaster. Finally, the principles of heterosis have been applied to obtain other hybrid crops, such as radishes, tomatoes, barley, wheat, and zucchini.

2. WHEAT. By selective breeding of more than 50 varieties of wheat (usually *Triticum aestivum* or *T. durum*), in which hundreds of genes were placed in desired combinations, improved yields of 30 to 100 per cent have been obtained. For example, genes from a semidwarf variety of Japanese wheat were introduced into Mexican wheat and made the following improvements: (1) shorter and stiffer straw, which means less growth is needed before flowering and greater wind resistance; (2) greater adaptability to the region; and (3) greater resistance to disease and insects. In another example, the yield of Swedish winter wheat

increased 25 per cent after winter-hardy varieties were crossed with those offering high yield, and the progeny were selected for winter hardiness and yield. Also selection for early ripening has extended the northern range of spring wheat.

3. RICE. Hybridization and selection has produced what has been called miracle rice. The usual Indica variety of rice grown in the tropics has wide leaves that shade the leaves of other plants as well as the lower leaves of the same plant. An Indica type that has short upright leaves and stem, called Dee-geo-woo-gen, was crossed to another type that is not tasty, called Peta. Subsequent selection yielded miracle rice (IR8). The yield of miracle rice in the tropics is four to six times the average yield of the Indica variety.

4. SUGAR BEETS. The sugar content of sugar beets has been increased from 6 per cent in 1818 to over 20 per cent today by means of selective breeding.

5. ORNAMENTAL PLANTS. Ornamental plants exist in a profusion of varieties produced by selective breeding.

## 25.3 Plant usefulness has been increased by polyploidy, introgression, and mutagenesis.

Additional genetic variability of plants has been induced through single-species polyploidy, multiple-species polyploidy, introgression, and mutagenesis. Examples are given of the use of each of these methods to produce new types of plants.

1. SINGLE-SPECIES POLYPLOIDY. This kind of polyploidy can occur naturally or it can be induced by mutagens. Moderate increases in the number of chromosome sets, such as a doubling or quadrupling, are usually accompanied by increased cell and plant size, and sometimes by more intense flower color. Such plants grow slower, however, and owing to the formation of chromosomally unbalanced gametes, have reduced fertility.
    a. *Apples*. The Gravenstein and Baldwin apples are triploids that are propagated asexually by budding or grafting.
    b. *Sugar beets*. Triploid sugar beets are high in sugar content and are resistant to mold.
    c. *Watermelon*. A cross between the tetraploid (44 chromosomes) watermelon ♀ and the diploid (22 chromosomes) watermelon ♂ produces a fruit with fewer seeds than the diploid produces. These seeds are triploid (33 chromosomes) and when grown produce watermelons with few or no seeds. (It is amusing to note that one company selling triploid seed is called the American Seedless Watermelon Seed Corporation.) In a recent year, 10 per cent of all watermelons grown in Formosa and the

United States were triploids. This fraction is expected to increase since new triploid varieties are disease-resistant and early-maturing.

   d. *Oranges*. Some seedless oranges are produced by pollinating normally cross-fertilizing diploid females with pollen from a tetraploid.

   e. *Red clover* and *alfalfa* have improved varieties as a result of polyploidy.

   f. *Tulips*. Triploid tulips are propagated asexually.

   g. *Ornamental plants* such as *narcissus*, *petunia*, and *cyclamen* have polyploids with brighter flower colors.

2. MULTIPLE-SPECIES POLYPLOIDY. Several advances have made it easier to obtain fertile progeny from interspecific crosses. The initial hybrid cell can be obtained by fertilization if the ovary's mechanism for rejecting foreign pollen has been suppressed by drugs. Growth of the hybrid is improved by culturing it *in vitro*, and fertility is aided by treating the plantlet with colchicine to double the chromosome number. In this way, a relatively high proportion of survivors can be obtained from such wide crosses as wheat × barley, barley × rye, and wheat × wild grass.

   a. *Truticale*. This is a relatively new cereal grain obtained by crossing wheat with rye. It combines the high yield of wheat and the drought and disease resistance of rye. Whereas wheat is a relatively poor source of lysine for human nutrition, *Triticale* produces 2 per cent more lysine than wheat. By 1973, over 1 million acres of *Triticale* were planted worldwide.

   b. *Strawberry*. The strawberry now cultivated (*Fragaria grandiflora*) arose in Europe in the mid-1700s as the polyploid of a hybrid between North and South American species.

   c. *Loganberry*. Hybridization of a raspberry and a blackberry produced the loganberry in California in the 1880s.

3. INTROGRESSION. Ordinary wheat is a multiple-species polyploid susceptible to leaf rust disease. It is possible to cross rust-resistant goat grass (*Aegilops*) with wheat and to carry out artificial introgression. That is, a series of crosses of the hybrid back to wheat is carried out. The result is wheat that contains segments of goat grass chromosomes that carry genes for leaf-rust resistance.

4. MUTAGENESIS. X rays and chemical mutagens have been used to induce mutations in such crop plants as maize, barley, and oats. Selection of X-ray-induced mutants in barley, for example, has resulted in such beneficial effects as earlier ripening, stiffer straw, and a higher yield. Since so few mutants are advantageous, however, screening of large organisms (such as crop plants or domesticated animals) for them can be very expensive. On the other hand, selection of mutants for increased production of antibiotics, alcohol, or other organic compounds by microorganisms, such as bacteria or fungi, has been very successful. Finally, the possibility exists that plants (and animals) may be infected with organisms or DNA's whose protein products increase the usefulness of the host. Such genetic manipulations are discussed in Section 27.12 in relation to genetic engineering.

## 25.4 The usefulness of many domesticated animals has increased due to selective breeding.

A variety of animals have been domesticated and and bred for many different purposes. We will mention the result of selectively breeding some animals.

1. DOGS. The farmer of 10,000 years ago had already domesticated the dog. Since then people in different parts of the world have selectively bred their dogs for a variety of purposes. When groups of people were well spaced apart, as the American Indians were, each had its own breed of dog. Dogs were bred for guarding and protecting, for hunting, for herding, for companionship, for eating, and as beasts of burden. The more than a hundred breeds recognized today all belong to the same species and testify to the great genetic variability already present and arising in a species in the wild.

2. CATTLE. All the main animals reared for meat were domesticated 10,000 years ago. In London, by 1710 a young bull weighed on the average 370 pounds; by 1795, after selection, it weighed on the average over 800 pounds. Today a Chianina bull of Italy weighs 4000 pounds. Inbreeding in cattle is facilitated by artificial insemination. Since female cattle have so few offspring in their lifetime, it is not practical to select them for future breeding on the basis of the quality of the progeny they produce. A bull, however, can sire a very large number of progeny by means of artificial insemination. Sperm can be diluted, stored in a deep freeze for many years, and still be viable when thawed. The bulls used for breeding are usually those that produce desirable kinds of female offspring. Inbreeding may be achieved through the fertilization of several generations of cows with semen from the same bull.

Two specific results of selective breeding of cattle are cited. The butterfat content from Friesian cows in the Netherlands has been increased about 20 per cent in this century. A superior new breed of beef cattle, the Santa Gertrudis, has been obtained by selecting among the descendants of crosses between two old breeds, the English shorthorn and the Indian Brahma. The new breed is able to thrive in hot, dry areas as well as in hot, humid ones. It is also resistant to ticks, insects, and parasites; and grows well when fed only on grass.

3. SHEEP. In Australia, selective breeding has increased the weight of wool per sheep from about 5 pounds in the 1880s to about 8.5 pounds in the 1960s.

4. POULTRY. Selective breeding has doubled the average annual egg production per hen from about 125 in 1933 to about 250 in 1965.

5. FISH. Tropical fish are selectively bred. In Japan, fancy carp or Koi have been bred for more than 2000 years. Koi breeding has become a major industry as well as one of Japan's leading hobbies.

6. PIGEONS, HORSES, CATS, AND RABBITS. These animals also come in many breeds as a result of selective breeding.

**25.5 One of agriculture's main problems is to maintain or obtain the genetic variability needed for the future adaptability of desirable plants and animals.**

A large pool of different genetic alternatives is needed among plants and animals if they are to be maintained or improved in the future. Consider from this standpoint the case of a disease-resistant variety of wheat. Such wheat is usually resistant to 6 to 10 different plant pathogens each of which required breeding into the wheat dozens to hundreds of different genes. The disease resistance erodes, however, in 5 to 10 years as a result of mutations in the pathogens and host. During this time a new wheat variety needs to be synthesized, using the genetic variability already present in other strains, to reestablish disease resistance. Of course, once disease resistance is lost insecticides and fungicides are required to obtain an adequate crop.

Consider next the case of maize in 1971. At that time about 80 per cent of maize planted in the United States carried the genes for male sterility. These genes made the corn plant susceptible to a fungus that caused the stalk and leaves of the plant to rot. This *corn leaf blight* destroyed about 15 per cent of the U.S. crop of 700 million bushels, some farms losing more than 50 per cent of their crop. Fortunately, by 1971 blight-resistant seed was available, although some farmers switched to other types of crops to avoid any loss.

Even though local crops of wheat or maize are often adapted to local conditions, all varieties tend to share a large number of genes in common to produce certain commercially desirable traits. Since nowadays all the plants in a crop are often genetically uniform, once a disease starts, it will tend to spread to all plants, causing the catastrophic loss that the prehistoric farmer was able to avoid. Moreover, such a disease can readily pass on to neighboring farms and often to neighboring areas, to produce a national or international disaster. If plant geneticists and pathologists are alert to such possibilities, new varieties may be bred that avoid or stop such catastrophic infections. Much of the genetic variability that would provide resistance to new varieties of pathogens or other improvements was already present in the old, locally adapted, genetically heterogeneous varieties. Unfortunately, these varieties were often lost when they were replaced by new genetically homogeneous varieties.

Since 1967 plant geneticists have instituted national and international programs to conserve the genetic diversity present in wild as well as cultivated varieties of plants. As a result of their efforts, some seed banks have been started. But unless such programs are more adequately funded than they are at present, we shall be continuing to lose the genetic variability we are sure to need in the future. Animal breeders have a more difficult time than plant breeders in maintaining the genetic variability of different strains. Cattle breeders, however, have established many artificial insemination centers that store a wide selection of bull semen; the bulls are exchanged among them every few years to maintain the variability of the strains.

## GENETICS AND ECOLOGY

Ecology deals with the relationship between organisms and their total environment. Genetics is related to ecology in that any ecological change, intended or not, may have important genetic consequences; and genetic principles and methods may be used to make purposeful ecological changes. These relationships are discussed and illustrated in the next two sections.

**25.6** Many unintended ecological changes have genetic consequences important to human beings.

Many ecological changes have genetic consequences, some of which are more important than others. Our attention will be restricted to ecological changes that were unintentionally produced by human beings and that affect human beings in a fairly direct manner. Some examples follow.

1. EXTINCTIONS. As the result of hunting and fishing practices, many species of African game, porpoises, whales, and so on, are presently facing extinction. Many animal species have already disappeared due to human activity. When an island is A-bombed fewer species reinhabit it than were present originally. Most accidental extinctions are undesirable since they disturb the ecological balance in unplanned ways. They also reduce the genetic variability of living forms. Since variety is one of the spices of life, an earth without lions, elephants, porpoises, and so on, is, at minimum, a less interesting planet.

2. ACCIDENTAL TRANSPORTATIONS. Because of the increased ease of transport, species are sometimes unintentionally transferred from one part of the earth to another. These transportations sometimes produce a harmful genetic invasion, as occurred, for example, after the introduction into the United States of the gypsy moth, Japanese beetle, chestnut blight, and Dutch elm disease. Killer bees, accidentally introduced into South America, are spreading northward.

The spread of potentially harmful organisms is avoided by public health measures such as quarantine and disease-preventing pills and injections, and by restrictions on the transport of live plants and animals from one country to another (and sometimes from one state to another within the United States). The use of rat guards on ships, the fumigation of aircraft and passengers, and the sterilization of interplanetary spacecraft are additional examples of precautions taken to minimize the accidental transportation of undesirable organisms.

3. RESISTANCE TO ANTIBIOTICS AND PESTICIDES. Antibiotics and pesticides are used to destroy or at least control organisms harmful to human beings or their food supply. When an undesirable organism exists in large numbers and reproduces rapidly, its population already contains, or comes to contain through spontaneous mutation, genetic alternatives that confer some degree of resistance to antibiotics or pesticides. In the absence of the controlling agent, resistant individuals are less fit than the wild-type ones and are either eliminated or survive in small numbers. When the population is exposed to the controlling agent, however, natural selection acts to establish the resistant forms as the wild type since under these conditions they are the fittest. It should be emphasized that resistant mutants result from mutations that are occurring more or less at random, spontaneously, and not in response to exposure to a controlling agent. Some examples of genetic resistance to chemical control are listed.

a. *Bacteria*. Antibiotics work because they attack some feature of the bacterium that distinguishes it from its host, say a human being. Such features include the bacterial cell wall, DNA polymerase, transcriptase, and ribosome. Bacteria can become resistant to an antibiotic through mutations that change the feature that makes them susceptible to the antibiotic, for example, their ribosomal proteins (Section 5.6), or through the acquisition of drug resistance particles (Section 9.6). In recent years drug-resistant forms of the bacteria that cause venereal diseases have become more prevalent.

b. *Fungi*. Fungicide-resistant forms of various fungi, including those that cause athlete's foot, have arisen.

c. Flies. Insecticide-resistant houseflies and fruit flies including *Drosophila* (a pest in tomato fields), have arisen following long-term exposure to such pesticides as DDT and its relatives.

d. *Mosquitoes*. Various mosquitoes are the vectors of parasites that cause certain diseases in human beings. For example, *Anopheles* transmits the protozoans causing malaria; *Culex* transmits the roundworm causing elephantiasis; and *Aedes* transmits the virus causing yellow fever. Insecticide-resistant forms of each of these have arisen following massive applications of insecticides and the resultant selection of resistant strains. For example, despite the use of insecticides and quinines, 6 million cases of malaria were recorded in 1976 in India alone. The incidence of malaria is increasing there at least 14 per cent per year, owing in part to the increased resistance of mosquitoes to insecticides.

e. *Rats*. Strains of rats that are resistant to certain rodenticides are appearing wherever such chemicals are used.

Immunity to different kinds of controlling chemicals requires mutants at different loci. Accordingly, the future control of infectious diseases and pests by means of chemicals depends upon the discovery of new chemical control agents, and the use of old and new chemical agents in novel and transient combinations.

4. HUMAN MUTAGENESIS. Human beings are unintentionally exposed to a wide variety of physical and chemical mutagens present in their

environment as a result of human actions. Various examples are given and the consequences are discussed in Chapter 27.

**25.7** Genetic principles and methods are applied in effecting many intentional ecological changes.

Intentional ecological changes for the purpose of adding useful organisms or of destroying pests are not uncommon. Some of the changes that apply genetic principles or methods are discussed.

1. ADDING USEFUL ORGANISMS. Fish, game birds, and mammals are often placed in new habitats to suit human needs; vast areas have been covered by reforestation with introduced species of trees. In order for such projects to be successful, the genetic constitution of an introduced species must (a) be adaptive to its new habitat, (b) contain sufficient genetic variability for natural selection to maintain future adaptiveness, and (c) not upset a favorable ecological balance in the region.

2. BIOLOGICAL CONTROL OF DISEASE. Genetically modified pathogens are used to combat certain diseases, such as poliomyelitis, in human beings. In these cases, persons are purposely infected with a mutant, nonpathogenic form of a microbe or virus so that antibodies are made that will combat any subsequently invading pathogenic form of the same organism.

3. BIOLOGICAL CONTROL OF PESTS. Insect pests can be biologically controlled or eradicated in two ways.
  a. *Induced sterility*. The adult insect has few if any somatic cells that divide. Therefore, doses of X rays can be given that will completely sterilize the adult without killing it. Sterility results from the dominant lethal mutations that are induced in cells of the germ line. These dominant lethals are due to chromosomal imbalances produced after germ-line cells containing induced gross chromosomal changes undergo nuclear division (Section 12.8).
  Sterile adults have been used to control or eradicate insect pests (such as blowflies) present in restricted areas (such as on an island). The pest is grown in the laboratory in tremendous numbers; the females are sterilized and then released in the infected area. Wild males that mate with these sterile females suffer complete gametic wastage. Control by this method is more efficient when the wild insect population is small. For example, a program was started in 1976 to eradicate the screwworm fly, whose victims are mostly cattle, from Mexico and the southwestern United States. This program entails the production of 300 million sterile screwworm females each week for a few years.
  b. *Baited traps*. Many female insects synthesize male-attracting substances that can be detected by males from great distances. Traps baited with females or simply with a sex attractant substance are being used to destroy the males.

4. ADDING PATHOGENIC ORGANISMS. A pest may be purposely exposed to an organism that produces a disease in the pest in the hope of controlling or eradicating the pest. This was planned for the rabbit, a pest in Australia. Rabbits are not native to Australia. In 1859, about two dozen wild English rabbits were deliberately released on a sheep ranch in Victoria, Australia. In the next six years, 20,000 rabbits were killed on that ranch alone. A hundred years later, several hundred million of these rabbits had spread all over southern Australia and had done great damage to agriculture and ranching. This population explosion was due to the absence of natural enemies or predators in Australia. It was then discovered that the European rabbit was nearly always killed when infected by a myxoma virus that is common in America (where it has a relatively mild effect on the American rabbit). This led to an experimental test in 1950 of the effect of this virus on the European rabbits in Australia. The virus was accidentally released into the wild population in 1954, at which time the virus had a rabbit kill rate of 99.8 per cent. The kill rate has since fallen to 90 per cent for two reasons. First, the rabbit population contains enough genetic variability for natural selection to retain more resistant genotypes. Second, the virus has mutated to a less virulent form that permits infected rabbits to live longer. (Since less virulent viruses have a selective advantage, this evolution is expected to occur routinely.) Increased longevity of the rabbit has helped the spread of the virus by mosquitoes.

The rabbit–myxoma story should caution us about the potential dangers of introducing species to new habitats. It also shows that biological control of a pest by a pathogenic organism is likely to be only partially effective, owing to *coevolution* of the host and the pathogen.

## SUMMARY AND CONCLUSIONS

Genetics has made, is making, and will continue to make important contributions to the progress of biological sciences such as agriculture and ecology.

The breeding schemes used for domesticated crops and animals have changed over the course of thousands of years. The success of such programs rested upon the presence in the population of sufficient genetic variability so that suitably fit organisms could be naturally or artificially selected. Specific improvements made through genetics are described for several plants, including maize, wheat, and rice; and in several animals, including cattle, sheep, and poultry. The future contribution of genetics to agriculture will depend, at least in part, upon how much of the genetic variability already present in nature can be preserved.

The relation between genetics and ecology was examined from two standpoints. Considered first were the genetic consequences of unintended ecological changes such as extinctions, accidental transportations, and resistance to chemical controls. Considered second were the genetic principles and methods applied to make intentional ecological changes such as adding useful organisms to an area, the biological control of diseases and pests, and adding pathogenic organisms to control a pest. Purposeful changes in ecology should be made keeping in mind that the fitness of an organism depends not only upon its nonorganismal environment but upon its organismal environment, with which it will coevolve.

## QUESTIONS AND PROBLEMS

**25.1** Why were the original farmers women?
**25.2** Are plants a better source for our nutrition than animals? Explain.
**25.3** Give three examples of how the use of genetic principles helped improve agricultural crops.
**25.4** Why don't farmers save money by planting the seeds from their hybrid corn for their next crop?
**25.5** Are mutagenic agents useful in improving agricultural crops? Explain.
**25.6** List five species of animals that have been improved through selective breeding.
**25.7** Why are cows ordinarily not selectively bred to breed better cows?
**25.8** What is the purpose of seed banks?
**25.9** Are mutations that confer resistance to antibiotics or pesticides more common today than they were 1000 years ago? Explain.
**25.10** What care needs to be taken before introducing a species to a new habitat?
**25.11** Describe three methods of controlling undesirable organisms by biological means.
**25.12** Give an example of coevolution involving: **a.** maize.   **b.** rabbits.   **c.** human beings.

*Chapter 26*

# Applications to Behavior, and Social and Political Issues

IN THIS CHAPTER we first take up some of the applications and implications of genetics to animal and human behavior. We then consider the social and political applications of genetics with respect to races and intelligence, politics, and law and religion.

## APPLICATIONS TO BEHAVIOR

The behavior of an organism is the result of the combined action of its environment and genotype. The relative importance of these two components differs for different behaviorial traits (Section 17.1). Our present concern is the importance of the genetic contribution to behavior. From the principles derived in the preceding portions of this book we expect that the genetic basis for behavior will often lie in proteins that affect sensory organs, the nervous system, the endocrine system, or effector organs such as muscles and glands. To adequately understand the basis of individual acts of behavior, therefore, we would like to identify the genes involved, their chromosomal location, the nature of their protein products, how these function, and the multiple phenotypic effects they are expected to produce (Section 17.5). We do not yet know all these things about any specific type of behavior, although we know some of these things about many kinds of behavior. For example, we have already discussed how transcription and translation are regulated by hormones (Section 20.5), many of which have effects on social or sexual behavior (Section 19.5). We know little or nothing, however, about the genes producing a specific hormone, where they are located, their precise mechanism of controlling the activity of other specific genes, and how such regulation produces the behavioral effect. In the next few sections we will summarize what pieces of the total desired knowledge we have at present about the molecular basis of learning and memory, the genetic basis of certain kinds of normal and abnormal behavior in other animals, and the genetic basis of mental retardation, abnormal nervous behavior, and intelligence in human beings.

**26.1** Learning and memory depend upon the synthesis of both specific and nonspecific RNA's and proteins.

*Learning* refers to the capacity of an organism to react in a new or changed way as the result of experience. *Memory* refers to the capacity to store and subsequently retrieve learned information. The learning that occurs during a series of trials performed over a short interval is said to involve *short-term memory*. After short-term memory is invoked, a period of *consolidation* or *fixation* must follow before the learning can become part of *long-term memory*. For example, a rat may be given a series of trials in which it has a choice of two kinds of behavior, one of which is followed by an electric shock. Short-term memory enables the rat to avoid the behavior that is followed by shock; after a period of consolidation, the rat retains this behavioral tendency indefinitely as part of its long-term memory. Cyclical behavior, in which a particular function is repeated at periodic intervals, also involves long-term memory.

Studies of sea hares, goldfish, mice, and rats indicate that learning and memory are intimately associated with the transcription of unique sequences of DNA and with repeated (perhaps rDNA) sequences. For example, single neurons of the cerebral cortex have been studied in right-handed rats that are forced to use the left hand to obtain food. Early in the learning process neurons that serve only the learning side synthesize only small amounts of RNA rich in A and U; later in the learning period these neurons synthesize a relatively large amount of RNA rich in G and C (as is rDNA). This result suggests that short-term memory and fixation of long-term memory require the synthesis of different types of RNA. In a different experiment, when rats are placed in a learning situation requiring balance, specific neurons shift the A/U ratio of the RNA they synthesize. The change in the base ratio of the RNA synthesized, and hence the types of RNA synthesized, seems to be specific for the type of learning behavior required.

Since RNA synthesis is implicated in learning and memory, it is reasonable to expect protein synthesis to be implicated also, since a portion of this RNA is probably mRNA. Support for this expectation has been obtained in studies of sea hares, goldfish, mice, and rats. For example, protein synthesis has been studied in a long-term learning situation where rats are forced to change their handedness to obtain food. During training, an increase in amount of a brain-specific protein occurs in certain specific nerve cells of the hippocanthus region, the portion of the brain specific for this learning process. When antibodies made against this protein are injected, they accumulate in the hippocanthal nerve cells of learning rats and prevent additional learning. On the other hand, neutralized antibodies have no such effect. On the basis of these and other studies it seems as though protein synthesis is needed for long-term memory but is probably not needed for short-term memory, and that some of the proteins involved are produced only by nervous tissue.

Although learning and memory involve RNA and protein synthesis, we do not know how these molecules are related to these behavioral changes. To discover how they are related will require increased knowledge of the chemistry and physiology of neurons and of the connective tissue cells, the *neuroglia*, that surround the neurons and support them physically and chemically.

**26.2** Different behavioral traits in animals have been associated with alleles at one, two, or many loci.

SINGLE-GENE EFFECTS

1. *In mice*. More than 90 different loci are known to have alleles that produce a defect in the nervous system of the mouse, and consequently have behavioral effects. For example, homozygotes for a certain recessive allele are *waltzers*. Such mice have inner-ear abnormalities, including the degeneration of the semicircular canals and of the cochlea. The semicircular canal defect causes them to run in circles for long periods of time and shake their heads; the cochlea defect results in deafness. As one might expect, almost any allele that causes a change in phenotype has, through its multiple phenotypic effects, a behavioral effect that can be detected if the tests performed are sufficiently extensive and delicate. For example, a considerable array of behavioral differences occur between albino and normal mice due in general to albinos hesitating to respond to environmental changes in incline, white light intensity, surface roughness, sound, or odor.
2. *In* Drosophila. Various single-gene effects on behavior are also known in *Drosophila*, some of which affect the nervous system directly. As in the mouse, body color alleles also affect behavior. For example, the X-limited allele for yellow body color, *y*, has as one of its multiple phenotypic effects a decreased efficiency in courting behavior. Male courtship includes a period of wing vibration; the *y* male vibrates his wings less often and for shorter periods than the normal gray-bodied male.

Analysis of the action of alleles having behavioral effects is facilitated in *Drosophila* by the study of genetic mosaics. Mosaics that express different alleles in different parts of the body occur frequently when heterozygous individuals contain a special unstable genotype. If the individual heterozygous for a behavioral allele is simultaneously marked by being heterozygous for a closely linked gene having an obvious phenotypic effect, say on body color, the anatomical site of action of the behavioral allele can often be localized. The focus of action of the behavioral gene is identified by determining from a series of mosaics which part of the fly must express the marker gene in order to express the behavioral allele. This localization is illustrated diagrammatically in Figure 26-1. Mosaic 1 shows that the behavioral allele is expressed when the marker gene is expressed in the lower two thirds of the abdomen. Mosaic 2 shows that the behavioral gene is expressed when the marker gene is expressed in the right half of the body. Mosaic 3 shows that when only the upper two thirds of the abdomen expresses the marker gene, the

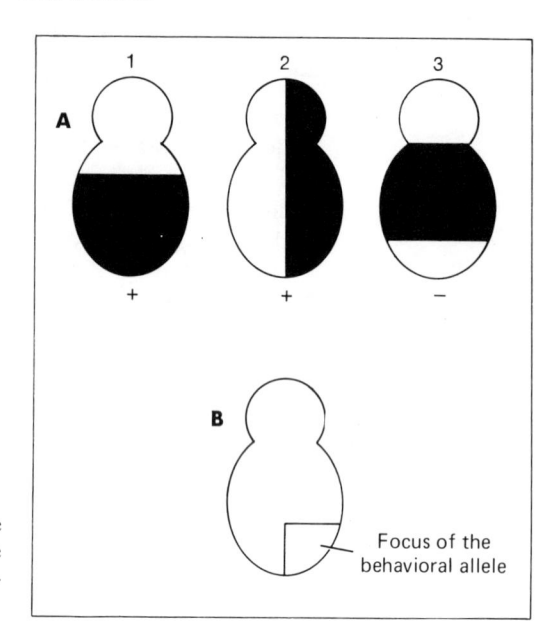

FIGURE 26-1. The use of mosaics (row A) to delimit the site of action (B) of a behavioral allele. The behavioral allele is expressed (+) or is not expressed (−) when the accompanying genetic marker is expressed (shaded areas).

behavioral gene is not expressed. The focus of action of the behavioral gene is, therefore, the right lower one third of the abdomen. Since it is now possible in *Drosophila* to obtain mutant alleles that affect behavior at many loci, to map them, and to identify their focus of action in the body, we are encouraged to think that by means of future genetic, biochemical, and physiological work we will eventually be able to completely dissect the sequential receptor–transmitter–effector steps that occur in certain simple types of behavior.

TWO-GENE EFFECTS. In the honeybee, growing individuals are encased in a wax cell of the comb, where they are sometimes killed by infection by the bacterium *Bacillus larvae*. In one strain of bees, the diploid workers take the cap off the wax cell and remove the dead individuals; workers of another strain neither uncap the wax cell nor do they remove the dead individual it contains if the cell is uncapped for them. Hybrid workers obtained by crossing these two strains neither uncap nor remove. Workers obtained by crossing the hybrids to the uncap-and-remove strain are one of four types: (1) uncappers and removers, (2) uncappers, (3) removers, and (4) neither uncappers nor removers. These results and others indicate that uncapping is due to a recessive allele at one locus and removing is due to a recessive allele at a different locus. The hive-cleaning behavior depends, therefore, upon homozygosity for two recessive pairs of genes. No biochemical difference has been found so far between the two strains that would explain their differences in hive cleaning.

MULTIGENE EFFECTS. When artificial selection is carried out for a large number of generations, a gradual change is sometimes observed in a particular kind of behavior. In such cases it seems likely that the selected behavior is dependent upon the action of many pairs of genes. For example, gradual change is found when rats are selected for fast or slow ability to solve a maze to obtain a reward. Gradual change is also found when *Drosophila* is selected for positive or negative response to gravity. In this case it has been possible to show that the X, 2, and 3 chromosomes contain genes affecting response to gravity. In neither example, however, do we know at present precisely how many loci are involved or how they interact.

**26.3** Mental retardation and abnormal nervous behavior in human beings can result from changes at single loci as well as from whole chromosome abnormalities.

As in other animals, normal mentality and normal nervous behavior in human beings require the normal structure and function of sensors, neurons, and effectors. We also expect in human beings that the usual precondition for normal behavior is the normal functioning of many genes on many chromosomes. A few of the single loci or groups of loci that are needed for normal behavior have been positively identified through abnormalities in nervous behavior that result from the presence of abnormal genetic alternatives. Some of the abnormal genetic constitutions that produce behavioral changes are described.

PHENYLKETONURIA. Phenylalanine is one of the amino acids that we require in our food. More than one half of the phenylalanine in our food is normally converted to tyrosine, another essential amino acid, principally in the liver. Some tyrosine is, in turn, converted by a series of chemical reactions into melanin pigment. Persons who are homozygous for a particular recessive, autosomal allele fail to make phenylalanine hydroxylase, the enzyme that converts phenylalanine to tyrosine. As a result of the absence of this enzyme, phenylalanine accumulates in the blood and some of it is converted to phenylpyruvic acid. Both phenylalanine and phenylpyruvic acid accumulate in the cerebrospinal fluid and, in children up to six years of age, produce some unknown effect that results in mental retardation. Affected individuals are said to have phenylketonuria (Sections 22.4 and 22.6). They have an IQ about 30, often have convulsions and an abnormal electroencephalogram pattern, and are generally lighter in pigmentation in their skin, hair, and eyes, owing to the lack of melanin production. Mental retardation in homozygotes can be averted, however, if during the first six years of life, only enough phenylalanine is included in their diet for protein synthesis. (In this case appreciable amounts of phenylalanine and phenylpyruvic acid cannot accumulate in their cerebrospinal fluid.) After about six years of age, when the brain is about 90 per cent of adult size, they can be given a normal diet. Genetically phenylketonuric mothers, however, need to restrict their phenylalanine intake during pregnancy to prevent a high concentration of phenylalanine from entering the circulation of the fetus and causing its mental retardation.

Most states in the United States require a test of newborns for phenylketonuria. Heterozygotes for phenylketonuria can often be detected by a phenylalanine tolerance test, since many heterozygotes metabolize phenylalanine more slowly than do normal homozygotes. Such a test is valuable in counseling parents about the chance of having phenylketonuric children (which is, of course, 25 per cent when two heterozygotes marry).

LESCH–NYHAN SYNDROME. The purines used to make nucleotides can be synthesized from non-purine-containing raw materials, or are present in the diet, or are salvaged when a person's own nucleotides are degraded. When an X-limited recessive gene is homozygous in a female or is present in a male, however, the individual cannot use already formed purines due to the absence of the enzyme hypoxanthine-guanine phosphoribosyltransferase (HGPRT). One of the biochemical consequences of this deficiency is the production of excess uric acid, one of the chemical breakdown products of purines; this excess causes extreme symptoms of gout. Other biochemical consequences, unknown at present, result in mental deficiency, muscular spasticity, and a compulsive, aggressive behavior toward others (spitting, biting, and hitting) as well as toward themselves (self-mutilation by chewing their lips and fingers). Affected persons usually die in childhood.

Affected individuals can be diagnosed before they are born. Tests are also available to identify women who are heterozygous for this recessive lethal allele.

TAY–SACHS DISEASE. Many membranes contain fatty substances called lipids. When lipids occur in excess they are normally broken down during metabolism. When any one of 10 known enzymes is missing, however, lipids build up and produce lipid storage diseases. Because of the excess lipid in nervous tissue, many lipid storage diseases produce mental deterioration. Tay–Sachs disease is one such disease; it occurs in homozygotes for an autosomal recessive allele. Affected individuals do not synthesize hexosaminidase A. Although affecteds are normal as infants, they become progressively abnormal, suffering mental degeneration, blindness, and loss of muscular control. Death usually occurs at three to four years of age.

Affected individuals can be diagnosed before they are born, and heterozygotes can be detected from their lower-than-normal level of hexosaminidase A. The defective allele is most common among Jewish persons of northern European descent, in whom it occurs with a frequency of about 0.025. The high frequency of this recessive lethal allele in the population may be due to a relatively high resistance that heterozygotes have to tuberculosis. If so, the Tay–Sachs allele would be heterotic only in populations suffering from tuberculosis, just as the sickle-cell allele is heterotic only in a falciparum malaria environment.

HUNTINGTON'S CHOREA. This disease is due to an autosomal dominant allele with complete penetrance. The metabolic defect is unknown and symptoms first appear at an average age of 35, although the range is mainly from 15 to 65. Because of the late onset of the disease, which is invariably fatal, persons carrying the defective allele often have children before they are aware of their own affliction. Affecteds suffer a progressive atrophy of the nervous system. The early symptoms include limb twitching (chorea), grimacing, and bodyweaving, followed over a period of years by impairment of speech and intellect, insanity, and death. More than 7000 people in the United States have Huntington's chorea. The disease has received wide attention because the folk singer Woody Guthrie was a victim. He had inherited the defective gene from his mother; it is not known if any of his children are carriers.

TESTICULAR FEMINIZATION. XY genetic males that carry a single copy of a particular abnormal allele are transformed into phenotypic females. Such females possess blind-ended, small vaginas, full development of the mammary glands, and a normal life span. They are raised as girls, are feminine in bearing and temperament, have a sex drive normal for women, and sometimes marry men. These females have testes that are usually undescended and are, of course, sterile. Such females usually seek medical care when their testes try to descend or because they do not menstruate. The disease is called *testicular feminization*. Affected individuals given injections of male sex hormone do not respond by becoming more male-like. Since no cure is available, physicians do not ordinarily inform affecteds that they have an XY constitution.

GROSS CHROMOSOMAL IMBALANCES. The gross sex chromosomal imbalance that occurs in X0, XYY, XXY, and other individuals who have extra X's can cause behavioral abnormalities, usually expressed in mental retardation (Section 19.6). Trisomy for chromosome 21 also causes mental retardation (Section 12.4), as does trisomy for 13, 17, 18, or 22. We have also noted that the heterozygous deficiency of the short arm of chromosome 5 that results in the cri-du-chat syndrome also produces mental deficiency (Section 12.10).

We can conclude from all of the preceding that the X chromosome and the autosomes contain single loci whose abnormal alleles, recessive or dominant, can produce abnormal mental, sexual, or other behavior. Gross chromosomal imbalances also produce such abnormal behavior. We are a long way from understanding how these abnormal genotypes produce abnormal behavior, and a still longer way from understanding how the normal alleles contribute to normal behavior.

# APPLICATIONS TO SOCIAL AND POLITICAL ISSUES

The remainder of this chapter considers the genetics of races and intelligence, politics, and law and religion.

# RACES AND INTELLIGENCE

**26.4** The genetic variability of human beings makes it possible for different races to be adapted to different environments.

As is the case within every species of animals and plants (Section 24.1), there is a large amount of genetic variability in human beings. Genetic variability enables different individuals or groups within a species to adapt to different environments (Section 24.5), and it serves the same function in human beings. Genetic constitutions that make human beings obviously less fit in some environments may be adaptive in other environments. For example, persons who are red-green colorblind can more readily see through camouflage than can persons with normal vision. During wartime, therefore, such colorblind persons have an adaptive advantage. Abnormally small people have an advantage in jobs where space is restricted. Genetic constitutions that produce phenotypes at the extremes of what is considered normal may be advantageous in some environments. For example, tallness is an advantage in basketball and much weight in football and weight lifting. Different genetic constitutions that produce phenotypes differing in less extraordinary ways are also adaptive in different environments. For example, white skin color is advantageous in northern and southern temperate zones because it permits the absorption of sunlight needed for the synthesis of vitamin D. Black skin is advantageous near the equator because it prevents the synthesis of too much vitamin D, which is toxic in overdose amounts. If the genes for skin color are the only criteria used to differentiate races, white and black races are adapted to different environments.

**26.5** A given environment can be modified so that different races have essentially equal fitness.

Black and white skins lose at least some of their adaptiveness with respect to vitamin D synthesis when the environment is modified. For example, white persons can live near the equator by becoming suntanned and/or by wearing protective clothing; black persons can live in northern and southern temperate zones by increasing their exposure to sunlight and/or by eating a vitamin D-enriched diet. There is every reason to believe that most, if not all, adaptive differences between individuals or races can be removed by changing the environment they share. Even the detrimental genotypes for Turner's syndrome, phenylketonuria, and diabetes can be made to phenocopy a normal genotype through appropriate changes in the environment. It should be just as possible to modify the environment so that persons or races whose genotypes would otherwise produce somewhat less detrimental phenotypes will have essentially the same fitness as genetically different persons or races.

**26.6** We do not know whether our present black and white races differ genetically in test intelligence.

It is clear from the preceding that in different environments the black and white races have different fitnesses, and that in an appropriate environment we expect these two races will have essentially equal fitness. Recently, much social and political controversy has arisen regarding the answer to the question: "Do blacks and whites differ genetically with respect to intelligence?" Since (1) both the black and white races contain a great deal of genetic variability, (2) intelligence is a behavioral trait due to the action of many gene pairs, and (3) the environment to which the members of each race are exposed varies considerably, great variations must undoubtedly occur in the expression of intelligence within each race. The question asks whether, as a whole, these two races differ in intelligence for genetic reasons. The answer requires that we specify what intelligence is, and that we test both races for it under the same environmental conditions.

It can be claimed that intelligence includes creativity, motivation, and personality. Intelligence may also be expressed in musical, artistic, or athletic talent; in an attitude of altruism or pacifism; in facility or eloquence in speech; or in the ability to survive in a hostile environment. The most commonly used tests of intelligence measure none of these components. The most widely used test, the *intelligence-quotient* (*IQ*) test, is designed to predict how successful an individual will be as a scholar in a white, middle-class society. This test is accurate and also indicates future success in such a culture. When white and black children are given such tests, black children do significantly poorer than white children.

There is no evidence, however, that the environments of tested black and white children are the same. IQ tests have been given to white children raised in black families, and black children raised in white families. There is good reason to believe that the environments, including color prejudice, are not the same in these two cases. Other IQ comparisons have been made between black and white children raised in orphanages since infancy. Even if one granted that the orphanage environment was the same for all children, one could wonder whether the white and black mothers (of comparable ages) who gave up their children were genetically random samples of all white and black parents. Even if such samples were random, there is no assurance that the uterine environment in which the two types of babies developed were equivalent.

The fetus is not immune to damage while it is developing. Since 1941 it has been known that pregnant women who contract German measles while their fetuses are developing eyes and ears often produce babies that have cataracts or are deaf. Such babies also have almost a 50 times greater chance of being autistic, a behavioral defect. What is the effect on the fetal nervous system of other infectious diseases contracted by the mother? We do not know. We have already mentioned the effect on the fetus of the drug thalidomide (Section 17.4). Drug addiction during pregnancy can produce babies who are addicted. Other, medically approved drugs are known to have harmful effects on the fetus. Recently, a list of more than

80 such drugs was compiled. Included are barbiturates, many antibiotics, amphetamines, and salicylates (the principal ingredient of aspirin). An excess of vitamins is harmful to a fetus. Even moderate consumption of alcohol during pregnancy produces adverse effects on a fetus, as does smoking. Some of the drugs in this list are already known to adversely affect the fetal nervous system. Others in the list have not yet been adequately studied in this respect. Now that we are becoming increasingly aware of the damage the fetus can suffer, we can expect a growing list of substances that can harm its nervous system, and hence its intelligence. Finally, we should bear in mind the likely possibility that different diets presently accepted as normal may have various effects on the development of particular features of the nervous system and on intelligence.

From the preceding we learned that many nongenetic intrauterine factors may harm a child's intelligence. Until we are reasonably certain that black mothers and white mothers have provided their babies with the same intrauterine environment, we will not be able to decide whether any difference between blacks and whites in any test of intelligence has a genetic basis. Until that time, which is not in the near future, we can continue to improve (1) our understanding of intelligence and our ways of testing for it, and (2) the prenatal and postnatal environment of all children to maximize their genetic potential for intelligence.

# POLITICS

**26.7** The science of genetics was suppressed for a generation in the Soviet Union for political reasons.

Communism holds that the ideal society is classless. To foster this attitude, Stalin was interested in supporting the view that all persons are or can become intrinsically equal. At the same time, there was a great need to improve the productivity of agriculture in the Soviet Union. The breeding methods of mendelian genetics were being used successfully in the Soviet Union, but the rate of progress was too slow for Stalin even though it was as rapid as could be expected. In the mid-1930s, an agronomist, T. D. Lysenko, forwarded the claim that rapid and stable improvements will result from alterations in the environment of plants and animals. Some of the attempts to produce permanently improved strains of plants and animals in one or a few generations employed exposure to high or low temperatures, grafting of one plant onto another, and the injection of the blood of one strain of animal into another strain. Lysenko did not believe in mendelian genes, but rather in a genetic material that diffused throughout the whole organism, collected in the germ cells, and mixed at the time of fertilization. Environmentally induced somatic improvements were supposed to become somatic genetic improvements and eventually germ line genetic improvements. Since, on this hypothesis, the genetic material was readily changed to any desired end, and could even generate new genera in one generation, it was considered unimportant to use genetically uniform, select lines. The environment was considered important, not the genetic background. Mendelian genetics was denounced as an anticommunist ideology fostered by enemies of the Soviet Union; Lysenko's views, however, were consistent with Stalin's concepts concerning the development of communism and the needs of Soviet agriculture.

The controversy between Lysenko (and his followers) and the Soviet mendelian geneticists ended in 1948 when Lysenko gained the official support of the Soviet Communist Party and of Stalin; he became the director of the Institute of Genetics of the Academy of Sciences of the USSR. Research in or teaching of mendelian genetics was suppressed. Lysenko's opponents lost their jobs, were imprisoned, or disappeared. Even such an eminent scientist as N. I. Vavilov died in prison. By 1954, the year of Stalin's death, every scientific position in agriculture was held by a follower of Lysenko, and mendelian genetics had all but ceased to exist in the Soviet Union.

It became apparent during the next decade that Lysenko's theories had failed to increase agricultural productivity in the Soviet Union as fast as promised. Meanwhile, the application of the principles of mendelian genetics in the United States and other countries had produced much improvement. Moreover, the dramatic (and sometimes farcical) claims of the Lysenkoists could not be repeated experimentally by scientists in other countries. Many of Lysenko's successes could be explained by the survival of favorable mendelian genotypes, fortuitously present in the unselected material that was subjected to selective environmental conditions. (Numerous reported successes were clearly prefabrications by supporters who knew they should obtain the expected results, and who knew their results would never be checked.) Under Krushchev, some Soviet mendelian geneticists were again allowed to do research.

By 1962, it was apparent that the Soviet Union had also fallen behind in molecular and microorganismal genetics. By 1965, Lysenko's power began to wane; under Brezhnev and Kosygin, he lost all his political and scientific influence. Genetics is now an actively pursued science in the Soviet Union, apparently free of direct governmental control.

The Lysenko affair is important not only because the scientific pursuit of the principles of genetics was actively suppressed in the Soviet Union for a generation, but because it illustrates the economic disadvantage that can result when a government preferentially supports a pseudoscience that advances its official doctrines. Science and its applications flourish best when scientific research is aided on the basis of scientific merit alone.

**26.8** The proposed "Aryan" master race of Hitler was impractical to attain and a biologically unsound goal.

Interest in the improvement of the genetic material of human beings, *eugenics*, grew in Germany in the period from 1900 to 1930. The most commonly proposed eugenic tool was that of controlled breeding. Individuals with superior genotypes were to be encouraged to have children by other individuals of superior genotypes, whereas persons with inferior genotypes were to be discouraged from having children. As this view became more and more popular, many Germans, including some German geneticists, came to believe that controlled breeding was the best way to improve their race. Hitler took advantage of this interest to foster the spread of his so-called "Aryan" master race. The Aryan phenotype was supposed to consist of a single set of ideal traits. Breeding that perpetuated and spread these traits was to be encouraged, whereas other breeding was to be discouraged. A law was passed in 1933 for the sterilization of the unfit and unwanted; eventually, more than 200,000 people were sterilized. To help attain racial purity, Hitler set up state-run breeding farms where Aryan women could bear the children of Aryan men, and the state undertook the care of Aryan children. In addition, Hitler had more than 6 million "non-Aryans," mostly Jews but including Gypsies and others, exterminated in gas chambers.

The plan for a pure Aryan master race was based in part upon a popular misunderstanding about the eugenic effect of selective breeding. Selective breeding against detrimental genes will improve the genetic constitution of a population. Progress will be most rapid when the detrimental allele is dominant and selection is against the heterozygote (Figure 23-1, curve $a$). When the detrimental allele is recessive, however, selection occurs against its homozygote. In this event, progress will be rapid only when the frequency of the detrimental allele is high (Figure 23-1, curves $b$ and $c$). When the allele's frequency is low, selection against its homozygote becomes less and less efficient in successive generations. Many of the traits considered undesirable in Aryans are quantitative traits that result from the action of many gene pairs, many of which are doubtless recessive. Since the selective breeding of human beings would depend largely upon the occurrence of homozygotes for recessives, it would take very many generations to remove such alleles from the population. Therefore, besides being implemented in a barbarous manner, Hitler's attempt to genetically purify the race by selective breeding was impossible in the short run and impractical in the long run.

The view that a single, homozygous, master genotype is the ideal one for any race, whether it has an Aryan or other phenotype is, moreover, contrary to our understanding of the adaptiveness of populations. No single genotype provides maximum fitness in all the ecologically different parts of the territory occupied by a population, race, or species. The most successful species are composed of cross-fertilizing, genetically heterogeneous populations. Accordingly, even if the genotype that produced the Aryan phenotype had been attained, it is unlikely that this genotype would have been the one most adaptive for all the other traits needed by a so-called master race.

Hitler's false doctrine of Aryan supremacy should caution us against present or future eugenic proposals for or against different races. The strength of the human species rests not in the superiority of a white, black, or yellow race, but in its genetic heterogeneity.

# LAW AND RELIGION

Various laws have genetic implications, for example, those that protect us from mutagenic radiations and chemical substances in our atmosphere, food, clothing, and workplaces. We consider in the next two sections how laws control our marriages, and hence our breeding behavior; and how genetics is used by the law to establish human parentage.

**26.9** Certain ecclesiastical and civil laws concerning marriage have genetic implications.

Ecclesiastical laws and ethical systems that deal with marriage often modify or regulate the breeding pattern of human beings. For instance, Confucius suggested that two persons with the same surname should not marry; the Roman Catholic Church prohibits marriage of its clergy; and the Orthodox Eastern Church prohibits marriage by monks, nuns, and bishops. Many religions prohibit the marriage of lay persons more closely related than first cousins. (They also prohibit the marriage of certain persons related only by marriage. Such prohibition must have had a social, not genetic, intent.)

In most European countries, civil law prohibits the marriage of persons who are more closely related than first cousins. Like church laws on marriage, civil laws oscillate in strictness regarding inbreeding. Whereas the first law making incest a crime was passed in Scotland in 1567, the first such law was not passed in England until 1908. In England, incest means sexual intercourse between a man and his granddaughter, daughter, sister, or mother; or between a woman and her four equivalent relatives. By 1960, twenty specific marriages were prohibited in England, half of which are between persons known to have genes in common and half that do not. In 1976 a government committee in Sweden recommended that incest—defined in that country as sexual intercourse between parent and child, or brother and sister—no longer be considered a crime.

Until the late 1880s, there were no civil laws in the United States prohibiting consanguineous marriages, that is, marriages between persons known to be related by descent. By 1959, however, 20 states prohibited marriages between two persons known to be related in descent to any degree; and 14 others prohibited marriages between relatives of certain specified closeness.

Several states have laws prohibiting the marriage of persons in public asylums who are epileptic, insane, or mentally retarded. Some states permit marriage of such women when they are over 45 years of age.

Most of the preceding ecclesiastical or civil laws that restrict marriage have the effect, if not the intent, of averting the birth of persons with undesirable phenotypes that may result from inbreeding or from the breeding of physically or mentally handicapped persons.

On the other hand, laws prohibiting marriage between blacks and whites favor inbreeding. In 1664, Maryland was the first of the United States to prohibit such a marriage. Many others followed suit. Since by 1790 about 20 per cent of the U.S. population was black, these laws prevented much crossbreeding. Although many states repealed such laws starting in 1840, there were still 19 states prohibiting white–black marriages in 1964. In 1967, all such laws were declared unconstitutional by the Supreme Court. So, at present, when the black population is about 10 per cent of the total, there is increased opportunity for random mating between blacks and whites.

**26.10** Genetics can be used in some cases to verify or disprove the maternity or paternity of children.

Thousands of cases of disputed paternity or maternity are brought before the United States law courts each year. Most of these are cases of disputed paternity, wherein the purpose is to judge whether a particular man is the parent of a particular child and, therefore, responsible for the child's support. Other cases include the identification of newly born babies who have been mixed up in the hospital or of children separated from one or both parents by war or social upheaval.

Genetics can be used either negatively, to exclude parentage, or positively, to assign parentage. An underlying assumption is that mutations occur too rarely to offer a reasonable alternative explanation. This is acceptable since the law does not require absolute proof, it only requires reasonable certainty. In science, explanations that have as little as a 1 per cent chance of being correct are accepted as being reasonable. Since mutations usually occur much more rarely, they are usually excluded from legal explanations of why a child's genotype differs from that of its reputed parents.

Paternity can thus be excluded when the child carries a mendelian allele that is carried neither by the known mother nor by the alleged father. The most suitable genes for such a determination are those which produce qualitative, fully penetrant traits that are not readily phenocopied, such as genes for blood type. The larger the number of traits having a known genetic basis for which the child, mother, and suspected father can be tested, the greater the likelihood that paternity can be excluded for a man who is not the father. Paternity can also be excluded if a son and the alleged father have Y chromosomes that differ in size, or if the child has a chromosomal rearrangement absent in both parents.

A positive assignment of paternity can be made when there are only two possible fathers and one can be genetically excluded from paternity. Positive assignment can also be made if both alleged father and child carry the same reciprocal translocation, half-translocation, rare dominant gene (such as for polydactyly), or a rare blood-type allele. Finally, positive assignment may be made if a rare combination of many nonalleles (none of which by itself need be rare) is found in both the child and the alleged father.

# APPLICATIONS TO BEHAVIOR

Genetic evidence to determine parentage is accepted in the courts of many countries and states. While evidence excluding paternity or maternity is accepted in many courts, many others do not recognize the positive assignment of paternity.

## SUMMARY AND CONCLUSIONS

To understand the genetic basis of individual acts of behavior, we need to identify the genes involved, their locations and products, and how these products affect metabolism. We have some of this information about different normal and abnormal kinds of animal and human behavior. For example, we have started to understand how transcription and translation are related to memory and learning, although we know almost nothing about the specific genes required or how their products are utilized to produce such behavior. In animals we have discovered examples of behavior determined principally by one, two, or many gene pairs, but know nothing about the gene products and how they affect behavior. In human beings more or fewer bits of the desired information are available for the various behavioral genetic diseases discussed.

The importance of both the genotype and the environment in the expression of different or similar traits in different races is emphasized. We cannot decide whether black and white races differ genetically in test intelligence because the two races have not been shown to have equivalent environments from conception to the time of testing.

The relation of genetics to politics is discussed with respect to the earlier suppression of genetics in the Soviet Union and the barbaric eugenic program in Germany under Hitler's rule.

The legal aspects of genetics include ecclesiastical and civil laws that discourage or encourage inbreeding, and the use of genetics in courts to solve cases of disputed parentage.

## QUESTIONS AND PROBLEMS

26.1 *Movies*: "Chinatown." What are the genetic implications of this movie?

26.2 *Movies*: **a.** "Walkabout." **b.** "Oliver Twist." What are the genetic implications of one or both movies?

26.3 *Book*: *Dunes*. What are the genetic implications of this book?

26.4 How is gene action related to learning and memory?

26.5 "Every gene has an effect on behavior." Explain whether you agree or disagree.

26.6 What advantages has *Drosophila* over the mouse as an experimental organism for the elucidation of the genetic basis for simple types of nervous behavior?

26.7 Propose one study that might help us learn the biochemical basis for hive-cleaning behavior in the bee.

26.8 Describe the biochemical basis for mental retardation caused by phenylketonuria. How can phenylketonurics phenocopy normal?

26.9 What genetic diseases are associated with each of the following? **a.** abnormal purine utilization. **b.** abnormal lipid storage. **c.** self-mutilation. **d.** limb twitching. **e.** sex transformation.

26.10 Why do most geneticists show no interest in attempting to prove that human races differ genetically in intelligence?

**26.11** Would the intrauterine environment be equivalent if random samples of both black and white babies were grown in black women as well as in white women? Explain you answer.

**26.12** Are the nuclei in a white zygote and a black zygote surrounded by equivalent cytoplasmic environments? Explain.

**26.13** A relation is sometimes traced between the Inquisition, which was particularly active in Spain, and the decline of Spain as a great power. What genetic basis do you think there may be for this decline?

**26.14** Do you think we should pass a law outlawing Lysenko-like genetics? Explain.

**26.15** The practice of cutting off the tail in puppies is commonly used in certain breeds of dogs. In these breeds some puppies are born with short or no tails, and this condition reappears in their offspring. Is this valid evidence of the inheritance of an acquired trait? Explain.

**26.16** Should we pass legislation that encourages marriages that will improve the genetic material of the human species? Explain.

**26.17** Do you think incest would be more prevalent if it were not against the law? Explain.

**26.18** Do you think all courts should accept the genetic proof and disproof of parentage? Why?

*Chapter 27*

# Applications to Medicine and Public Health

THIS CHAPTER CONCLUDES our consideration of the applications and implications of genetics. We consider here the relation between genetics and such medical and public health topics as (1) immunology; (2) aging, death, and cancer; (3) environmental mutagenesis; and (4) genetic diseases and genetic engineering.

## APPLICATIONS OF IMMUNOLOGY

Through the formation of antibodies, genes play an important role in blood transfusions, tissue transplantations, and certain diseases.

**27.1** ABO blood types must be compatible for safe blood transfusion.

Persons who carry gene $I^A$ have a particular sugar at the end of a large molecule. Allele $I^B$ causes these molecules to end with a different sugar, whereas allele $I^O$ adds no sugar. These sugars occur at the surface of red blood cells where, when present, they are recognized as normal and do not act as antigens. An individual who does not make either of these sugar compounds (blood type O, Figure 27-1) makes antibodies against both probably in response to their presence in infecting bacteria. An individual who makes only one (blood type A or B) forms antibodies against the other, and one who makes both (blood type AB) forms antibodies against neither.

Blood transfusions are safe if the entering blood of the donor is not clumped by antibodies of the recipient. (The antibodies of the donor are so quickly diluted in the recipient that they do not cause clumping of the

FIGURE 27-1. The antigens and antibodies present in different genotypes for ABO blood groups.

| Genotype | Blood Group or Antigens | Antibodies |
|---|---|---|
| $I^A I^A$ or $I^A I^O$ | A | B |
| $I^B I^B$ or $I^B I^O$ | B | A |
| $I^A I^B$ | AB | O |
| $I^O I^O$ | O | AB |

recipient's blood.) Referral to Figure 27-1 will make it apparent that persons of AB blood type are universal recipients because they can clump no incoming blood, and that persons of O blood type are universal donors because they have neither antigen causing incoming blood to be clumped. Each blood type can, of course, receive a transfusion of the same type of blood as well as of other types that do not carry an antigen to an antibody in the recipient.

## 27.2 The blood types of mother and fetus may be incompatible.

Antibodies of the mother can sometimes pass through the placenta into the bloodstream of a developing fetus. About one baby per thousand births suffers from a slight anemia due to an ABO mother–fetus incompatibility. For example, when a mother of O blood type is carrying a fetus of A or B blood type, some of her anti-A and anti-B antibodies can enter the fetus, where they can clump and, thereby destroy, fetal erythrocytes.

A similar incompatibility between mother and fetus can occur for a different erythrocyte antigen. Persons who are homozygous $rr$ do not have the antigen Rh made by $Rr$ or $RR$ persons (and by Rhesus monkeys). Accordingly, $rr$ persons are said to be $Rh^-$ and $RR$ or $Rr$ persons are $Rh^+$. Since $rr$ persons are not normally exposed to Rh antigen, they normally do not carry anti-Rh antibodies. Women who are $rr$ will synthesize anti-Rh antibodies if they accidentally receive $Rh^+$ blood either by transfusion or while giving birth to an $Rr$ child. In a subsequent pregnancy, anti-Rh antibodies can travel across the placenta and into the fetus; if the fetus is $Rr$, a severe anemia may occur that is sometimes lethal. Newborns can be rescued from such anemia by whole-blood replacement. Women who are $rr$ are, therefore, not given transfusions of $Rh^+$ blood; and, at the time of birth of an $Rh^+$ baby, they are injected with some anti-Rh antibodies to destroy any fetal erythrocytes that may have entered their bloodstream before such cells can trigger the synthesis of a large quantity of anti-Rh antibody.

When mother and fetus are incompatible with respect to both the ABO and Rh blood types, the ABO incompatibility offers protection against the Rh incompatibility. In this case, any fetal red blood cells that leak into the mother's circulation will tend to be destroyed by already present anti-A or anti-B antibodies before they can elicit the formation of anti-Rh antibodies.

**27.3** Incompatibility of tissues has a genetic basis.

Whether a transplanted tissue is to be tolerated or rejected seems to depend upon the alleles present at two loci in human beings. Each locus has a dozen or so different alleles, each of which specifies a different antigen. Individuals will make antibodies against any of these antigens (and hence any of these alleles) present in a graft that they themselves do not possess. Since most people differ by at least one allele at these loci, most transplants elicit antibody formation and are rejected. Rejection can be avoided if transplant recipients are given drugs that suppress the lymphocytes forming antibodies. (These drugs also suppress other lymphocytes that fight disease.)

Occasionally, persons make antibodies against their own tissues. Such autoimmunity can result in a variety of diseases in which antibodies are formed against the thyroid gland, adrenal glands (Addison's disease), heart (rheumatic fever), various other tissues (rheumatoid arthritis), or even one's own DNA (systemic lupus erythematosis). In some autoimmune diseases, the antibodies are probably induced by antigens in infecting organisms that resemble human antigens. Other autoimmune diseases may have a genetic basis, since lines of mice and chickens have been bred that have high frequencies of such diseases.

# APPLICATIONS TO AGING, DEATH, AND CANCER

**27.4** Normal development and differentiation include the programming of aging and death for the somatic line.

Since present-day people, mice, and insects are direct descendants of an unbroken ancestral reproductive chain, the cells of their germ lines are apparently able to divide an unlimited number of times. Several kinds of evidence indicate, however, that their somatic lines cannot be perpetuated indefinitely.

1. The death of somatic tissues and organs during development is normally programmed in a variety of organisms (Section 14.5).
2. Aging and death of the somatic cells of the adult are also normally programmed, as indicated by the following observations.
    a. Despite the increase in average life expectancy in human beings and other animals due to advances in nutrition and medicine, there is no evidence that life span can be lengthened indefinitely. In other words, somatic death seems to be an intrinsic feature of organisms. The limitation in life span is

associated with the aging process, which in human beings includes a decreased efficiency and atrophy of the muscular and nervous systems, loss of flexibility of collagen (a connective tissue protein), and a decreased frequency of cell division.

b. As indicated in Section 10.3, cancerous cells can apparently divide indefinitely in cell culture. A subculture of normal fibroblasts obtained from human embryos, however, can double its size in a particular culture medium only $50 \pm 10$ times before subculturing fails. The number of doublings possible decreases with age of the fibroblast; if the fibroblasts are obtained from an individual over 20 years old, the number of doublings is only $30 \pm 10$. Shorter-lived species have a lesser capacity for fibroblast division in cell culture. Embryos of chicken, rat, mouse, and hamster have a doubling number of about 15, this number being considerably smaller for fibroblasts obtained from the adult forms.

c. That the restricted life span for subculturing a normal somatic cell culture is programmed intrinsically is demonstrated by the following experiment. Fibroblasts without Barr bodies (originally isolated from males) that have doubled about 40 times are cultured together with fibroblasts with Barr bodies (isolated from females) that have doubled only about 10 times. After 25 more doublings, only cells with Barr bodies are present in the subcultures. Apparently, the more divisions a somatic cell undergoes, the greater the chance that it will fail to divide—this chance becoming 100 per cent by the 45th to 50th doubling.

d. Young cells can be serially transplanted in young hosts more times than old cells. This shows that aging of a cell is not caused by some nutritional or hormonal deficiency in its environment.

e. Finally, when human fibroblast-like cells that are young (dividing) are fused with those that are senescent (postdividing), DNA synthesis in the young nucleus is inhibited. This inhibition is probably due to a cytoplasmic factor (of nuclear origin?) in the senescent cell.

**27.5** Aging and death may result from a genetic program for the somatic increase and/or accumulation of protein damage.

Aging and death are manifested by the increased likelihood with time that a cell will fail to function properly or to undergo division, or both. Such metabolic failure may be due primarily to the malfunctioning of the cell's membranes and enzymes, and, therefore, primarily to damage to the cell's proteins. Aging and death are expected to occur if protein damage increases and/or accumulates with time. There are two general ways that the genetic material can program the occurrence of more protein damage in somatic cells than in germ cells.

1. CHANGES IN GENOTYPE. Cells in the somatic line of certain species are programmed to lose part or all of their nuclear genetic material (Section 14.5). Such losses may include genes that prevent, repair, or remove protein damage. In species where the genetic content is the same in the germ and somatic lines, the germ-line cells may be better able to repair mutants than somatic cells. For example, while homologous

chromosomes are synapsed during meiosis, chromosome breaks may be more readily restituted, the number of repeated regions in a chromosome may be rectified (Section 21.5), and mutants may be repaired by using as template the normal allele in the homologous chromosome. More mutations having effects on proteins may occur in somatic cells than germ cells due to differences in gene action in the two kinds of cells, as indicated next.

2. CHANGES IN GENE ACTION. The genotype may specify that (a) genes that prevent, repair, or remove protein damage shall be less active in the somatic line than in the germ line, and/or (b) genes that increase or accumulate protein damage shall be active in the somatic line and inactive in the germ line. One specific consequence such differential gene action could have would be a greater frequency of mutations in the somatic line than in the germ line. Certain mutations would have special significance for somatic aging and death since the protein damage they produce in turn generates more protein damage. These mutations include those that occur in genes coding for enzymes, other proteins, and RNA's that are needed for replication, transcription, translation, and repair. They also include mutations in genes that control the functioning of these genes.

Certain lines of evidence are consistent with protein damage as the basis for aging and death. (Such evidence does not prove, however, that the protein-damage hypothesis is correct.)

1. As expected, damaged protein increases with age. This is shown by (a) the accumulation of altered enzymes in aging fibroblasts, and (b) the increased frequency with which amino acid analogs are incorporated.
2. Base analogs cause protein missynthesis and premature aging.
3. There is some indication that changes in gene action occur as cells age. For example, the ability of fibroblasts to acetylate their histones decreases with age (hence aging may prevent the proper activation of DNA). Also, while mRNA ages without being translated, the length of its poly A decreases. (The specific gene-action consequences of this are, however, unknown.)
4. As expected, mutations are associated with aging. Evidence for this includes:
    a. Nicks in DNA increase with age in the brain tissues of mice and dogs, and single-stranded regions in DNA increase with age in the mouse liver.
    b. The amount of nucleolus organizer rDNA decreases with age, perhaps due to deletions.
    c. Gross chromosomal changes accumulate more rapidly in mammals with shorter life spans.
    d. Gross chromosomal mutations increase as fibroblasts age. (These mutations may be the result of protein damage produced by previously occurring mutations.)
    e. The ability to repair UV-damaged DNA is proportional to the longevity of different mammals.
    f. Mutants with a deficiency in DNA repair also age prematurely.
    g. Radiation and chemical mutagens shorten life span.
    h. Antimutagens increase longevity.

The protein-damage hypothesis is just a general mechanism for programmed aging and death. Protein damage may be the basis for various hypotheses that have been advanced as the specific cause of aging. These hypotheses include (a) the loss of the ability of cells to divide, (b) progressive loss of immunity, (c) atrophy of muscles and nerves, (d) decay of connective tissue proteins such as collagen, (e) increasing amounts of highly reactive compounds, and (f) the synthesis of specific aging proteins. At present we do not know how genetic material is specifically related to any of these hypotheses for aging and death.

**27.6** Cancer may be due to abnormal changes in the content or expression of genetic material.

Whereas aging and death seem to be normal outcomes of the differentiation of somatic cells, cancer seems to be the result of the abnormal differentiation of normal cells. Cancer cells are often characterized by:

1. An abnormally high rate of cell division. This is associated with a change in cell surface properties so that the cancer cell is not, like normal cells, inhibited from dividing by being in contact with other cells.
2. The production of progeny cells that are also cancerous.
3. A lesser response than normal cells to the organism's mechanisms of intercellular regulation.
4. The partial or complete loss of differentiation accompanied by changes in energy metabolism.
5. Metastasis, the ability to spread to other regions of the body.

There are at least 100 different types of cancer. The apparent permanency and transmissibility that they all have in common implies that a cancer cell results from a permanent modification in gene content or action. It is possible that different cancers originate by mutation (say, by the addition, subtraction, or relocation of normally present or infectious genetic material), or by a change in transcription (by a permanent abnormal turning off or turning on of genes that are normally present), or by a change in translation (of normal mRNA's or of RNA viruses). It is not surprising, therefore, that cancer can be induced by exposure to mutagenic agents such as high-energy radiations and certain chemical substances as well as by various viral infections.

It is probable that different cancers arise by different means. Two of the various hypotheses for the genetic basis of cancer that are actively being tested experimentally at present are as follows.

1. The *virogene–oncogene hypothesis* proposes that cancer genes, which may have originated in viruses, are normally present in cells in inactive form. Cancer induction simply requires the activation of these genes.
2. The *provirus–protovirus hypothesis* proposes that cancer induction requires a viral infection, since the genes needed to produce cancer are normally absent or are present but defective.

Both hypotheses may apply to different cancers. Still other mechanisms may be involved in initiation, promotion, and maintenance of cancer. For example, cancer cells may arise frequently but only become established when the organism's immunity systems fail.

In human beings, specific types of cancer are associated with specific chromosomal or genetic abnormalities. For example, persons with retinoblastoma have a deletion in chromosome 13; persons with chronic myeloid leukemia have, in their cancer cells but not other cells, a segment missing from chromosome 22 that has apparently been translocated to chromosome 9; persons with xeroderma pigmentosum, who are homozygotes for a recessive autosomal allele that fails to synthesize an enzyme needed to excise pyrimidine dimers, often develop skin cancer when exposed to sunlight. Human cancers are known to be produced by X rays and radium; and certain human cancers seem to contain viruses or RNA's specific for viruses. In no case, however, do we yet know how the genetic changes entailed are related to the production of any cancer.

# APPLICATIONS TO ENVIRONMENTAL MUTAGENESIS

The mutational load in human beings is being increased by exposure to environmental physical and chemical mutagens. Mutants induced in somatic cells are, of course, restricted to the person in whom they occur. Since a person will usually have a normal allele in addition to the mutated one, most somatic mutants only produce their detrimental effects in heterozygous condition. When such mutants are intragenic, these effects are relatively minor. When such mutants produce gross deletions or acentric or dicentric chromosomes, however, they may cause appreciable damage when the cells containing them divide.

Mutations in the germ line may be transmitted to the next generation. The earlier that mutation occurs in the germ line, the greater will be the proportion of the germ cells carrying the new mutant. The new, usually detrimental, mutant will probably first occur in heterozygous condition, where it can be selected against; it may persist, however, and become homozygous in a subsequent generation, where it will usually produce even greater detriment. We will consider the somatic and germ line effects first of environmental radiation mutagens and then of environmental chemical mutagens.

**27.7** Our environment exposes us to physical mutagens.

Mutations are chemical changes that require energy, and the more energetic a radiation is the more likely it is to induce mutations. Less energetic radiations (such as visible light) are usually converted to heat and produce fewer mutations than more energetic radiations (such as UV light) whose energy is converted to heat and also excites molecules by shifting the position of an electron within a molecule. High-energy radiations (such as X rays and gamma rays, as well as bombardment by particles such as alpha particles, beta particles, neutrons, protons, and other atomic particles such as occur in cosmic rays) not only heat and excite molecules but leave most of their energy in cells in the form of electrically charged particles, or *ions*, produced by *ionization*. These ions are electrons, torn free of atoms, that can move at great speed to cause other atoms to lose electrons. Each electron lost from one atom is eventually gained by another atom. Since every loss or gain of an electron causes a chemical change, such changes in genetic DNA or RNA often produce mutations.

The less energetic the radiation is, the less able it is to penetrate cells before its energy is absorbed. Thus, visible and ultraviolet light do not penetrate deeper than our skin. Accordingly, UV-induced mutations occur only in the skin of exposed individuals (such mutations and skin cancer being especially frequent in persons with xeroderma pigmentosum), producing somatic, not germinal damage.

Since ionizing radiations penetrate deeply, they can cause somatic as well as germinal mutations. The unit of dose of ionizing radiation is called the *rem*. The frequency of mutations resulting from single chromosome breaks or other single gene changes increases in direct proportion as rem dosage increases. There is, therefore, no safe dosage of ionizing radiation for mutations of this type. When a given dose is administered quickly, moreover, it is often likely that more gross rearrangements resulting from the simultaneous presence of two or more breaks will occur than when the same dose is given over a long period of time. In the latter case, some breaks may restitute before others required for a gross rearrangement are produced.

Because of various uncertainties it is thought that the number of rem required to double our spontaneous mutation frequency, which is of the order of 1 mutation per 100,000 gametes, lies between 20 and 200. It is calculated that in the United States the average exposure to ionizing radiation is about 5 rems per 30 years, the average length of a generation. Of this amount only about 80 per cent, or 4 rems, reach the gonads. At present, therefore, ionizing radiation is probably producing less than 25 per cent of our spontaneous mutation frequency.

About 60 per cent of our radiation exposure is due to natural radiation (cosmic rays, radioactive rocks, or foods). Almost all of the remaining 40 per cent is man-made (radioactive fallout from atomic or nuclear bombs, from nuclear power and other industries, the greatest amount coming from radiation used in medicine and dentistry). Our present radiation exposure is not considered to be a threat to the survival of the human species. But since almost all mutants are harmful, and can only be removed by genetic death, any steps taken to reduce our radiation exposure are desirable.

Many actions have already been taken to decrease radiation exposure. These include stopping:

1. The drinking of and bathing in radioactive waters (fashionable in the early part of this century).
2. The licking of brushes used to paint radium on watch dials.
3. The use of X-ray fluoroscopy to fit shoes (common in the 1930s).
4. X-ray visualization of one's heart beating for purposes of amusement.

5. Exposure of technicians and radiologists to X-ray scatter while treating patients.
6. Exposure of medical and dental patients to any more ionizing radiation than is needed for purposes of diagnosis or treatment.
7. The use of X rays in certain medical practices (such as the visualization of the fetus; treatment of children for swollen tonsils or adenoids; treatment of adults suffering from ankylosing spondylitis—a fusion of the vertebrae).
8. The testing of atomic or nuclear bombs by the major powers. It should be noted that whole-body exposure to 300 rems has a 50 per cent chance of killing. Such doses are readily exceeded by atomic and nuclear bombs.

In recent years guidelines have been established to further reduce radiation exposure of persons working in atomic power industries, medicine, and dentistry. The use of X rays to detect breast cancer in women has just recently been restricted in the United States to women having the greatest likelihood of having the disease. Medical and dental radiation exposure may be further reduced by as much as 50 per cent by avoiding excessive repetitive X rays, by shielding from X-ray scatter areas of the body not being examined or treated, and by improving the machinery and procedures used in radiation diagnosis and therapy as well as the training of the personnel involved.

Finally, it should be noted again that energetic radiations produce cancer. Increased frequencies of cancer have been found among various groups mentioned earlier—persons exposed to too much sunlight, radium dial painters, early radiologists, children irradiated *in utero* or treated for swollen glands, persons irradiated to cure ankylosing spondylitis, and persons exposed to atomic bomb explosions. Although we cannot be sure such cancers are due to radiation-induced mutations, they can be considered to be due to permanent changes in either gene content or gene action.

## 27.8 Our environment exposes us to chemical mutagens.

Since genetic DNA is chemically essentially identical in all cellular organisms, a penetrating radiation that produces mutations in one organism can doubtless do so in all others. The same situation does not exist with regard to chemical substances. Since different organisms differ in form and metabolism, a chemical substance that is mutagenic to bacteria may be inactivated in this regard before it can reach the nuclear DNA of a *Drosophila*, mouse, or human being. The converse is also true. Metabolism may convert a nonmutagenic chemical substance into a mutagenic one in a mammal but not in a bacterium. Since we do not use human beings to test experimentally whether a given chemical substance is mutagenic, we must rely on work done with human tissue cultures or with other organisms, keeping in mind that there will be some false positive and false negative results. Since the larger the experimental organism, the more expensive it is to test for mutagenesis, screening for chemical mutagens is usually best done with bacteria, subsequently testing bacterial mutagens in *Drosophila*, mice, or human tissue cultures.

All penetrating radiations are both mutagenic and carcinogenic. Many chemical mutagens are also carcinogenic. The intimate relationship between mutagenicity and carcinogenicity has led to two kinds of experiments. One kind discovers new chemical mutagens that can subsequently be tested for carcinogenicity, say in mice. The other kind tests whether already known chemical carcinogens in mammals are also mutagens.

Some environmental chemical substances known to be carcinogenic and/or mutagenic are listed in Figure 27-2. Some of these substances are used in industry and agriculture, including benzene and benzene derivatives, mercury compounds, vinyl chloride, asbestos, and pesticides. Others occur in our food, including nitrates and nitrites, cyclamate and saccharine, caffeine, and excess vitamin C; or in our cosmetics, including permanent hair dyes. Still others are used to treat diseases—methotrexate (vs. psoriasis), hycanthone (vs. schistosomiasis, a parasitic fluke infection), isoniazid (vs. tuberculosis), and alkylating drugs (vs. cancer).

Ionizing radiation and/or alkylating drugs can cure cancer because they are mutagens that produce many gross rearrangements involving deficiency, duplication, acentric, or dicentric chromosomes. Since cancer cells divide more often than noncancer cells, such gross rearrangements more often produce gross chromosomal imbalance, and hence death, in the progeny of cancer cells than of normal cells. These same mutagens can also initiate new cancers. Such new cancers, whose origin is often delayed

FIGURE 27-2. Environmental mutagens and carcinogens.

| Substance | Mutagenic | Carcinogenic |
|---|---|---|
| Industry and Agriculture | | |
| benzene | + | + |
| modified fused ring hydrocarbons (benzpyrene, benzanthracene) | + | + |
| mercury compounds | + | |
| vinyl chloride | + | + |
| asbestos | | + |
| pesticides | | |
| (DDT, dieldrin, aldrin) | + | |
| (DBCP) | | + |
| fungicide (captan) | + | |
| plant growth inhibitor (maleic hydrazide) | + | |
| Food and Cosmetics | | |
| nitrates and nitrites modified to | | |
| nitrous acid | + | } + |
| nitrosamines | + | |
| cyclamate | + | + |
| saccharine | | + |
| caffeine | + | |
| vitamin C (ascorbic acid) in excess | + | |
| permanent hair dyes | + | + |
| Medicine | | |
| methotrexate | + | |
| hycanthone | + | |
| isoniazid | + | + |
| alkylating drugs (Melphalan, Chlorambucil, Thiotepa, N mustards) | + | + |

for years, are only a problem in a small fraction of persons cured of a cancer that otherwise would have killed them.

Nitrates and nitrites (common in pastrami, hot dogs, and bacon) and certain benzene derivatives are converted into carcinogens after they enter the mammalian cell. They are also mutagenic in bacteria after such conversion but not before. It should be noted that once the public has become accustomed to the presence of certain chemical substances in the environment it is very difficult to restrict their use if they are subsequently found to be mutagenic and/or carcinogenic. This has been true for cigarettes (whose tars are carcinogenic), cyclamate and saccharine, and permanent hair dyes (which are the basis of a quarter of a billion dollar per year industry serving more than 33 million people in the United States).

Many new and old chemical substances will need to be tested for mutagenesis. Although we expect chemical mutagens to most readily affect somatic cells, we need to keep in mind that they may also produce significant numbers of germinal mutations.

# APPLICATIONS TO GENETIC DISEASES AND GENETIC ENGINEERING

Research achievements in genetics have had such a great impact on medicine that Nobel prizes in the category of physiology or medicine have been awarded for genetic discoveries 10 times, eight of these within the past 20 years. These advances have had many applications to many areas of medicine—for example, the use of ionizing radiations and chemical mutagens in diagnosis and therapy; the molecular basis and treatment of diseases caused by one's own genes or by infection; the analysis of pedigrees and populations with respect to disease; disease resistance and immunology. All these have already been discussed here, at least briefly.

We now wish to discuss certain actual or potential applications of genetics to medicine and public health that are or may become particularly important and that have ethical implications. The ethics involved can be noted only briefly, since a complete discussion of ethical issues is beyond the scope of this book.

**27.9** Many genetic diseases can be detected before as well as after birth.

At least 30 inborn errors of metabolism, or *genetic diseases*, can be detected in the growing fetus by means of *amniocentesis*, that is, by puncturing the amnion and obtaining a sample of the fluid bathing the fetus. The sample is obtained by withdrawing some of the fluid through a hypodermic needle passed through the vagina or the abdominal wall and into the amniotic cavity. This fluid contains cells cast off by the fetus as well as cells of the amnion, both of which have the same genotype. After these cells are cultured, they can be examined microscopically for chromosomal abnormalities and can also be tested biochemically for the presence or absence of certain enzymes or other metabolic features.

Amniocentesis is usually employed only when the fetus has a high risk of genetic disease. When the fetus demonstrates a chromosomal defect such as trisomy, or a deficiency for an enzyme such as HGPRT (causing Lesch–Nyhan syndrome) or hexosaminidase A (causing Tay–Sachs syndrome), the usual practice is to terminate pregnancy, that is, to induce abortion. Induced abortions reduce the frequency of detrimental alleles in the population. Some pertinent areas of direct concern to the general public are the availability of amniocentesis, legal and religious restrictions on therapeutic abortions, who shall make the decision to abort, and how serious the genetic impairment need be before a therapeutic abortion is performed.

Genetic diseases that are detectable through amniocentesis can also be tested for at birth or later. In addition, whole populations of relatively high-risk groups of people can be tested at birth for genetic diseases such as phenylketonuria, homocystinurea, and galactosemia. Since the personnel and funds required for such testing are limited, the question arises: What groups shall be screened for which genetic diseases? When a disease is rare in a given population group, the cost of detecting a case (by testing every member of the population) becomes very high per detected case. Should only high-risk groups therefore be tested for a given genetic disease? For example, should only Jews be screened for Tay–Sachs disease while only blacks are screened for sickle-cell trait? It has been argued, on the other hand, that the racial problems produced by selective screening outweigh its genetic advantage. For example, some people claim that blacks are stigmatized by being selectively screened for sickle-cell trait, and are discriminated against because their rare cases of Tay–Sachs disease are not discovered. Should whites also be tested for sickle-cell trait?

## 27.10 The phenotypic effects of many genetic diseases can be treated with medicines or corrected by surgery.

The harmful phenotypic effects of genetic disease can be prevented or alleviated in several ways. We have already noted how a controlled diet can prevent the mental retardation of phenylketonurics. Other birth defects that have a genetic basis can be remedied surgically; these include such defects as cleft palate, pyloric stenosis, clubfoot, bilateral retinoblastoma, and certain types of congenital heart disease. Most persons with these birth defects can lead essentially normal lives after surgical repair. Persons with other genetic diseases can be treated with medicines. For instance, the blood of persons with hemophilia will clot normally if a special cryoprecipitate is injected frequently. Since only the main phenotypic effects of detrimental alleles are being treated dietetically or medically, other, usually harmful, multiple phenotypic effects may remain. The reduction of harmful phenotypic effects slowly but surely increases the persistence and the frequency of these alleles in the population (Section 24.2).

The treatment of genetic disease can be very expensive. For example, it costs thousands of dollars a year to supply cryoprecipitate to a hemophiliac. Other treatments that require transplantation of eye parts, bone marrow, or of whole organs such as the kidney or heart are limited by the availability of donors (usually siblings). Hopefully, the use of cadavers as organ donors will become more widespread in the future. How heroic shall the attempt be to alleviate the phenotypic damage of genetic disease? In recent years, the tendency has been not to treat individuals who, after treatment, would be seriously retarded mentally or paralyzed. For example, spina bifida, the failure of the spinal cord to close normally, can be repaired surgically. But after surgery most children who had extreme cases of spina bifida remain paralyzed or mentally retarded, or both. Public and medical discussion is needed in order to establish guidelines that will help determine who is to be treated medically and in what way.

**27.11** Genetic diseases that cannot be prevented or cured at present may be amenable to treatment in the future.

General increases in genetic diseases can be prevented in large part by the avoidance of needless exposure to radiation and chemical mutagens. Specific genetic diseases in families can be avoided through selective breeding, even though the frequency of rare recessive alleles in the population will be reduced only slowly. Selective breeding is assisted by the identification of heterozygotes; by genetic counseling, which gives prospective parents an estimate of the risk of having children with certain genetic diseases; and by the use of artificial insemination, which employs semen (capable of being stored in sperm banks) of genetically normal men in place of the sperm of a husband who carries a gene for a genetic disease. The fertilization and early development of human beings can presently be carried out *in vitro* and can be followed by implantation and normal development. It is likely that in the future it will be possible to implant such developing embryos in foster mothers. A fetus could thus be protected from somatic damage due to a genetic disease carried by the mother. Conversely, a woman who carried a detrimental gene could nurture an implanted fetus produced by fertilizing a donor egg with her husband's sperm.

The genetic cure of genetic disease requires the addition or removal of genes to repair imbalances, or the addition of normal alleles to counteract (and preferably replace) defective alleles. All the techniques needed to produce genetic cures in somatic cells of human beings are already available and have been used in prokaryotes, in other eukaryotes, or in cultured cells. These techniques include the following:

1. TRANSGENOSIS. Cultured human fibroblast cells that are $gal^-$ have been infected by transducing $\phi\lambda$ carrying the $gal^+$ region of the *E. coli*

chromosome. The $gal^+$ region introduced was both replicated and functional for 40 days, or eight successive divisions. In other experiments the *E. coli* gene for $\beta$-galactosidase attached to a defective $\phi\lambda$ chromosome was functional when introduced into $\beta$-galactosidase-deficient cultured human fibroblast cells either by phage infection or by uptake of naked phage DNA.

2. TRANSFORMATION. Human cells from a genetically defective patient can be grown in cell culture and then exposed to normal human DNA. Genetically normal, transformed cells can then be reimplanted in the donor. The success of live transplantations in rats indicates that the reimplantation of transformed liver cells could cure phenylketonuria in human beings.

3. TRANSDUCTION. Viruses can be used to transduce normal human DNA into genetically defective human cells. We expect many viruses that normally infect human beings can be used to produce specific transductions (Section 8.7).

4. CELL FUSION. Normal genes can be introduced into genetically abnormal cells by fusing normal and abnormal cells. For example, the gene for a particular enzyme present in a chick erythrocyte has been introduced by somatic cell hybridization into a line of mouse fibroblast cells genetically deficient for the enzyme. This gave rise to a line of mouse fibroblast cells that actively produced the enzyme in question.

5. RADIATION SURGERY. Microscopic beams of penetrating radiation can be used to damage, and hence destroy, whole chromosomes or parts of chromosomes. Such surgery might be used to partially cure chromosome overdoses. For example, a trisomic explanted cell may be cured by radiation surgery and then be implanted under circumstances that favor the replacement of defective cells by repaired cells.

6. GENE SYNTHESIS. Any transcribed gene can be synthesized *in vitro* by reverse transcription of its RNA transcript. It is now also possible to synthesize any desired duplex DNA *de novo*. For example, the gene sequence promoter-Tyr tDNA-terminator has recently been synthesized *de novo* using enzymes and individual deoxyribonucleotides. This gene sequence was functional when introduced in *E. coli*. Synthesized genes may prove to be useful in the future in the study and repair of mutations.

Whether any of the above techniques for preventing or curing genetic diseases becomes feasible for use in human beings in the near future depends upon public support of basic and applied medical research. Care must be exercised to avoid the political or social misuse of either selective breeding or of experimental techniques that change the genetic material in our somatic or germ lines.

## 27.12 For better (or worse) genetic engineering can revolutionize life on earth.

*Genetic engineering* refers to a directed intervention with the content and/or organization of an organism's genetic material. In addition to the genetic engineering techniques described in the last section, others are already known or being developed that will make possible the construction of large numbers of specific types of human beings and other organisms. These new genetic engineering techniques employ the cloning of cells, nuclei, or molecules.

NUCLEAR AND CELL CLONING. In amphibians, the nucleus of a fertilized egg can be replaced by the properly prepared nucleus of a somatic cell. Development then proceeds normally, but in accordance with the genotype of the nucleus introduced. This nuclear transplantation experiment can be repeated using different cells of the same tissue as nuclear donors. The result is a clone of individuals that are genetically identical with respect to nuclear genes. Nuclear transplantation (and hence nuclear cloning) should also be possible in human beings (although it is expected to be more difficult because human egg cells are much smaller). It is also possible to obtain whole carrot plants by culturing single somatic cells. Cloning single cells of the same carrot would produce a clone of genetically identical carrots. Although very unlikely to be feasible, single cloned human cells may be induced to develop into individuals, thus producing clones of identical human beings. Cloning techniques applied to humans could produce as many persons of a specific type as desired. Who would make the decision as to who would be cloned? Cloning techniques applied to other animals or to plants are clearly advantageous and present no ethical problems.

MOLECULAR CLONING. It is now possible to transplant genes from almost any organism into a microorganism, where they will produce clones of genes as the result of chromosome replication and cell division. Figure 27-3 illustrates one of several mechanisms that have been successfully used for this purpose. A drug-resistance plasmid with the following characteristics is employed: it is normally present in many copies in its *E. coli* host; its duplex ring DNA chromosome carries only one palindrome within which staggered nicks are produced by a particular restriction nuclease; the nicked sequence is not part of this plasmid's genes for either DNA replication or for drug resistance. After nuclease treatment of isolated plasmid DNA, partial strand separation produces a duplex linear chromosome with complementary single-stranded ends. After nicking by the same nuclease, any isolated foreign DNA that contains two (or more) copies of the same palindrome will produce duplex chromosome segments with homologous, complementary, single-stranded ends. To insert the foreign DNA into the drug-resistance chromosome, the two kinds of nicked DNA are mixed so they can hybridize with each other, after which nicks are sealed with ligase. Drug-sensitive *E. coli* is then treated so that the recombinant, circular

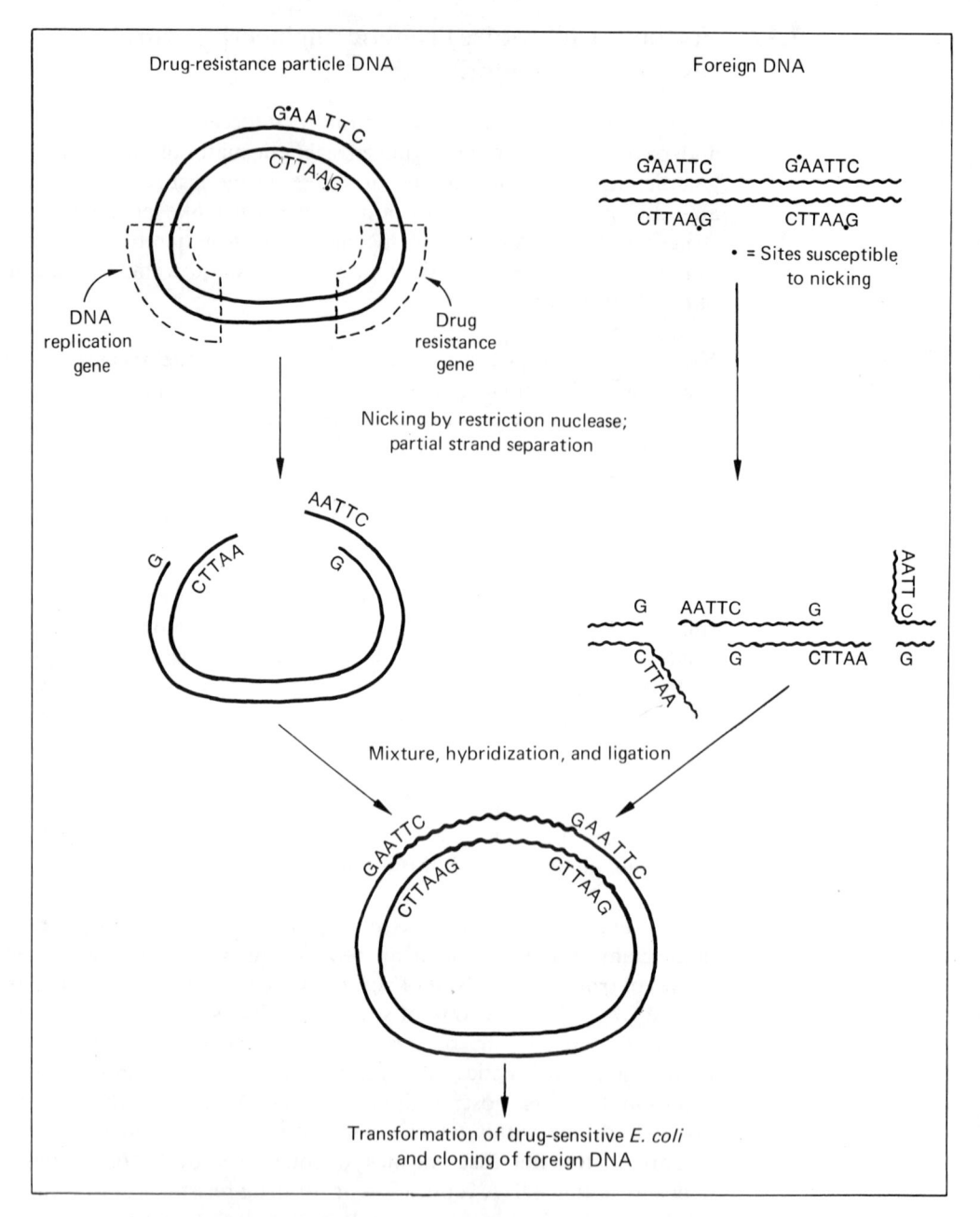

FIGURE 27-3. Foreign DNA can be inserted into a drug-resistance particle and cloned in E. coli.

duplex can enter and transform the cell. Only those cells which are, in fact, transformed survive subsequent drug treatment. Survivors form clones, each of whose members contains many copies of the same recombinant plasmid chromosome and, therefore, many copies of the same piece of any incorporated foreign DNA. Clones can then be screened for particular foreign genes that they may carry.

Since different restriction nucleases make staggered nicks in different palindromes, a given piece of DNA can be divided into pieces in several different ways. It is apparently possible, therefore, to clone any, or almost any, foreign gene in *E. coli*. Genes cloned in *E. coli* include those derived from other bacteria, *Drosophila*, *Xenopus* (nucleolus organizer DNA), and human beings (the genes for insulin and somatostatin). Some of these genes are transcribed (nucleolus organizer DNA, somatostatin DNA), and some mRNA's are translated (somatostatin). We can clone, therefore, not only DNA genes, but their transcripts and the proteins these code for.

Molecular cloning has great potential benefits. It can be used to determine the functions of different segments of nuclear or organelle DNA. It can be used to inexpensively synthesize large quantities of antibiotics, hormones, and enzymes. It is conceivable that human intestinal bacteria could be made to digest cellulose, in which case the threatening biological energy shortage would be greatly postponed. Once molecular cloning has been made applicable to eukaryotes, it will be easier to add specific genetic material to plants and animals, including human beings. For example, it might become possible to add to crop plants the genes needed to fix atmospheric nitrogen. This would eliminate their need for nitrates in fertilizer.

Molecular cloning may produce great benefits for human beings, but it may also entail considerable hazard. All kinds of cloning can be misused for political and social purposes, as can any genetic procedure aimed at selectively improving human beings and other organisms. Furthermore, molecular cloning is potentially a great biological hazard. Certain molecular recombinants that can be cloned in *E. coli* or other microorganisms might produce disastrous effects if released accidentally. Two potential hazards are (1) the production of strains of organisms, not found in nature, that have been cloned for genes that provide either drug resistance or the ability to produce toxins; and (2) the cloning of DNA, from tumor viruses or other animal viruses or bacteria, whose accidental release may increase the incidence of cancer. A conference of scientists and nonscientists from more than 15 countries was held in 1975 on this topic; the participants agreed that caution should be excessive rather than minimal. The conference recognized the great benefits achievable through molecular cloning, but recommended that such work be done only under conditions of strict containment in which the host strains used can live only under stringent laboratory conditions. Certain experiments were considered too hazardous to perform even with the most careful containment. The guidelines for cloning DNA proposed by the U.S. National Institutes of Health in 1976 have been widely adopted by research biologists here and abroad. Some scientists admit that these guidelines represent overkill on the safety side. It seems reasonable that the public alarm that such research work has aroused be allowed to dictate a level of safety beyond that which scientists consider necessary. If experiments now planned reveal in a year or two that the cloning of recombinant DNA is not hazardous, regulations can be relaxed somewhat. Meanwhile, through 1977, the U.S. federal government has failed

# 390 CONSEQUENCES OF GENETICS

to pass national recombinant DNA regulations that would assure that such research in industry was following safe guidelines.

## SUMMARY AND CONCLUSIONS

Genetics has made and will continue to make important contributions to medicine and public health. With regard to immunology, genetics helps us understand the basis for successful and unsuccessful transfusions and transplantations, and for certain blood incompatibility diseases. Genetics also provides us with an orientation to or an understanding of the normal and abnormal programs entailed with aging, death, and cancer.

We are exposed to a wide variety of physical and chemical mutagens in our environment. Many of these are also carcinogenic. Whereas physical mutagens can produce many germ-line mutations, most chemically induced mutations probably occur in somatic cells. Since most mutations are harmful, it is desirable to minimize our exposure to man-made physical and chemical mutagens.

Many present-day genetic diseases can be detected before birth. Purposeful abortion of affecteds reduces the population frequency of the genes responsible. Genetic diseases present at birth can be treated with medicines and surgery. Such treatment increases the population frequency of the genes responsible. Genetic diseases that cannot be prevented or cured at present may, in the future, be avoided by selective breeding or be genetically cured by genetic engineering. Genetic engineering techniques for cloning cells, nuclei, and especially DNA, RNA, and protein molecules hold great promise for human betterment (but not without some possible risk). Abortions, treatment of genetic diseases, and genetic engineering all entail ethical problems.

Future progress in basic research in genetics and future application of its principles to other sciences and nonscientific fields for the benefit of all people will depend upon many factors. It is clear that attainment of such advances will require the cooperative effort of geneticists, of experts in the various nongenetic areas concerned, and of an informed public.

## QUESTIONS AND PROBLEMS

27.1 *Riddle*: What human female can donate to a human male, organs that will not be rejected after transplantation?

27.2 *Songs*: **a.** "Does It Make My Brown Eyes Blue?" **b.** "Five Foot Two, Eyes of Blue." What is the genetic application or implication of either or both songs?

27.3 *Songs and Movies*: **a.** "My Heart Belongs to Daddy." **b.** "Frankenstein." What is the main problem in either or both cases?

27.4 *Movies and TV*: **a.** "Planet of the Apes." **b.** "Mr. Ed." What genetic explanations are needed to make one or both believable?

25.7 *Book*: **a.** *Brave New World*. **b.** *The Boys from Brazil*. What are the genetic applications and implications of either or both books?

27.6 From what ABO blood types can persons of the following blood types receive transfusions? **a.** A. **b.** B. **c.** AB. **d.** O.

# APPLICATIONS TO MEDICINE 391

**27.7** A baby has MN, AB, and Rh-positive blood types. The mother has M, B, and Rh-negative blood. What blood types are possible for the father?

**27.8** A hospital has three babies mixed up. Which baby belongs to the parents who are MN, A, Rh positive, and N, O, Rh positive? **a.** N, AB, Rh positive. **b.** MN, O, Rh negative. **c.** M, B, Rh positive.

**27.9** How can you explain the observation that $I^A I^O \male \times I^O I^O \female$ marriages produce 25 per cent fewer $I^A I^O$ children than do the reciprocal marriages?

**27.10** Some societies, for example the Japanese, are reducing the size of their families. What effect do you expect this to have on the occurrence of: **a.** Rh incompatibility? **b.** trisomy 21? **c.** cousin marriage? **d.** homozygotes for a rare recessive allele?

**27.11** Describe one evidence that aging and death is genetically programmed.

**27.12** How are proteins related to aging?

**27.13** Name one genetic and one chromosomal abnormality associated with cancer.

**27.14** Distinguish between permissible doses and safe doses of ionizing radiation.

**27.15** Name three exposures to ionizing radiation in the past that are now considered foolish or undesirable. Name one present exposure that is foolish or undesirable.

**27.16** Suppose it was proven that saccharine caused cancer in the frequencies indicated below. At what incidence would you outlaw its use, and why? **a.** 1 in 1 billion. **b.** 1 in 100 million. **c.** 1 in 100,000. **d.** 1 in 10,000.

**27.17** What would happen to the hair dyeing industry if our first lady went gray? Is this desirable? Explain.

**27.18** Should a genetically deformed baby ever have the right to sue its parents for genetic negligence? Explain.

**27.19** Should the parents of a half-translocation Down's syndrome baby be permitted to sue a physician for malpractice if he assured them they would not again have such a child, and they did? Explain.

**27.20** Should persons who contract cancer due to working conditions be permitted to sue their employers? Explain.

**27.21** Should sunlamps be outlawed? Explain.

**27.22** What effect may SST's and Freon have in common that has genetic implications?

**27.23** Should gene-splicing experiments be controlled at the national level? Why?

**27.24** Should the money now used to explore space be used for genetic research? Why?

**27.25** Which alternatives are correct? A man of O blood type marries a woman of AB blood type. **a.** None of their children will have the same blood type as their parents. **b.** They could have grandchildren of A, B, AB, or O types. **c.** This couple carries three different alleles of a gene. **d.** A given child cannot, if it marries only once, have all four types among its offspring. **e.** Their children will show the results of segregation and their grandchildren the consequences of independent segregation.

**27.26** Which alternatives are correct? Blood transfusions can always be made safely between twins of the same sex: **a.** if the twins are monozygotic. **b.** if the twins are dizygotic. **c.** if the parents have the same blood type. **d.** if the parents have different blood types. **e.** a. plus d. **f.** b. plus c.

## CONSEQUENCES OF GENETICS

**27.27** *Crossword Puzzle.*

ACROSS

1. parade rest
7. His, Arg, or Lys
17. antigout and antispindle drug
19. prefix: twice
20. city in Russia
21. Fr.: he
22. Sp.: another
23. managed under odds
26. anger
28. start of limerick (*Hint*: see text's dedication pedigree)
32. river
33. without (comb. form)
34. Hindu sounds of assent
35. symbol for tin
36. Fr.: fire
37. Noah's boat
38. formerly
42. half the width of an em
44. social disease
45. life (comb. form)
46. equal (comb. form)
47. limerick continued
54. exclamation
55. try; dissertation
56. small tick; a small amount
57. form of "to be"
59. breakfast, lunch, or dinner
62. turmeric; extinct train delivery service (abbr.)
63. continent (abbr.)
66. obstruct; female parent
67. limerick continued
70. last mo. of yr.
72. booze-induced violent delirium
73. encore; twice (variant of 19 across)

74. construct
76. __ __ Tittle
79. god of love; Cupid
81. provided with openings
84. digraph; wind
85. limerick continued
89. wages
91. eye; sphere
92. endoplasmic reticulum (abbr.)
94. dimer-inducing light
95. rigor mortis or drink will ___
99. 1150
101. chromosome with two exchanges (abbr.)

103. form of "to be"
105. crew; troops
106. kind of window
107. gull relative
108. end of limerick
114. city in Illinois
115. to utter a convulsive sigh
116. fit
117. continent (abbr.)
118. color; dull
120. electrical engineer (abbr.)
121. bromouracil or aminopurine
125. made possible by Roentgen
126. steady; firm

DOWN

1. make less basic
2. ___ one's nerve (become scared)
3. elevated train
4. current
5. Olympic put
6. one of two
7. sym___ (type of life style)
8. indefinite article
9. zone of action
10. assist
11. southern France
12. shakes one's head in assent
13. conjunction
14. prefix: airplane
15. tribes
16. legal document
18. Watson and ___
24. electrical units
25. gone by
27. DNA's
29. nerve (comb. form)
30. Rough Rider's initials
31. ___ one ___ (get drunk)
37. alternative forms of a gene
38. whirlpool
39. reporter's question
40. megacorporation (abbr.)
41. jungle cat
43. Japanese theater
44. by way of
46. that is (abbr.)
47. fruit preserve; in a ___
48. baseball position
49. requested
50. presentiment-like
51. coins of Morocco
52. gypsum
53. Russ.: yes
54. goes with Laurel or Weinberg
57. Des Moines, ___
58. Biblical: strike
60. end of a list
61. Doctor: "Say ___."

62. automaton
64. tailless Amphibia
65. stained moray
66. April 15, taxes are ___
68. newspaper story
69. alternative
71. $1.00/___
75. dove's call
77. Greek M
78. red wine
80. Great Barrier ___
81. U.S. chief exec.
82. poetic: over
83. hesitant sound in speech
85. cross $AA$ ___ $aa$
86. "... when I put out __ __."
87. Organization formerly headed by J. E. Hoover
88. expand
89. Chinese coin
90. a passport endorsement
93. *Of Human* ___
96. Australian birds
97. Roman fiddler
98. Fr.: one priest
99. perhaps
100. *Sphaerium*; Roman circular robe
102. coal tar product
104. "___ ___ song of sixpence..."
106. erythro___
107. preposition
108. zenith
109. cautious
110. Charon's river
111. suited to __ __
112. placekicker or ballet dancer
113. fasting period
119. digraph
122. article
123. organization of former drinkers (abbr.)
124. pound (abbr.)

# Glossary

**Abortive transduction.** Transduction in which genetic material is not integrated (or replicated) but may be otherwise functional.
**Acentric.** Lacking a centromere.
**Adaptive value.** *See* Fitness.
**Alkylation.** Addition of any radical of the methane series or a derivative of that series.
**Allele.** One of the alternative forms of a gene that may occupy a particular locus in a chromosome.
**Allele frequency.** The number of loci at which a given allele is found in a population, divided by the total number of loci at which it could occur.
**Allolactose.** Effector for *lac* repressor protein.
**Ambiguous code.** Genetic code in which a codon has more than one meaning during translation.
**Aminoacyl-tRNA synthetase.** Enzyme that attaches a specific amino acid to a specific tRNA.
**Amniocentesis.** Technique of obtaining amniotic fluid for the prenatal detection of fetal disorders.
**Amplification.** Synthesis of all or part of a chromosome set in extra copies which are functional but temporary.
**Anabolism.** Constructive metabolism that involves the synthesis of complex molecules from simpler ones.
**Antibody.** *See* Immunoglobulin.
**Anticodon.** Complement of a codon; part of the structure of all tRNA's and of 16S and 18S rRNA's.
**Antigen.** Foreign giant molecular substance which combines with the immunoglobulin whose production it induces.
**Antigen-binding site.** One of two identical regions in an immunoglobulin which can bind identical antigens.
**Antimutagen.** Agent that reduces the mutation rate induced by a mutagen.
**Antiparallel.** Two parallel molecules that point in opposite directions.
**Assortative mating.** Nonrandom mating in which like types tend to pair.
**Attachment site.** Locus in a bacterial chromosome where a phage chromosome normally integrates.

# GLOSSARY

**Autoradiograph.** Photograph that shows the location of radioactive substances in cells or tissues, obtained by exposing a photographic emulsion in the dark to radioactive emissions from the preparation, and then developing the latent image.

**Autosome.** Chromosome other than a sex chromosome.

**Auxotroph.** Organisms whose metabolism requires supplements to the basic food medium required by the wild-type organism.

**Backcross.** Cross of one individual with another whose genotype is the same as an ancestor of the first individual.

**Bacteriophage.** Bacterial virus; phage.

**Balanced lethals.** Two recessive lethal alleles whose hybrid survives.

**Barr body.** See Sex chromatin.

**Base analog.** Compound that resembles a nucleic acid base and can therefore sometimes be incorporated in nucleic acid in place of the normal base.

**Basic amino acids.** Amino acids, including Arg, Lys, and His, which have a net positive charge in neutral solutions.

**Biological fitness.** See Fitness.

**Bridge.** Chromosome region between the two centromeres of a dicentric which is being pulled toward both poles of a spindle at anaphase.

**cAMP.** Cyclic adenosine monophosphate—a nucleotide.

**CAP.** cAMP acceptor protein.

**CAP interaction site.** Region of the promoter that seems to contain the base sequence which cAMP·CAP recognizes.

**Catabolism.** Destructive, often energy-yielding, metabolism that involves the breakdown of complex molecules into simpler substances.

**Cell.** Smallest membrane-bound unit of protoplasm produced by independent reproduction.

**Cell membrane.** The protoplasmic cover of the cell.

**Cell wall.** Polysaccharide-rich, rigid layer outside the cell membrane of plant cell.

**Centric.** Having a centromere.

**Centriole.** DNA-containing structure detected within some centrosomes.

**Centromere.** Portion of chromosome that undergoes directed movement when attached to spindle tubules.

**Centrosome.** Structure that serves as a pole at either end of the spindle in animal cells.

**Charged tRNA.** tRNA carrying an amino acid.

**Chloroplast.** Green plastid; site of photosynthesis in green plants.

**Chromatid.** Thread in a chromosome which is visible in the light microscope and contains a single DNA duplex.

**Chromatid break.** Break through one of the chromatids present in a replicated chromosome.

**Chromatin.** Partly clumped, tangled mass of interphase nuclear chromosomes.

**Chromatin reconstitution.** Reassembly of the DNA and proteins in chromatin.

**Chromosome.** Fiber completely or partially composed of genetic nucleic acid.

**Chromosome arm.** Chromosome limb; a part of a rod chromosome to one side of the centromere.

**Chromosome break.** Break that scissions the chromosome.

**Chromosome segregation.** Separation of the members of a pair of homologous chromosomes so that only one member is present in any postmeiotic nucleus.

**Chromosome set.** Group of one or more chromosomes containing one chromosome of each kind normally present.

**Clone.** All the cells or individuals descended asexually from a single cell or individual.

**Cloning.** Making identical copies of individuals, cells, nuclei, genes, RNA's, or proteins by an asexual, biological process.

**Codon.** Three successive nucleotides (or bases) in mRNA which specify an amino acid or polypeptide chain termination.

**Coevolution.** Interdependent evolution of two species, for example that of a pathogen and its host.

**Competence.** Ability to be transformed.

**Complementary chains.** The two polynucleotides whose base pairing produces a regular double helix.

**Complementary genes.** Two or more nonalleles whose phenotypic effects are needed to produce a single trait.

**Complete transduction.** Transduced genetic material is integrated.

**Concatemer.** Chromosome linearly repeated two or more times.

**Congenital.** Existing at birth. Congenital malformations usually exist before birth.

**Conjugation.** Process of sexual reproduction in unicellular organisms in which two organisms of opposite mating type temporarily pair and transfer genetic material.

**Consanguinity.** State of being descended from a common ancestor.

**Consolidation.** Period required for short-term memory to become fixed as long-term memory.

**Constant (C-type) position effect.** Position effect whose phenotype is constant or uniform.

**Constant region.** Portion of each L or H chain of an immunoglobulin whose sequence is relatively invariant in different immunoglobulins.

**Constitutive heterochromatin.** Chromatin that is always heterochromatic; composed of highly repetitive, late-replicating DNA.

**Continuous trait.** *See* Quantitative trait.

**Controlling elements or genes.** Transposable genes that regulate gene action.

**Cortex.** Outer layer of cells or plasm.

**Crossing over.** Process of reciprocal exchange between the nonsister chromatids of homologous chromosomes.

**Crossover.** Recombinant product of crossing over.

**Crossover unit.** Linkage map unit equal to 1 crossover per 100 postmeiotic products.

**C-type position effect.** Position effect whose phenotype is constant or uniform.

**Cytoplasm.** Protoplasm of the cell outside the nucleus.

**Deficiency.** Absence of a chromosome segment.

**Degenerate code.** Code that has more than one codon for the same amino acid.

**Deletion.** Deficiency.

*de novo* **synthesis.** Made without using a template.

**Deoxyribonucleic acid.** DNA.

**Deoxyribose.** Five-carbon sugar characteristically found in DNA.

**Depolymerization.** Process of breaking a polymer into smaller units, usually into monomers.

**Derepressed operon.** Negatively regulated operon which is being transcribed.

**Determination.** Fixation of the way in which a tissue will subsequently differentiate.

**Development.** Orderly sequence of changes that occur during the life history of an organism.

**Dicentric.** Chromosome or chromatid with two centromeres.

**Differential polyploidy.** Polyploidy of one genome but not another in the same cell.

**Differentiation.** Sequence of changes that results in the specific structures and functions of cells and tissues.
**Dimer.** Product of union of two monomers.
**Diploid.** Having two representatives of each type of chromosome or locus.
**Disassortative mating.** Opposite of assortative mating.
**Discontinuous trait.** *See* Qualitative trait.
**Disjunction.** Separation of chromosomes at anaphase of either mitosis or meiosis.
**DNA.** Deoxyribonucleic acid.
**DNA modification.** Programmed changes made in DNA, usually after it is synthesized.
**DNA polymerase.** Enzyme that synthesizes DNA using DNA as a template.
**Dominance.** Condition in which (1) the phenotype of a heterozygote resembles one homozygote more than the other; (2) a rare allele is expressed phenotypically when heterozygous; or (3) the phenotype is the same when a gene is heterozygous as when it is present alone in single dose.
**Dominant.** Trait showing dominance; the allele or individual responsible for dominance.
**Dominant lethal.** Genetic material that is invariably lethal in single dose.
**Dosage compensation.** Regulation of gene action so that the same phenotypic effect is produced by a particular gene in single or double dose.
**Double helix.** Two strands coiled about each other.
**Drug-resistance particle.** Duplex ring DNA plasmid or episome which codes for resistance to one or more antibiotics.
**Duplicate loci.** Two nonallelic loci that have the same genetic meaning.
**Duplication.** Occurrence of a chromosome segment twice in the same chromosome or chromosome set.
**Effector.** Small molecule that combines with a regulatory protein and thereby enhances or inhibits its functioning.
**Elongation factor.** Protein needed to lengthen a polypeptide chain that is being synthesized on a ribosome.
**Endoplasmic reticulum.** System of membranes that forms sheets and vesicles in the cytoplasm of eukaryotes.
**Enzyme.** Protein catalyst of a metabolic chemical reaction.
**Episome.** Dispensable chromosome that can replicate autonomously or as part of a host chromosome.
**Epistasis.** Interference of one allele with the detection of the phenotypic expression of a nonallele.
**Erasure.** Removal of a setting.
**Erythrocytes.** Red blood cells.
**Euchromatin.** Relatively uncoiled or unclumped interphase chromatin.
**Euchromatization.** Process that makes constitutive heterochromatin appear like euchromatin.
**Eugenics.** Improvement of the genetic constitution of the human species through selective breeding.
**Eukaryotes.** Organisms whose cells contain a nucleus.
**Evolutionary tree.** Diagram whose branch lengths show the mutational distances between the same protein in different species, or the evolutionary distance between species.
**Exchange union.** Joining between broken ends of chromosomes, chromatids, or nucleic acids which produces a gene order different from the prebreakage order.
**Expressivity.** Degree of phenotypic expression of a penetrant gene.

**F.** Episomal sex particle in *E. coli*.

**F derivative.** F particle part of which is replaced by a segment of another (usually *E. coli*) chromosome.

**Facultative heterochromatin.** Euchromatin that becomes heterochromatized during interphase in some tissues but not in others.

**Fertilization.** Union of two gametes to form a zygote.

**Fitness (*w*).** Adaptive value of a genotype; the relative ability of an organism to transmit its genes to the next generation.

**Fixation.** (1) Period of consolidation required for short-term memory to become long-term memory. (2) Attaining an allele frequency of 1.0 or 0.0.

**Follicle cells.** Layer of cells surrounding an oocyte.

**Founder principle.** A new population started by colonizers that contain only a portion of the genetic variability of the parent population.

**Frame-shift mutation.** *See* phase-shift mutation.

**Functional mosaic.** Hybrid that expresses one gene in some parts and its allele in other parts.

**Gamete.** Cell used in fertilization.

**Gene.** Smallest, independently functional unit of genetic material. Some genes function by binding proteins, others function by coding for proteins, tRNA's, or rRNA's.

**Gene dosage.** Number of times that a given gene is present in a cell or chromosome set.

**Gene frequency.** *See* Allele frequency.

**Generalized transduction.** Unrestricted transduction; transduction of any segment of a donor chromosome, mediated by a virus whose chromosome is completely replaced by donor DNA.

**Genetic background.** Genes in a genotype other than the gene(s) under consideration.

**Genetic code.** mRNA codons and their corresponding meanings in protein synthesis.

**Genetic death.** Failure to reproduce.

**Genetic disease.** Disease due to a genetic abnormality in an organism's own genetic material.

**Genetic donor.** Male.

**Genetic drift.** Nondirectional change in allele frequencies due to chance variations.

**Genetic engineering.** Directed intervention with the content and/or organization of an organism's genetic material.

**Genetic environment.** All the organisms in the immediate environment of an organism.

**Genetic load.** Genetic burden (mutations, etc.) that reduces the fitness of an individual or population.

**Genetic marker.** Alternative genetic constitution (usually an allele) used to identify a genetic region (usually a locus) in successive generations.

**Genetic material.** Nucleic acid that replicates or that was made using a nucleic acid template that is known to be used as a template to synthesize nucleic acid.

**Genetic recipient.** Female.

**Genetics.** Study of the properties, functions, and significance of nucleic acids that specify their own replication; the science concerned with genetic material.

**Genotype.** Genetic constitution of an organism.

**Germ line.** Cells from which gametes are derived.

**Globin.** Protein portion of hemoglobin or myoglobin.

**Glucosylation.** Addition of glucose to compounds such as the organic bases in DNA.

**Gonad.** Sex organ; site of gamete formation.

**Gynandromorph.** Sexually mosaic individual with some portions of the body typically male and others typically female.

**Half-translocation.** One half of a reciprocal translocation.

**Haploid.** Having a single representative of each type of chromosome or locus.

**Hardy–Weinberg principle.** Allele and genotypic frequencies remain constant in a large randomly mating population in the absence of migration, mutation, and selection.

**Hb.** Hemoglobin.

**Heme.** Iron-containing, oxygen-transporting, group attached to each chain of hemoglobin or myoglobin.

**Hemoglobin.** Heme-containing, oxygen-transporting protein in red blood cells.

**Heterochromatin.** Relatively coiled or clumped, interphase chromatin; generally or always inactive in transcription.

**Heterochromatization.** Process that converts euchromatin into facultative heterochromatin.

**Heteroduplex.** Double-stranded nucleic acid molecule whose strands come from different sources.

**Heterosis.** *See* Hybrid vigor.

**Heterozygous.** Carrying two (or more) different alleles of a single gene.

**Histone cluster.** Combination of two molecules each of histones $IIb_1$, $IIb_2$, III, and IV.

**Histones.** Proteins, usually found in combination with nuclear DNA, that are rich in basic amino acids and have a MW of 11,000 to 21,000.

**Homoduplex.** Duplex whose strands are exactly complementary.

**Homologous.** Genetically corresponding.

**Homologous chromosomes.** Chromosomes that pair during meiosis.

**Homozygous.** Having identical alleles at two (or more) corresponding loci.

**Hybrid.** Heterozygous.

**Hybridization.** (1) Combining molecularly different complements into duplex nucleic acid. (2) Crossing genetically different individuals.

**Hybrid nucleic acid.** Duplex containing one DNA and one RNA strand; or a DNA–DNA or an RNA–RNA duplex whose two strands are not exactly complementary.

**Hybrid vigor.** Heterosis; the hybrid is more fit than either homozygote.

**Hydrogen bond.** Weak chemical bond that results from the sharing of a proton ($H^+$) between two nitrogens, two oxygens, or between one of each.

**Hydrolysis.** Splitting of a molecule by adding the elements of water.

**Hydrophilic.** Water loving.

**Hydrophobic.** Water shunning.

**Hypostasis.** The condition of the locus whose phenotypic expression is interfered with because of epistasis.

**Ig.** Immunoglobulin.

**Immunoglobulin (Ig).** Molecule composed of two pairs of polypeptides which is capable of binding two identical antigens.

**Inbreeding.** Crossing of closely related individuals.

**Independent segregation.** Nonhomologous chromosomes or chromosome segments segregate independently.

**Induced enzyme formation.** The presence of a nutrient in the medium induces the synthesis of large amounts of the enzymes needed to catabolize the nutrient.

# GLOSSARY

**Inducer.** Chemical or physical agent causing induction.

**Induction.** (1) Determination of the developmental fate of one tissue by another. (2) Stimulation of enzyme synthesis by a specific inducer.

**Infrasex.** Supermale or superfemale.

**Initiation factor.** Any one of several proteins needed to start protein synthesis.

**Initiator codon.** First translated codon in mRNA; AUG.

**Initiator substance.** Substance (sometimes a positive regulatory protein) that is required to initiate nucleic acid replication.

**Integration.** Stable incorporation into the linear genetic sequence.

**Intemperate.** Virulent; causing lysis.

**Interphase.** All stages of the cell cycle other than nuclear division.

**Intersex.** Individual with sexual characteristics intermediate between those of males and females.

**Introgression.** Incorporation of genes of one species into the genetic constitution of another species by means of interspecific hybridization and backcrossing.

**Inversion.** Chromosome that contains a segment which has been turned through 180°.

*in vitro.* Biological processes studied outside the whole organism. (Literally, in glass.)

*in vivo.* Within the living organism.

**Isochromosome.** Chromosome with two identical arms.

**Isolating mechanism.** Any barrier that prevents successful mating between two or more related groups of organisms.

**Joint.** Place where two duplexes are held together by base pairing between their single-stranded ends.

**Kinetoplast.** Mitochondrion-like organelle found in certain parasitic protozoans.

**Klinefelter's syndrome.** Abnormal human male phenotype characteristics of an XXY chromosome constitution.

**Leader sequence.** The portion of mRNA that precedes the place where translation starts.

**Learning.** Capacity of system to react in a new or changed way as the result of experience.

**Lethal.** Genetic condition that causes the premature death of its carrier.

**Leucoplast.** White plastid.

**Ligase.** Enzyme that repairs nicks in duplex DNA.

**Linkage.** The greater-than-chance association of two or more nonalleles.

**Load.** *See* Genetic load.

**Locus.** Site in the chromosome occupied by a gene.

**Long-term memory.** Memory that persists after a period of fixation.

**Lysis.** Destruction of a cell by the rupture of its cell membrane.

**Lysogen.** Bacterium harboring a temperate phage.

**Lysogenic.** State of being a lysogen; bacterial state which, after induction, is followed by lysis.

**Magnification.** Process that permanently increases the number of reiterated loci.

**Map unit.** Unit used to measure relative distance between linked genes.

**Medulla.** Inner cell layer surrounded by the cortex.

**Meiosis.** Two spindle-using, nuclear divisions which reduce chromosomes from the paired to the unpaired condition (e.g., diploid to haploid).

**Meiotic mutant.** Mutant that produces an abnormality in the meiotic mechanism.

**Memory.** Capacity to store and retrieve learned information.

**Mendelian genes.** Genes distributed to progeny nuclei by means of a spindle.

**Messenger (m) RNA.** Transcript that contains a protein-specifying base sequence.

**Metabolism.** Sum total of the chemical processes in living cells.
**Methylase.** Enzyme that adds a methyl group to compounds such as the organic bases in nucleic acid.
**Migration.** Cause of gene flow between populations due to immigration and emigration.
**Minor bases.** Derivatives of the four usual bases in RNA which are present in small numbers in tRNA's.
**Missense mutation.** Mutation that converts a codon for one amino acid into a codon for a different amino acid.
**Mitochondrion.** Cytoplasmic organelle that is the main site of cellular respiration.
**Mitosis.** Spindle-using nuclear division which produces two identical daughter nuclei.
**Mold.** Template.
**Molecular recombinant.** Individual whose genetic material is derived from two or more different sources.
**Monomer.** Single unit of a polymer (e.g., amino acids and nucleotides); one subunit of a protein which is composed of two or more identical polypeptides.
**Monozygotic.** Two or more individuals derived from the same fertilized egg.
**Mosaicism.** Having two or more different genotypes or phenotypes in the same part or individual.
**Mutagen.** Physical or chemical agent which greatly increases the frequency of mutation.
**Mutagenic.** Causing mutation.
**Mutant.** Cell or organism produced by mutation.
**Mutation.** More-or-less permanent, uncoded, relatively rare change in the kind, number, or sequence of nucleotides in genetic material.
**Mutational distance.** Minimal number of nucleotides that need to be mutated in order to convert a nucleic acid or protein of one organism into one of another organism.
**Mutator.** Allele that increases the mutation rate.
**Myoglobin.** Monomeric, heme-containing protein in vertebrate muscle cells.
**Natural selection.** Selection as it occurs in nature.
**Negative control.** Gene action inhibited by a regulatory protein.
**Neuron.** Nerve cell.
**Nick.** Single-strand break in double-helical nucleic acid.
**Nondisjunction.** Failure of chromosomes to separate at anaphase of either mitosis or meiosis.
**Nongenetic environment.** Immediate environment of an organism exclusive of other organisms.
**Nonhistone proteins.** Acidic proteins found in nuclear chromosomes.
**Nonhomologous chromosomes.** Chromosomes belonging to different pairs of chromosomes.
**Nonhomologous genes.** Genes located in different pairs of homologous chromosomes.
**Nonmendelian genes.** Genes distributed to progeny cells by means other than a spindle.
**Nonrestitutional union.** *See* Exchange union.
**Nonsense codon.** Codon for the termination of translation.
**Nonsense mutation.** Mutation that converts a codon for an amino acid into a chain-terminating codon.
**Nonsister chromatids.** Chromatids that belong to different members of a pair of homologous chromosomes.

**Nuclear membrane.** The double-layered boundary of the nucleus.
**Nuclease.** Any enzyme that breaks down nucleic acids.
**Nucleic acid.** Polymer of nucleotide monomers.
**Nucleolus.** Product of the functioning of the nucleolus organizer.
**Nucleolus organizer.** DNA gene sequence that specifies the formation of a nucleolus; genetic region in a nuclear chromosome that codes for the two larger types of rRNA.
**Nucleoplasm.** Protoplasm of the nucleus.
**Nucleotide.** Compound composed of an organic base, a sugar, and a phosphate.
**Nucleotide sharing.** The same nucleotide has more than one biological meaning, sometimes being part of two different genes.
**Nucleus.** Large, spheroid organelle that contains the main chromosomes of a cell.
**Nullosomic.** Lacking both members of a pair of chromosomes.
**Nurse cells.** Cells that function to nourish an oocyte.
**Oocyte.** Cell that undergoes meiosis to form the egg.
**Operator.** Negatively controlled gene that regulates the transcription of the structural genes in an operon.
**Operon.** Duplex DNA sequence in prokaryotes which contains an operator gene that regulates the transcription of one or more polypeptide-coding genes.
**Organelle.** Any one of several membrane-bound structures found within a cell.
**Organism.** Individual characterized by its interdependent genetic material and protein.
**Organismal (biological) evolution.** Gene-based changes that occur during the history of organisms.
**Origin.** Locus where chromosome replication or chromosome transfer starts.
**Outbreeding.** Opposite of inbreeding.
**Palindrome.** A base sequence in one strand is reversed in its complement.
**Pedigree.** Diagram presenting a line of ancestors; a genealogical tree.
**Penetrance.** Proportion of individuals of a specified genotype that show the expected phenotype.
**Peptide bond.** Chemical union required to link an amino acid to another amino acid or to a chain of amino acids.
**Persistence.** Number of generations an allele is present in a population before it is lost.
**Phage (bacteriophage).** Bacterial virus.
**Phage cross.** Recombination between two phages in a multiply infected host.
**Phase-shift mutation.** Frame-shift mutation; an addition or loss of genetic nucleotides (any nonmultiple of three) that results in mRNA codons being read out of phase (or frame).
**Phenocopy.** Environmentally induced phenotype which mimics the phenotype produced by a different genotype.
**Phenotype.** Collection of traits possessed by a cell or organism that results from the interaction of the genotype and the environment.
**Pilus.** Hair-like projection from donor bacterium which is converted into a conjugation tube during conjugation.
**Plasma cell.** Cell of immune system that secretes antibodies.
**Plasmid.** Dispensable chromosome that replicates autonomously, but cannot be integrated into a chromosome of the host.
**Plastids.** Self-replicating cytoplasmic organelles of various types in plant cells.
**Polar body.** Discarded, almost cytoplasm-free cell produced by a meiotic division of an oocyte.
**Polarity.** Having a direction.
**Poly A.** Polymer of A-containing nucleotides.

**Polydeoxyribonucleotide.** Deoxyribonucleic acid, or DNA.
**Polymer.** Large molecule composed of many like subunits, or monomers.
**Polymerase, nucleic acid.** Enzyme that synthesizes nucleic acid from nucleotide monomers.
**Polymerization.** Process of making a polymer through the union of many monomers.
**Polynucleotide.** Polymer of nucleotide monomers; a nucleic acid.
**Polyoma.** DNA virus that causes tumors in rodents.
**Polypeptide.** Polymer of amino acid monomers.
**Polypeptide polymerase.** Enzyme for forming peptide bonds, that is, for polymerizing amino acids into polypeptides.
**Polyploid.** Having more than two haploid sets of chromosomes.
**Polyribonucleotide.** Ribonucleic acid, or RNA.
**Polyribosome.** Polysome; linear array of ribosomes attached to a single messenger RNA.
**Polysomic.** Cell or individual with one or more extra copies of a chromosome.
**Population.** All the interbreeding members of a group of one kind of organism.
**Position effect.** Change that sometimes occurs in the functioning of a gene upon changing its position in the genetic material.
**Positive control.** The action of a gene enhanced by a regulatory protein.
**Primer.** Single-stranded piece of nucleic acid whose end is used as a start point for nucleic acid synthesis.
**Prokaryotes.** Organisms whose cells lack a nucleus.
**Promoter.** Gene that binds transcriptase preparatory to transcription.
**Prophase.** First phase of a mitotic or meiotic division.
**Prophase I(II).** Prophase of the first (second) meiotic division.
**Protamine.** Relatively small protein, very rich in Arg, found in combination with nuclear DNA in certain cells.
**Protein.** Giant molecule composed of a single poypeptide or of two or more associated polypeptide subunits synthesized using organismal machinery.
**Proteinoids.** Proteins formed when amino acids are polymerized by exposure to dry heat or by other nonorganismal means.
**Protoplasm.** The metabolizing substance of a cell.
**Prototroph.** Organism that has no nutritional requirements in addition to those of the wild type.
**Puff.** Band of a multichromatid chromosome which is swollen due to transcription.
**Purine.** Double-ring type of base found in nucleic acid.
**Pyrimidine.** Single-ring type of base found in nucleic acid.
**Pyrimidine dimer.** Result of the union of two pyrimidines.
**Qualitative trait.** Discontinuous trait; trait whose alternatives can be described without being measured since they fall into discrete categories.
**Quantitative trait.** Continuous trait; trait whose alternatives require measurement in order to be described since they vary over a continuous scale.
**Race.** Population whose genetic constitution is significantly different from other populations of the same species.
**Radioautography.** Localization of radioactive substance by its ability to expose a coating of photographic emulsion.
**Random mating.** Any individual of one sex has an equal probability of mating with any individual of the opposite sex.
**rDNA.** Gene that codes for rRNA.
**Reading.** Sequential translation of codons.
**Recessive.** Trait partially or completely hidden by the phenotypic effect of a dominant allele; the allele or individual that has such a phenotypic effect.

# GLOSSARY

**Recessive lethal.** Invariably lethal when present in homozygous condition.
**Reciprocal cross.** Second cross identical to the first but with the sexes of the parents interchanged. A♀ × B♂ and B♀ × A♂ are reciprocal crosses.
**Reciprocal translocation.** Mutual exchange of segments between non-homologous chromosomes or of unequal segments between homologous chromosomes.
**Recombinant.** Product of recombination.
**Recombination.** Change in the sequence or grouping of genetic nucleotides. Changes in sequence include making and breaking concatemers, integration and excision, and crossing over. Changes in grouping include the separation of sister chromosomes in prokaryotic and eukaryotic cell divisions; segregation and independent segregation in meiosis; and conjugation, cell fusion, and fertilization.
**Recombination map.** Linkage map whose unit is recombination frequency.
**Redundancy.** Repeated nucleotide sequences or genes.
**Regulator gene.** Gene that codes for a regulatory protein.
**Regulatory protein.** Product of a regulator gene used to regulate the action of another gene.
**Reiteration.** Redundancy.
**Replication eye.** Strand-separated region of a double helix undergoing replication.
**Replicon.** Single, complete unit of nucleic acid replication.
**Repressible operon.** Negatively regulated operon in which an effector facilitates the binding of the repressor to the operator.
**Repressor.** Protein that inhibits the transcription of one or more genes.
**Repressor gene.** Gene that codes for a repressor.
**Reproduction.** Formation of more copies of the same kind.
**Reproductive isolation.** Absence of interbreeding between members of different populations.
**Responding gene.** Controlling gene whose expression is regulated by a signaling gene.
**Restitutional union.** Joining between broken chromosome ends which restores their prebreakage order.
**Restricted transduction.** Transduction of specific, restricted regions of the bacterial chromosome which are adjacent to an episome integration site; mediated by an excised episome whose genome is partially replaced by the DNA being transduced.
**Restriction enzyme.** Nuclease that degrades improperly modified duplex DNA.
**Reticulocyte.** Immature red blood cell.
**Reverse transcription.** Synthesis of DNA using an RNA template.
**Reversion.** Mutation back to the original condition.
**Ribonucleic acid.** RNA.
**Ribose.** Five-carbon sugar characteristically found in RNA.
**Ribosomal (r) RNA's.** Transcripts that combine with ribosomal proteins to form ribosomes.
**Ribosome.** Cellular particle composed of two unequal subunits, each made up of roughly equal parts of rRNA and protein, which is the site of protein synthesis.
**RNA.** Ribonucleic acid.
**RNA polymerase.** Enzyme that synthesizes RNA using RNA as a template.
**RNA replicase.** Enzyme that replicates RNA.
**Rod.** Chromosome having ends; a nonring chromosome.
**rRNA, 5S.** Smallest of the rRNA's, found in all but mitochondrial ribosomes.

**S.** Svedberg unit; used to denote the sedimentation velocity of a giant molecule.
**Segration.** Separation of the members of a pair of genes or chromosomes, for example, during meiosis.
**Selection.** Process that increases (positive selection) or decreases (negative selection) the probability of reproduction.
**Selection coefficient (s).** Equal $1 - w$; measure of unfitness of a genotype.
**Self-fertilization.** Fusion of the male and female gametes from the same individual.
**Self-replication.** The reproduction of a molecule that is aided by information contained in the molecule.
**Semiconservative synthesis.** Synthesis of duplex nucleic acid in which the parental strands separate and each forms a duplex with its complementary daughter strand.
**Semilethal mutant.** Mutant that produces more than 90 per cent but less than 100 per cent mortality before adulthood.
**Sense codon.** Codon that specifies an amino acid.
**Sense strand.** DNA strand used as a template for transcription.
**Setting.** Fixing the level of potential activity at a locus (sometimes that of a controlling gene).
**Sex chromatin.** Barr body; facultative heterochromatin of the X chromosome of normal mammalian females but not males.
**Sex chromosomes.** Chromosomes that differ in number or morphology in different sexes and contain genes determining sex type.
**Sex mosaic.** Individual with some parts typically male and others typically female.
**Sex particle.** Dispensable episome or plasmid that induces conjugation.
**Short-term memory.** Transient memory that will be lost if not fixed as long-term memory.
**Siblings.** Children (offspring) of the same parents.
**Sickle-cell anemia.** Lethal anemia that occurs in homozygotes for the recessive lethal allele.
**Sigma ($\sigma$) factor.** Polypeptide needed for transcriptase to initiate transcription.
**Signaling gene.** Controlling gene whose expression regulates a responding gene.
**Sister chromatids.** Chromatids in the same chromosome; identical chromatids.
**Site.** Portion of a macromolecule that has a particular function; often the locus of a gene.
**Somatic.** Pertaining to nongerm tissue, cell, or protoplasm.
**Spacer DNA.** Nontranscribed DNA of unknown function located between transcribed DNA's.
**Speciation.** Species formation.
**Species.** All populations that can interbreed with each other but which maintain a genetic constitution different from all other such groups.
**Spindle.** Structure composed of small tubules employed to move chromosomes during mitosis and meiosis.
**Spreading effect.** Position effect that extends to genes in the vicinity of a break point of a chromosomal rearrangement, presumably as a result of the spread of heterochromatization starting at the break point.
**Subvital mutant.** Mutant that has a significantly reduced viability but greater than 10 per cent chance of survival to adulthood.
**Supercoil.** *See* Superhelix.
**Superfemale.** A female with 3 X chromosomes and two sets of autosomes.
**Superhelix.** Supercoil; double-helical nucleic acid which is itself coiled into a helix of larger dimensions.

**Supermale.** A sterile *Drosophila* male with one X chromosome and three sets of autosomes, whose sexual phenotype is more extreme than normal maleness.

**Synapsis.** Side-by-side pairing of homologous duplex nucleic acids or chromosomes.

**Tailoring.** Modification of RNA by shortening, lengthening, methylation, and so on.

**tDNA.** Gene that codes for a tRNA.

**Temperate phage.** Phage that has both lysogenic and lytic stages.

**Template.** Giant molecule that provides the information for synthesizing a complement.

**Terminal redundancy.** Rod chromosome whose two ends contain the same nucleotide or gene sequence.

**Termination factor.** Protein needed to release a newly synthesized polypeptide chain from tRNA.

**Terminator codon.** Nonsense codon.

**Termination factor.** Protein needed to release a newly synthesized polypeptide chain from tRNA.

**Testcross.** Cross between an individual of unknown genotype and an individual homozygous for the recessive genes in question.

**Tetraploid.** Cell or individual with four sets of chromosomes.

**Tetrasomic.** Cell or individual with four homologous chromosomes.

**Transcriptase.** DNA-dependent RNA polymerase.

**Transcriptase interaction site.** Region of the promoter that seems to contain the base sequence which transcriptase recognizes as a site for initiation of transcription.

**Transcription.** Synthesis of RNA using a DNA template.

**Transcription-silent DNA.** Nontranscribed DNA, of unknown function, such as that which precedes rDNA's and genes for histones; "spacer" DNA.

**Transcription terminator.** DNA gene whose sequence terminates transcription.

**Transduction.** Genetic recombination in which the transfer of genetic material between cells is mediated by a plasmid or episome.

**Transfer (t) RNA.** Transcript that accepts an amino acid and transports it to the ribosome–mRNA complex for use in protein synthesis.

**Transformation.** Genetic recombination in which naked DNA from one cell can enter and integrate in another cell.

**Transgenosis.** Transfer of genes of one organism to a widely different species in which they are expressed.

**Transition.** Mutation in which there is a substitution of one purine by another, or one pyrimidine by another, or one base pair by another which retains the orientation of purine and pyrimidine.

**Translation.** Synthesis of protein from information in mRNA.

**Transposition.** Shift of a gene or chromosome segment to a new locus in the chromosome set.

**Transversion.** Mutation in which there is a substitution of a purine by a pyrimidine or the reverse, or a substitution of one base pair by another which reverses the orientation of purine and pyrimidine.

**Trimer.** Product of union of three monomers.

**Triploid.** Cell or individual with three sets of chromosomes.

**Trisomic.** Cell or individual that contains three homologous chromosomes.

**Turner's syndrome.** Characteristics of an abnormal human female who has an X0 chromosome constitution.

**Underreplicated.** The part of a chromosome set that did not undergo amplification.

**Unrestricted transduction.** *See* Generalized transduction.
**Variable region.** The portion of the L or H chain of an immunoglobulin which has many different sequences in a single individual.
**Variegated (V-type) position effect.** The position-effect phenotype which is inconstant, variegated, or mottled.
**Virulent.** Causing lysis; intemperate.
**Virus.** Infective organism composed of nucleic acid and protein which can only reproduce within a cell.
*w.* Fitness.
**Wild type.** The type commonly found in nature.
**Zygote.** Product of fertilization.

# Bibliography

THE FIRST TITLES are general references that have wide interest or application to genetics. These are followed by chapter references selected because they (1) provide a general presentation of the subject matter, (2) are key papers in the discovery or application of major genetic principles, or (3) provide entry into the current literature.

*General References*

ADELBERG, E. A. (Editor). 1966. *Papers on bacterial genetics*, second edition. Boston: Little, Brown and Company.
CAIRNS, J., STENT, G. S., and WATSON, J. D. 1966. *Phage and the origins of molecular biology.* Cold Spring Harbor, N.Y.: Cold Spring Harbor Laboratory of Quantitative Biology. (Covering mostly 1945–1966.)
CARLSON, E. A. 1966. *The gene; a critical history.* Philadelphia: W. B. Saunders.
CORWIN, H. O., and JENKINS, J. B. (Editors). 1976. *Conceptual foundations of genetics. Selected readings.* Boston: Houghton Mifflin Co.
DAVIS, B. D., DULBECCO, R., EISEN, H. N., GINSBERG, H. S., WOOD, W. B., and MCCARTY, M. 1973. *Microbiology*, second edition. New York: Harper & Row, Inc.
DUNN, L. C. 1965. *A short history of genetics.* New York: McGraw-Hill Book Company. (Covers the period 1864–1939.)
HAYNES, R. H., and HANAWALT, P. C. (Editors). 1968. *The molecular basis of life.* San Francisco: W. H. Freeman and Company, Publishers. (Readings from *Scientific American*.)
HERSKOWITZ, I. H. 1965. *Genetics*, second edition. Boston: Little, Brown and Company. (A supplement contains Nobel prize lectures through 1962 dealing with genetics.)
HERSKOWITZ, I. H. 1977. *Principles of genetics*, second edition. New York: Macmillan Publishing Co., Inc. (Contains a more complete discussion and a longer list of references for most of the topics in this text.)
KENNEDY, D. (Editor). 1965. *The living cell.* San Francisco: W. H. Freeman and Company, Publishers. (Readings from *Scientific American*.)
KING, R. C. 1974. *A dictionary of genetics*, second edition revised. New York: Oxford University Press.
*Nobel lectures in molecular biology 1933–1975.* 1977. New York: Elsevier North-Holland Publishing Co.
PETERS, J. A. (Editor). 1959. *Classical papers in genetics.* Englewood Cliffs, N.J.: Prentice-Hall, Inc.

SRB, A. M., OWEN, R. D., and EDGAR, R. S. (Editors). 1970. *Facets of genetics.* San Francisco: W. H. Freeman and Company, Publishers. (Readings from *Scientific American.*)

STEIN, G. S., STEIN, J. L., and KLEINSMITH, L. J. (Editors). 1976. *Molecular genetics.* San Francisco: W. H. Freeman and Company, Publishers. (Readings from *Scientific American.*)

STENT, G. S. (Editor). 1965. *Papers on bacterial viruses*, second edition. Boston: Little, Brown and Company.

STERN, C. 1973. *Principles of human genetics*, third edition. San Francisco: W. H. Freeman and Company, Publishers.

STURTEVANT, A. H. 1965. *A history of genetics.* New York: Harper & Row, Inc.

SUTTON, H. E. 1975. *An introduction to human genetics*, second edition. New York: Holt, Rinehart and Winston, Inc.

TAYLOR, J. H. (Editor). 1965. *Selected papers on molecular genetics.* New York: Academic Press, Inc.

TOMIZAWA, J. (Editor). 1972. *Selected papers in biochemistry.* Vol. 1, *Bacterial genetics and temperate phage*; Vol. 2, *Virulent phage.* Baltimore: University Park Press.

WATSON, J. D. 1976. *Molecular biology of the gene*, third edition. Menlo Park, Calif.: W. A. Benjamin, Inc.

WOODWARD, D. O., and WOODWARD, V. W. 1977. *Concepts of molecular genetics.* New York: McGraw-Hill Book Company.

ZUBAY, G. L., and MARMUR, J. (Editors). 1973. *Papers in biochemical genetics*, second edition. New York: Holt, Rinehart and Winston, Inc.

*Chapter 1*

FRAENKEL-CONRAT, H., and WILLIAMS, R. C. 1955. Reconstitution of tobacco mosaic virus from its inactive protein and nucleic acid components. Proc. Nat. Acad. Sci., U.S., 41: 690–698. Reprinted in *Classic papers in genetics*, Peters, J. A. (Editor). Englewood Cliffs, N. J.: Prentice-Hall, Inc., 1959, pp. 264–271; and Bobbs-Merrill Reprint Series. Indianapolis: Howard W. Sams Company, Inc.

GIERER, A., and SCHRAMM, G. S. 1956. Infectivity of ribonucleic acid from tobacco mosaic virus. Nature, Lond., 177: 702–703. Reprinted in *Papers in biochemical genetics*, Zubay, G. L. (Editor). New York: Holt, Rinehart and Winston, Inc., 1968, pp. 16–18.

GUTHRIE, G. D., and SINSHEIMER, R. L. 1963. Observations on the infection of bacterial protoplasts with deoxyribonucleic acid of bacteriophage $\phi$X174. Biochim. Biophys. Acta, 72: 290–297.

HERSHEY, A. D., and CHASE, M. 1952. Independent functions of viral protein and nucleic acid in growth of bacteriophage. J. Gen. Physiol., 36: 39–54. Reprinted in *Papers on bacterial viruses*, second edition, Stent, G. S. (Editor). Boston: Little, Brown and Company, 1965, pp. 87–104; and Bobbs-Merrill Reprint Series. Indianapolis: Howard W. Sams Company, Inc. (This "Hershey–Chase" experiment was the first strong evidence for DNA as the genetic material of T phages.)

MIRSKY, A. E. 1968. The discovery of DNA. Scient. Amer., 218 (No. 6): 78–88, 140. Scientific American Offprints. San Francisco: W. H. Freeman and Company, Publishers.

*Chapter 2*

*Chromosome structure and function.* Cold Spring Harbor Sympos. Quant. Biol., 38 (1973).

CRICK, F. H. C. 1957. Nucleic acids. Scient. Amer., 197 (No. 3): 188–200, 278, 280. Scientific American Offprints. San Francisco: W. H. Freeman and Company, Publishers.

DAVIDSON, J. N. 1977. *The biochemistry of the nucleic acids*, eighth edition. New York: Academic Press, Inc.

HERSKOWITZ, J. 1970. The DOUBLE talking HELIX blues. (A classical record. Available from: Vertebral Disc, 86 Hunting Lane, Sherborn, Mass., 01770. Price $1.00.)

HOLLEY, R. W. 1966. The nucleotide sequence of a nucleic acid. Scient. Amer., 214 (No. 2): 30–39, 138. Scientific American Offprints. San Francisco: W. H. Freeman and Company, Publishers.

MAXAM, A. M., and GILBERT, W. A new method for sequencing DNA. Proc. Nat. Acad. Sci., U.S., 74: 560–564.

MIESCHER, F. 1871. On the chemical composition of pus cells. Translated in *Great experiments in biology*, Gabriel, M. L., and Fogel, S. (Editors). Englewood Cliffs, N.J.: Prentice-Hall, Inc., 1955, pp. 233–239. (The discovery of nucleic acid.)

SPIEGELMAN, S. 1964. Hybrid nucleic acids. Scient. Amer., 210 (No. 5): 48–56, 150. Scientific American Offprints. San Francisco: W. H. Freeman and Company, Publishers.

WATSON, J. D. *The double helix*. 1968. New York: Atheneum. (A personalized account of the discovery of the double-helical organization of DNA.)

WATSON, J. D., and CRICK, F. H. C. 1953. Molecular structure of nucleic acids. A structure for deoxyribose nucleic acid. Nature, Lond., 171: 737–738. Reprinted in *Classic papers in genetics*, Peters, J. A. (Editor). Englewood Cliffs, N.J.: Prentice-Hall, Inc., 1959, pp. 241–243; and Bobbs-Merrill Reprint Series. Indianapolis: Howard W. Sams Company, Inc.

WATSON, J. D., and CRICK, F. H. C. 1953. Genetical implications of the structure of deoxyribonucleic acid. Nature, Lond., 171: 964–969. Reprinted in *Papers on bacterial genetics*, second edition, Adelberg, E. A. (Editor). Boston: Little, Brown and Company, 1966, pp. 127–132; and Bobbs-Merrill Reprint Series. Indianapolis: Howard W. Sams Company, Inc.; and *The biological perspective, introductory readings*, Laetsch, W. M. (Editor). Boston: Little, Brown and Company, 1969, pp. 126–131.

WATSON, J. D., and CRICK, F. H. C. 1953. The structure of DNA. Cold Spring Harbor Sympos. Quant. Biol., 18: 123–131. Reprinted in *Papers on bacterial viruses*, second edition, Stent, G. S. (Editor). Boston: Little, Brown and Company, 1965, pp. 230–245; and *Papers in biochemical genetics*, Zubay, G. L. (Editor). New York: Holt, Rinehart and Winston, Inc., 1968, pp. 28–36.

## Chapter 3

DRESSLER, D. 1975. DNA replication: portrait of a field in mid passage. Sympos. Soc. Gen. Microbiol., 25: 51–76.

KORNBERG, A. 1968. The synthesis of DNA. Scient. Amer., 219 (Oct.): 64–70, 75–78, 144.

KORNBERG, A. 1974. *DNA synthesis*. San Francisco: W. H. Freeman and Company, Publishers.

WEISSMANN, C., and OCHOA, S. 1967. Replication of phage RNA. Progr. Nucleic Acid Res. Mol. Biol., 6: 353–399.

ZINDER, N. D. (Editor). 1975. *RNA phages*. Cold Spring Harbor, N.Y.: Cold Spring Harbor Laboratory.

## Chapter 4

BALTIMORE, D. 1970. RNA-dependent DNA polymerase in virions of RNA tumour viruses. Nature, Lond., 226: 1209–1211. (Reverse transcription.)

COHN, W. E., and VOLKIN, E. (Editors). 1977. *mRNA: the relation of structure to function*. New York: Academic Press, Inc.

LOSICK, R., and CHAMBERLIN, M. (Editors). 1976. *RNA polymerase*. Cold Spring Harbor, N.Y.: Cold Spring Harbor Laboratory.

PERRY, R. P. 1976. Processing of RNA. Ann. Rev. Biochem., 45: 605–629.

STEWART, P. R., and LETHAM, D. S. 1977. *The ribonucleic acids*, second edition. New York: Springer-Verlag. (m, t, and rRNA's.)

TEMIN, H. M. 1972. RNA-directed DNA synthesis. Scient. Amer., 226 (No. 1): 24–33, 122. (Reverse transcription.)

TILGHMAN, S. M., CURTIS, P. J., TIEMEIER, D. C., LEDER, P., and WEISSMANN, C. 1978. The intervening sequence of a mouse β-globin gene is transcribed within the 15S β-globin mRNA precursor. Proc. Nat. Acad. Sci., U.S., 75: 1309–1313.

*Transcription of genetic material.* Cold Spring Harbor Sympos. Quant. Biol., 35 (1971).

## Chapter 5

CRICK, F. H. C. 1966. The genetic code: III. Scient. Amer., 215 (No. 4): 55–62, 150. Scientific American Offprints. San Francisco: W. H. Freeman and Company, Publishers.

*The mechanism of protein synthesis.* Cold Spring Harbor Sympos. Quant. Biol., 34 (1970).

MORGAN, A. R., WELLS, R. D., and KHORANA, H. G. 1966. Studies on polynucleotides, LIX. Further codon assignments from amino acid incorporations directed by ribopolynucleotides containing repeating trinucleotide sequences. Proc. Nat. Acad. Sci., U.S., 56: 1899–1906.

NIRENBERG, M. W., and MATTHAEI, J. H. 1961. The dependence of cell-free protein synthesis in *E. coli* upon naturally occurring or synthetic polyribonucleotides. Proc. Nat. Acad. Sci., U.S., 47: 1588–1602. (The first cracking of the code.)

RICH, A. 1963. Polyribosomes. Scient. Amer., 209 (No. 6): 44–53, 178. Scientific American Offprints. San Francisco: W. H. Freeman and Company, Publishers.

RICH, A., and RAJBHANDARY, U. L. 1976. Transfer RNA: molecular structure, sequence, and properties. Ann. Rev. Biochem., 45: 805–860.

SHINE, J., and DALGARNO, L. 1974. The 3'-terminal sequence of *Escherichia coli* 16S ribosomal RNA: complementarity to nonsense triplets and ribosome binding sites. Proc. Nat. Acad. Sci., U.S., 71: 1342–1346.

SMITH, W. P., TAI, P.-C., THOMPSON, R. C., and DAVIS, B. D. 1977. Extracellular labeling of nascent polypeptides traversing the membrane of *Escherichia coli*. Proc. Nat. Acad. Sci., U.S., 74: 2830–2834. (Membrane-bound polyribosomes are making secretory proteins.)

## Chapter 6

DRAKE, J. W. 1969. *An introduction to the molecular basis of mutation.* San Francisco: Holden-Day, Inc.

DRAKE, J. W., and BALTZ, R. H. 1976. The biochemistry of mutagenesis. Ann. Rev. Biochem., 45: 11–37.

HANAWALT, P. C., and HAYNES, R. H. 1967. The repair of DNA. Scient. Amer., 216 (No. 2): 36–43, 146. Scientific American Offprints. San Francisco: W. H. Freeman and Company, Publishers.

MULLER, H. J. 1922. Variation due to change in the individual gene. Amer. Nat., 56: 32–50. Reprinted in *Classic papers in genetics*, Peters, J. A. (Editor). Englewood Cliffs, N.J.: Prentice-Hall, Inc., 1959, pp. 104–116.

MULLER, H. J. 1927. Artificial transmutation of the gene. Science, 66: 84–87. Reprinted in *Classic papers in genetics*, Peters, J. A. (Editor). Englewood Cliffs, N.J.: Prentice-Hall, Inc., 1959, pp. 149–155; and *Great experiments in biology*, Gabriel, M. L., and Fogel, S. (Editors). Englewood Cliffs, N.J.: Prentice-Hall, Inc., 1955, pp. 260–266.

SETLOW, R. B. 1968. The photochemistry, photobiology, and repair of polynucleotides. Progr. Nucleic Acid Res. Mol. Biol., 8: 257–295.

TOPAL, M. D., and FRESCO, J. R. 1976. Complementary base pairing and the origin of substitution mutations. Nature, Lond., 263: 285–289. (Hypothesizes that AA, GG, and GA pairs may also occur, leading to transversions.)

## Chapter 7

POTTER, H., and DRESSLER, D. 1977. On the mechanism of genetic recombination. The maturation of recombination intermediates. Proc. Nat. Acad. Sci., U.S., 74: 4168–4172.

RADDING, C. M. 1973. Molecular mechanisms in genetic recombination. Ann. Rev. Genet., 7: 87–111.

SIGNER, E. 1971. General recombination. In *The bacteriophage lambda*, Hershey, A. D. (Editor). Cold Spring Harbor, N.Y.: Cold Spring Harbor Laboratory, pp. 139–174.

STREISINGER, G., EMRICH, J., and STAHL, M. M. 1967. Chromosome structure in phage T4, III. Terminal redundancy and length determination. Proc. Nat. Acad. Sci., U.S., 57: 292–295.

## Chapter 8

AVERY, O. T., MACLEOD, C. M., and MCCARTY, M. 1944. Studies on the chemical nature of the substance inducing transformation of pneumococcal types. J. Exp. Med., 79: 137–158. Reprinted in *Papers on bacterial genetics*, Adelberg, E. A. (Editor). Boston: Little, Brown and Company, 1960, pp. 147–168; and *The biological perspective, introductory readings*. Laetsch, W. M. (Editor). Boston: Little, Brown and Company, 1969, pp. 105–125.

CAMPBELL, A. 1964. Transduction. In *The bacteria*, Vol. 5, Gunsalus, I. C., and Stanier, R. Y. (Editors). New York: Academic Press, Inc., pp. 49–89.

DEGNEN, G. E., MILLER, I. L., EISENSTADT, J. M., and ADELBERG, E. A. 1976. Chromosome-mediated gene transfer between closely related strains of cultured mouse cells. Proc. Nat. Acad. Sci., U.S., 73: 2838–2842.

HAYES, W. 1968. *The genetics of bacteria and their viruses*, second edition. New York: John Wiley & Sons, Inc., pp. 574–649.

SCHAEFER, P. 1964. Transformation. In *The bacteria*, Vol. 5. Gunsalus, I. C., and Stanier, R. Y. (Editors). New York: Academic Press, Inc., pp. 87–153.

ZINDER, N. D. 1958. "Transduction" in bacteria. Scient. Amer., 199: 38–43.

## Chapter 9

BACHMANN, B. J., LOW, K. B., and TAYLOR, A. L. 1976. Recalibrated linkage map of *Escherichia coli* K-12. Bact. Rev., 40: 116–167. (The most complete map to date.)

CAMPBELL, A. 1969. *Episomes*. New York: Harper & Row, Inc.

CLOWES, R. C. 1973. The molecule of infectious drug resistance. Scient. Amer., 228 (No. 4): 18–27, 124.

HAYES, W. 1968. *The genetics of bacteria and their viruses*, second edition. New York: John Wiley & Sons, Inc., Chap. 21, pp. 620–649.

HERSHEY, A. D. (Editor). 1971. *The bacteriophage lambda*. New York: Cold Spring Harbor Laboratory.

JACOB, F., and WOLLMAN, E. L. 1958. Episomes, added genetic elements. (In French.) C.R. Acad. Sci., Paris, 247: 154–156. Translated and reprinted in *Papers on bacterial genetics*, Adelberg, E. A. (Editor). Boston: Little, Brown and Company, 1960, pp. 398–400.

JACOB, F., and WOLLMAN, E. L. 1961. Viruses and genes. Scient. Amer., 204 (No. 6): 92–107.

LANDY, A., and ROSS, W. 1977. Viral integration and excision: structure of the lambda *att* sites. Science, 197: 1147–1160.

LEDERBERG, J., and TATUM, E. L. 1946. Gene recombination in *Escherichia coli*. Nature, Lond., 158: 558. Reprinted in *Classic papers in genetics*, Peters, J. A. (Editor). Englewood Cliffs, N.J.: Prentice-Hall, Inc., 1959, pp. 192–194. (The discovery of sexuality in bacteria.)

LWOFF, A. 1966. Interaction among virus, cell, and organism. Science, 152: 1216–1220. (Nobel prize lecture on the history, significance, and molecular biology of lysogeny.)

## Chapter 10

BAJER, A. S., and MOLÉ-BAJER, J. 1973. *Spindle dynamics and chromosome movements*. New York: Academic Press, Inc. (In both mitosis and meiosis.)

BRACHET, J., and MIRSKY, A. E. (Editors). 1961. *The cell*, Vol. 3, *Meiosis and mitosis*. New York: Academic Press, Inc.

BRITTEN, R. J., and KOHNE, D. E. 1970. Repeated segments of DNA. Scient. Amer., 222 (No. 4): 24–31, 130.

HSU, T. C. 1973. Longitudinal differentiation of chromosomes. Ann. Rev. Genet., 7: 153–176.

MENDEL, G. 1866. Experiments in plant hybridization. Translated in *Principles of genetics*, fifth edition, Sinnott, E. W., Dunn, L. C., and Dobzhansky, Th. New York: McGraw-Hill Book Company, 1958, pp. 419–443; in *Genetics, the modern science of heredity*, Dodson, E. O. Philadelphia: W. B. Saunders Company, 1956, pp. 285–311; and in *Classic papers in genetics*, Peters, J. A. (Editor). Englewood Cliffs, N.J.: Prentice-Hall, Inc., 1959, pp. 1–20. (Original proofs of segregation and independent segregation of gene pairs.)

MENDEL, G. 1867. Part of a letter to C. Nägeli. Supplement I in *Genetics*, second edition, Herskowitz, I. H. Boston: Little, Brown and Company, 1965. (A summary of his discovery of segregation and independent segregation.)

NOVIKOFF, A. B., and HOLTZMAN, E. 1976. *Cells and organelles*, second edition. New York: Holt, Rinehart and Winston, Inc.

PRESCOTT, D. M. 1976. *Reproduction of eukaryotic cells*. New York: Academic Press, Inc. (Nuclear and cell division.)

YUNIS, J. J. 1977. *Molecular structure of human chromosomes*. New York: Academic Press, Inc.

## Chapter 11

BRIDGES, C. B. 1916. Non-disjunction as proof of the chromosome theory of heredity. Genetics, 1: 1–52, 107–163.

CREIGHTON, H. S., and McCLINTOCK, B. 1931. A correlation of cytological and genetical crossing-over in *Zea mays*. Proc. Nat. Acad. Sci., U.S., 17: 492–497. Reprinted in *Classic papers in genetics*, Peters, J. A. (Editor). Englewood Cliffs, N.J.: Prentice-Hall, Inc., 1959, pp. 155–160; and in *Great experiments in biology*, Gabriel, M. L., and Fogel, S. (Editors). Englewood Cliffs, N.J.: Prentice-Hall, Inc., 1955, pp. 267–272.

McKUSICK, V. A., and RUDDLE, F. H. 1977. The status of the gene map of the human chromosomes. Science, 196: 390–405.

MORGAN, T. H. 1910. Sex limited inheritance in *Drosophila*. Science, 32: 120–122. Reprinted in *Classic papers in genetics*, Peters J. A. (Editor). Englewood Cliffs, N.J.: Prentice-Hall, Inc. 1959, pp. 63–66.

MORGAN, T. H. 1911. Randon segregation versus coupling in Mendelian inheritance. Science, 34: 384. Reprinted in *Great experiments in biology*, Gabriel, M. L., and Fogel, S. (Editors). Englewood Cliffs, N.J.: Prentice-Hall, Inc. 1955, pp. 257–259.

STURTEVANT, A. H. 1913. The linear arrangement of six sex-linked factors in *Drosophila*, as shown by their mode of association. J. Exp. Zool., 14: 43–59. Reprinted in *Classic papers in genetics*, Peters, J. A. (Editor). Englewood Cliffs, N.J.: Prentice-Hall, Inc., 1959, pp. 67–78.

## Chapter 12

BORGAONKAR, D. S. 1977. *Chromosomal variations in man: a catalog of chromosomal variants and anomalies*. New York: A. R. Liss, Inc.

EPHRUSSI, B., and WEISS, M. C. 1969. Hybrid somatic cells. Scient. Amer., 220 (No. 4): 26–35, 146.

MULLER, H. J. 1954. The nature of the genetic effects produced by radiation. In *Radiation biology*, Hollaender, A. (Editor). New York: McGraw-Hill Book Company, pp. 351–473.

RUSSELL, L. B. 1962. Chromosome aberrations in experimental animals. Progr. Med. Genet., 2: 230–294.

SMITH, G. F., and BERG, J. M. 1976. *Down's anomaly*, second edition. New York: Churchill Livingstone.

## Chapter 13

COHEN, S. 1970. Are/were mitochondria and chloroplasts microorganisms? Amer. Scientist, 58: 281–289.
FOUTS, D. L., MANNING, J. E., and WOLSTENHOLME, D. R. 1975. Physicochemical properties of kinetoplast DNA from *Crithidia acanthocephali, Crithidia luciliae,* and *Trypanosoma lewisi.* J. Cell Biol., 67: 378–399.
GELVIN, S., HEIZMANN, P., and HOWELL, S. M. 1977. Identification and cloning of the chloroplast gene coding for the large subunit of ribulose-1,5-bisphosphate carboxylase from *Chlamydomonas reinhardi.* Proc. Nat. Acad. Sci., U.S., 74: 3193–3197.
LINNANE, A. W., LUKINS, H. B., MOLLOY, P. L., NAGLEY, P., RYTKA, J., SRIPRAKASH, K. S., and TREMBATH, M. K. 1976. Biogenesis of mitochondria: Molecular mapping of the mitochondrial genome of yeast. Proc. Nat. Acad. Sci., U.S., 73: 2082–2085.
SAGER, R. 1972. *Cytoplasmic genes and organelles.* New York: Academic Press, Inc.
STEWART, P. R., and LETHAM, D. S. 1977. *The ribonucleic acids,* second edition. New York: Springer-Verlag. (chl and mit RNA's.)

## Chapter 14

BROWN, D. D., and DAWID, I. B. 1968. Specific gene amplification in oocytes. Science, 160: 272–280. Reprinted in *Papers on regulation of gene activity during development,* Loomis, W. F., Jr. (Editor). New York: Harper & Row, Inc., 1970, pp. 201–209.
DRAKE, J. (Editor). 1973. *The genetic control of mutation.* Genetics, 73 (Suppl.): 205 pp.
JACOB, F. 1966. Genetics of the bacterial cell. Science, 152: 1470–1478. (Nobel prize lecture, discusses the replicon.)
JACOB, F., BRENNER, S., and CUZIN, F. 1963. On the regulation of DNA replication in bacteria. Cold Spring Harbor Sympos. Quant. Biol., 28: 329–348. Reprinted in *Papers on bacterial genetics,* second edition, Adelberg, E. A. (Editor). Boston: Little, Brown and Company, 1966, pp. 403–436.
LAIRD, C. D. 1973. DNA of *Drosophila* chromosomes. Ann. Rev. Genetics, 7: 177–204.
MILLER, O. L., JR., and BEATTY, B. R. 1969. Extrachromosomal nucleolar genes in amphibian oocytes. Genetics, 61 (Suppl. 1/2): 133–143.
PAVAN, C., and DA CUNHA, A. B. 1969. Gene amplification in ontogeny and phylogeny of animals. Genetics, 61 (Suppl. 1/2): 289–304.

## Chapter 15

DICKSON, R. C., ABELSON, J., BARNES, W. M., and REZNIKOFF, W. S. 1975. Genetic regulation: the *lac* control region. Science, 187: 27–35.
FELSENFELD, G. 1978. Chromatin. Nature, Lond., 271: 115–122.
JACOB, F., and MONOD, J. 1961. Genetic regulatory mechanisms in the synthesis of proteins. J. Mol. Biol., 3: 318–356.
KORNBERG, R. D. 1974. Chromatin structure: a repeating unit of histones and DNA. Science, 184: 868–871.
LI, H. J., and ECKHARDT, R. A. (Editors). 1976. *Chromatin and chromosome structure.* New York: Academic Press, Inc.
LODISH, H. F. 1976. Translational control of protein synthesis. Ann. Rev. Biochem., 45: 39–72.
MACLEAN, N. 1976. *Control of gene expression.* New York: Academic Press, Inc.
PTASHNE, M., and GILBERT, W. 1970. Genetic repressors. Scient. Amer., 222 (No. 6): 36–44, 152.
STEIN, G. S., STEIN, J. S., and KLEINSMITH, L. J. 1975. Chromosomal protein and gene regulation. Scient. Amer., 232 (No. 2): 46–57, 114.

## Chapter 16

BARR, M. L. 1959. Sex chromatin and phenotype in man. Science, 130: 679–685.

BEUTLER, E., YEH, M., and FAIRBANKS, V. F. 1962. The normal human female as a mosaic of X-chromosome activity; studies using the gene for G-6-PD-deficiency as a marker. Proc. Nat. Acad. Sci., U.S., 48: 9–16.

LUCCHESI, J. C. 1973. Dosage compensation in *Drosophila*. Ann. Rev. Genet., 7: 225–237.

LYON, M. F. 1961. Gene action in the X-chromosome of the mouse (*Mus musculus* L). Nature, Lond., 190: 372–373. Reprinted in *Papers on regulation of gene activity during development*, Loomis, W. M., Jr. (Editor). New York: Harper & Row, Inc., 1970, pp. 181–183. (The original presentation of the single-active-X hypothesis.)

MCCLINTOCK, B. 1968. Genetic systems regulating gene expression during development. Develop. Biol., Suppl. 1 (1967): 84–112. (In maize.)

MULLER, H. J. 1950. Evidence of the precision of genetic adaptation. *The Harvey lectures* (1947–1948), Ser. 43: 165–229, Springfield, Ill.: Charles C Thomas. Excerpted in *Studies in genetics*, Muller, H. J. Bloomington, Ind.: Indiana University Press, 1962, pp. 152–171. (Dosage compensation in *Drosophila*.)

PROKOFYEVA-BELGOSKAYA, A. A. 1947. Heterochromatization as a change of chromosome cycle. J. Genet., 48: 80–98. (The pioneer paper in this area.)

## Chapter 17

CROW, J. F., and TEMIN, R. G. 1964. Evidence for partial dominance of recessive lethal genes in natural populations of *Drosophila*. Amer. Nat., 98: 21–33.

HADORN, E. 1961. *Developmental genetics and lethal factors*. New York: John Wiley & Sons, Inc.

SANG, J. H. 1963. Penetrance, expressivity and thresholds. J. Heredity, 54: 143–151.

## Chapter 18

BATESON, W. 1909. *Mendel's principles of heredity*. Cambridge: Cambridge University Press.

MATHER, W. B. 1964. *Principles of quantitative genetics*. Minneapolis: Burgess Publishing Company.

WRIGHT, S. 1963. Genic interaction. In *Methodology in mammalian genetics*, Burdette, W. J. (Editor). San Francisco: Holden-Day, Inc., pp. 159–192.

## Chapter 19

BRIDGES, C. B. 1925. Sex in relation to chromosomes and genes. Amer. Nat., 59: 127–137. Reprinted in *Classic papers in genetics*, Peters, J. A. (Editor). Englewood Cliffs, N.J.: Prentice-Hall, Inc., 1959, pp. 117–123.

HANNAH-ALAVA, A. 1960. Genetic mosaics. Scient. Amer., 202: 118–130.

HICKS, J. B., and HERSKOWITZ, IRA. 1977. Interconversion of yeast mating types. II. Restoration of mating ability to sterile mutants in homothallic and heterothallic strains. Genetics, 85: 373–393.

LEVITAN, M., and MONTAGU, A. 1977. *Textbook of human genetics*, second edition. New York: Oxford University Press.

SIMPSON, J. L. 1976. *Disorders of sexual differentiation*. New York: Academic Press, Inc. (In human beings.)

## Chapter 20

BRACK, C., and TONEGAWA, S. 1977. Variable and constant parts of the immunoglobulin light chain gene of a mouse myeloma cell are 1250 nontranslated bases apart. Proc. Nat. Acad. Sci., U.S., 74: 5652–5656.

COOPER, M. D., and LAWTON, A. R., III. 1974. The development of the immune system. Scient. Amer., 231 (No. 5): 59–72, 146.

DATTA, A., DE HARO, C., and OCHOA, S. 1978. Translational control by hemin is due to binding to cyclic AMP-dependent protein kinase. Proc. Nat. Acad. Sci., U.S., 75: 1148–1152.

DAVIDSON, E. H. 1977. *Gene activity in early development*, second edition. New York: Academic Press, Inc.
DAVIS, B. D., DULBECCO, R., EISEN, H. N., GINSBERG, H. S., WOOD, W. B., JR., and MCCARTY, M. 1973. *Microbiology*, second edition. New York: Harper and Row, Inc. (See the section on immunology.)
EDELMAN, G. M. 1970. The structure and function of antibodies. Scient. Amer., 223 (No. 2): 34–42, 128.
GORENSTEIN, C., and WARNER, J. R. 1976. Coordinate regulation of the synthesis of eukaryotic ribosomal proteins. Proc. Nat. Acad. Sci., U.S., 73: 1547–1551.
GURDON, J. B. 1968. Transplanted nuclei and cell differentiation. Scient. Amer., 219 (No. 6): 24–35, 144.
GURDON, J. B. 1975. *Control of gene expression in animal development*. Cambridge, Mass.: Harvard Univ. Press.
LOOMIS, W. F., JR. (Editor). 1970. *Papers on the regulation of gene activity during development*. New York: Harper & Row, Inc.
STILES, C. D., LEE, D.-L., and KENNEY, F. T. 1976. Differential degradation of messenger RNAs in mammalian cells. Proc. Nat. Acad. Sci., U.S., 73: 2634–2638.
TOMKINS, G. M. 1974. Regulation of gene expression in mammalian cells. Harvey Lectures, 69: 37–65. (By hormones.)
YAMAMOTO, K. R., and ALBERTS, B. M. 1976. Steroid receptors: elements for modulation of eukaryotic transcription. Ann. Rev. Biochem., 45: 721–746.

## Chapter 21

FITCH, W. M., and MARGOLIASH, E. 1967. Construction of phylogenetic trees. Science, 155: 279–284. Reprinted in *Papers on evolution*, Ehrlich, P. R., Holm, R. W., and Raven, P. H. (Editors). Boston: Little, Brown and Company, 1969, pp. 450–462.
FOX, S. W., and DOSE, K. 1972. *Molecular evolution and the origin of life*. San Francisco: W. H. Freeman and Company, Publishers.
HOROWITZ, N. H., and HUBBARD, J. S. 1974. The origin of life. Ann. Rev. Genet., 8: 393–410.
KAFATOS, F. C., EFSTRATIADIS, A., FORGET, B. G., and WEISSMAN, S. M. 1977. Molecular evolution of human and rabbit $\beta$-globin mRNA's. Proc. Nat. Acad. Sci., U.S., 74: 5618–5622. (Mutational distance based on mRNA's.)
LANCELOT, G., and HÉLÈNE, C. 1977. Selective recognition of nucleic acids by proteins: The specificity of guanine interaction with carboxylate ions. Proc. Nat. Acad. Sci., U.S., 74: 4872–4875.
MILLER, S. L., and ORGEL, L. E. 1973. *The origins of life on the earth*. Englewood Cliffs, N.J.: Prentice-Hall, Inc.
OPARIN, A. I. 1964. *The chemical origin of life*. Springfield, Ill.: Charles C Thomas.
WONG, J. T.-F. 1976. The evolution of a universal genetic code. Proc. Nat. Acad. Sci., U.S., 73: 2336–2340.

## Chapter 22

HARDY, G. H. 1908. Mendelian proportions in a mixed population. Science, 28: 49–50. Reprinted in *Classic papers in genetics*, Peters, J. A. (Editor). Englewood Cliffs, N.J.: Prentice-Hall, Inc., 1959, pp. 60–62; in *Great experiments in biology*, Gabriel, M. L., and Fogel, S. (Editors). Englewood Cliffs, N.J.: Prentice-Hall, Inc., 1955, pp. 295–297; and in *Evolution*, Brousseau, G. E., Jr. (Editor). Dubuque, Iowa: William C. Brown Company, 1967, pp. 48–50.
LERNER, I. M., and LIBBY, W. J. 1976. *Heredity, evolution, and society*, second edition. San Francisco: W. H. Freeman and Company, Publishers.
SPIESS, E. B. 1977. *Genes in populations*. New York: John Wiley & Sons, Inc.
STERN, C. 1943. The Hardy–Weinberg law. Science, 97: 137–138.

## Chapter 23

ALLISON, A. C. 1956. Sickle cells and evolution. Scient. Amer., 195: 87–94.
AYALA, F. J. 1976. *Molecular evolution*. Sunderland, Mass.: Sinauer Associates, Inc.

BIBLIOGRAPHY

BODMER, W. F., and CAVALLI-SFORZA, L. L. 1976. *Genetics, evolution, and man.* San Francisco: W. H. Freeman and Company, Publishers.
DOBZHANSKY, TH. 1970. *Genetics of the evolutionary process.* New York: Columbia University Press.
FISHER, R. A. 1930. *The genetical theory of natural selection.* Oxford: Clarendon Press.
GOWEN, J. W. (Editor). 1952. *Heterosis.* Ames, Iowa: Iowa State College Press.
KIMURA, M., and OHTA, T. 1974. On some principles governing molecular evolution. Proc. Nat. Acad. Sci., U.S., 71: 2848–2852.
WRIGHT, S. 1932. The roles of mutation, inbreeding, crossbreeding and selection in evolution. Proc. 6th Intern. Congr. Genet., Ithaca, pp. 356–366. Reprinted in *Evolution*, Brousseau, G. E., Jr. (Editor). Dubuque, Iowa: William C. Brown Company, 1967, pp. 68–78.

## Chapter 24

BENNETT-CLARK, H. C., and EWING, A. W. 1970. The love song of the fruit fly. Scient. Amer., 223 (No. 1): 84–92, 136. (Courtship song as a reproductive isolating mechanism.)
COOK, L. M., ASKEW, R. R., and BISHOP, J. A. 1970. Increasing frequency of the typical form of the peppered moth in Manchester. Nature, Lond., 227: 1155. (Correlated with the introduction of smoke control.)
DOBZHANSKY, TH. 1970. *Genetics of the evolutionary process.* New York: Columbia University Press.
EHRLICH, P. R., HOLM, R. W., and RAVEN, P. H. (Editors). 1969. *Papers on evolution.* Boston: Little, Brown and Company.
MORTON, N. E. 1960. The mutational load due to detrimental genes in man. Amer. J. Human Genet., 12: 348–364.

## Chapter 25

BAER, A. S. (Editor). 1977. *Heredity and society: readings in social genetics*, second edition. New York: Macmillan Publishing Co., Inc.
BORLAUG, N. E. 1970. The green revolution, peace and humanity. Nobel prize talk reprinted in: CIMMYT Reprint and Translation Series, No. 3, Jan. 1972.
*Conservation of germplasm resources: An imperative.* 1978. ix + 118 pp. National Academy of Sciences, Printing and Publishing Office.
DAY, P. R. 1977. Plant genetics: increasing crop yield. Science, 197: 1334–1339.
FORD, E. B. 1971. *Ecological genetics.* London: Chapman & Hall Ltd.
KIHARA, H. 1975. Plant genetics in relation to plant breeding research. Seiken Zihô (Rep. Kihara Inst. Biol. Res.), 25–26: 25–40.
SIGURBJÖRNSSON, B. 1971. Induced mutations in plants. Scient. Amer., 224 (No. 1): 86–95, 122. (Radiation and chemicals used to obtain rare beneficial mutations.)

## Chapter 26

ALLEN, G. E. 1975. Genetics, eugenics and class struggle. Genetics, 79: 29–45.
ANSELL, G. B., and BRADLEY, P. B. (Editors). 1973. *Macromolecules and behavior.* Baltimore: University Park Press.
BENZER, S. 1973. Genetic dissection of behavior. Scient. Amer., 229 (No. 6): 24–37, 148.
BODMER, W. F., and CAVALLI-SFORZA, L. L. 1976. *Genetics, evolution, and man.* San Francisco: W. H. Freeman and Company, Publishers. (Discusses intelligence.)
EHRMAN, L., and PARSONS, P. A. 1976. *The genetics of behavior.* Sunderland, Mass.: Sinauer Associates, Inc.
HELLER, J. H. 1969. Human chromosome abnormalities as related to physical and mental dysfunction. J. Hered., 60: 239–248. Reprinted in *Genetics and society*, Bresler, J. B. (Editor). Reading, Mass.: Addison-Wesley Publishing Company, 1973; and *Heredity and society: readings in social genetics*, second edition, Baer, A. S. (Editor). New York: Macmillan Publishing Co., Inc., 1977.

KABACK, M. M. (Editor). 1977. *Tay-Sachs disease: screening and prevention.* New York: A. R. Liss, Inc.
MEDVEDEV, Z. A. 1969. *The rise and fall of T. D. Lysenko*, translated by I. M. Lerner. New York: Columbia University Press.
MEDVEDEV, Z. A. 1977. Soviet genetics: new controversy. Nature, Lond., 268: 285–287. (About a prominent Soviet geneticist who is against studying human genetics.)
MILUNSKY, A., and ANNAS, G. F. (Editors). 1976. *Genetics and the law.* New York: Plenum Publishing Corp.
STINE, G. J. 1977. *Biosocial genetics.* New York: Macmillan Publishing Co., Inc.

*Chapter 27*

AMES, B. N. 1974. A combined bacterial and liver test system for detection and classification of carcinogens as mutagens. Genetics, 78: 91–95. (A liver extract is used to activate some carcinogens; the bacteria, to test for mutagenicity.)
AUSUBEL, F., BECKWITH, J., and JANSSEN, K. 1974. The politics of genetic engineering: who decides who is defective. Psychology Today, 8 (No. 1): 30–43, 120.
COHEN, S. N. 1975. The manipulation of genes. Scient. Amer., 233 (No. 1): 24–33, 132.
CROCE, C. M., and KOPROWSKI, H. 1978. The genetics of human cancer. Scient. Amer., 238 (No. 2): 117–125, 162.
DULBECCO, R. 1967. The induction of cancer by viruses. Scient. Amer., 216 (No. 4): 28–36, 146. (Experimental cancers induced by polyoma virus and SV40.)
HART, R. W., SETLOW, R. B., and WOODHEAD, A. D. 1977. Evidence that pyrimidine dimers in DNA can give rise to tumors. Proc. Nat. Acad. Sci., U.S., 74: 5574–5578.
HAYFLICK, L. 1968. Human cells and aging. Scient. Amer., 218 (No. 3): 32–37, 150.
HILTON, B., CALLAHAN, D., HARRIS, M., CONDLIFFE, P., and BERKLEY, B. (Editors). 1973. *Ethical issues in human genetics.* New York: Plenum Publishing Corp.
HOLLAENDER, A. (Editor). 1971–1976. *Chemical mutagens*, Vols. 1–4. New York: Plenum Publishing Corp.
HOLLIDAY, R., HUTSCHSCHA, L. I., TARRANT, G. M., and KIRKWOOD, T. B. L. 1977. Testing the commitment theory of cellular aging. Science, 198: 366–372.
MERTENS, T. R. (Editor). 1975. *Human genetics: readings on the implications of genetic engineering.* New York: John Wiley & Sons, Inc.
Recombinant DNA research guidelines. Draft environmental impact statement. 1976. Federal Register, 41 (No. 176), Sept. 9, Part III, pp. 38426–38483.
WATSON, J. D. 1976. *Molecular biology of the gene*, third edition. Menlo Park, Calif.: W. A. Benjamin, Inc. (Discusses cancer.)

# Answers to Selected Questions and Problems

**1.1** **a.** genetics.  **b.** chromosome.  **c.** polymer.  **d.** peptide.  **e.** enzyme.
**1.2** T or λ bacteriophages.
**1.4** By the same criteria we use now.
**1.5** No. The existence of God is a belief not subject to scientific proof or disproof.
**2.1** **a.** nucleotide.  **b.** pyrimidine.  **c.** purine.  **d.** nuclease.  **e.** redundancy.
**2.2** **a.** T T A C G G C T A.  **b.** U U A C G G C U A.
**2.6** They would have the same base ratio, same melting point, would hybridize best with each other, would be circular in electron micrographs, and so on.
**2.7** T = 30 per cent; A = 30 per cent; C = 20 per cent; G = 20 per cent.
**2.11** Test for double-strandedness, U, or ribose.
**2.12** "I'm Wilkins."
**3.1** **a.** template.  **b.** palindrome.  **c.** ligase.  **d.** primer.  **e.** methylase.
**3.2** DNA replication eyes.
**3.12** "I'm all eyes."
**4.1** **a.** transcription.  **b.** operon.  **c.** nucleolus.  **d.** tailoring.  **e.** promoter.
**4.2** Single-stranded (sense) DNA. Complementary, single-stranded RNA.
**4.3** Both require cutting, rearranging, adding.
**4.4** Promoter, operator, terminator.
**4.5** At least the five needed to make transcriptase.
**4.6** 1. Recognize replicase.  2. Recognize transcriptase (by promoter).  3. Recognize (repressor) protein (by operator).  4. Recognize ribosome.  5. Transcribe to mRNA.  6. Transcribe to tRNA.  7. Transcribe to rRNA. You will learn of others later.
**4.10** Single-stranded RNA. Complementary, single-stranded DNA.
**5.1** **a.** translation.  **b.** anticodon.  **c.** ribosome.  **d.** ambiguity.  **e.** degeneracy.
**5.2** When it does not contain A.
**5.3** The initiator codon, AUG.
**5.4** GAT.
**5.5** Nothing, since it has no initiator codon.
**5.8** Since the code is universal, some monkey proteins would probably be synthesized.
**5.9** They are cut (f or fMet is removed) and pressed (folded).
**5.14** "Take me to your leader."
**5.15** "Nonsense."
**5.16** "Where's the CAT?"
**6.1** **a.** phenotype.  **b.** mutation.  **c.** transversion.  **d.** photorepair.  **e.** mutagen.

# ANSWERS

**6.4** When a radioactive P in a phosphate decays.
**6.6** They certainly do when the new base is limited in supply.
**6.10** Nuclease, DNA, polymerase, ligase.
**6.11** Those whose changes are too small to deform the DNA duplex; those that produce a change in the same position in both complements.
**6.13** No. Repair is genetically programmed.
**6.14 a.** By breaking dimers enzymatically.   **b.** By excision repair, assuming its complement is present as a result of replication being under way.
**6.17** It proofreads newly synthesized DNA and removes bases added in error.
**7.1** **a.** concatemer.   **b.** virulent.   **c.** joint.   **d.** allele.   **e.** diploid.   **f.** recombination.
**7.3** The one formed by base pairing between complementary single-stranded ends of duplexes.
**7.7** *M N* and *m n*. The parental ones: *M n* and *m N*.
**7.8** 3.
**8.1** **a.** clone.   **b.** auxotroph.   **c.** synapsis.   **d.** integration.   **e.** transduction.
**8.3** The auxotroph. It is genetically simpler.
**8.4** **c** (due to spontaneous mutation from rough to smooth).
**8.5** Make one or two wrong purine–pyrimidine pairs (C A; G T).
**8.6** Yes. If it occurs in the donor strand, the transforming region may be lost by repair.
**8.7** The auxotroph's culture medium contains Met and Thr, the prototroph's presumably does not. Spin down bacteria and see which will grow when placed in the other's medium. If only one grows, it is the prototroph.
**8.11** The abortive transductant leaves a single, erratic trail; the complete transductant leaves a repeatedly branched trail.
**9.1** **a.** temperature.   **b.** episome.   **c.** plasmid.   **d.** lysogenic.   **e.** excision.
**9.2** By transduction of F by $\phi$P1.
**9.3** If the host DNA were heavy and the phage DNA radioactive, only the required phage would be heavy and radioactive.
**9.6** It enters linearly and no ligase is present in the phage head.
**9.7** Collect lysogens for one, two, and three different phage episomes.
**9.9** $\phi\lambda$ helps free $\lambda$ DNA entry, perhaps by forming a joint between the phage and the free $\lambda$ DNA's.
**9.11** Obtain an Hfr after infection of F⁻ with a derived F carrying *lac*.
**9.16** "I have undergone a sex transformation."
**10.1** **a.** mitosis.   **b.** meiosis.   **c.** chromatid.   **d.** spindle.   **e.** prophase.   **f.** chromatin.
**10.2** A rod chromosome.
**10.3** Sister chromatids; identical twins.
**10.4** When it is facultative heterochromatin.
**10.9** 40 micrometers in a ring or an equal-armed rod chromosome.
**10.10** The former has twice the DNA content and chromosomes each with two chromatids.
**10.14** Sperm.
**10.16** Interphase I has no S stage and the haploid number of chromosomes.
**10.17 a.** No.   **b.** Yes. After meiosis II.
**10.19** Metaphase II has half the number of tubules leading from the centromere to the poles.
**10.20 a.** 2.   **b.** 4.   **c.** 8.   **d.** 16.   **e.** $2^n$.
**10.21 a.** 4.   **b.** 8.   **c.** 16.
**10.23** e.
**10.24** d.
**10.25** b.
**10.26** Mitosis.

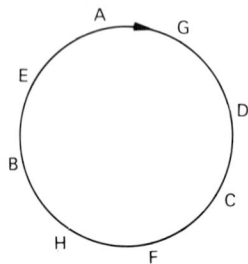

# ANSWERS

**10.27** 2, 3, 1, 6, 5, 4.
**10.28** **a.** Fertilization and first mitotic cell division. **b.** 1 **c.** At least the top chromosome should be J-shaped.
**10.29** Meiosis in ♀. 8.
**11.1** **a.** autosome. **b.** albinism. **c.** nondisjunction. **d.** crossover.
**11.2** Pairs move to metaphase I; crossovers.
**11.3** **d.**
**11.4** $X^C X^c A a$; normal.
**11.5** Parents: $X^C X^c A a$ and $X^C Y A a$; son: $X^c Y a a$.
**11.6** 25 per cent.
**11.7** **c.**
**11.8** **c.**
**11.9** 10 map units.
**11.11** Case A: $B d/b D$, the two loci are 8 map units apart. Case B: the two loci are segregating independently because they are either on nonhomologous chromosomes or far apart on the same chromosome.
**11.13** **a.** $B d$, 0.46; $b D$, 0.46; $B D$, 0.04; $b d$, 0.04. **b.** $B d$, 0.50; $b D$, 0.50.
**11.14** b, c, e.
**12.1** **a.** polyploid. **b.** trisomic. **c.** dicentric. **d.** inversion. **e.** bridge.
**12.2** Many cells condense their chromosomes but do not complete mitosis.
**12.3** d, e.
**12.4** Chromosomes are unpaired at meiosis, producing chromosomally unbalanced gametes.
**12.5** **a.** haploid. **b.** monosomic diploid. **c.** trisomic diploid.
**12.6** **d.**
**12.7** b, c.
**12.11** Two monocentric rings, one (double-length) dicentric ring.
**12.12** **a.** Four normal chromatids, 2 with and 2 without the inversions. **b.** Two normal chromatids, 1 with and 1 without the inversion; plus two duplication-deficiency chromatids, 1 dicentric and the other acentric.
**12.13** **a.** Initiates a bridge–breakage–fusion–bridge cycle. **b.** Same, if centric fragment is included in the future oocyte; if discarded, however, future oocyte is normal.
**13.1** **a.** mendelian. **b.** plastid. **c.** kinetoplast. **d.** centrosome.
**13.2** centrosomal (centriolar) DNA.
**13.7** $2 \, mt^+ \, nea^+$ and $2 \, mt^- \, nea^+$.
**13.8** Isolate mit DNA, treat with active or inactive RNA nuclease. Observe fragmentation only with active RNA nuclease.
**13.9** It is probably coded in the nucleus and translated in the cytoplasm.
**13.14** b.
**14.1** **a.** replicon. **b.** repressor. **c.** amplify. **d.** acentric.
**14.2** When it transduces a sex particle into a female.
**14.3** In the amphibian oocyte.
**14.4** Replicase will not work on foreign DNA; replication can be controlled at the replication initiation locus.
**14.5** Replication initiation requires phage genes not located at the replication initiation locus.
**14.6** The mutant involves the polymerase locus and perhaps its promoter; the normal but not the mutant polymerase acts as a repressor of the polymerase locus.
**14.11** It is a mutation when the cell is not programmed to lose its nucleus; it is not a mutation when the cell is programmed to do so.
**15.1** A bead made up of a cluster of histones.
**15.2** It makes the transcriptase phage-specific.
**15.3** When glucose is abundant, lac enzymes are not needed and the *lac* operon is not transcribed.
**15.5** *lac* promoter, *lac* operator, $z^+$, $\alpha$, $\beta$, $\beta'$, $\omega$, $\sigma$, CAP-coding, cAMP-regulating, $i^+$.
**15.9** Length and nature of leader sequence and ribosome-binding sequence; additional ribosome attachment or drop-off sites.
**15.11** **a.** heterochromatin. **b.** bands.

**15.12** DNA, RNA, histones or protamine, nonhistone protein.

**15.17** d, because histone DNA is only transcribed then and histone DNA probably needs to be specifically activated by stage-specific nonhistone protein.

**16.2** XXY, XXXY, XXXXY, XXXYY, etc.

**16.3** **a.** The amount produced using 1 X. **b.** The amount produced using 2 X's.

**16.5** In all cells where there is no Barr body—including all cells in early development and in the germ line.

**16.9** Perhaps additional heterochromatin competes for a limited supply of histones used for heterochromatization.

**16.13** Both seem to be heterochromatic; both are transposable; both repress adjacent euchromatic genes.

**16.14** Transposition of controlling genes to and from a color locus; segregation of innately colored and colorless plastids.

**16.15** c.

**17.1** **a.** dizygotic. **b.** phenocopy. **c.** dominant. **d.** recessive. **e.** lethal.

**17.5** Yes. One may be XY; the other, X0.

**17.10** No. Transcription and translation of a gene have nothing to do with the dominance or recessiveness of its allele.

**17.13** The recessive is less detrimental in the tetraploid because it is less often expressed there.

**17.14** A recessive lethal is involved. Compare the numbers of abortions and progeny produced here with those from a cross between nonyellows.

**17.15** Yes. The white seedlings will die.

**17.18** **a.** Feeble-mindedness is due to a recessive allele with complete penetrance.
**b.** Short-sightedness is due to a dominant, autosomal allele with complete penetrance.

**17.19** Before they split into two groups of cells.

**18.3** A hindrance, since the presence of the recessive is masked in the heterozygote.

**18.5** The walnut parents were heterozygotes for two independently segregating pairs of genes. Walnuts have both dominants; rose or pea each have one, different, dominant; single has no dominants.

**18.7** **a.** $\frac{1}{8}$. **b.** $\frac{1}{32}$. **c.** $\frac{1}{64}$.

**18.10** Both are $Ww\ Yy$.

**18.11** White parent is $Ww\ YY$; green parent is $ww\ yy$.

**18.12** **a.** First parent is 18 inches, second parent is 16 inches. **b.** 19 inches. **c.** 15 inches. **d.** $\frac{1}{16}$.

**18.13** **a.** Curve becomes narrower. **b.** Curve becomes wider. **c.** Curve becomes narrower (and shifts position toward the selected extreme).

**18.14** $Tt\ L^M L^N;\ Tt\ L^N L^N;\ tt\ L^M L^N;\ tt\ L^N L^N$.

**18.15** **a.** light. **b.** mulatto. **c.** mulatto. **d.** mulatto.

**18.16** $\frac{1}{128}$.

**18.17** **a.** $AA\ bb$ and $aa\ BB$. **b.** Both are $Aa\ Bb$.

**19.1** **a.** supermale. **b.** intersex. **c.** medulla. **d.** gonad. **e.** mosaic.

**19.11** **a.** none. **b.** 1. **c.** 2.

**19.14** See if he contains antigens for antibodies made against the Y chromosome.

**19.16** Environment differentially affecting a constant genotype.

**19.17** daughter could be $X^c\ 0$, owing to maternal nondisjunction.

**19.18** The son is $X^C\ X^c\ Y$, owing to paternal nondisjunction.

**19.19** The hormones of the male twin caused sexual differentiation to be abnormal in the female twin.

**19.20** **a.** Both are gynandromorphs. **b.** In A: The X lost from the left half of the body was wild type, leaving an X containing alleles for white eyes and lighter body color. In B: A wild-type X chromosome was lost later in development, leaving an X containing an allele for white eyes.

**19.21** A. Nondisjunctional $X^w\ X^w\ Y$ female. B. Wing mutation. C. Nondisjunctional $X^{w^+}\ 0$ (sterile) male. D. Contamination, since nondisjunction *and* mutation are needed to explain the fly, an explanation that is too unlikely.

**20.1** **a.** hormone. **b.** induction. **c.** antibody. **d.** heme. **e.** antigen.

# ANSWERS 425

**20.2** They cannot jump over the fence.
**20.7** Perhaps the leader or ribosome-binding sequence changed.
**20.14** Yes, if they are long enough.
**20.17** Because a clone produces mRNA's that code for the same variable region but different constant regions of an antibody chain.
**21.1** Forming the first organism was a rare event, after which organisms evolved and spread so rapidly that conditions favorable for another origin no longer existed.
**21.2** Chemical stability and self-replication.
**21.4** Complementarity.
**21.5** Proteinoids have a nonorganismal origin.
**21.7** Nucleic acid.
**21.8** Yes, as required for auxotrophs to become prototrophs, for motion, for coordination, for self-regulation.
**21.9** Gains and losses of small amounts, since these disturb metabolism least.
**21.10** With evolutionary trees.
**21.12** Completely and/or partially redundant DNA's that may have a regulatory function.
**22.2** Apparently, yes.
**22.3** 48 per cent.
**22.4** 0.1.
**22.5** **a.** 0.1. **b.** 0.8. **c.** 0.5. **d.** 0.3.
**22.7** $p = q = 0.5$.
**22.8** **a.** 52 per cent. **b.** 76 per cent. **c.** No effect.
**22.9** **a.** 0.18. **b.** $BB = 0.81, bb = 0.01$.
**22.11** $T =$ about 0.45, $t =$ about 0.55.
**22.12** About $\frac{2}{5}$.
**22.15** 25 per cent.
**23.1** **a.** selection. **b.** fitness. **c.** heterosis. **d.** migration.
**23.6** 0.66.
**23.9** 1.3.
**23.10** 0.02.
**23.11** **a.** 0.9. **b.** 0.99.
**23.13** **a.** 0.00018. **b.** 0.0001. **c.** 0.0002.
**23.15** 0.8.
**24.7** No. Genetic death means the failure to reproduce.
**24.8** Probably more, because it persists longer and has more chance to produce detriment (say in old age) not affecting reproduction.
**24.13** The new species arose as a multiple-species polyploid of the other two species.
**25.1** They were more often home.
**25.4** The genotypes of these seeds contain undesired recombinations.
**25.5** Only if the expense is affordable.
**25.9** No. Not unless the chemical under discussion is itself a mutagen.
**26.10** They have no assurance that the environment of different races is equivalent.
**26.17** Not appreciably, according to a Swedish committee, since there is a strong social pressure against incest.
**26.18** Yes. This is one of the fruits of genetics.
**27.1** An X0 female to her XY monozygotic twin brother.
**27.7** MN or N, A or AB, and Rh positive.
**27.8** **b.**
**27.14** Mutations can be produced by doses that are socially or medically permissible.
**27.17** It would probably disappear. Yes, from the consumer's standpoint; no, from industry's.
**27.19** Yes, such permission was recently granted.
**27.22** They are reported to destroy ozone. Atmospheric ozone protects us from the genetic effects of the sun's UV.
**27.23** Yes. So that such industrial and academic experimentation will follow the same guidelines.
**27.25** a, b, c.

**27.26** a, e.

**27.27**

| A | T | E | A | S | E | ■ | B | A | S | I | C | A | M | I | N | O | A | C | I | D |
|---|---|---|---|---|---|---|---|---|---|---|---|---|---|---|---|---|---|---|---|---|
| C | O | L | C | H | I | C | I | N | E | ■ | B | I | ■ | O | R | E | L | ■ | ■ | E |
| I | L | ■ | ■ | O | T | R | O | ■ | C | O | P | E | D | ■ | D | ■ | R | A | G | E |
| D | O | N | T | T | H | I | N | K | T | H | A | T | I | M | S | T | O | N | E | D |
| I | S | E | R | ■ | E | C | T | ■ | O | M | S | ■ | ■ | ■ | ■ | I | ■ | S | N | ■ |
| F | E | U | ■ | A | R | K | ■ | E | R | S | T | W | H | I | L | E | ■ | E | N | ■ |
| Y | ■ | R | L | ■ | V | D | ■ | ■ | ■ | H | ■ | B | I | O | ■ | ■ | I | S | O | ■ |
| ■ | ■ | J | O | E | L | S | A | I | D | T | O | R | A | Y | M | O | N | D | E | ■ |
| H | A | ■ | E | S | S | A | Y | ■ | M | I | T | E | ■ | N | ■ | A | ■ | I | S | ■ |
| A | M | E | A | L | ■ | K | ■ | R | E | A | ■ | S | A | ■ | D | ■ | D | A | M | ■ |
| R | ■ | T | H | E | R | E | I | S | O | N | L | Y | O | N | E | Y | O | U | ■ | I |
| D | E | C | ■ | S | ■ | D | T | ■ | B | I | S | ■ | ■ | U | ■ | E | R | E | C | T |
| Y | A | ■ | M | ■ | P | ■ | E | R | O | S | ■ | P | O | R | E | D | ■ | ■ | O | E |
| ■ | ■ | B | U | T | O | F | M | E | T | H | E | R | E | A | R | E | T | W | O | ■ |
| P | A | Y | ■ | O | R | B | ■ | E | ■ | ■ | E | R | ■ | ■ | E | ■ | I | ■ | ■ | B |
| U | V | ■ | S | T | I | F | F | E | N | U | S | ■ | M | C | L | ■ | D | C | O | ■ |
| ■ | I | S | ■ | E | ■ | ■ | ■ | M | E | N | ■ | B | A | Y | ■ | T | E | R | N | ■ |
| A | S | I | W | A | S | N | A | T | U | R | A | L | L | Y | C | L | O | N | E | D |
| P | A | N | A | ■ | T | ■ | T | O | S | O | B | ■ | A | B | L | E | ■ | ■ | S | A |
| E | ■ | G | R | A | Y | ■ | E | E | ■ | ■ | B | A | S | E | A | N | A | L | O | G |
| X | R | A | Y | E | X | P | E | R | I | M | E | N | T | ■ | S | T | A | B | L | E |

# Index

Page numbers in *italics* refer to figures; those in **bold face** refer to the glossary.

a and α mating type, 275
A. *See* adenine
$A_1$ gene, 243, *244*
Å (angstrom), 18, 22
abdomen, 359, 383
abnormality, sexual, 281
A-bomb, 352
ABO blood type, 150, 251, 373, *374*
abortion, 178, 281, 384
abortive transduction, *119*, 120, *121*, **395**
acclimation, 234
acentric, *179*, 182, **395**
*Acetabularia*, 194
acetylation, 230, 377
achondroplastic dwarf, 293, 327
acridine, 96, *97*
activation, sliding, hypothesis, 297, *298*, 300, *301*
  of transcription, 232
active repressor, *223*
adaptation, 334, 364
adaptive, race, 335
  value. *See* fitness
Addison's disease, 375
adenine (A), 16, *17*, *18*, 25
  methylated, *51*, 53
adenoids, 381
adenosine triphosphate, 197
adenovirus, *35*
adrenal gland, 375
adsorption, 308
adult, *Drosophila*, 211
  globin, *296*, *298*
  hemoglobin (Hb-A), 297
*Aedes*, 353

*Aegilops*, 349
*Aerobacter* DNA content, *23*
African game, 352
agar, 29, 113, *114*
age, and Down's syndrome, 176, *177*
  and heterochromatization, 240
  mental, 176
  and sex abnormality, 282
aging, 52
  and death, 375–378
agriculture and genetics, 343–352
alanine (Ala), *71*
albinism, 150, 159, *160*, 162, *163*, 265
  mouse, 359
alcohol, 347, 349, 366
aldrin, *382*
alfalfa, 349
algae, 4, 5, *191*, 195
alkylating agents, 92, *382*
alkylation, **395**
allele, 105, **395**
  abnormal, and differentiation, 293
  and behavior, 359
  deleterious, in populations, *324*
  frequency, 315, *316*, 318, **395**
    changes in, 323–330
    disequilibriated, 328–330
    and mutation, 326
  neutral, 329
  rare, 318, 327
alligator DNA content, *23*
allolactose, 224, **395**
alpha, -amanitin, *61*
  chain, 58
  mating type, 275
  particles, 379
alphabet, Greek, 295
altruism, 365
ambiguous code and ambiguity, 73, 75, **395**
American Indian, 350

amino acid, 7, 70
  analog, 377
  basic, 38, **396**
  sequence similarities, 311
  structure, *71*
amino group, *51*, 70, *71*
amino terminus, 70, *80*
aminoacyl-tRNA synthetase, 76, *77*, 228, 233, **395**
aminopurine, 92
amniocentesis, 383, **395**
amphetamines, 366
amphibian, 4, 159, 193
  chromosome, *62*, *63*
  DNA content, *23*
  nuclear cloning in, 387
  nucleolus organizer, 214, *216*
  o⁺ protein, 292
*Amphiuma* DNA content, *23*
amplification, 211, *214*, *215*, **395**
  and differentiation, 289
  and heterochromatin, 242
Ana. *See* anaphase
anabolism, 6, 7, **395**
  and operon, 224
analog, 377
anaphase (Ana), *138*, *139*, 140
  I and II, *148*, 164, *165*
  abnormal, 173
Ancon sheep, 293
anemia, Cooley's, 264
  fetal, 374
  sickle-cell, 254–257
angstrom (Å), 18, 22
animals, domesticated, 350
  tetraploid, 174
  *See also* names of specific animals
ankylosing spondylitis, 381
annelid DNA content, *23*
*Anopheles*, 353
ant, 279

427

# INDEX

anthocyanin pigment, 243
antibiotics, 134, 349
  and cloning, 389
  and fetus, 366
  resistance to, *196*, 200, 353
antibody, *201*, 295, 298, 311, 375, **395**
  against blood antigens, 374
  and disease, 354
  and immunology, 373, *374*
  and learning, 358
  synthesis, 298–301
anticodon, 68, *69*, 77, *78*, **395**
antigen, 298, *374*, 375, **395**
  -binding site, *299*, **395**
antimutagen, 98, 377, **395**
antiparallel, 20, **395**
apple, 348
*Arabidopsis, 191*
arabinose, 88, 308
arginine (Arg), *71*, 229
arm, 142, 293
arrow, 159
art, 365
arteries, 217
arthritis, 375
artificial insemination, 350, 352, 385
Aryan, 367
asbestos, *382*
ascorbic, acid, *382*
*Ascaris,* 217
*Asidea* DNA content, *23*
asparagine (Asn), *71*
aspartic acid (Asp), *71*
*Aspergillus* DNA content, *23*
aspirin, 366
assortative mating, 318, **395**
athlete, 364, 365
  foot, 353
atmosphere of earth, 307
atom, 380
  bomb, 380, 381
atrophy and aging, 378
attachment site, 126, **395**
Australia, 350, 355
autism, 365
autoimmunity, 375
autoradiograph, *30*, *229*, **396**
autosome, 158, **396**
  -X translocation, *239*
auxin, 294
auxotroph, 114, **396**

B (any usual base), *51*
baby, 251
*Bacillus, 52,* 117, 120
  and bees, 360
  DNA content, *23*
backbone broken, 88
backcross, **396**
background, genetic, 252
bacon, 383
bacteria, 4, 5, *6*
  antibiotic-resistant, 353
  competent, 115
  conjugation. *See* conjugation
  chromosomes, 33, *36, 37*

DNA cloned, 389
DNA content, *23*
  intestinal, 251
  lysogenic, 126
  mutation-testing in, 381
bacteriophage (phage, $\phi$), *11*, **396**
  attachment to host, *12*
  coat, *11, 12*
  core, *12*
    DNA, *12*
  f2, *35*
  head, *12,* 104
  hexagonal plate, *12*
  immunity to, 126
  incubation, 13
  infection, 12
  λ, 13, *129,* 209
    chromosome, *31,* 33
    DNA, 23, *28*
    life cycle, 123, *124, 125*
    map, *208*
    repressor, 209
    and rho, 226
    and transduction, *127*
    and transgenosis, 385
  M13, *31*
  MS2, *35,* 73
  Mu-1, 218, 243
  P1, 117, 118, 119, 126
  PBS1, PBS2, *52*
  protein, *12*
  Qβ, 45
  R17, 25, *27*
  sheath, *12*
  striations, *12*
  T2, *23, 24, 35, 52*
  T4, *52,* 110
  T6, *52*
  T7, *31*
  T-even, *52,* 102, *103*
  T series, *12,* 13
  tail and tail fibers, *12*
  temperate, 123, *124*
  and transduction, 117, *119*
  and translation regulation, 228
  virulent, 105
  X174, *11,* 64, 109
    DNA, 23, *31, 32*
    nucleotide sharing in, 227
balance, ecological, 354
balanced lethal, 325, **396**
baldness, 266
Baldwin apple, 348
band, 143, *144,* 172. *173*
  and interband, 213
bank, seed or sperm, 350, 351, 385
barbiturates, 366
barley, *343,* 347, 349
Barr body, 237, *238,* 240, 373, **396**
base, abnormal, 92
  absorbing UV, *26,* 29
  analog, 92, 377, **396**
  complementary, *18,* 25
  composition of DNA, *24*
  derivatives of, in tRNA, *26*
  organic, 16

methylated, *51*
modification of RNA, 59
  and mutation, 88
  pairing, 25, *26,* 44
  pairs, *78,* 91
  ratio determination, 29
  rearranged, 59
  sequences, 24, *25, 26*
basic amino acid, 38, **396**
basketball, 364
bean, *343*
bee, 279, 352, 360
beet, sugar, 348
behavior, and evolution, 336
  and genetics, 357–363
  normal, 361
  and sex type, 281
benzanthracene, *382*
benzene as mutagen and carcinogen, *382*
benzpyrene, *382*
beta, chains, 58
  -galactosidase, 386
  particles, 379
*bio,* 127
biochemical sequences, origin of, 310
biological fitness, 323, *324,* **396**
biology, 9
bird, 4, *23,* 159, 335
birth, control, 345
  defects, 320, *321*
bishop, 369
*Biston,* 335
biting, 362
blackberry, 349
blacks, 267, 364, 370, 384
blastula, 292
blight, corn leaf, 351
blood, incompatibility, 374
blood transfusions, 373
blood type, 150, 374
  ABO, 251, 373, *374*
  MN, 264, 265, 315, *316*
blowflies, 354
blue-green algae, 5, *191*
body, color and behavior, 359
  parts coordinated, 292–294
  -weaving, 363
boiling DNA, *26*
bombs, atomic or nuclear, 380
bond, disulfide, 299
  peptide, 70
bone, 293, 385
*Bonellia,* 286
*Brachy,* 293
Brahma, 350
brain, 176, 358, 377
breaks, *34,* 88, 102, *182*
  chromatid, 180, *181*
  chromosome, 178, *179*
breast, 281, 282, 381
breeding, 335, 385
bridge, *179,* 291, **396**
  -breakage-fusion-bridge cycle, 180, 182
British peppered moth, 335

# INDEX

bromouracil (BU), 92
brother-sister mating, 319
budding, 337, 348
bugs, 190
building blocks, 6
bull, 350
bullet, 88
butterfly, 159

C. *See* cytosine
C bands, 143, *144*, 172
C strand, 22, 30, 46
cabbage, 174, *175*
cadaver, 385
caffeine, *382*
calf thymus, *229*
California, 349
camouflage, 364
cAMP, 223, 294, **396**
camphor, 134
canal, cytoplasmic, *217*
cancer, 65, 120, 192, 259, 376
    breast, 381
    and cloning, 389
    cure of, by mutagens, 382
    and genetics, 378–379
*Candida, 313*
CAP, 221, **396**
    interaction site, *222*, 223, **396**
capillaries, 255
capsule, *114*, 115
captan, *382*
carbolic acid (phenol), 9, 11
carbon, 7, 8
    dioxide, 194, 307
    monoxide, 307
carboxyl group, *70, 71*
*Carcharias* DNA content, *23*
carcinogen and mutagen, *382*
carp, *23*, 230, 350
carriers, 328
carrot, 289, 387
cartilage, 293
cassava, *343*
cassette, 276
*Cassiopeia* DNA content, *23*
cat, *252*, 253, 351
catabolism, 6, *7*, 224, **396**
catalysis and catalyst, 7, *8*, 308, 310
cataract, 365
"cat-cry" syndrome, 187
cattle, 350, 354
cell, 4, 5, **396**
    cloning, 387
    cycle, *138*
    cytoplasmic canal, 291
    daughter, 141
    division, 376, 378
    follicle, *217*, 291
    fusion and genetic disease, 386
    membrane, 4, *5*, 11, **396**
        bacterial, 137
        and chromosome, 190
        and DNA, 13, 121, 200, *201*, 298
        and hormone, 294
        and secretion, *82*

motility, 200
nerve, 295
nurse, *217*, 291
plasma, 298
pole, 148
red blood. *See* erythrocyte
root, 289
sex, 144
size, 348
somatic, 61, 141
surface and cancer, 378
target, 294
wall, 11, **396**
white blood, 290, 295, 298
cellulose, 389
centric, **396**
centriole, *5*, 13, 202, **396**
centromere, 138, 145, **396**
    broken, *283*
    and crossing over, 169
    programmed, *152*
    and redundancy, 143
centrosome, 202, **396**
cereals, *343*
cerebrospinal fluid, 361
cerebrum or cerebellum, 176, 358
chain, complementary, *18*, **397**
    globin, 295, *296*
    light (L) and heavy (H), 298, *299*
    nucleotide, 16
charged tRNA, *69, 80*, **396**
checkerboard, 160
chemical, messenger and sex type, 286
    mutagen, 92, 381–383
    and cancer, 378
    selection, 308
chestnut blight, 352
Chianina bull, 350
chickpea, *343*
chicken, 4, 111, 200, 232, 233, *313*, 376
    and autoimmunity, 375
    base ratio, *24*
    and cell fusion, 386
    Creeper, 293
    DNA content, *23*
    globin, 297
    wing, 293
*Chlamydomonas*, 195, 275
Chlorambucil, *382*
chloramphenicol, 134
*Chlorella, 191*
chloroplast, 5, 13, 81, *191, 192, 194*, **396**
    DNA, 193, 195, 210
    ribosomes, *74*
chondrodystrophic dwarf, 293, 327
chordates, DNA content of, *23*
chorea, 363
chorion, 291
chromatid, 138, **396**
    break, 180, *181*, **396**
    exchanges, *146*
    sister, 213, 146
chromatin, 143, 228, *229*, **396**
    reconstitution, *232*, **396**

sex, 237, *238*
chromosome, 5, *6*, 13, **396**
    abnormal, 361, 363
    acentric, *179*, 214
    addition, 175
    amphibian, *62, 63*
    amplified, 212, *214, 215*
    arm, 142, **396**
    arrangement, 218
    balance, 175
    bands and interbands, 172, *173*, 213, 311
    breaks, 88, 178–*187*, **396**
        two, 180, *182*, 183
        and virus, 218, 219
    bridge, *179*
    and cell membrane, 190
    changes, and aging, 377
        and evolution, 310
        gross, in nuclear, 172–187
        in unbroken, 173, 178
        and X rays, 354
    circular, *37, 38*
    condensation, *38*
    defect and amniocentesis, 384
    degradation, 175
    deletion, *88*, 377, **397**
    dicentric, *179*, 219
    distribution, 137, 141
    divided human, *39*
    duplex number in, 38
    eukaryotic, 142
    exchanges, *146*
    eyes, 49, *50*
    folding, 36, *38*, 229, *231*
    giant interphase, 213
        *See also* chromosome, larval salivary gland
    homologous, 142
    human, 36, *39, 140, 173*
    iso-, *283*
    larval salivary gland, 211, *212, 213, 214, 215*, 217, 240, 242
    loss, 174, 175, 281, 289
        and virus, 219
    metaphase, cytology of, 173
    mosaic, 175
    movement, 138
    multichromatid, 211, *212, 213*, 240
    nick, *34*, 102, 377
    nondisjunction or nonseparation, 164, *165*
    nonhomologous, 144
    nonsister, 146
    nuclear, number, 142
    set of, 36
    number reduced, 144
    organization, *31–38*
    and parentage, 370
    puffs, 240
    rearrangements, gross, in population, 332
    and sex type, 283, *284*
    somatic, and antibody, 299
    redundancy, 143
    replication, 43–53, 102

# INDEX

chromosome (cont.)
  repulsion, 38, 145
  ring, 128, *129, 181,* 182
    acentric, *216*
    and virus, 219
  RNA, and folding of DNA, *38*
    phage, 27
  rod, 36, *50,* **405**
  segregation, 149, **396**
  set, 13, *23,* **396**
    extra, 173
  sex, 158, *159*
  shape, 182
  stained, *30,* 172
  sticky, 152
  structure and organization, 16, 289
  supercoiling, 36, *37, 38,* 229
  three-armed, 212
  tips, 167, 169
  X and Y, 158, *159*
  X$^S$ and X$^L$, *159*
  Y$^S$ and Y$^L$, *159*
chronic myeloid leukemia, 379
*Chrysanthemum,* 337
church, 369
*cI, 208,* 209
cigarettes, 383
cilia, 202, 286
circulatory system, 217
circus, 282
civil law, 369
clay, 308
cleavage, 292
cleft palate, 384
clergy, 369
clone, 113, *114, 118,* **396**
cloning, 389, **397**
clover, red, 349
clubfoot, 251
coat color, 252
coccid, 212
cochlea, 359
code, genetic, 72, 73
codon, 67, *68, 69, 71, 72,* **397**
  anti-, 68, *69, 77, 78,* **395**
  initiator, 68, *69,* **401**
  sense, missense, and nonsense, 72, *75, 89,* 90
  terminator, 72, *75*
coelenterate DNA content, *23*
coevolution, 355, **397**
coffee, 344
colchichine, 174, *175,* 349
colcimide, 174
collagen, 376, 378
colorblindness, red-green, *161, 162, 166,* 236, 268, 282
  adaptive, 364
communism, 366
compartmentalization, 7
competence, 115, **397**
complementary, bases, *18*
  chains, *18,* **397**
  genes, *265,* 266, **397**
complete transduction, *119,* 120, **397**

complexity, biochemical, evolved, 309
concatemer, 103, *104, 105, 124, 125, 199,* **397**
conformation similarities, 311
Confucius, 369
congenital, 320, *321,* **397**
conjugation, 113, 123, *128*–134, **397**
connective tissue and aging, 378
consanguinity, 369, **397**
consolidation, 358, **397**
constant (C-type) position effect, 240, **397**
constant (C) region, 299, **397**
constitutive heterochromatin, 213, *215,* 241, **397**
contact, cell, and DNA replication, 210
  inhibition, 378
continuous (quantitative) traits, 268–273, **397**
control, negative, 208, *223*
  positive, 208, 221, *223,* 226
controlling genes, 243, **397**
Cooley's anemia, 264
coordination of body parts, 292–294
corals, 4
corn leaf blight, 351
  *See also* maize
cortex, *280,* **397**
cosmetics, mutation, and cancer, *382*
cosmic rays, 307, 380
cotton, 174
couplet, 190
courtship behavior, 359
cousin marriage, 319, 320, *321,* 369
cow, 230
crab DNA content, *23*
creativity, 365
Creeper chicken, 293
cri-du-chat syndrome, 187, 363
Crick, F. H. C., 20
crops, prehistoric, *343*
cross, test or back-, *257,* **396**
crossing over, 158, 166–*169,* **397**
  and antibody, 299
  and duplications, 310, *311*
crossover 166, *167, 169,* **397**
  frequency, 168
  unit, 168, **397**
crustacean DNA content, *23*
cryoprecipitate, 384, 385
C-type position effect, 240, **397**
*Culex,* 353
curing genetic diseases, 384, 385
curve, distribution, *269,* 270, 271, *272*
cyanoacetylene, 307
cyclamate, *382*
cyclamen, 349
cycle, bridge-breakage-fusion-bridge, 180, 182
  lysogenic, *124,* 125, 209
  lytic, 209
cyclical behavior, 358
cysteine (Cys), *71*
cytochrome *c,* 312, *313*
cytokinin, 294

cytology, 159, 172, 241
  and transcription, 228
cytoplasm, *5,* 210, **397**
cytoplasmic, bridges, 291
  canal, *217*
  DNA, 200, *201*
  ribosomes, *74*
cytosine (C), 16, *17, 18,* 25
  derivatives, 51, *52*
  dimers, 93
  and UV, *93*

DBCP, *382*
DDT, 353, *382*
deafness, 359, 365
death, 320, *321,* 333
  and aging, 375–378
Dee-geo-woo-gen rice, 348
deficiency, 108, 180, *182,* **397**
  mental, 187, 269, 318, 361, 369
degenerate code, 72, **397**
degeneracy, 72, *78*
dehydration, *7,* 308
deletion, *88,* 377, **397**
*Delphinium,* 338
*de novo* synthesis, 43, 60, **397**
dentistry, 380, 381
deoxyribonucleic acid (DNA), **397**
  abnormal, 52
  amplification, 211
  antiparallel strands in, *20*
  AT-rich regions of, *28, 222,* 223
  backbone, *21*
  bacterial, *82*
  as genetic material, 114
  base, composition, *24*
    sequences, 24, *25*
  beads, 230, *231*
  binding, 115
  boiled, *26*
  cell membrane, 121, 200, *201,* 298
  centrosomal, 202
  chloroplast, 193, 195
  circular, *31, 32,* 197
  circularization, *124,* 125
  cloned, 387, *388,* 389
  coiling, 19, 20
  conformation, 51
  content of various organisms, *23*
  cooled, *26,* 27, 28
  cytoplasmic, 200, *201*
  differs from RNA, 19
  dispensable, 134
  -DNA hybrids, 289
  double helix, 19, *21,* 28
  duplex, mapping of, *28*
  extruded, 117
  fingerprinted, 51
  folding, *38,* 229, *231*
  gaps repaired, *48*
  GC-rich regions of, *222,* 223
  as genetic material, 9, 13
  giant, *61*
  globin, 232

# INDEX

heated, 26, 28
hybrid molecules and hybridization, 29, 109
immunity against, 375
inserted, 65
integrated, 116
kinetoplast, 200, 201, 210
kinetosome, 202
knot-free, 33
labeled, 47
and learning, 358
m, 61
major and minor spaces around, 20, 21
and memory, 358
methylases, 51
mitochondrial, 196, 197, 199, 291
modification, 51, **398**
molecular model, 21
nicks repaired, 48
nonessential, 123, 126
nuclear, and chl DNA, 210
nucleolus organizer, 193, 213, 216
   amplified, 214
packaging, 231
packing of, 33, 34, 35
penetration, 115
polarity in, 20
as a polymer, 18
polymerase, 47, 48, 65, **398**
   and mutation, 98
   proofreads, 98
   stockpiled, 291
r, 61
radioactive, 105
recombinant, cloning, 388
redundant or repeated, 22, 31, 61, 62, 63, 143
released from nucleus, 121
repair, 94–97
replication, 46, 47, 48, 49, 50
   and cell contact, 210
   synchronous, 211
ring, 134, 199, 214, 216
-RNA hybrid, 290
and RNA viruses, 65
semiconservative replication, 46–50
with single-stranded ends, 31, 33
single-stranded, with double-helical regions, 25, 26
spacer, 61, 62, 63
splicing, 387, 388
stockpiled, 291
strand separation, 26, 27, 28
supercoiling, 229
synthesis, and histone, 230
   and DNA repair, 95
t, 61
terminal redundancy, 31
transcription-silent and -active, 61, 62, 63, 214
and transformation, 115
underreplicated, 211
untwisting, 34, 47
viruses, 11, 31, 63

X-ray diffraction, 22
yolk, 202, 291
See also chromosome
deoxyribose, 18, 19, **397**
   in evolution, 308
   and mutation, 88
dephosphorylation, 231
depolymerization, 7, **397**
derepressed operon, 226, **397**
desert, 336
detergent, 10
determination, 293, **397**
   of sex in eukaryotes, 275–286
detriment due to inbreeding, 320
development, 289–302, **397**
   early, 291–292
diabetes, 254
dicentric, 179, 181, **397**
dictionary, tRNA's as, 69
dieldrin, 382
diet, 269, 366
differential polyploidy, 212, **397**
differentiation, 289–302, **398**
   and cancer, 378
   of gonad, 280
diffraction, X-ray, 21, 22
digits, 253
dimer, 7, **398**
   excision, 95, 379
   pyrimidine, 93
   repaired, 94, 95
dinucleotide, 7
dipeptide, 7, 70
diploid, 105, **398**
dipnoan DNA content, 23
Diptera, 210
disassortative mating, 320
discontinuous (qualitative) trait, 268, **398**
disease, biological control of, 354
   genetic, 383, 385
   lipid-storage, 362
   resistance, 347, 349, 351
disequilibrium in allele frequency, 323–330
disjunction, **398**
disomic, 176
distance, mutational, 312, 313
distribution curve, 269, 270, 271, 272
disulfide bond, 299
divergent spindle, 152, 153
division, cytoplasmic, 141
dizygotic twins, 249, 250, 251
DNA. See deoxyribonucleic acid
dog, 293, 313, 350
   DNA 23, 377
domains of supercoiling, 38
domesticated animals, 350
dominance, 263, 269, 270, 271, **398**
   and allele frequency, 317
   and populations, 317
   and selection, 324
dominant, 256, 257, **398**
   lethal, 259
donkey, 313

donor, bacterial, 113
   organ, 385
dosage, compensation, 236–241, **398**
   and optimum effect, 258
double-cross method, 345, 346
double helix, 19, 20, 94, **398**
doubling dose, radiation, 380
doubling number, 376
Down's syndrome, 176, 186, 187, 285
*Drosophila*, adult, 276
   behavior, 359, 360
   chromosomes, 212, 213, 277
   DNA, cloned, 389
      content, 23
      replication, 50
   dosage compensation, 240
   egg chamber, 217
   as founders, 337
   giant chromosomes, 211, 212, 213, 215, 217, 240, 242
   and gravity, 361
   meiosis, 149, 152
   mutation rate, 218
   mutation-testing in, 381
   as pest, 353
   polyploid, 174
   population, 332, 333
   position effect, 241–243
   sex determination, 276–279
   stages, 211, 276
   transformed, 117
   transgenosis, 191
   and transposition, 243
   wing, 266, 267
   X chromosome, 213
drift, genetic, 323, 329, 330
drug, 254, 349, 365
   mutation, and cancer, 382
   and rejection, 375
   -resistance, particle, 134, **398**
      plasmid, 387, 388
duck, 23, 313
duplex, hybrid, 29
duplicate loci, 265, 267, **398**
duplication, 180, 183, **398**
   and crossing over, 310, 311
   in evolution, 310
   and redundancy, 310, 311
Dutch elm disease, 352
dwarf, 269, 293, 327
dye, acridine, 97

ear, 252, 281
   maize, 345, 346
earth, 3, 4, 307
earthquake, 336
earthworm, 285
ecclesiastical law, 369
echinoderm DNA content, 23
ecology, 336
   and genetics, 352–355
ectoderm, 293
effector, 223, 224, 233, 294, 357, 361, **398**

egg, 4, 61, 149
    chamber, *217*
    chicken, 111, 350
    exceptional, *165*
    fertilized, 144, 249
Egypt, 321
elasmobranch DNA content, *23*
electric shock, 358
electroencephalogram, 361
electron, 88, 380
    micrograph, 6, *10, 11, 28, 81, 103, 129, 194*
    microscope and mapping, *108*
electrophoresis, 333
elephantiasis, 353
elimination of harmful mutant, 259
elongation factor, protein, 79, **398**
embryo, 282, 293, 299
    and globin, 295, *296, 298*
emigrants, 329
encephalitis virus, 11
endoplasmic reticulum, 4, *5,* 81, *82,* 197, **398**
energy, 6
    for protein synthesis, 79
engineering, genetic, 349, 387
England, 335, 369
English shorthorn, 350
engulfment, 291
enhancer, active or inactive, *223,* 224
enzyme, 7, *8,* 398
    hybrid, 240, 262
    induced, 224
    modification, 51
    pigment, 252
    restriction, 52, *53, 135*
    ribosome-bond, 76
    virus-specific, *52*
    See also names of specific enzymes
environment, and behavior, 357
    genetic, 251
    and IQ, 365
    and mutagens, 379–383
    nongenetic, 252
    and phenotype, 249–259
    and quantitative trait, 268
    and sex, 275, 285–286
*Ephestia, 191*
epilepsy and marriage, 369
episome, 126, *130,* 208, 209, **398**
epistasis-hypostasis, *265,* 266, **398**
Epstein–Barr virus, *191,* 192
equator, 364
equilibrium, 317, 327
erasure, 242, **398**
errors in repair, 96
erythrocyte, 216, 217, *232,* 236, 237, **398**
    and cell fusion, 386
    shape, 254, *255*
*Escherichia coli,* 11, 117–120, 123–134
    base ratio, *24*
    chromosome, *37,* 49
    DNA, *23, 35*

and DNA cloning, 387, *388,* 389
    linkage map, *131, 132*
    operons, 221–227
    as phage host, *12, 103*
    RNA polymerase, *43*
    transcriptase, 262
    transcription in, 57, 58
    transformed, 117
estrogen, 294
ethics, 383, 386, 387, 389
ethyl ethanesulfonate, *92*
euchromatin, *231,* **398**
    and crossing over, 168
    replicon, 210
    and transcription, 228
euchromatization, 243, 297, *298,* 300, **398**
eugenics, 367, **398**
*Euglena,* 194
eukaryote, *5,* **398**
    chromosomes, 142, *230*
    DNA replication, 49
    gene action regulation, 228–234
    mutation rate, 218
    and nonmendelian genes, 190
    and position effect, 239
    replicons, 209
    sex determination, 275–286
    translation regulation, 233–234
    See also names of specific eukaryotes
European strawberry, 349
evaporation, 308
evolution, 3, 14, 307, 332
    of genetic material, 310–313
evolutionary tree, 312, *313,* **398**
exchange, programmed, 151
    union, 179, **398**
excision, *124, 126*
    of dimers, *95,* 379
    faulty, *127, 133,* 379
    -splicing hypothesis, 301
expense of treatment, 385
explosives, 347
expressivity, 252, **398**
extinction, 352
extreme phenotype, *269,* 270, 271, *272*
eye, 176, 187, 259, 361, 385
    replication, 47, *48*

$F_1$, *160*
F derivative, *133,* **399**
F particle, *130,* 208, **399**
F repressor, 208
$F^+$ and $F^-$ sex types, *130*
facultative heterochromatin, 212, 237, **399**
falciparum malaria, 256, 325
fallout, 380
farmers, 343
female, 128, *129, 130*
    apparent, XY, *284*
    chromosome, *140*
    as mosaic, 237, 239
    super-, *277*
    X0 and XXX, 281

feminization, testicular, 363
ferns, 4
ferritin, 199
fertilization, *149,* 159, **399**
    double, 282
    multiple, 173
    self-, *319*
    and translation, 233
fertilized egg, 144
fertilizer, 344, 389
fetal hemoglobin (Hb-F), 296
fetus, 178
    damage to, 365
    and globin, 295, *296,* 297, *298*
    and meiosis, 148
    -mother incompatability, 374
    transplanted, 385
    X-rayed, 381
Feulgen stain, *238*
fidelity in replication, 98
fibroblast, 121, *181, 238,* 290
    aging, 376, 377
    and cell fusion, 386
    divisions, 376
    and transgenosis, 385
filter or filtration, 29, 117, *118*
fingers, 138, 253, 362
firehose, 33
fish, 4, 350
*Fissurella* DNA content, *23*
fitness ($w$), 323, *324,* **399**
fixation, 358, **399**
flagellum, 202
flora, 251
flower, color, 267, 268, 348
    conformation, RNA, 25, *27, 31*
flowering plants, evolution of, 310
fluke, 382
fluorescence, 172
fluoroscopy, 380
fly, 333, 353, 354
follicle cell, *217,* 291, **399**
food, as barrier, 336
    mutation, and cancer, 380, *382*
foot, 251
football, 364
formaldehyde, 307
Formosa, 348
founder principle, 337, **399**
*Fragaria,* 349
frame-shift mutation, *90, 96, 97,* **399**
freckles, 252
freezing, 308
frequency, allele, 315, 318
Friesian cow, 350
frog, *23,* 174, 217
frontal lobes, 176
fruit fly, drug-resistant, 353
    See also *Drosophila*
functional mosaic, 237, 239, **399**
fungi, 4, 218, 349, 351
    DNA content, *23*
    drug-resistant, 353
fungicide, mutation, and cancer, 353, *382*

# INDEX

fungus gnat, 217
furazolidone, 134
future, genetic diseases in the, 385

G. *See* guanine
G bands, 172, *173*
G1 and G2 period, 137, *138, 160*
*gal*, 127, 227, 386
galactosemia, 384
galaxies, 3
gamete, 144, *263,* 316, *317,* **399**
gamma rays, 379
garden pea, 142
gastrulation, 292
gel, *25,* 333
gene, 45, **399**
    action, and aging and death, 377
        regulation, 221–234
    activation, *298, 300, 301*
    amplification, 211
    balance, 175, 178, 276
        and sex, 285
    behavioral, 359
    chloroplast, 195
    cloned, 387, *388,* 389
    complementary, *265*
    controlling, 243
    destruction, 207, 216–218
    detrimental, in population, 332, *333*
    dosage, 258, **399**
    flow, 329
    frequency, 323–330, **399**
    function, 46
        evolution, 312
    globin, *295*
    Ig, *300*
    intercellular effects, 293
    *lac* repressor, 224
    lethal, 257, *259*
    mendelian and nonmendelian, 190
    multiple phenotypic effects, 254
    polymerase recognition, 45
    product, 247, 292
    for protein, 58, *59*
    regulator, 208, *223,* 292
    replication origin, 47, *48*
    responding, 244, 276
    ribosome-binding, *225*
    segregation, 149
    signaling, *244,* 276
    single, and behavior, 359
    single copy, transcription, 290
    splicing, 387. *388*
    synthesis, and genetic disease, 386
        programmed, 207–216
    transposable, 243
    underreplicated, 211
    *See also* genes; mutant
generalized transduction, 113, *118,* 120, **399**
genes, for antibiotic resistance, 196, 200
    duplicate, *265,* 267
    heterozygous, and traits, *269, 270*
    linked, 167

many, affecting trait, 268–273
and meiosis, 151
and mental retardation, 361
and mitosis, 141
multiple, and behavior, 361
nonhomologous, 158
and parentage, 370
X-limited, 236–241
*See also* gene; mutant
genetic, background, 252, **399**
    code, 68, 70, 309, **399**
    death, 333, **399**
    disease, 383, 385, **399**
    donor, 128, **399**
    drift, 323, 329, 330, **399**
    engineering, 369, 387, **399**
    environment, 251, **399**
    equilibrium, 317
    heterogeneity, 368
    load, 333, 334, **399**
    marker, 105, *106,* 254, **399**
    material, 3, 8, 13, 56, 57, **399**
        information content of, 22
        origin and evolution, 307–313
    recipient, 128, **399**
    recombination. *See* recombination
    transduction. *See* transduction
    transformation. *See* transformation
    variability, 345, 351, 364
genetics, 14, **399**
    and agriculture, 343–352
    and behavior, 357–363
    and cancer, 378–379
    consequences of, 341–390
    and ecology, 352–355
    and law, 369
    and medicine, 373
    misuse of, 386
    and politics, 363–371
    and public health, 373
    and religion, 369
    and sociology, 363–371
    suppressed, 366
genitalia, 280
genotype, 89, **399**
    and aging and death, 376
    and behavior, 357
    and clone, 113
    and environment, 249–259
    frequency, *316,* 318, *319*
    variability, 329, 332–335
geography, 336
germ line, 141, **399**
    aging and death, 375
    breaks in, 180
    and dehydrogenase, 263
    and dosage compensation, 240
    mutants, 379
German measles, 365
Germans, 368
giant DNA, *61*
giant molecules, *7*
gibberellic acid, 294
Giemsa stain, 172
giraffes, 232

gland, pituitary, 294
*Glaucocystis, 191*
globin, 254, 290, 295, *296,* **399**
    mRNA, *232,* 234
glucose, *52*
    6-phosphate dehydrogenaise, 236
glucosylation, **400**
glutamic acid (Glu), *71,* 254
glutamine (Gln), *71*
glycine (Gly), 68, *71*
goat grass, 349
goldfish, 358
gonad, *280,* **400**
grafting, 337, 348
grass, 349, 350
Gravenstein apple, 348
gravity, 361
Greek alphabet, 295
grid, 161
grimacing, 363
growth, 6
guanine (G), 16, *17, 18, 25*
    methylated, 51
guidelines for cloning, 389
Guthrie, W., 363
gynandromorph, 278, *279,* **400**
Gypsies, 368
gypsy moth, 352

H. *See* hydrogen
H (heavy) strand or chain, 197, 298, *299*
hair, 266, 282, 361, *382*
half-translocation, 183, *184,* 185, *186, 187,* **400**
hamster, 376
handedness in rats, 358
hands, 138
haploid, 105, *277,* **400**
Hardy–Weinberg principle, 316, 317, **400**
Hawaii, 337
hazard of genetic engineering, 389
Hb, **400.** *See* hemoglobin
health and genetics, 373
heart, 375, 380, 384, 385
heavy (H) strand or chain, 197, 298, *299*
height, 268, 269, 281, 282, 283
    adaptive, 364
HeLa, *198,* 199
*Helix,* 285
helix, *10*
    double, DNA as a, 19, *21*
heme, 254, 297, **400**
hemoglobin, 232, 262, 290, **400**
    A (Hb-A), 254
    $A_2$ (Hb-$A_2$), 297
    evolution, 311, 312
    F (Hb-F), *296, 297*
    S (Hb-S), 254
    synthesis, 295–298
hemophilia, 384, 385
*Hemophilus, 36, 51, 53*
herpes simplex virus, *31,* 234

# INDEX

heterochromatin, 143, 213, *215*, 218, *231*, 237, 241, **400**
  constitutive, 143, *144*
  and crossing over, 167
  facultative, 143, 212
  replicon, 210
  and transcription, 228
heterochromatization, 236–244, *285*, **400**
  and antibody, 300, *301*
  and globin, 297, *298*
heteroduplex, *116*, 117, **400**
heterogeneity, genetic, 368
heterosis. *See* hybrid vigor
heterozygosity and races, 334
heterozygote, 149
  and heterosis, 325
  for reciprocal translocation, *184*, *185*
heterozygous, 149, **400**
hexosaminidase A, 363, 384
Hfr sex type, 130, *131*, 209
HGPRT, 362, 384
hippocanthus, 358
hips, 282
histidine (His), *71*
histone, 38, 193, 196, 200, 210, **400**
  acetylation, 377
  classes, *230*
  cluster, 230, *231*, **400**
  and redundancy, 230
  stockpiled, 291
  and transcription, 232
Hitler, A., 367
hitting, 362
hive, 360
*ho* and *HO*, 275
hobbies, 350
homocystinurea, 384
homoduplex, *116*, 117, **400**
homologous, **400**
  chromosomes, 142
  regions in protein, *299*
homozygosis and mating scheme, 318, *319*, 320
homozygous and homozygote, 149, **400**
honeybee, 360
hormones, 233, 278, 293, *294*
  and behavior, 357
  and cell division, 376
  and cloning, 389
  and phenocopy, 281
  and sex type, 280
horse, *23*, *313*, 351
hot dogs, 383
housefly, 353
human being, base composition of, *24*
  DNA content, 23
  monosomic or trisomic, 176, *177*
  and polyploidy, 173, 174
  sex types in, 280–285
  with translocations, *185*
  and trees, *313*
human, cells transformed, 117
  chromosomes, *140*, *144*, *173*, 176

mental retardation, 318, 361, 369
  pedigree symbols, *251*
  sex chromosomes, *159*–166, 237–238, 281–*285*
  mutagenesis, 353, 379–383
humidity, 336
Huntington's chorea, 362
hybrid, 149, **400**
  corn, 345, *346*
  and crossing over, 338
  enzyme, 240, 262
  and isolation, 336
  mated, *149*
  molecular, 29
  nucleic acid, 29, **400**
  single- or double-cross, *346*, 347.
  vigor, 325, 345, *346*, 362, **400**
hybridization, **400**
  interspecific, 337
  and mapping, 198
  nucleic acid, 289, 290
  rate, 143
hycanthone, *382*
hydration, *7*
hydrocortisone, 294
hydrogen (H), 7, 8, 307
  bond, *18*, *47*, **400**
  cyanide, 307
hydrolysis, **400**
hydrophilic, 70, *71*, **400**
hydrophobic, 70, *71*, 81, **400**
hydroxylamine, *92*
Hymenoptera, 279
hypostasis-epistasis, *265*, 266, **400**
hypoxanthine (I), 26, *78*
  -guanine phosphoribosyl transferase (HGPRT), 362, 384

$i^+$, 222, 224, *225*
ice, 336
identical twins, 249, *250*, *251*
idiocy, juvenile amaurotic, *328*
Ig, **400**
immigrants, 329
immunity, 126, 298, 353, 375, 378
immunoglobulin (Ig), 298, *299*, **400**
immunology, 373–375
inbreeding, 318, 320, 321, 350, **400**
incest, 369
income and food, *344*
incubation period, 9
independent segregation, *150*, *264*, **400**
India, 353
Indian Brahma bull, 350
Indian corn. *See* maize
Indica rice, 348
individual as founder of species, 337
induced enzyme formation, 224, **400**
inducer, 224, **401**
induction, *293*, **401**
industry and cloning, 390
infant death, 320, *321*
infection, experiments, 9, *10*
  and sex type, 128

infectious bovine rhinotracheitis virus, 52
infective DNA and RNA in eukaryotes, *191*
influenza virus, 11, 110
information flow, *67*
infrasex, 277, **401**
initiation, factor, **401**
  proteins, 79
initiator, codon, 68, 69, **401**
  substance, 208, 209, **401**
injury and polyploidy, 174
insanity and marriage, 369
insect, 4, 23
  pests, 345, 350
insecticides, 353
insemination, artificial, 350, 352, 385
Insertion Sequences, 243
insulin, 254, 389
integration, *124*, *126*, **401**
  of DNA, *116*, 191
intelligence, 268
  quotient (IQ), 365
  and races, 364–366
intemperate, 105, **401**
interaction, between loci, 262–272
  of gene products, 247
interarm inversion, *182*
intercellular effects of genes, *293*
interconversion of mating type, 275
interlocked rings, *199*
interphase, *138*, *139*, 143, **401**
  I and II, 147, *148*, *154*
intersex, 277, 281, **401**
intestine, 290
intragenic mapping, 109, *111*
introgression, 338, 348, 349, **401**
inversion, 88, *180*, *182*, **401**
*in vitro* and *in vivo*, 43, 349, **401**
ions and ionization, 380
IQ, 281, 364
island, A-bombed, 352
isochromosome, *283*, **401**
isolating, mechanism, **401**
  reproductive, 335, 336
isoleucine (Ile), *71*
isoniazid, *382*

Japan, 350
Japanese, 320, *321*, 347
  beetle, 352
jellyfish, 4, *23*
Jews, 362, 368, 384
joint, 102, *104*, 105, 125, **401**
juvenile amaurotic idiocy, *328*

kangaroo, 240, *313*
kappa, *191*
kernel, 243, *244*, 251
kidney, 197, 200, 385
killer, *191*
  bees, 352
kinetoplast, 13, *192*, 200, *201*, **401**
  DNA, 210

# INDEX

kinetosome, *192*, 202
Klinefelter's syndrome, 282, 284, **401**
Koi, 350
Kosygin, A., 367
Krushchev, N., 367

L (light) strand or chain, 298, *299*
labeled DNA, *47*
*lac*, 221, 224
lactic dehydrogenase, 262
lactose, 224
  operon, 221, *223*, 224, *225*
ladder, DNA as a, *18*
language, biological, 67, *68*
larkspur, 338
larva, 210, *211*, 217, 242
law, 369, 384
leader sequence, *59, 60*, 62, *75*, 226, *227*, **401**
learning and memory, 358, **401**
leaves, 9, *10*
leg, 293
legumes, *343*
*Leishmania*, 200
lensless mouse, 293
lentil, *343*
Lesch–Nyhan syndrome, 362, 384
lethal, *259, 324*, **401**
  balanced, 325
  synthetic, 259
leucine (Leu), *71*
leucocyte, 290, 295, 298
leucoplast, 193, **401**
leukemia, 379
lichens, 335
life, 3
  span, 375
ligase, *48*, 65, *95, 96*, **401**
  and cloning, 387, *388*
ligation of broken ends, 178
light, blue, 94
light bulb filament, 38
light (L) chain or strand, 197, 298, *299*
lightning, 307
*Limnea*, 141
limpet DNA content, *23*
linkage, 107, *269*
  autosomal, *160*
  sex, 158, 163
  X-, *161, 162, 163*
lipid, 7, 81, 362
lips, 362
liquid transport, 217
lithium, *22*
liver, 173, 200, 232, 377, 386
load, genetic, 333, 334, **399**
  mutational, 379
localization of gene products, 292
loci, chloroplast, 195
  segregated, 149
  sequenced, 107, *108*
  X- and Y-limited, 159, *164*
locus, 105, 249–259, **401**
  multiple, heterosis, 326

silent or active, 276
  single, heterosis, 325
locust base ratio, *24*
loganberry, 349
lollipop, *12*
London, 350
longevity, 377
  of mRNA, 234
long-term memory, 358, **401**
looking glass, 159
lungfish DNA content, *23*
lupus erythematosis, 375
lymphocyte, 120, 193, 200, *201*, 298, 300
  suppressed, 375
Lysenko, T. D., 366
lysine (Lys), *71*, 229, 349
lysis, 105, **401**
lysogen, 209, **401**
lysogenic, 126, **401**
lytic cycle, 105, *124*, 125, 209

M (mitotic) period, *138*
macrophage, 290
maggot, 210
maize (corn), 142, 251
  chloroplast, *194*
  controlling genes, 243
  disease-resistance, 351
  evolution, 338
  heterosis, 345, *346*
  hybrid, 345, *346*
  leaf blight, 351
  and spindle, 152, *153*
  and X rays, 349
malaria, *191*, 256, 353
  and heterosis, 325
male, 128, *129*
  apparent XX, *284*
  chromosomes, *142*
  parasitic, 286
  super-, *277*
  XYY and XXY, 282
maleic hydrazide, *382*
mammals, 4, *23*, 292
man. *See* human being
mannose, 88
map, circular, 109, 195, 196, 198
  crossover, 168
  electron microscope, *108*, 109
  human X linkage, *170*
  intragenic, 109, *111*
  λ DNA, 208
  linkage, *131*
  of mitochondrial genes, *198*
  recombinational, *108*, 109, *110*
  unit, 107, *108*, **401**
maps, congruence between, *108*
mapping, AT-rich regions, *28*
  by hybridization, *29, 30*, 198
marriage, 7, 369
  cousin, 319, 320, *321*
Mars, 159
Maryland, 370

maternal, age and chromosomal loss or gain, 177
  materials stockpiled, 292
maternity proven, 370
mating, assortative, 318
  disassortative, 320
  nonrandom, 318–321
  random, 315, 370
  sibling, 319
  systems, 315–321
  type, 195, 275
maze, 361
mDNA, *61*
mean phenotype, *269*
measles, 365
medicine and genetics, 373, 384
medulla, *280*, **401**
meiosis, 137, 144–153, **401**
  I and II, *147, 148*
  in females, 148
  programmed, 151
  and recombination, 148
meiotic mutant, **401**
melanin, 150, 159, 267, 361
Melphalan, *382*
membranes, 4, 7, 81, 198
memory and learning, 358, **401**
Mendel, G., 190, 266
mendelian genes, 190, **401**
menstruation, 281
mental, ability, 269
  age, 176
  deficiency, 187, 269
  or penal institution, 282
  retardation, 318, 361, 369
mercury compounds, *382*
mesoderm, 293
messenger (m) RNA, 59, **401**
  antibody, 299
  artificial, 73
  globin, *232*, 234, 297
  longevity, 227, 234
  stockpiled, 291
  suppression, 233
  tailoring, *60*
  translation, 68
Met. *See* methionine or metaphase
metabolism, 6, **402**
  and mutation rate, 218
metaphase (Met), *138, 139, 140, 142*, 143
  I and II, *147, 148, 150, 154*
  and translation, 233
metastasis, 378
meteorite, 307
methane, 307
methionine (Met), 68, 69, *71*, 114, 117, *118*
  f, 68, 69
methotrexate, *382*
methyl, 19
methylase, *51*, 87, **402**
methylated, A, *53*
  RNA, 59, 63
methylation, *51*, 62, 87, *230*

# INDEX

Mexico, 347, 354
*Microbracon,* 279
microscope, light, 138, 141, 228
   phase-contrast, *238*
migration, 323, 329, 330, **402**
miracle rice, 348
missense mutation, 89, **402**
mitochondrion, *5,* 13, 64, 81, *191, 192, 201,* **402**
   DNA, 196, *197*
   ribosomes, *74*
   stockpiled, 291
mitomycin, 94
mitosis, 137–144, *148,* **402**
   frequency, 218
   germ-line, 141
   programmed, 141
   somatic, 141
MN blood type, 150, 264, 265, 315, *316*
mold, 44, 345, **402**
molecular, cloning, 387
   recombinant, **402**
molecules, giant, 6, *7*
mollusk DNA content, *23*
mongolism, 176
monk, 369
monkey, 64, *313,* 374
monomer, 6, *7,* **402**
monosomic, 176, 281
monozygotic, 249, *250, 251,* **402**
mosaic, 175, 241, **402**
   and behavior, 359, *360*
   functional, 237
   human, 282
   sex, 278, *279*
mosquitoes, 353
mosses, 4, 285
mother-fetus incompatability, 374
mother, foster, 385
moths, 159, 174, 278, *313,* 335
mountain, 336
mouse, 237, 238, *239,* 241, 289
   and autoimmunity, 375
   and behavior, 359
   *Brachy,* 293
   cell divisions, 376
   and cell fusion, 386
   DNA content, *23*
   encephalomyocarditis virus, *31*
   learning in, 358
   lensless, 293
   mutation-testing in, 381
   nicked DNA of, 377
   transformation in, 117
mouth, 176
mRNA. *See* messenger RNA
mulatto, 267
multiple phenotypic effects, 359
multiple-species polyploidy, 174, *175,* 337, 338, 349
muscle, 249, 295, 311
music, 365
mustard, nitrogen, *382*
mutagen, 92, **402**
   and aging, 377
   and agriculture, 348, 349
   anti-, 98, 377
   and cancer, 378, 382
   chemical, 381–383
   human effects of, 353
   radiation, *93*
mutagenesis, and agriculture, 349
   environmental, 353, 379–383
mutagenic, **402**
mutant, 89, **402**
   adaptive, 258, 334
   and aging, 377
   base sequence, 89
   beneficial, 259
   detection, 89
   detrimental, 90
   elimination of, 259
   germ line, 379
   irreparable, 95
   neutral, 259
   recessive or dominant, 256–258
   in a redundant gene, 91
   silent, 91
   somatic, 379
   spontaneous, 97
   undetected, 91
   and viability, *259*
   *See also* gene; genes
mutation, 87–98, **402**
   and cancer, 378
   and chromosome loss, 175
   and DNA polymerase, 98
   and load, 333, 334
   nonrecurrent or recurrent, 326
      phase-(or frame-)shift, *90, 96, 97*
   and polyploidy, 173
   in populations, 323, 326–328
   produces raw material for selection, 326
   programmed, 207, 218
   and radiation, 379
   rate, 218
   and recombination, 101
   reverse, 326
   and selection, 327
   sense, missense, or nonsense, 89
   spontaneous, 97, 107, 380
   and twins, 250, *251*
mutational, distance, 312, *313,* **402**
   load, 379
   sites, intragenic, 110, *111*
mutator, **402**
mutilation, 281, 362
mycins, 134
myoglobin, 311, 312, **402**
myxoma virus, 355

narcissus, 349
natural selection, 310, **402**
natural selection and populations, 323–326
nature vs. nurture, 249
neck, webbed, 281, 283
*Necturus* DNA content, *23*
negative control, 208, *223,* **402**
nerve cell or neuron, 295, 358, 359, 361
nervous system, mouse, 359
Netherlands, 350
neuroglia, 359
neurons, 295, 358, 359, 361
*Neurospora, 313*
neutral, alleles, 329
   mutant, 259
neutrons, 379
newborns, 362
Newcastle disease virus, *63*
nick, **402**
   staggered, *125,* 130, *135*
nightshades, 337
nitrates, 344
   nitrites, mutation, and cancer, *382*
nitrogen, 7, 8
   fixation, 389
   mustard, *94,* 382
nitrosamines, *382*
nitrous acid, *92, 382*
Nobel prizes, 22, 383
nondisjunction, *165, 166,* **402**
   and age, 178
   and sex abnormality, 282
nongenetic environment, 252, **402**
nonhistone proteins, 38, *232,* **402**
nonhomologous chromosomes or genes, **402**
nonidentical twins, 250, *251*
nonmendelian genes, 190–202, **402**
nonrandom mating, 318–321
nonrestitutional union, 179, **402**
nonsense, codon, *72, 75,* **402**
   mutation, 89, **402**
nonsister chromatids, 146, **402**
noodles, 5
norepinephrine, *294*
nose cone, 120
notochord, 293
N-terminus, *299*
nuclear, bomb, 380, 381
   chromosome, 13
   replication, *50*
   cloning, 387
   membrane, 4, *5,* 138, **403**
   power, 380
nuclease, 24, 47, 33, *34,* 65, *95,* 118, *201,* **403**
   DNA, 52
   phage-specific, 228
   restriction, *135*
   and cloning, 387, *388*
   RNA, *48,* 59, 76
   and transformation, 117
nucleic acid, 7, 8, 9, **403**
   as antigen, 298
   conformation in virus, *31*
   crosslinked, 94
   dependency upon protein, 309
   as genetic material, 3–15
   hybrid, 29
   information, *67*
   nonreplicational, 44
   organization, 16–31

origin of, 308
  and evolution of, 307
polymerase, 43
replication, 43
single-stranded, with double-helical regions, 25, *26*
structure, 16
See also deoxyribonucleic acid; ribonucleic acid
nucleolus, *5*, 61, 138, **403**
  free, 214
  in oocyte, 214
  organizer, 61, *62*, *63*, 193, 214, *216*, **403**
    rDNA, *61*, 193, 214, *216*, 377, 389
nucleoplasm, **403**
nucleotide, *7*, 16, **403**
  origin of, 308
  ribose, *43*, *44*
  sharing, 46, *64*, 227, **403**
  and transcription, 58
  whole, changes, *88*
nucleus, 4, *5*, 229, **403**
  division, 137, *138*
  polar, 216
  receiving protein, 230
  -released DNA, 121
  transplanted, 210, 289
nullosomic, 178, **403**
nun, 369
nurse cells, *217*, 291, **403**
nurture vs. nature, 249

$o$, replication origin gene, 47, *48*, *49*
$o^+$, 222
$o^+$ protein, 291, 292
oat, 349
obesity, 176
odor, 359
oil, selection for, 345
omega chain, 58
oncogene, 378
one-arm inversion, *182*
onion, *139*
oocyte, 61, 148, *217*, **403**
  human, 177
  and nucleoli, 214
  stockpiling in, 291
operator, 58, *59*, 222, 224, *225*, **403**
  mutant, 90
operon, 58, *59*, 221–227, **403**
  complex, 226
  *gal*, 227
  *lac*, 221, *223*, 227
  in viruses, 63
*Ophryotrocha*, 286
optimum phenotypic effect, 258
orange, 344, 349
organ donor, 385
organelle, *5*, *192*, **403**
organic, base, 16
  soup, 308
organism, **403**
  complex, evolution of, 309
  DNA content of, 22

origin of, 3, 309
structure of, 4
organismal evolution, **403**
*ori*, *208*, 209
orientation, defective, 281
origin, **403**
  of genetic material, 307–310
orphanage, 365
Orthodox Eastern Church, 369
ostrich, 4
outbreeding, 320, **403**
ovalbumin, 290
ovary, *280*, *282*
ovulation, 281
oxygen, 7, 8, 19, 254, 256, 297, 307

$p$ or $q$ allele frequencies, 315, *316*
$p^+$, 221, *222*, 225
$P_1$, *160*
$^{31}P$ and $^{32}P$, 88
pacifism, 365
pain, 255
palindrome, 53, 124, *125*, *126*, *127*, 130, *135*, 222, 224, 225, **403**
  and cloning, 387, *388*
palm prints, 176
parainfluenza virus, 111
paralysis, 385
*Paramecium*, *191*
parasite, 289, 336, 350
parent and heterochromatization, 242
parentage, assigned or excluded, 370
pastrami, 383
paternity proven, 370
paws, 252
pea, 230, *343*
peanut, *343*
pedigree, *253*, *320*, **403**
  of causes, *256*
  symbols, *251*
penal or mental institution, 282
penetrance, 252, 327, **403**
penguin, *313*
penicillin, 134
peptide bond, 70, *72*, **403**
peroxide, 93
perrenials, 337
persistence, 334, 385, **403**
personality, 365
pesticides, 253, *382*
pests, 352, 354
Peta rice, 348
petunia, 349
phage. See bacteriophage
phage cross, **403**
Pharoahs, 321
phase-shift mutation, *90*, *96*, *97*, **403**
phenocopy, 253, 281, 364, **403**
phenol (carbolic acid), 9, 11
phenotype, 87, 127, **403**
  and chromosome changes, 176
  and clone, 113
  and dominance, 263
  and environment, 249–259
  frequency, *316*, 318
  multiple effects on, 359

selected, 323
sex, *280*
and single loci, 249–259
and two or more loci, 262–272
variable, 252
variegated, 241, 242
phenotypic, classes, *269*, *270*
  effects, 254, 255
  ratios, 263, *265*, 268
phenylalanine (Phe), *71*, 318, 361
  hydroxylase, 361
phenylketonuria, 318, *320*, 325, 361, 384, 386
phenylpyruvic acid, 361
phocomelia, 254, 266, 293
phosphate, 16, *17*, 344
  and chromosome folding or coiling, 38
  and mutation, 88
phosphoric acid, origin of, 307
phosphorus, 8
phosphorylation, 230, *231*
photographic emulsion, 29, *30*
photorepair, 94
photosynthesis, 193
physiology, 6
  and isolation, 336
pig, *313*
pigeon, *313*, 351
pigment, 150, 243, 335, 361
pilus, *128*, **403**
pituitary gland, 294
placenta, 374
*Plagusia* DNA content, 23
planets, 3, 4
*Planococcus*, 212
plant, breeders, 345
  evolution, 310
  growth inhibition, *382*
  hormone, 294
  ornamental, 348, 349
plasma, cell, 298, **403**
  membrane, *201*
plasmid, 126, 130, 134, **403**
  drug-resistance, 387, *388*
plastids, 193, **403**
*Pneumococcus*, *114*, 117
polar body or nucleus, 216, **403**
polarity, 20, *44*, 78, **403**
  and reunion, 178
pole cell, 148
poliomyelitis (polio) virus, 11, *35*, *191*, 354
politics and genetics, 363–371
pollen, 195, 345
pollution, 335
poly A, *43*, 45, 59, 61, 62, 65, 230, 377, **403**
poly Phe, 73
polydactyly, 150, *253*, 266
polydeoxyribonucleotide (DNA), 8, **404**
polymer, *7*, 11, **404**
  as antigen, 298
polymerase, DNA, 47, *48*
  nucleic acid, 43, **404**

polymerase (cont.)
  polypeptide, 68, 76, 79, **404**
  recognition gene, 45
  RNA, *48*
  stockpiled, 291
polymerization, 7, **404**
  of amino acids, 70
polynucleotide, 7, 8, **404**
polyoma, *31*, 33, *34*, 64, *191*, 192, **404**
  DNA, *23*, 35
polypeptide, 7, **404**
  growing, *80*
  as hormone, 294
  origin of, 308
  polymerase 68, 76, 79, **404**
  proteins needed to make a, 79
  signal, 81, 195
  synthesis speed, 234
  *See also* protein
polyploid, 173, 212, 310, 348, **404**
polyploidy and speciation, 337–338
polyribonucleotide, 8, **404**
polyribosome (polysome), 79, *80*, *81*, *82*, 230
polysomic, 282, **404**
population, 315, **404**
  explosion, rabbit, 355
  gene frequencies, factors affecting, 323–330
  genetic variability of, 332–335
  genotypes, 315–321
  natural, 332, *333*
Porifera DNA content, *23*
porpoise, 352
position effect, 239, 241, 243, **404**
positive control, 208, 221, *223*, 226, **404**
postnatal globin, *296*, *298*
potato, 337, *343*
poultry, 350
pox virus, *191*
predator, 335, 336
pregenetic evolution, 307
primer, 47, 65, **404**
primitive streak, 234
prints, palm and sole, 176
Pro. *See* prophase or proline
proboscis, 286
program, aging and death, 375
  gene synthesis and destruction, 207
  mutation, 207
prokaryote, 5, *6*, **404**
  DNA replication, *49*
  gene action regulation, 221
  mutation rate, 218
  and position effect, 239
  replicon, 207, 208
proline (Pro), 68, *71*
promoter, 58, *59*, *60*, 222, 225, **404**
  globin, *295*
  mutant, 90
  regulation of, 221
  proofreading by polymerase, 98
prophase (Pro), *138*, *139*, 143, **404**
  I and II, *148*, **404**

protamine, 38, **404**
protein, 7, 9, *10*, **404**
  as antigen, 298
  basic and acidic, 38, 104, 138
  brain-specific, 358
  chloroplast, 193
  coat, *103*
  D and E, *64*
  damage, aging and death, 376
  dependent upon nucleic acid, 309
  entering nucleus, 230
  evolution of, 307
  gene for, 46
  homologous regions in, *299*
  initiator, 208
  interacting, 262
  L7, L12, L18, L25, 75
  learning and memory, 358
  modification, 230
  nonhistone, 38, *232*
  origin of, 307, 308
  regulatory, 208
  repressor, 208
  ribosome, *74*, 75
  S12, 75
  secreted or exported, 81, *82*
  specific aging, 378
  specific tissue, 295–301
  starvation, 200
  subunits, *10*, *11*
  suppressing or activating translation, 233
  synthesis, *67*, *68*
  tailoring, 68, 81
  and translation regulation, 228
  *See also* polypeptide
proteinoids, 308, **404**
protons, 379
protoplasm, 4, 5, **404**
*Protopterus* DNA content, *23*
prototroph, 114, *118*, **404**
protovirus, 378
protozoan, 4, *191*, 353
provirus-protovirus hypothesis, 378
*Pseudomonas*, 120
pseudouracil, *26*
pseudoscience, 367
psoriasis, 382
puberty, 148
public health and genetics, 373
puff, **404**
pupa, *211*, 242
pure, 149
purine (Pu), 16, 17, *51*, *53*, **404**
  and behavior, 362
  dimers, *94*
  as mutagens, 98
  origin of, 307
pyloric stenosis, 384
pyrimidine (Py), 16, *17*, *52*, *53*, **404**
  dimer, 93, **404**
  origin of, 307
  substitutions, *52*
  and UV, *93*

$q$ or $p$ allele frequencies, 315, *316*
Q bands, 172
qualitative (discontinuous) trait, 268, **404**
quantitative evolution, 310
quantitative (continuous) trait, 268–273, **404**
sex as, 278
quarantine, 352
queen, 279
quinine, 353

$r^+$, 225, 226, 227
R bands, 172
R group, *70*, *71*
rabbit, *313*, 351, 355
race, 334, 335, **404**
  becomes species, 336
  and intelligence, 364–366
  master, 367
  Mongoloid, 176
radiation, and aging, 377
  and cancer, 378
  dose, 380, 381
  guidelines, 381
  and mutation, 379
  penetration, 307
  and polyploidy, 174
  surgery, 386
radioactive DNA, *25*, 46
radioactivity, *30*, 143
radioautography, *229*, **404**
radiology, *381*
radish, 174, *175*, 347
radium, 379, 380
random mating, 315, 370, **404**
raspberry, 349
rat, 352, 353, 358, 376
ratio, genotypic, *160*
  phenotypic, *160*, 163, 263, *265*, 268
  of X to autosomes, *277*
rays, cosmic, 307
rDNA, 214, *216*, 377, **404**
reading, **404**
rearrangement. *See* chromosome rearrangement
recessive, 256, *257*, **404**
  and allele frequency, 317
  lethal, 259, **405**
  in populations, 332, *333*, *334*
recipient, bacterial, 113
reciprocal, cross, **405**
  translocation, *183*, *184*, *185*, 186, **405**
  and sex type, 283
recognition sequence, 45
recoil, 88
recombinant, 101, **405**
  DNA, cloning, *388*
  molecular, *104*, *105*
recombination, 101, *102*, **405**
  advantages of, 101
  between bacteria, 113–134
  and breakage, 102, 123

in eukaryotes, 137–153, 158–169
frequency, 107, *108*
map, *108*, 109, *110*, **405**
unit, 107, *108*
during meiosis, 148
of mit DNA, 199
between phages, 105, *106–110*
programmed, 137
and RNA viruses, 110
stabilized, 338
between viruses, 101–111
reconstitution of chromatin, *232*
red blood cell. See erythrocyte
red-green colorblindness, *161, 162, 163, 166,* 236, 268, 282
redundancy, 143, *183*, **405**
DNA, 61, *62, 63,* 193
and duplication, 310, *311*
in evolution, 312
and histone, 230
reforestation, 354
regulation of gene action, 221–234
regulator gene, 208, *223,* 292, **405**
regulatory protein, 208, **405**
reiteration, **405**
rejection of tissue, 375
religion, 369, 384
rem, 380
reovirus, *31,* 45
repair of dimers, 94, *95*
replicase, RNA, 45
replication, bidirectional, *49, 50*
of chromosomes, 43
of DNA, 43, 46, *47, 48, 49, 50*
and cell contact, 210
synchronous, 211
eye, 47, *48,* 51, **405**
of genetic material, 14
late, 210
and mutant repair, 95, *96*
origin gene, 47, *48*
programmed, 207
of RNA, 43, *44*
of RSV, *65*
semiconservative, 46, *47, 48,* 131
time, 144
replicon, 207, 209, **405**
repressible operon, 226–227, **405**
repressor, **405**
active or inactive, *223,* 224
F, 115
gene, **404**
λ, 209
protein, 208
reproduction, 6, **405**
reproductive, barrier, 336
isolation, 335, 336, **405**
reptile, 4, 23, 159
research on cloning, 390
resistance, antibiotic, 353
disease, 257, 351
to parasites and pests, 350
to pesticide, 353
responding gene, *244,* 276, **405**
restitutional union, 179, **405**

restricted transduction, 113, 123, 127, **405**
restriction, enzyme, 52, *53, 135,* **405**
nuclease and cloning, 387, *388*
retardation, mental and physical, 176, 282, 361
reticulocyte, *232,* **405**
retinoblastoma, 259, 379, 384
reverse transcriptase, 56, *65, 67,* **405**
reverse transcription, 56, 64, 386
reversible chemical reaction, 7, *8*
reversion, **405**
Rh blood type or Rhesus, 150, 374
rheumatic fever, 375
rheumatoid arthritis, 375
rho, 226
ribonucleic acid (RNA), 8, *10*
backbone, *17*
bases sequenced, 24
cellular, is not genetic material, 14
complementary, *48*
*de novo,* 60
differs from DNA, 19
-DNA hybrid, 290
double-stranded, *31,* 45
flower form, 25, *27*
and folding of DNA, *38*
4S, 5S, 16S, 23S, 59
as genetic material, 9 *10*
giant, *60*
and hybrid molecules, 29
as intermediate, 56
learning and memory, 358
modification, 59
phage, 25, *27*
as a polymer, 16, *17*
polymerase, *43, 44, 48,* **405**
stockpiled, 291
as primer, 47
replicase, 45, 64, **405**
replication, 43, *44*
single-stranded, with double-helical regions, 25, *26*
spliced, *60*
tailoring, 59, *60*
30S, 200S, 61
transcripts, 59
viral, preferential translation of, 79
viruses, 11, *31,* 62, 64
ribosomal (r) RNA, 59, *75,* **405**
5S, 25, *27, 74,* 76, 291, **405**
16S, 18S, 23S, 28S, *74, 75*
stockpiled, 291
ribose, 16, *17, 19, 43,* **405**
in evolution, 308
and mutation, 88
nucleotides, 47
ribosome, *5,* 59, 68, *69, 82*
attachment to, *77*
and drop off, 228
-binding gene, *225*
-bound enzymes, 76
chloroplast, *74,* 193
cytoplasmic, *74*
DNA, *61*

functioning, 74
mitochondrial, *74*
phage-modified, 228
proteins, *74,* 75
RNA's in, *74*
sizes, *74, 80, 82*
stockpiled, 291
subunits, *74*
and translation regulation, 233
ribulose-1, 5-bisphosphate carboxylase, 194
rice, *343,* 348
rickettsiae, *191*
rifle, 88
ring chromosome, *31, 32, 181,* 182
rocket, 120
rocks, 380
Rocky Mountain spotted fever, *191*
rod, 36, *50,* **405**
rodenticide, 353
Roman Catholic Church, 369
root, 174, 289
rough clone, *114*
roundworms, 174, 217, 353
Rous sarcoma virus (RSV), *65,* 191
rubber, 33, 347
rust, 345, 349
rye, 174, 349

*s,* 324
S period, 137, *138,* 143, 147, 230
S unit, 59, **406**
saccharine, *382*
*Saccharomyces,* 23, 275, *313*
salamanders, 174
salicylate, 134, 366
salivary gland chromosome, 211, *212, 213, 215,* 217, 240, 242
salmon base ratio, 24
*Salmonella,* 120, *121*
Santa Gertrudis cattle, 350
saw fly, 279
schistosomiasis, 382
*Sciara,* 217
science, pseudo-, 367
scope of genetics, 14
Scotland, 369
screening for genetic diseases, 384
screwworm, *313,* 354
sea hare, 358
sea urchin, *23,* 24, 174, 292
seasons and isolation, 336
secretion, *82*
seeds, 4, 217, 351
segregation, 137, *264,* **406**
of chromosomes and genes, 149
in hybrid, 338
independent, *150*
and translocations, *184*
selection, 323, **406**
and agriculture, 345
chemical, 308
coefficient, *324,* **406**
and gene dosage, 258
against hybrids, 259

selection (cont.)
  and mutation, 327
  natural, 310
  and oil, 345
  for quantitative trait, 272
self-fertilization, 319, **406**
self-mutilation, 362
self-replication, 13
semicircular canal, 359
semiconservative synthesis, 46, 47, 48, **406**
semilethal mutant, **406**
senescence, 376
sense, codon, 72, **406**
  mutation, 89
  strand, 58, **406**
sensors, 357, 361
sequence similarities, 311
sequencing, bases, 24, 25
  loci, 169
serine (Ser), 71
sets of chromosomes, 173
setting, 242, 243, **406**
sex, attractant, 354
  bacterial, 128–135
  cell, 144
  chromatin, 237, 238, **406**
  chromosome, 158, 159, **406**
  determination in eukaryotes, 275–286
  and heterochromatization, 242
  index, 277
  infectious, 128
  infra-, 277
  linkage, 158–166
  mosaic, **406**
  organ, 280
  and isolation, 336
  particle, 128, 129, **406**
  ratio, 191
  type, + or –, 128
    and environment, 275
  Hfr, 130
  *See also* names of particular organisms
sharing nucleotides, 46, 227
shark DNA content, 23
sheep, 293, 350
shell, 142
shellfish, 4
shock, electric, 358
shoes, 380
shoot, 174
short-term memory, 358, **406**
Siamese cat, 252, 253
siblings, 319, 385, **406**
sickle-cell, anemia, 150, 254–257, **406**
  trait, 384
sieve cell, 217
sigma, 58, 221, **406**
signal polypeptide, 81, 195
signaling gene, 244, 276, **406**
silk fibroin, 290
silks, 345
silkworm, 142

silver, 228, 229
Simian virus 40 (SV40), 64, 191, 192
single-species polyploidy, 173, 337, 348
sister chromatids, 146, **406**
sister-brother mating, 319
site, **406**
  antigen-binding, 299
  of integration, 126
  ribosome attachment and drop off, 228
  transcriptase interaction, 222
size and sex type, 286
skin, 252
  cancer, 379
  color, 267, 361, 364
sliding activation hypothesis, 297, 298, 300, 301
smoke, 335
smoking, 366
smooth clone, 114
smut, 345
snail, 23, 141, 285
snake, 313
sociology and genetics, 363–371
*Solanum*, 337
solar system, 3
sole prints, 176
somatic, 61, **406**
  cell fusion, 175
  line, aging and death of, 375
    breaks in, 180
    and dosage compensation, 240
  mutants, 379
  arearrangement and antibody, 299
  rRNA, 61
somatostatin DNA cloned, 389
somite, 234
soot, 335
sorghum, 343
sound, 359
soup, 4, 308
Soviet Union, 347, 366
soybean, 343
spacecraft, sterile, 352
spacer DNA, 61, 62, 63, **406**
speciation, 174, 175, 335–338, **406**
species, 4, 336, **406**
  and base ratios, 24
  formation, 174, 175, 335–338
  inter-, hybrids, 337, 338
  from races, 336
speech, 365
speed of polypeptide synthesis, 234
sperm, 148
  bank, 385
  base composition in, 24
  disomic, 178
  DNA content, 23
  exceptional, 166
  nucleus and stockpiling, 291
  storage, 350
spermatid, 148
spiders, 4
spina bifida, 385

spindle, 138, 141, 145, 165, 202, **406**
  and mendelian genes, 190
  orientation, 150, 153, 154
  shape, 152, 153
  tubule, 174
    protein stockpiled, 291
*Spirogyra*, 193
spiroplasma, 191
spitting, 362
spleen, 255
splicing nucleic acids, 60, 301, 387, 388
splint, 180
sponge, 4, 23
spontaneous mutation, 97
spreading effect, 239, 241, **406**
stain, chromosome, 172, 238
Stalin, J. 366
*Staphylococcus*, 120
starfish, 4
stars, 3
starvation, 174, 200
statistics, 271
sterility, 176, 281, 282, 283, 285, 363
  induced, 354
  and isolation, 336
  in population, 332, 333
sterilization, 368
steroid, 294
stillbirth, 320, 321
stockpiling, 291, 292
Stone Age, 344
strands, + and –, 44, 45, 63, 64
  complementary, 18
  separation, and UV, 94
straw, 347, 349
strawberry, 349
streptomycin, 75
subunit, ribosomal, 74
subvital mutant, **406**
sugar, 16, 19
  and ABO blood type, 373
  as antigen, 298
  beet, 348
  in evolution, 308
  and mutation, 88
  origin of, 307
sulfonamide, 134
sulfur, 7, 88
sun, 308
sunlight, 252, 307, 364
  as barrier, 336
  and cancer, 379
supercoiling or superhelix, 31, 33, 34, 37, 38, 47
superfemale, 277, **406**
supermale, 277, **407**
Supreme Court, 370
surgery, 281, 384, 386
survival of early development, 291–292
SV40, 64, 191, 192
Sweden, 344, 347, 369
sweet potato, 343
swelling, 255
sycamore, 191, 192

# INDEX

symbionts, 289
symbols, 159, *251*
synapsis, 115, *116*, 145, 151, **407**
  in hybrid, 338
  mismatched, 310, *311*
syndromes, 281, 282, 284, 285, 362
synthesis, antibody, 298–301
  gene, and genetic disease, 386
  hemoglobin, 295–298
  of nucleic acid, 43
  spontaneous, 307
synthetic lethal, 259
systemic lupus erythematosis, 375

*t*⁺, 225
T. *See* thymine
tadpole, 217
tail, *12*, 217, 252
tailoring, 59, 68, 81, 407
tan, 252, 364
target cell, 294
taro, *343*
tars, 383
tassels, 345
Tay–Sachs disease, 362, 384
tDNA, **407**
*Tectorius* DNA content, *23*
teleost DNA content, *23*
telophase (Tel), *138, 139,* 141
  I and II, *147, 148*
temperate, *Drosophila,* 218
  phage, 123, *124,* **407**
temperature, as barrier, 336
  and coat color, 252
  and mutation, 98
  and polyploidy, 174
template, *44,* 56, **407**
teosinte, 338
terminal redundancy, *105,* **407**
termination factor, 79, **407**
terminator, 58, *59, 60*
  codons, 72, *75,* **407**
territory, 335
test, cross, *257,* **407**
  intelligence, 364
testicular feminization, 363
testis, *280,* 282
testosterone, *294*
tetracycline, 134
tetramer, 262
tetraploid, 173, 237, *238, 277,* **407**
tetrasomic, **407**
thalassemia, 264
thalidomide, 254, 365
Thiotepa, *382*
Third World, 344
threonine (Thr), *71,* 114, 117, *118*
thumb, 266
thymine (T), *17, 18, 25*
  derivative, *52*
  dimers, *93*
thyroid, 297, 375
ticks, 350
tissue, culture, 376, 381
  incompatability, 375

inducing or responding, *293*
  transplantation, 375
toad, *23,* 199, 290
toadfish, 234
tobacco, 174
  mosaic virus (TMV), 9, *10, 31, 35, 191*
  and recombination, 111
toes, 253
tomato, 347, 353
tongue, 176
tonsils, 381
toxins and cloning, 389
*tra,* 278
trait, affected by more than one gene, 266
  continuous (quantitative), 268–273
  mental and physical, 251
  qualitative (discontinuous), 268
  secondary sexual, 280, 282
  in twins, 249, *250*
transcript, 56
transcriptase, 56, 60, *61, 82,* 221, 262, **407**
  interaction site, *222,* 223, **407**
  nuclear, *61*
  reverse, 56, *65*
  virus-specific, 64
transcription, 56–65, *69,* **407**
  activated, 232
  -active and -silent DNA, 61, *62, 63,* 214, *216,* **407**
  and antibody, 300
  and cancer, 378
  in chloroplast, 193
  coupled to translation, 79, *82*
  differential, 289
  and differentiation, 289, *290*
  and dosage compensation, 237, 238
  in eukaryotes, *60*
  and globin, 297
  and hormones, 294
  inhibited, 297
  *in vitro, 232*
  in mitochondria, 197
  one-complement, *58*
  in prokaryotes, 57
  regulation, 221–226, 228–233
  reverse, 56, 64, 214, 386
  of single copy genes, 290
  speed, 79
  terminator, *225,* 226, **407**
  two-complement, 64
  unit, 58, *59, 60*
  visualized, *82*
transduction, 117–121, 123–127, **407**
  abortive, *119,* 120, *121*
  complete, *119,* 120
  as an error, 119, 126
  in eukaryotes, 120
  generalized, 113, 117, *118,* 120
  and genetic diseases, 386
  restricted, 113, 123, 127, *130, 133*
transfer (t) DNA, *61*
transfer (t) RNA, 25, *26,* 59, **407**

as adapter, 69
base sequence, *26*
CCA terminus, 77
charged, *69, 80*
chloroplast, 193
cloverleaf structure, *26*
  and degeneracy, *78*
functions, 76, 77
of mitochondria, 178
multiple, 78
  Phe, 77
as primer, 65
shape, 76, 77
stockpiled, 291
and translation regulation, 228, 233
uncharged, *69*
transformation, 113–117, **407**
  and genetic disease, 386
  as a genetic recombination, 115, *116*
  sex, 278
transfusions, blood, 373
transgenosis, *191,* 192, 385, **407**
transition, 88, *91,* **407**
translation, 67–83, **407**
  and antibody, 300
  and cancer, 378
  in chloroplast, 193
  coupled to transcription, 79, *82*
  and differentiation, 290
  and globin, 297
  and hormones, 294
  and its code, 67
  in mitochondrion, 197
  regulation, 226–228, 233–234
  speed, 79
  of virus RNA, 79
translocation, half-, 183, *184,* 185, *186, 187,* 283
  and leukemia, 379
  reciprocal, *183, 184, 185,* 186, 238
transplantation, 210, 289, 375, 385
transportation, accidental, 352
transposable genes, 243
transposition, *88,* 243, 299, **407**
transversion, 88, 254, **407**
traps, baited, 354
treating genetic diseases, 384, 385
treatment expense, 385
tree, 332, 335, 354
  evolutionary, 312, *313*
trimer, 7, **407**
triplet, 67, *69,* 309
triploid, 173, 174, **407**
  apple, 348
  *Drosophila,* 276, *277*
  plants, 349
  sugar beet, 348
trisomic, 176, 281, 363, 384, **407**
*Triticale,* 349
*Triticum,* 347
tropics, 218, 348, 350
*trp,* 226, 227
*Trypanosoma,* 200, *201*
tryptophan (Trp), *71,* 224
  operon, 224, 227

tube, conjugation, 128, *129*, 131
tuberculosis, 24, 251, 362, 382
tubules, spindle, 138
tulips, 349
tumor, 64, 120
   *See also* cancer
tuna, *313*
tunicate DNA content, *23*
Turner's syndrome, 281, 283, **407**
turtle, *23, 313*
twin studies, 249
twine, 33
two-arm inversion, *182*
tyrosine (Tyr), *71,* 318, 361

U. *See* uracil
ultracentrifugation, 59
ultraviolet light (UV), 94, 195, 377
   and bases, *26,* 29
   and mutation, *93,* 379, 380
underreplicated, 211, 242, **407**
union, restitutional or nonrestitutional, 179
   sister, 179
unit, map, 107, *108,* 168
universe, 3, *4*
unrestricted transduction, **408**
uracil (U), 16, *17*
   derivatives of, *52,* 77
   dimers, 93
United States, *344,* 347
   maize, 351
   marriage laws, 369
   mutation rate in, 380
   National Institutes of Health, 389
   pests in, 352
   and screwworm, 354
   watermelon, 349
uric acid, 362
USSR, 347, 366
uterus, 251

vaccinia, *24,* 35
vagina, 363, 383
valine (Val), *71,* 254, 255
variability, genotypic, 329, 332–335
   and mutants, *259*
variable (V) region, *299,* **408**

variegated (V-type) position effect, 241, **408**
variegation, 241
Vavilov, N. I., 367
Venus, 159
vertebrae, 293, 381
vertebrates, DNA content of, *23*
vesicular stomatitis virus (VSV), 120
vigor, hybrid, 325, 345, *346*
vinyl chloride, *382*
virogene-oncogene hypothesis, 378
virulent, 105, **408**
virus, 4, 6, **408**
   and cancer, 378
   DNA modified in, *52*
   and mutation, 219
   nucleic acid conformation in, *31*
   and recombination, 101
   reconstituted, 9, *10*
   RNA, as mRNA, 62
      translation, 79
   transcription in, 62
   and translation control, 234
   *See also* names of specific viruses
vitamins, 364, 366, *382*
volcano, 307, 308, 336

*w, 324,* **408**
*w*⁺, 241, 242
W strand, 22, *30,* 46
waltzers, 359
war, 370
wasp, 279
watches, 380
water, *7,* 307, 308
   as barrier, 336
   nucleic acid in, 9, 11
   radioactive, 380
watermelon, 348
Watson, J. D., 20, 22
wax, 360
weight lifting, 364
Weinberg, W., 316, 317
whale, 352
wheat, *77, 174, 343,* 344, 347, 349, 351
white blood cell, 290, 295, 298
whites, 267, 364, 370
wild type, 114, 258, *276,* **408**
Wilkins, M. H. F., 20, *21*

wind, 336, 345
wing, 266, *267,* 268, 359
wingless chicken, 293
winter wheat, 347
woods, 335
wool, 350
workers, 279, 360
World War II, 347
worm, 4, *23,* 286

xanthine dehydrogenase, 333
*Xanthomonas,* 52
X chromosome, 158, *159,* 284
   human, linkage map, *170*
   inactivation, 239
   -limited loci, 159, *164*
   -linkage, *161, 162, 163*
   ring, 284
   translocation, 238
X$^S$ and X$^L$ chromosome, *159,* 283, *284, 285*
X0 female, 281
*Xenopus,* 199, 202, 389
xeroderma pigmentosum, 379, 380
X- and Y-linked loci, *164*
X rays and agriculture, 349
X-ray, diffraction, 21, *22*
   induced mutation, 93, 379
   induced sterility, 354
XXX and XXY, *177,* 281, 282
X-Y translocation, *285*

Y$^L$ or Y$^S$, *159,* 283, *284, 285*
Y chromosome, 158, *159*
   -limited loci, 159, *164*
   and parentage, 370
   size, 284
yam, *343*
yeast, *77, 191,* 200
   base ratio, *24*
   sex in, 275
   tRNA, *26*
yellow fever, 353
yolk, 4, 13, 141, *192,* 202, 291
Y-X translocation, *285*

*Zea,* 338
zucchini, 347
zygote, 144, **408**